Foundations of Engineering Mechanics

Series Editors

Vladimir I. Babitsky, School of Mechanical, Electrical and Manufacturing Engineering, Loughborough University, Loughborough, Leicestershire, UK

Jens Wittenburg, Karlsruhe, Germany

More information about this series at http://www.springer.com/series/3582

György Szeidl · László Péter Kiss

Mechanical Vibrations

An Introduction

Springer

György Szeidl
Institute of Applied Mechanics
University of Miskolc
Miskolc-Egyetemváros, Hungary

László Péter Kiss
Institute of Applied Mechanics
University of Miskolc
Miskolc-Egyetemváros, Hungary

ISSN 1612-1384 ISSN 1860-6237 (electronic)
Foundations of Engineering Mechanics
ISBN 978-3-030-45076-2 ISBN 978-3-030-45074-8 (eBook)
https://doi.org/10.1007/978-3-030-45074-8

This Springer imprint is published by the registered company Springer Nature Switzerland AG
The registered company address is: Gewerbestrasse 11, 6330 Cham, Switzerland

Preface

The present book is based on the lectures delivered for those Hungarian and foreign M.Sc. students who have to take the course Mechanical Vibrations. When we began to write it we faced with the following problem: what level of preliminary knowledge could be expected. Hungarian students in mechanical engineering with a B.Sc. degree have some familiarity with tensor algebra and analysis. However, this is, in general, not the case for foreign students because of the differences in curricula. We had to compromise. Mainly in Chap. 1 which is a summary of the fundamental results of dynamics. In order to avoid the difficulties caused by the differences in the preliminary knowledge of the students we preferred matrix notations to tensorial ones. In spite of that the parts of the text which are bounded with two thin horizontal lines contain some materials which are presented in vectorial and tensorial notations. For the first reading, these parts of the text can be left out of consideration.

As we have mentioned above, Chap. 1 is a sort of Introduction which is devoted to the fundamentals of dynamics. The students are, in principle, familiar with the content of Chap. 1. As regards the content of the chapter Introduction, the necessary kinematic relations with an emphasis on rigid body motions are all covered. Then the equivalence of the effective forces to the external forces, the principle of impulse and momentum as well as the principle of work and energy are presented in the text. Proofs are, sometimes, omitted or are left for problems to be solved.

Chapter 2 deals with impact. First, we consider central impact. Since the graphical solutions are very illustrative the use of Maxwell's diagrams is also shown. As regards the issue of eccentric impact, we clarify the way how to reduce the equations providing the solution into such a form which coincides with the equations valid for central impact. The chapter contains various exercises (for which the solutions are also included in the text) and is closed by the problems that are left for the independent work of the students.

Some simple but important vibration issues are presented in Chap. 3 which is, among others, a short introduction to the vibratory problems of single degree of freedom systems. Undamped and damped free vibrations and forced vibrations are considered. As regards the applications, some simple machine foundation problems are modeled and solved. It is worth emphasizing that the machine foundation problems are not simple at all and the one degree of freedom model we have presented is only aimed at drawing the attention of the readers to these important problems.

Chapter 4 is an Introduction to multidegree of freedom systems. We establish Lagrange's equation of the second kind for a system of particles first. They can, however, be applied to establishing the equations of motion for such systems which involve rigid bodies as well. Special emphasis is laid on spring–mass systems with two degrees of freedom. Solutions are presented for various free and forced vibration problems. It is also shown how to tune a system to avoid resonance.

The general theory of multidegree of freedom system is considered in Chapter 5. The eigenvalue problem that provides the eigenfrequencies for a vibrating finitedegree of freedom system is also presented. It is shown what properties these eigenvalue problems have including the fundamental characteristics of the eigenvalues and eigenvectors. Some simple solution procedures are suggested. The concept of the Rayleigh quotient is introduced. The case of forced vibrations is investigated at the end of the chapter.

Chapter 6 covers some phenomena of the rotating motion. We discuss the most important properties of a flywheel which can be used to store kinetic energy and to make the rotational motion smoother by reducing the speed of fluctuations. Stability problems caused by a change in the load torque is also investigated. If the shaft is not rigid further problems occur. We present Laval's theorem and the gyroscopic effect of the rotational motion.

Chapter 7 is devoted to the vibration problems of systems with infinite degrees of freedom. First, we consider the longitudinal vibrations of rods. Then solutions are presented for the transverse vibration of a string and for the torsional vibrations of rods with circular cross section. This is followed by the analysis of transverse vibrations of beams.

Chapter 8 is concerned with the eigenvalue problem of ordinary differential equations. We present the definition of the Green functions and reduce some eigenvalue problems to homogeneous Fredholm integral equation with the Green function as kernel. A solution algorithm is suggested by the use of which numerical solutions are given for some vibration problems of circular plates subjected to constant radial in plane load and for the vibratory behavior of beams loaded by an axial force.

Eigenvalue problems of ordinary differential equation systems are discussed in Chap. 9. The concept of the Green function matrix is introduced. By utilizing the Green function matrices the eigenvalue problems described by ordinary differential equation systems can be reduced to eigenvalue problems governed by homogeneous Fredholm integral equation systems. The solution algorithm presented in Chap. 8 is generalized for such eigenvalue problems. The applications are related to the vibration problems of Timoshenko beams.

Vibration problems of curved beams with a centerline of constant radius are governed by degenerated differential equation systems. Chapter 10 provides a definition for the Green function matrices concerning the degenerated differential equation systems. These matrices are determined for pinned-pinned, fixed-fixed, and pinned-fixed heterogeneous curved beams. By utilizing the Green function matrices the eigenvalue problems that describe the vibratory behavior of these beams are reduced to Fredholm integral equation systems. Numerical solutions are also presented in graphical format.

Appendix A is a very short introduction to tensor algebra. It presents those tools which are needed to understand some proofs in Chap. 1 and Appendix C which is a collection of solutions to some selected problems. Appendix B gives the definition of the Dirac function.

Miskolc-Egyetemváros, Hungary

György Szeidl
László Péter Kiss

Acknowledgments In preparing the manuscript, serious assistance was provided by Dr. Gábor Csernák who read the first version of the manuscript and suggested various changes in order to make it better. We gratefully acknowledge his assistance.

In structuring and correcting the manuscript we got valuable assistance from Mrs. Sudhany Karthick at Springer. Her help is highly appreciated.

Since this book is a textbook it is important to acknowledge our debts to our teachers. Imre Kozák introduced us to the modern literature on applied mechanics. We could always turn to him for advice when difficulties arose in our research work.

We are also grateful to Miss Lisa Lui (Zhejiang Guanbao Industrial Co.), Mr. Tony Griffiths of lathes.co.uk and Mr. Brian Le Barron (Vibratech TVD 180 Zoar Valley Rd. Springville, New York, USA) for allowing us to include some photos in the book—see Figs. 3.22, 4.11 and 4.24.

László Péter Kiss is especially grateful to his parents and sister, simply for everything they mean to him. There is further dedication to his little nephew, Balázs.

Gyorgy Szeidl owes his wife Babi a debt of gratitude for her support and encouragement through her persistent asking the question "Did you deal with your book today?". This always turned his attention to writing instead of dealing with less important things. Special thanks are due to his daughter Agnes and son Adam for their continuous support.

There is also an opportunity here to make some remarks for editing the text of the book:

(a) Vectors and tensors are typeset in boldface. Matrices are also typeset in boldface but the letters are underlined.
(b) Equations are numbered sequentially in each chapter.
(c) The formulae that we regard important are, in general, framed.
(d) Bibliographies are presented at the end of the chapters. The main text is followed by Appendices and Index.

Contents

Chapter 1
Introduction

1.1 Model Creation

When one studies a dynamical phenomenon there arises the question what model should be applied. It is a fundamental expectation that the model established should reflect those features of the phenomenon in question which are the most important for the examination to be carried out.

When making a model the following facts should, therefore, be taken into account:

(a) what are the main objectives of the examinations (calculations) since the model should be based on them;
(b) what are the limits of the applied simplifications when the model is established.

For instance:

- A particle is such a body the motion of which can be described by the motion of one point of the body considered. The earth is a particle when its motion is considered around the sun.
- The rigid body is such a body for which the distance between any two but different points within the body remains constant during the motion. This means that a rigid body does not deform under the action of the applied forces.
- A solid body is capable of deformation.
- A mass distribution can be

(i) discrete (then the considered dynamical system consists of particles and has a finite degree of freedom—the degree of freedom is the minimum number of coordinates required to describe completely the motion of a particle or body) or
(ii) continuous (then the dynamical system has infinite degrees of freedom).

- As regards the material behavior that can be either

(i) linear (the springs in the system follow Hooke's law) or
(ii) non-linear.

G. Szeidl and L. P. Kiss, *Mechanical Vibrations*, Foundations of Engineering Mechanics,
https://doi.org/10.1007/978-3-030-45074-8_1

- The geometrical (kinematic) relations between the displacements and deformations can also be (i) linear or (ii) non-linear.
- The constraints in the dynamical system can be either integrable (or holonomic) or non-integrable (nonholonomic)—these concepts are to be discussed later.

1.2 Kinematic Relations—A Summary

1.2.1 Motion of a Material Point (Particle)

Figure 1.1 shows the Cartesian coordinate system (xyz)—the unit vectors of the coordinate axes are denoted by $\mathbf{i}_\ell$ $(\ell = x, y, z)$—the particle with mass m and its path on which s is the arc coordinate. The position vector of the particle is $\mathbf{r}(t) = \mathbf{r}[s(t)]$ where t is time. The function $\mathbf{r}(t)$ is called motion law of the particle. The unit tangent and normal of the path are denoted by $\mathbf{t}$ and $\mathbf{n}$, respectively. Let $\mathbf{v}$ be the velocity of the particle. It is obvious that

$$\mathbf{v} = \frac{\mathrm{d}\mathbf{r}}{\mathrm{d}t} = \frac{\mathrm{d}\mathbf{r}}{\mathrm{d}s}\frac{\mathrm{d}s}{\mathrm{d}t} = v(s)\mathbf{t}(s), \tag{1.1a}$$

where

$$v(s) = \mathrm{d}s/\mathrm{d}t \tag{1.1b}$$

is the speed of the particle. Given the velocity of the particle

$$\mathbf{a} = \frac{\mathrm{d}\mathbf{v}}{\mathrm{d}t} = \frac{\mathrm{d}v}{\mathrm{d}t}\mathbf{t} + v\frac{\mathrm{d}\mathbf{t}}{\mathrm{d}s}\frac{\mathrm{d}s}{\mathrm{d}t} \tag{1.2a}$$

is the acceleration. Since

$$\frac{\mathrm{d}\mathbf{t}}{\mathrm{d}s} = \frac{\mathbf{n}}{\rho} \tag{1.2b}$$

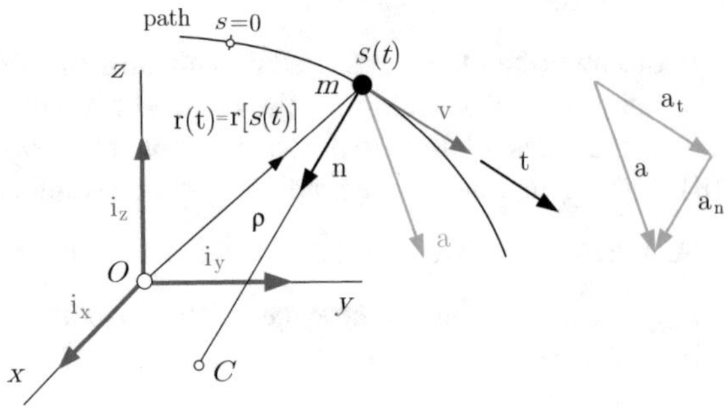

Fig. 1.1 Resolution of the acceleration for particle m

in which ρ is the radius of curvature, Eq. (1.2a) can be rewritten in the following form:

$$\mathbf{a} = a_t \mathbf{t} + a_n \mathbf{n}, \tag{1.2c}$$

where

$$a_t = \frac{\mathrm{d}v}{\mathrm{d}t} \quad \text{and} \quad a_n = \frac{v^2}{\rho} \tag{1.2d}$$

are the acceleration components tangential to the path and perpendicular to it.

If the motion of the particle is not constrained it has three degrees of freedom.

Typical units: [m/s] for the velocity and $\left[\mathrm{m/s^2}\right]$ for the acceleration.

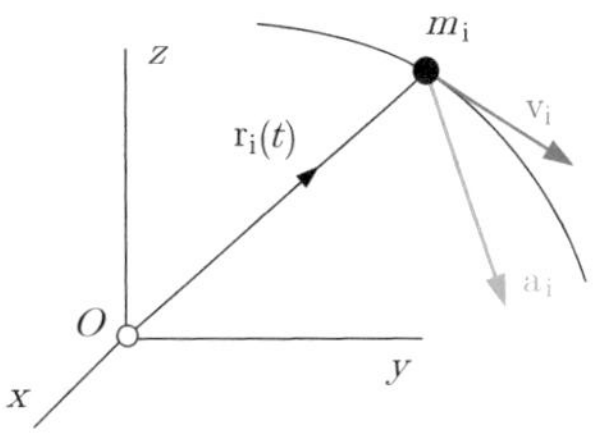

Fig. 1.2 The i-th particle in a system of particles

Considering a system of particles with masses m_i $(i = 1, \ldots, N)$—see Fig. 1.2 for details—the velocity and acceleration for the i-th particle can be given by

$$\mathbf{v}_i = \dot{\mathbf{r}}_i = \frac{\mathrm{d}\mathbf{r}_i}{\mathrm{d}t} \tag{1.3a}$$

and

$$\mathbf{a}_i = \dot{\mathbf{v}}_i = \frac{\mathrm{d}\mathbf{v}_i}{\mathrm{d}t}, \tag{1.3b}$$

where $\mathbf{r}_i(t)$ is the motion law of the i-th particle.

1.2.2 Motion of a Rigid Body

Figure 1.3 shows, among others, the rigid body $\mathcal{B}$ and the points $A \neq B \neq G$ of the body with position vectors $\mathbf{r}_A, \mathbf{r}_B, \mathbf{r}_G$. The position vector of the point B with respect to the point A (the vector from A to B) is $\mathbf{r}_{AB}$. Velocities and accelerations at the points A and B are denoted by $\mathbf{v}_A$, $\mathbf{a}_A$ and $\mathbf{v}_B$, $\mathbf{a}_B$. Further let $\boldsymbol{\omega}$ [1/s] and $\boldsymbol{\alpha}$ [1/s^2] be the angular velocity and angular acceleration of the body.

Let t be a given (fixed) point of time. Then

$$\mathbf{v} = \mathbf{v}(\mathbf{r}), \qquad \mathbf{r} \in \mathcal{B}$$

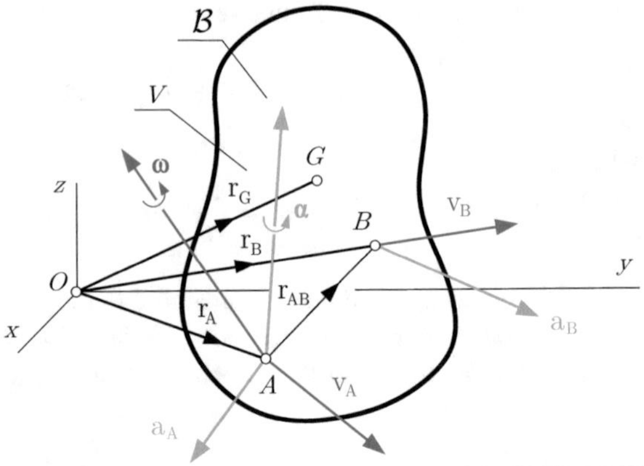

Fig. 1.3 Characteristic values for the states of velocity and acceleration in a rigid body

is the velocity distribution within body $\mathcal{B}$—referred to as the velocity state of the body at time t. As is well known—see [1–3], for instance,

$$\boxed{\mathbf{v}_B = \mathbf{v}_A + \boldsymbol{\omega} \times \mathbf{r}_{AB} = \mathbf{v}_A + \mathbf{r}_{BA} \times \boldsymbol{\omega}\,, \quad \mathbf{r}_{AB} = \mathbf{r}_B - \mathbf{r}_A = -\mathbf{r}_{BA}\,.} \tag{1.4}$$

Hence, $\mathbf{v}_A$ and $\boldsymbol{\omega}$ uniquely determine the velocity state of the body $\mathcal{B}$.

For a given point of time t, the function

$$\mathbf{a} = \mathbf{a}(\mathbf{r})\,, \quad \mathbf{r} \in \mathcal{B}$$

is the acceleration distribution within body $\mathcal{B}$—referred to as the acceleration state of the body at time t. Since

$$\dot{\mathbf{r}}_{AB} = \dot{\mathbf{r}}_B - \dot{\mathbf{r}}_A = \mathbf{v}_B - \mathbf{v}_A = \boldsymbol{\omega} \times \mathbf{r}_{AB} \quad \text{and} \quad \dot{\boldsymbol{\omega}} = \boldsymbol{\alpha} \tag{1.5}$$

it follows that

$$\boxed{\mathbf{a}_B = \dot{\mathbf{v}}_B = \mathbf{a}_A + \boldsymbol{\alpha} \times \mathbf{r}_{AB} + \boldsymbol{\omega} \times (\boldsymbol{\omega} \times \mathbf{r}_{AB})\,.} \tag{1.6a}$$

Consequently, $\mathbf{a}_A$, $\boldsymbol{\omega}$, and $\boldsymbol{\alpha}$ uniquely determine the acceleration state of the body $\mathcal{B}$.

Assume that the body is in a plane motion for which $\boldsymbol{\omega} = \omega_z(t)\mathbf{i}_z = \omega(t)\mathbf{i}_z$, $\boldsymbol{\alpha} = \alpha_z(t)\mathbf{i}_z = \alpha(t)\mathbf{i}_z$ and $\mathbf{r}_{AB} \perp \mathbf{i}_z$—$\mathbf{r}_{AB}$ is in the plane of motion. Making use of (A.1.10) we can write

$$\boldsymbol{\omega} \times (\boldsymbol{\omega} \times \mathbf{r}_{AB}) = \underbrace{(\boldsymbol{\omega} \cdot \mathbf{r}_{AB})}_{=0}\,\boldsymbol{\omega} - (\boldsymbol{\omega} \cdot \boldsymbol{\omega})\,\mathbf{r}_{AB} = -\omega_z^2 \mathbf{r}_{AB} = -\omega^2 \mathbf{r}_{AB}$$

thus

$$\boxed{\mathbf{a}_B = \mathbf{a}_A + \alpha \mathbf{i}_z \times \mathbf{r}_{AB} - \omega^2 \mathbf{r}_{AB}\,.} \tag{1.6b}$$

1.3 Fundamental Relations of Kinetics—A Summary

1.3.1 The First Moment of a Mass Distribution

Figure 1.4 shows (a) a system of particles (for which the mass distribution is not continuous) and (b) a rigid body denoted by $\mathcal{B}$ which occupies volume V (for body $\mathcal{B}$ the mass distribution is, in general, a continuous—or a piecewise continuous—function of the position vector, ρ [kg/m^3] stands for the density, $\boldsymbol{\rho}$ is the position vector of dm with respect to the material point B).

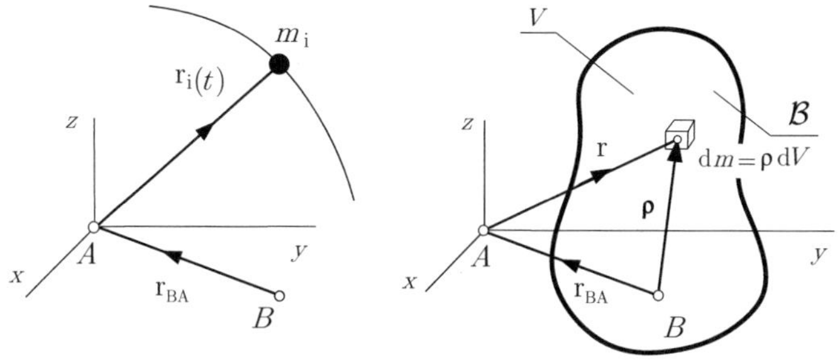

Fig. 1.4 A system of particles and a rigid body

The first moment of a mass distribution with respect to the point A (about the point A) is denoted by $\mathbf{Q}_A$ and is defined by the following relations:

$$\mathbf{Q}_A = \sum_{i=1}^{N} m_i \, \mathbf{r}_i \,, \qquad \mathbf{Q}_A = \int_V \mathbf{r} \, \rho(\mathbf{r}) \, \mathrm{d}V \,. \tag{1.7a}$$

The total mass is given by the relations

$$m = \sum_{i=1}^{N} m_i \,, \qquad m = \int_V \rho(\mathbf{r}) \, \mathrm{d}V \,. \tag{1.7b}$$

The unit of $\mathbf{Q}_A$ is $\left[\mathrm{kg\,m}\right]$.
If $\rho = \rho(\mathbf{r}) \neq$ constant then the body is inhomogeneous.
If $\rho = \rho(\mathbf{r}) =$ constant, i.e., ρ is independent of $\mathbf{r}$ then the body is homogeneous.
It is obvious that

$$\boxed{\mathbf{Q}_B = \int_V \boldsymbol{\rho} \, \rho \mathrm{d}V = \int_V (\mathbf{r} + \mathbf{r}_{BA}) \, \rho \mathrm{d}V = \int_V \mathbf{r} \, \rho \mathrm{d}V + \mathbf{r}_{BA} \int_V \rho \mathrm{d}V = \mathbf{Q}_A + m \, \mathbf{r}_{BA}} \tag{1.8}$$

is the first moment of the body with respect to the point B.

The mass center (or the center of gravity) is the point (denoted by G) for which

$$\mathbf{Q}_G = \mathbf{0}\,. \tag{1.9a}$$

For $G = B$ it follows from Eq. (1.8) that

$$\boxed{\mathbf{r}_{AG} = -\mathbf{r}_{GA} = \frac{\mathbf{Q}_A}{m}\,.} \tag{1.9b}$$

1.3.2 Momentum Distribution—Vector System of Momenta

A vector is called (a) free vector if its point of application can freely be selected, i.e., it is freely movable in space, (b) fixed vector if it has a unique point of application, and (c) sliding vector if it has a unique line of action on which its point of application can freely be selected. A system of vectors with fixed points of application is referred to as vector system. However, it may happen that some effects of the vectors that constitute the vector system mentioned are independent of their points of application[1] and in this respect they behave as if they were sliding vectors.

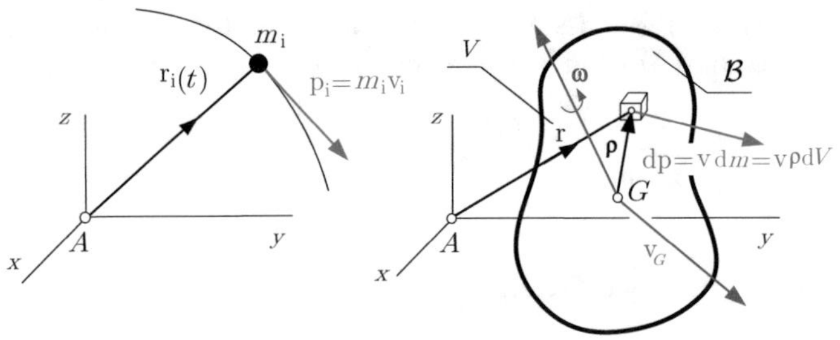

Fig. 1.5 Momentum vectors

The momentum vector is denoted by $\mathbf{p}$ and, for a single particle, is defined as $\mathbf{p} = m\mathbf{v}$ [kgm/s]—m and $\mathbf{v}$ are the mass and the velocity of the particle.

Resultants of [the momentum vectors $\mathbf{p}_i$] and {the infinitesimal momentum vectors $\mathrm{d}\mathbf{p} = \mathbf{v}\rho\mathrm{d}V$} are given by

$$\mathbf{p} = \sum_{i=1}^{N} \mathbf{p}_i = \sum_{i=1}^{N} m_i\,\mathbf{v}_i \qquad \mathbf{p} = \int_V \mathbf{v}\,\rho\mathrm{d}V = m\,\mathbf{v}_G\,. \tag{1.10}$$

Equation $(1.10)_2$ follows from the manipulation

[1]Forces acting on (a rigid body)[deformable] body are (sliding)[fixed] vectors.

$$\mathbf{p} = \int_V \mathbf{v}\,\rho \mathrm{d}V = \underset{(1.4)}{\uparrow} = \int_V (\mathbf{v}_G + \boldsymbol{\omega} \times \boldsymbol{\rho})\rho \mathrm{d}V = \underset{(1.9a)}{\uparrow} =$$

$$= \underbrace{\int_V \rho\,\mathrm{d}V}_{m}\mathbf{v}_G + \boldsymbol{\omega} \times \underbrace{\int_V \boldsymbol{\rho}\rho\,\mathrm{d}V}_{\mathbf{Q}_G=\mathbf{0}} .$$

We remark that some textbooks denote the momentum vector by **L**—see, for instance, book [1].

1.3.3 Moment of Momentum for a Rigid Body

Figure 1.6 shows body $\mathcal{B}$ in the absolute (not moving) coordinate system (xyz). The points A and G are those of the body, the point G is the mass center. The coordinate system $(\xi\eta\zeta)$ with origin at G and the coordinate system $(\xi'\eta'\zeta')$ with origin at A move together with the body in such a way that coordinate axes (a) x, ξ, ξ'; (b) y, η, η'; and (c) z, ζ, ζ' remain parallel to each other, i.e., the Greek coordinate systems moving with the body do not change their orientation. In Fig. 1.6, the position vector of the infinitesimal mass element $\mathrm{d}m$ with respect to the mass center is denoted by $\boldsymbol{\rho}$. The moment of momentum with respect to the point G is defined as

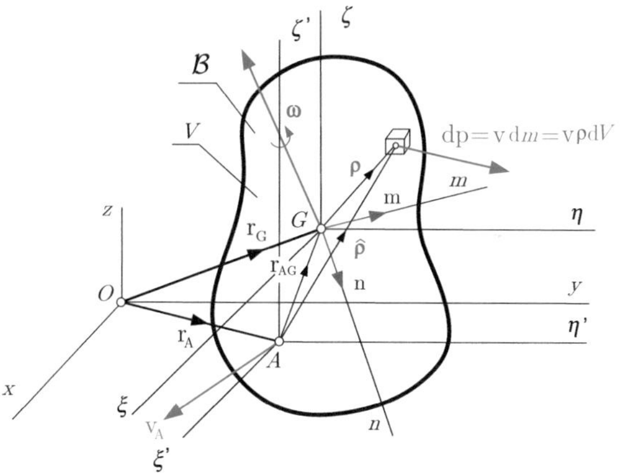

Fig. 1.6 Various coordinate systems (fixed and moving with the body)

$$\mathbf{H}_G = \int_V \boldsymbol{\rho} \times \mathbf{v}\rho\,\mathrm{d}V .$$

It follows from Eq. (1.4) that

$$\mathbf{v} = \mathbf{v}_G + \boldsymbol{\omega} \times \boldsymbol{\rho} .$$

Consequently,

$$\mathbf{H}_G = \int_V \boldsymbol{\rho} \times (\mathbf{v}_G + \boldsymbol{\omega} \times \boldsymbol{\rho})\, \rho\, \mathrm{d}V = \\ = \underbrace{\int_V \boldsymbol{\rho}\, \rho\, \mathrm{d}V}_{\mathbf{Q}_G = \mathbf{0}} \times \mathbf{v}_G + \int_V \boldsymbol{\rho} \times (\boldsymbol{\omega} \times \boldsymbol{\rho})\, \rho\, \mathrm{d}V\,.$$

Hence,

$$\mathbf{H}_G = \int_V \boldsymbol{\rho} \times (\boldsymbol{\omega} \times \boldsymbol{\rho})\, \rho\, \mathrm{d}V\,. \tag{1.11}$$

Remark 1.1 For our later consideration we shall introduce matrix notations. Let **a** and **b** be two vectors. Their matrices are written as

$$\underline{\mathbf{a}} = \underset{(3\times1)}{\underline{\mathbf{a}}} = \begin{bmatrix} a_x \\ a_y \\ a_z \end{bmatrix} \quad \text{and} \quad \underline{\mathbf{b}} = \underset{(3\times1)}{\underline{\mathbf{b}}} = \begin{bmatrix} b_x \\ b_y \\ b_z \end{bmatrix}. \tag{1.12}$$

Further let

$$\underline{\mathbf{a}}_\times = \underset{(3\times3)}{\underline{\mathbf{a}}_\times} = \begin{bmatrix} 0 & -a_z & a_y \\ a_z & 0 & -a_x \\ -a_y & a_x & 0 \end{bmatrix} \tag{1.13}$$

be a skew matrix which is given in terms of a_x, a_y, and a_z. It is obvious that there belongs a similar skew matrix to any other vector. Consider now the cross product

$$\mathbf{c} = \mathbf{a} \times \mathbf{b}\,. \tag{1.14a}$$

It is not too difficult to check that the column matrix $\underline{\mathbf{c}}$ can be calculated as

$$\underline{\mathbf{c}} = \underset{(3\times1)}{\underline{\mathbf{c}}} = \underset{(3\times3)}{\underline{\mathbf{a}}_\times}\,\underset{(3\times1)}{\underline{\mathbf{b}}} = \begin{bmatrix} 0 & -a_z & a_y \\ a_z & 0 & -a_x \\ -a_y & a_x & 0 \end{bmatrix} \begin{bmatrix} b_x \\ b_y \\ b_z \end{bmatrix}. \tag{1.14b}$$

Making use of the calculation rule for triple cross products from (A.1.10) we can transform Eq. (1.11) into the following form:

$$\mathbf{H}_G = \int_V \left(\rho^2\, \boldsymbol{\omega} - \boldsymbol{\rho}\, (\boldsymbol{\rho} \cdot \boldsymbol{\omega})\right) \rho \mathrm{d}V\,. \tag{1.15}$$

The matrices of $\boldsymbol{\rho}$, $\boldsymbol{\omega}$, and $\mathbf{H}_G$ in the coordinate system (ξ, η, ζ) are given by

$$\underset{(3\times1)}{\underline{\boldsymbol{\rho}}} = \begin{bmatrix} \xi \\ \eta \\ \zeta \end{bmatrix}, \quad \underset{(3\times1)}{\underline{\boldsymbol{\omega}}} = \begin{bmatrix} \omega_\xi \\ \omega_\eta \\ \omega_\zeta \end{bmatrix} \quad \text{and} \quad \underset{(3\times1)}{\underline{\mathbf{H}}_G} = \begin{bmatrix} H_{G\xi} \\ H_{G\eta} \\ H_{G\zeta} \end{bmatrix}. \tag{1.16}$$

A comparison of Eqs. (1.15) and (1.16) yields

$$\underline{\mathbf{H}}_G = \int_V \left\{ [\xi^2+\eta^2+\zeta^2] \begin{bmatrix} 1 & 0 & 0 \\ 0 & 1 & 0 \\ 0 & 0 & 1 \end{bmatrix} - \begin{bmatrix} \xi \\ \eta \\ \zeta \end{bmatrix} [\xi \; \eta \; \zeta] \right\} \rho \mathrm{d}V \, \underline{\omega} \,,$$

that is,

$$\underline{\mathbf{H}}_G = \int_V \begin{bmatrix} \eta^2+\zeta^2 & -\xi\eta & -\xi\zeta \\ -\eta\xi & \xi^2+\zeta^2 & -\eta\zeta \\ -\zeta\xi & -\zeta\eta & \xi^2+\eta^2 \end{bmatrix} \rho \mathrm{d}V \, \underline{\omega}$$

or

$$\underbrace{\begin{bmatrix} H_{G\xi} \\ H_{G\eta} \\ H_{G\zeta} \end{bmatrix}}_{\underset{(3\times1)}{\underline{\mathbf{H}}_G}} = \underbrace{\begin{bmatrix} J_\xi & -J_{\xi\eta} & -J_{\xi\zeta} \\ -J_{\eta\xi} & J_\eta & -J_{\eta\zeta} \\ -J_{\zeta\xi} & -J_{\zeta\eta} & J_\zeta \end{bmatrix}}_{\underset{(3\times3)}{\underline{\mathbf{J}}_G}} \underbrace{\begin{bmatrix} \omega_x \\ \omega_y \\ \omega_z \end{bmatrix}}_{\underset{(3\times1)}{\underline{\omega}}} , \tag{1.17a}$$

where $\underline{\mathbf{J}}_G$ is the matrix of inertia in which the diagonal elements

$$J_\xi = \int_V (\eta^2+\zeta^2)\, \rho \mathrm{d}V \,, \quad J_\eta = \int_V (\xi^2+\zeta^2)\, \rho \mathrm{d}V \,, \quad J_\zeta = \int_V (\xi^2+\eta^2)\, \rho \mathrm{d}V \tag{1.17b}$$

are the moments of inertia with respect to the axes ξ, η, and ζ while the off-diagonal elements

$$J_{\xi\eta} = J_{\eta\xi} = \int_V \xi\eta\, \rho \mathrm{d}V \,, \quad J_{\eta\zeta} = J_{\zeta\eta} = \int_V \eta\zeta\, \rho \mathrm{d}V \,, \quad J_{\zeta\xi} = J_{\xi\zeta} = \int_V \zeta\xi\, \rho \mathrm{d}V \tag{1.17c}$$

are referred to as products of inertia.

Let **n** and **m** be unit vectors at G—see Fig. 1.6. Further let **n** and **m** be perpendicular to each other. Then

$$J_n = \underset{(1\times3)}{\underline{\mathbf{n}}^T} \underset{(3\times3)}{\underline{\mathbf{J}}_G} \underset{(3\times1)}{\underline{\mathbf{n}}} \quad \text{and} \quad J_{nm} = J_{mn} = -\underset{(1\times3)}{\underline{\mathbf{m}}^T} \underset{(3\times3)}{\underline{\mathbf{J}}_G} \underset{(3\times1)}{\underline{\mathbf{n}}} = -\underset{(1\times3)}{\underline{\mathbf{n}}^T} \underset{(3\times3)}{\underline{\mathbf{J}}_G} \underset{(3\times1)}{\underline{\mathbf{m}}} \tag{1.17d}$$

are the moment of inertia with respect to the axis n, and product of inertia with respect to the axes n and m. Since the matrix of inertia $\underline{\mathbf{J}}_G$ is a symmetric matrix it holds that

$$\underset{(3\times3)}{\underline{\mathbf{J}}_G} = \underset{(3\times3)}{\underline{\mathbf{J}}_G^T} \,.$$

The eigenvalue problem

$$\underset{(3\times3)}{\underline{\mathbf{J}}_G} \underset{(3\times1)}{\underline{\mathbf{n}}} = J \underset{(3\times1)}{\underline{\mathbf{n}}} \,, \qquad \underset{(1\times3)}{\underline{\mathbf{n}}^T} \underset{(3\times1)}{\underline{\mathbf{n}}} = 1 \,, \tag{1.18}$$

in which J and $\mathbf{n}$ are the unknowns, has at least three real solutions for the vector $\mathbf{n}$. If the number of solutions for $\mathbf{n}$ is equal to three the solution vectors $\mathbf{n}_i$ $(i = 1, 2, 3)$ are mutually perpendicular to each other:

$$\mathbf{n}_i \cdot \mathbf{n}_j = \begin{cases} 1 & \text{if } i = j \\ 0 & \text{if } i \neq j \end{cases} \qquad (i, j = 1, 2, 3) \ . \tag{1.19}$$

Fig. 1.7 Principal directions

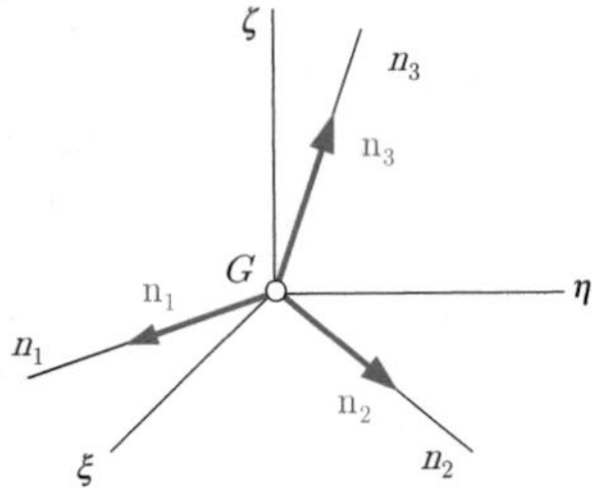

If the number of solutions for $\mathbf{n}$ is more than three then the number of solutions is infinite: one can, however, always select such three solutions which are mutually perpendicular to each other. The direction the solution for $\mathbf{n}_i$ determines is called principal direction. The eigenvalue (called principal moment of inertia) that belongs to the principal direction determined by $\mathbf{n}_i$ is denoted by J_{Gi}. It is well known that

$$\left.\begin{matrix} n_i \\ J_{Gi} \end{matrix}\right\} \text{ is referred to as principal } \left\{\begin{matrix} \text{axis of inertia.} \\ \text{moment of inertia.} \end{matrix}\right.$$

The matrix of inertia is a diagonal one in the coordinate system $(n_1 n_2 n_3)$ constituted by the principal axes:

$$\underset{(n_1 n_2 n_3)}{\underline{\mathbf{J}}_G} = \begin{bmatrix} J_{G1} & 0 & 0 \\ 0 & J_{G2} & 0 \\ 0 & 0 & J_{G3} \end{bmatrix} , \tag{1.20}$$

where we have assumed that

$$J_{G1} \geq J_{G2} \geq J_{G3} \, , \qquad (J_{Gi} > 0 - \underset{(3\times3)}{\underline{\mathbf{J}}_G} \text{ is positive definite!}) \ . \tag{1.21}$$

As regards the eigenvalue problem of symmetric tensors the reader is referred to Sect. A.2.6 in the Appendix (Fig. 1.7).

By using tensorial notations (1.15) can be written as

$$\mathbf{H}_G = \int_V \left(\rho^2\,\boldsymbol{\omega} - \boldsymbol{\rho}\,(\boldsymbol{\rho}\cdot\boldsymbol{\omega})\right)\,\rho\mathrm{d}V = \int_V \left(\rho^2\,\boldsymbol{1} - \boldsymbol{\rho}\circ\boldsymbol{\rho}\right)\,\rho\mathrm{d}V\,\cdot\boldsymbol{\omega},$$

where $\boldsymbol{1}$ is the unit tensor and $\circ$ stands for the operation sign of the dyadic product. Here

$$\boldsymbol{J}_G = \int_V \left(\rho^2\,\boldsymbol{1} - \boldsymbol{\rho}\circ\boldsymbol{\rho}\right)\,\rho\mathrm{d}V \tag{1.22}$$

is the tensor of inertia at G. Using the tensor of inertia we can give $\mathbf{H}_G$ as a product:

$$\mathbf{H}_G = \boldsymbol{J}_G\,\cdot\boldsymbol{\omega}\,. \tag{1.23}$$

The matrix $\underline{\mathbf{J}}_G = \underset{(3\times 3)}{\underline{\mathbf{J}}_G}$ of the tensor of inertia $\boldsymbol{J}_G$ is given by Eqs. (1.17).

Let A be a point of the rigid body ($A \neq G$). The moment of momentum about the point A—see Fig. 1.6 for details—is defined by the equation

$$\mathbf{H}_A = \int_V \hat{\boldsymbol{\rho}} \times \mathbf{v}\rho\,\mathrm{d}V\,. \tag{1.24}$$

It can be proved that the matrix of $\mathbf{H}_A$ can be calculated as

$$\boxed{\underset{(3\times 1)}{\underline{\mathbf{H}}_A} = m\,\underset{(3\times 3)}{\underline{\mathbf{r}}_{AG\times}}\,\underset{(3\times 1)}{\underline{\mathbf{v}}_A} + \underset{(3\times 3)}{\underline{\mathbf{J}}_A}\,\underset{(3\times 1)}{\underline{\boldsymbol{\omega}}} = \underset{(3\times 3)}{\underline{\mathbf{Q}}_{A\times}}\,\underset{(3\times 1)}{\underline{\mathbf{v}}_A} + \underset{(3\times 3)}{\underline{\mathbf{J}}_A}\,\underset{(3\times 1)}{\underline{\boldsymbol{\omega}}}}\,, \tag{1.25a}$$

where

$$m\underset{(3\times 3)}{\underline{\mathbf{r}}_{AG\times}} = \underset{(3\times 3)}{\underline{\mathbf{Q}}_{A\times}} = m\begin{bmatrix} 0 & -\zeta'_{AG} & \eta'_{AG} \\ \zeta'_{AG} & 0 & -\xi'_{AG} \\ -\eta'_{AG} & \xi'_{AG} & 0 \end{bmatrix}, \quad \underset{(3\times 1)}{\underline{\mathbf{v}}_A} = \begin{bmatrix} v_{A\xi'} \\ v_{A\eta'} \\ v_{A\zeta'} \end{bmatrix} \tag{1.25b}$$

$$\underset{(3\times 3)}{\underline{\mathbf{J}}_A} = \begin{bmatrix} J_{\xi'} & -J_{\xi'\eta'} & -J_{\xi'\zeta'} \\ -J_{\eta'\xi'} & J_{\eta'} & -J_{\eta'\zeta'} \\ -J_{\zeta'\xi'} & -J_{\zeta'\eta'} & J_{\zeta'} \end{bmatrix} \tag{1.25c}$$

in which

- ζ'_{AG}, η'_{AG}, ξ'_{AG} are the coordinates of the mass center in the primed coordinate system;
- $v_{A\xi'}$, $v_{A\eta'}$, $v_{A\zeta'}$ are the velocity components in the same coordinate system;
- $J_{\xi'}$, $J_{\eta'}$, $J_{\zeta'}$ are the moments of inertia with respect to the axis ξ', η', ζ'; and
- $J_{\eta'\xi'}$, $J_{\xi'\zeta'}$, $J_{\eta'\zeta'}$ are the corresponding products of inertia.

If the body rotates about the point A, which is regarded as a fixed point, then $\mathbf{v}_A = \mathbf{0}$, consequently Eq. (1.25a) simplifies to

$$\boxed{\underset{(3\times 1)}{\underline{\mathbf{H}}_A} = \underset{(3\times 3)}{\underline{\mathbf{J}}_A} \underset{(3\times 1)}{\underline{\boldsymbol{\omega}}}\,.} \tag{1.26}$$

It also holds that

$$\mathbf{H}_A = \mathbf{H}_G + \mathbf{r}_{AG} \times \mathbf{p}\,. \tag{1.27}$$

1.4 Parallel Axis Theorem

It can be proved that

$$\boxed{\underset{(3\times 3)}{\underline{\mathbf{J}}_A} = \underset{(3\times 3)}{\underline{\mathbf{J}}_G} + m \underbrace{\begin{bmatrix} \eta_{GA}^2 + \zeta_{GA}^2 & -\xi_{GA}\eta_{GA} & -\xi_{GA}\zeta_{GA} \\ -\eta_{GA}\xi_{GA} & \xi_{GA}^2 + \zeta_{GA}^2 & -\eta_{GA}\zeta_{GA} \\ -\zeta_{GA}\xi_{GA} & -\zeta_{GA}\eta_{GA} & \xi_{GA}^2 + \eta_{GA}^2 \end{bmatrix}}_{\underset{(3\times 3)}{\underline{\mathbf{J}}_{GA}}}} \tag{1.28a}$$

or

$$\boxed{\underset{(3\times 3)}{\underline{\mathbf{J}}_A} = \underset{(3\times 3)}{\underline{\mathbf{J}}_G} + m\underset{(3\times 3)}{\underline{\mathbf{J}}_{GA}},} \tag{1.28b}$$

where m is the mass of the body $\mathcal{B}$, while ξ_{GA}, η_{GA}, and ζ_{GA} are the coordinates of the point A with respect to the point G in the coordinate system $(\xi\eta\zeta)$. Equations (1.28) constitute the well-known parallel axis theorem.

1.5 Fundamental Theorems of Kinetics

1.5.1 Introductory Remarks

As is well-known kinetics is a study of motion and its causes. In the present section, we shall give a short summary of the most important concepts and theorems of kinetics. After successfully completing a course in dynamics, the students are expected to have a familiarity with the fundamental concepts and theorems of dynamics. On the other hand, a repetition is the mother of knowledge and refreshes the things you learnt earlier.

1.5.2 Force Systems

Figure 1.8a shows a system of N particles. The j-th particle exerts a force $\mathbf{F}_{ij}$ $(i, j = 1, \ldots, N)$ on the i-th particle: $\mathbf{F}_{ij} = -\mathbf{F}_{ji}$ (law of action and reaction), $\mathbf{F}_{ii} = \mathbf{0}$ (a particle does not exert a force on itself). *The forces* $\mathbf{F}_{ij}$ *are internal forces. The forces exerted on a particle by the bodies (and/or the particles) that do not belong to the system are referred to as external forces.* The resultant of the external forces acting on particle m_i is denoted by $\mathbf{F}_i$.

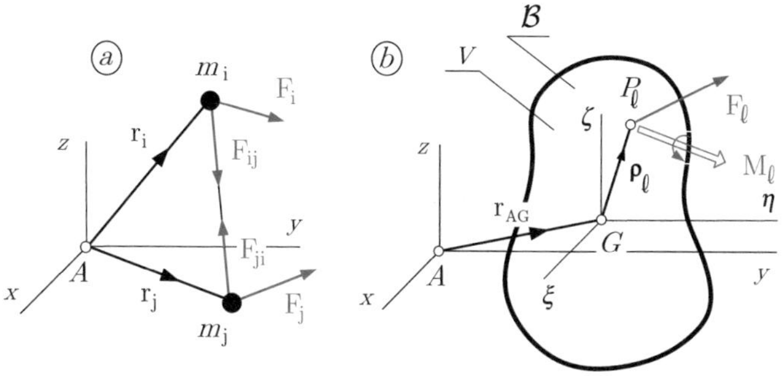

Fig. 1.8 Force systems acting on a system of particles and a rigid body

Figure 1.8b shows a rigid body $\mathcal{B}$. The body is loaded at the points P_ℓ $(\ell = 1, \ldots, K)$: an external force $\mathbf{F}_\ell$ and an external couple $\mathbf{M}_\ell$ are acting on the body at the point P_ℓ—the couples can have values of zero. The force couple system equivalent to the external loads [at A—system of particles] (at G—rigid body) is given by the equations

$$
\begin{aligned}
\mathbf{R} &= \mathbf{F} = \sum_{i=1}^{N} \mathbf{F}_i\,, & \mathbf{R} &= \mathbf{F} = \sum_{\ell=1}^{K} \mathbf{F}_\ell\,, \\
\mathbf{M}_A &= \sum_{i=1}^{N} \mathbf{r}_i \times \mathbf{F}_i\,, & \mathbf{M}_G &= \sum_{\ell=1}^{K} \boldsymbol{\rho}_\ell \times \mathbf{F}_\ell + \sum_{\ell=1}^{K} \mathbf{M}_\ell\,.
\end{aligned}
\tag{1.29}
$$

It is customary to denote the equivalent force couple system by putting the resultant and the moment about the point considered (called simply resultant moment) into a pair of square brackets in which they are separated from each other by a comma: $[\mathbf{R}, \mathbf{M}_A]$ or $[\mathbf{R}, \mathbf{M}_G]$.

For the rigid body $\mathcal{B}$, the moment of the external forces about the point A can be calculated as

$$
\mathbf{M}_A = \mathbf{M}_G + \mathbf{r}_{AG} \times \mathbf{R}, \tag{1.30}
$$

where the right side is the moment of the external forces about the point G plus the moment of the resultant $\mathbf{R}$—which is attached mentally to the point G—about the point A.

Remark 1.2 It is assumed here and later in Sect. 1.5.4 that the external force system is a special one since it includes the forces $\mathbf{F}_\ell$ and couples $\mathbf{M}_\ell$ only. This assumption does not violate the generality: the fundamental theorem of dynamics presented in the subsection cited above remains valid for any loading type: for example, in the case of distributed loads as well.

1.5.3 Systems of Effective Forces

The effective forces for [the system of particles] (the rigid body $\mathcal{B}$) shown in Figs. 1.8 and 1.9 are defined by the relations

$$[\mathcal{K}_i = m_i\,\mathbf{a}_i]\;(\mathrm{d}\mathcal{K} = \mathbf{a}\mathrm{d}m = \mathbf{a}\,\rho\mathrm{d}V).$$

Note that $\mathrm{d}\mathcal{K}$ is the intensity of the effective forces distributed on the volume V of the body.

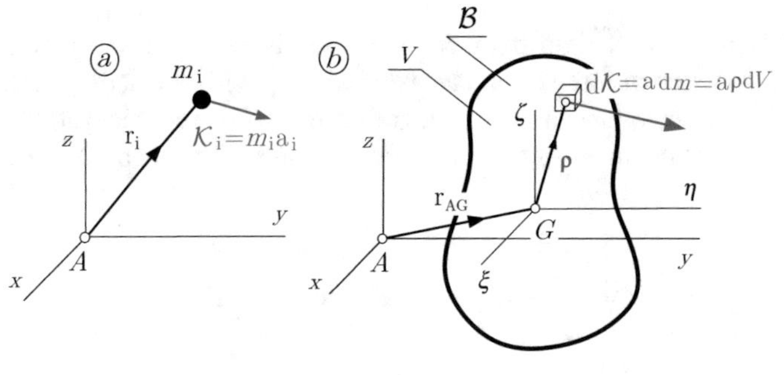

Fig. 1.9 Effective forces

The resultant of the effective forces for the system of particles is

$$\mathcal{K} = \sum_{i=1}^{N} \mathcal{K}_i = \sum_{i=1}^{N} m_i\,\mathbf{a}_i\,. \tag{1.31a}$$

The resultant of the effective forces acting on the rigid body $\mathcal{B}$ can be calculated as

$$\mathcal{K} = \int_V \mathbf{a}\,\rho\mathrm{d}V \underset{(1.6\mathrm{a})}{=\uparrow=} \int_V (\mathbf{a}_G + \boldsymbol{\alpha}\times\boldsymbol{\rho} + \boldsymbol{\omega}\times(\boldsymbol{\omega}\times\boldsymbol{\rho}))\,\rho\mathrm{d}V =$$

$$= \underbrace{\int_V \rho\mathrm{d}V}_{m}\,\mathbf{a}_G + \boldsymbol{\alpha}\times\underbrace{\int_V \boldsymbol{\rho}\,\rho\mathrm{d}V}_{\mathbf{Q}_G=0} + \boldsymbol{\omega}\times\left(\boldsymbol{\omega}\times\underbrace{\int_V \boldsymbol{\rho}\,\rho\mathrm{d}V}_{\mathbf{Q}_G=0}\right) = m\mathbf{a}_G$$

or simply

$$\mathcal{K} = m\,\mathbf{a}_G\,. \tag{1.31b}$$

Remark 1.3 Observe that we have taken into account that (a) $\int_V \rho \mathrm{d}V = m$ and have utilized the fact that the first moment of a rigid body with respect to the mass center vanishes $\mathbf{Q}_G = \mathbf{0}$.

A comparison of (1.10) and (1.31) yields

$$\boxed{\mathcal{K} = \frac{\mathrm{d}\mathbf{p}}{\mathrm{d}t} = \dot{\mathbf{p}}\,.} \tag{1.32}$$

The moment of the effective forces is, in general, denoted by $\mathcal{D}$.

For the system of particles, the moment of the effective forces about the fixed point A is

$$\mathcal{D}_A = \sum_{i=1}^{N} \mathbf{r}_i \times \mathcal{K}_i = \sum_{i=1}^{N} \mathbf{r}_i \times m_i \mathbf{a}_i = \sum_{i=1}^{N} \mathbf{r}_i \times \frac{\mathrm{d}\mathbf{p_i}}{\mathrm{d}t} = \underset{\mathbf{v}_i \times \mathbf{p}_i = \mathbf{0}}{\uparrow} = \\ = \frac{\mathrm{d}}{\mathrm{d}t} \sum_{i=1}^{N} \mathbf{r}_i \times \mathbf{p_i} = \frac{\mathrm{d}}{\mathrm{d}t} \mathbf{H}_A\,. \tag{1.33}$$

Remark 1.4 It is worth emphasizing that the sum

$$\sum_{i=1}^{N} \mathbf{r}_i \times \mathbf{p_i}$$

in the above equation is the moment of momentum with respect to the point A concerning the system of particles considered.

For the rigid body $\mathcal{B}$ shown in Fig. 1.9b, the moment of the effective forces about the mass center G is given by

$$\mathcal{D}_G = \int_V \boldsymbol{\rho} \times \mathrm{d}\mathcal{K} = \int_V \boldsymbol{\rho} \times \mathbf{a}\,\rho \mathrm{d}V = \\ = \underset{(1.6a)}{\uparrow} = \int_V \boldsymbol{\rho} \times (\mathbf{a}_G + \boldsymbol{\alpha} \times \boldsymbol{\rho} + \boldsymbol{\omega} \times (\boldsymbol{\omega} \times \boldsymbol{\rho}))\,\rho \mathrm{d}V = \\ = \int_V \boldsymbol{\rho}\,\rho \mathrm{d}V \times \mathbf{a}_G + \int_V \boldsymbol{\rho} \times (\boldsymbol{\alpha} \times \boldsymbol{\rho})\,\rho \mathrm{d}V + \int_V \boldsymbol{\rho} \times [\boldsymbol{\omega} \times (\boldsymbol{\omega} \times \boldsymbol{\rho})]\,\rho \mathrm{d}V\,. \tag{1.34}$$

Before examining the three integrals on the right side of the above equation we shall introduce the following matrix notations for the moment of the effective forces $\mathcal{D}_G$ (taken about G), the angular acceleration $\boldsymbol{\alpha}$, and the angular velocity $\boldsymbol{\omega}$:

$$\underline{\mathbf{D}}_G = \underset{(3\times1)}{\underline{\mathbf{D}}_G} = \begin{bmatrix} \mathcal{D}_{G\xi} \\ \mathcal{D}_{G\eta} \\ \mathcal{D}_{G\zeta} \end{bmatrix}, \quad \underline{\boldsymbol{\alpha}} = \underset{(3\times1)}{\underline{\boldsymbol{\alpha}}} = \begin{bmatrix} \alpha_\xi \\ \alpha_\eta \\ \alpha_\zeta \end{bmatrix},$$
$$\underline{\boldsymbol{\omega}}_\times = \underset{(3\times3)}{\underline{\boldsymbol{\omega}}_\times} = \begin{bmatrix} 0 & -\omega_\zeta & \omega_\eta \\ \omega_\zeta & 0 & -\omega_\xi \\ -\omega_\eta & \omega_\xi & 0 \end{bmatrix}. \tag{1.35}$$

- For the first integral on the right side of Eq. (1.34), it holds that

$$\underbrace{\int_V \boldsymbol{\rho}\, \rho \mathrm{d}V}_{\mathbf{Q}_G} \times \mathbf{a}_G = \mathbf{Q}_G \times \mathbf{a}_G \underset{(1.9a)}{=\uparrow} = \mathbf{0}\,. \tag{1.36a}$$

- Recalling Eqs. (1.11) and (1.17a), we can come to the conclusion that the final form of the second integral

$$\int_V \boldsymbol{\rho} \times (\boldsymbol{\alpha} \times \boldsymbol{\rho})\, \rho \mathrm{d}V$$

on the right side of Eq. (1.34) can be obtained from (1.17a) if we write α for ω. Consequently, equation

$$\underbrace{\begin{bmatrix} J_\xi & -J_{\xi\eta} & -J_{\xi\zeta} \\ -J_{\eta\xi} & J_\eta & -J_{\eta\zeta} \\ -J_{\zeta\xi} & -J_{\zeta\eta} & J_\zeta \end{bmatrix}}_{\underset{(3\times3)}{\underline{\mathbf{J}}_G}} \underbrace{\begin{bmatrix} \alpha_x \\ \alpha_y \\ \alpha_z \end{bmatrix}}_{\underset{(3\times1)}{\underline{\boldsymbol{\alpha}}}} = \underline{\mathbf{J}}_G\, \underline{\boldsymbol{\alpha}} \tag{1.36b}$$

is the matrix form of the second integral.
- As regards the third integral on the right side of Eq. (1.34) take into account that

$$\boldsymbol{\rho} \times [\boldsymbol{\omega} \times (\boldsymbol{\omega} \times \boldsymbol{\rho})] = \boldsymbol{\omega} \times [\boldsymbol{\rho} \times (\boldsymbol{\omega} \times \boldsymbol{\rho})]\,.$$

(The validity of this equation can be checked with ease if we make use of Eq. (A.1.10) for calculating the value of the triple cross products and take into account that $\boldsymbol{\omega} \times \boldsymbol{\rho}$, which we regard as a single factor when applying the expansion of the cross product, is perpendicular both to $\boldsymbol{\omega}$ and to $\boldsymbol{\rho}$.) Upon substitution of the above equation into the third integral, we get

$$\int_V \boldsymbol{\rho} \times [\boldsymbol{\omega} \times (\boldsymbol{\omega} \times \boldsymbol{\rho})]\, \rho \mathrm{d}V = \underbrace{\boldsymbol{\omega}\times}_{\underline{\boldsymbol{\omega}}_\times} \underbrace{\int_V \boldsymbol{\rho} \times (\boldsymbol{\omega} \times \boldsymbol{\rho})\, \rho \mathrm{d}V}_{\underline{\mathbf{J}}_G\, \underline{\boldsymbol{\omega}}}\,. \tag{1.36c}$$

Observe that the above equation shows the result in matrix notation too.

Making use of Eqs. (1.36) for the moment of the effective forces about the mass center, we get

$$\boxed{\underset{(3\times 1)}{\underline{\mathbf{D}}_G} = \underset{(3\times 3)}{\underline{\mathbf{J}}_G}\,\underset{(3\times 1)}{\underline{\boldsymbol{\alpha}}} + \underset{(3\times 3)}{\underline{\boldsymbol{\omega}}_\times}\,\underbrace{\underset{(3\times 3)}{\underline{\mathbf{J}}_G}\,\underset{(3\times 1)}{\underline{\boldsymbol{\omega}}}}_{\underset{(3\times 1)}{\underline{\mathbf{H}}_G}} = \underset{(3\times 3)}{\underline{\mathbf{J}}_G}\,\underset{(3\times 1)}{\underline{\boldsymbol{\alpha}}} + \underset{(3\times 3)}{\underline{\boldsymbol{\omega}}_\times}\,\underset{(3\times 1)}{\underline{\mathbf{H}}_G}}\,. \tag{1.37}$$

For completeness, here, we give the above equation in tensorial notation as well:

$$\mathcal{D}_G = \boldsymbol{J}_G \cdot \boldsymbol{\alpha} + \boldsymbol{\omega} \times \boldsymbol{J}_G \cdot \boldsymbol{\omega} = \boldsymbol{J}_G \cdot \boldsymbol{\alpha} + \boldsymbol{\omega} \times \mathbf{H}_G\,. \tag{1.38}$$

It can be proved that

$$\mathcal{D}_G = \frac{\mathrm{d}}{\mathrm{d}t}\mathbf{H}_G\,. \tag{1.39}$$

Let A be a point of the rigid body $\mathcal{B}$ ($A \neq G$). The moment of the effective forces about the point A—see Fig. 1.10 for details—is defined by the equation

$$\mathcal{D}_A = \int_V \hat{\boldsymbol{\rho}} \times \mathbf{a}\rho\,\mathrm{d}V\,. \tag{1.40}$$

It can be proved by using the definition given above that

$$\boxed{\underset{(3\times 1)}{\underline{\boldsymbol{D}}_A} = \underset{(3\times 3)}{\underline{\mathbf{Q}}_{A\times}}\,\underset{(3\times 1)}{\underline{\mathbf{a}}_A} + \underset{(3\times 3)}{\underline{\mathbf{J}}_A}\,\underset{(3\times 1)}{\underline{\boldsymbol{\alpha}}} + \underset{(3\times 3)}{\underline{\boldsymbol{\omega}}_\times}\,\underset{(3\times 3)}{\underline{\mathbf{J}}_A}\,\underset{(3\times 1)}{\underline{\boldsymbol{\omega}}}\,,} \tag{1.41a}$$

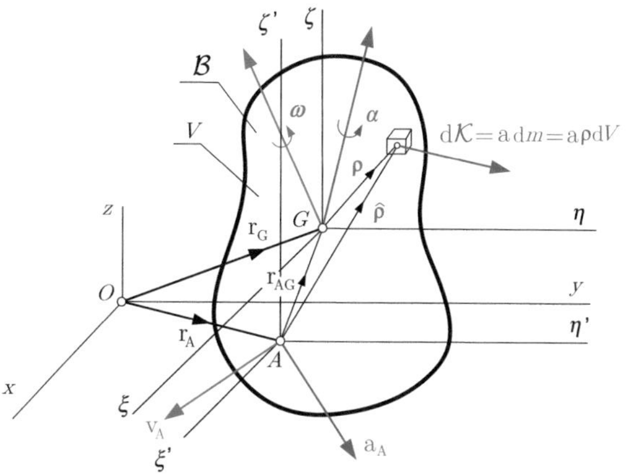

Fig. 1.10 Effective forces on a rigid body depicted for the calculation of $\underline{\boldsymbol{D}}_A$

where

$$\underset{(3\times3)}{\underline{\mathbf{Q}}_{A\times}} = m\underline{\mathbf{r}}_{AG\times} = m\begin{bmatrix} 0 & -\zeta'_{AG} & \eta'_{AG} \\ \zeta'_{AG} & 0 & -\xi'_{AG} \\ -\eta'_{AG} & \xi'_{AG} & 0 \end{bmatrix}, \quad \underline{\mathbf{a}}_A = \begin{bmatrix} a_{A\xi'} \\ a_{A\eta'} \\ a_{A\zeta'} \end{bmatrix}. \tag{1.41b}$$

Here we give, again for completeness, Eq. (1.41a) in tensorial notation as well:

$$\mathcal{D}_A = \underbrace{m\mathbf{r}_{AG}}_{\mathbf{Q}_A} \times \mathbf{a}_A + \boldsymbol{J}_A \cdot \boldsymbol{\alpha} + \boldsymbol{\omega} \times \boldsymbol{J}_A \cdot \boldsymbol{\omega} = \mathbf{Q}_A \times \mathbf{a}_A + \boldsymbol{J}_A \cdot \boldsymbol{\alpha} + \boldsymbol{\omega} \times \boldsymbol{J}_A \cdot \boldsymbol{\omega}. \tag{1.42}$$

If the body rotates about the point A, which is now regarded as a fixed point, then $\mathbf{v}_A = \mathbf{a}_A = \mathbf{0}$. Hence, Eq. (1.41a) simplifies to

$$\underset{(3\times1)}{\underline{\boldsymbol{D}}_A} = \underset{(3\times3)}{\underline{\mathbf{J}}_A}\ \underset{(3\times1)}{\underline{\boldsymbol{\alpha}}} + \underset{(3\times3)}{\underline{\boldsymbol{\omega}}_\times}\ \underset{(3\times3)}{\underline{\mathbf{J}}_A}\ \underset{(3\times1)}{\underline{\boldsymbol{\omega}}}\ . \tag{1.43}$$

Assume that the body $\mathcal{B}$ is in plane motion. Let the coordinate plane xy be the plane of motion. If the body is in plane motion the velocities and accelerations are all parallel to the plane of motion. Hence, the moment of momentum as well as the moment of the effective forces should be perpendicular to that plane. If there are no constraints to prevent the motion of the body $\mathcal{B}$ in the direction z (or which is the same perpendicularly to the plane of motion) the only way this condition can be satisfied is that the products of inertia $J_{\zeta\xi} = J_{\zeta\eta} = J_{\zeta'\xi'} = J_{\zeta'\eta'}$ all vanish. Since for the plane motion considered $\boldsymbol{\omega} = \omega_z\mathbf{i}_z = \omega\mathbf{i}_z$ and $\boldsymbol{\alpha} = \alpha_z\mathbf{i}_z = \alpha\mathbf{i}_z$ it follows that

$$\underset{(3\times3)}{\underline{\mathbf{J}}_G}\ \underset{(3\times1)}{\underline{\boldsymbol{\alpha}}} = \begin{bmatrix} 0 \\ 0 \\ J_\zeta\alpha \end{bmatrix} \quad \text{and} \quad \underset{(3\times3)}{\underline{\mathbf{J}}_G}\ \underset{(3\times1)}{\underline{\boldsymbol{\omega}}} = \begin{bmatrix} 0 \\ 0 \\ J_\zeta\omega \end{bmatrix}. \tag{1.44}$$

Consequently,

$$\underset{(3\times3)}{\underline{\boldsymbol{\omega}}_\times}\ \underset{(3\times3)}{\underline{\mathbf{J}}_G}\ \underset{(3\times1)}{\underline{\boldsymbol{\omega}}} = 0, \tag{1.45}$$

and Eq. (1.37) simplifies to

$$\mathcal{D}_G = J_\zeta\alpha\,\mathbf{i}_z\,. \tag{1.46}$$

If the body in plane motion rotates about the fixed point A we get in the same way that

$$\mathcal{D}_A = J_{\zeta'}\alpha\,\mathbf{i}_z. \tag{1.47}$$

1.5.4 *The Fundamental Theorem of Dynamics*

Assume that the rigid body $\mathcal{B}$ is loaded at the point P_ℓ ($\ell = 1, \ldots, K$) by an external force $\mathbf{F}_\ell$ and an external couple $\mathbf{M}_\ell$—see Fig. 1.11. They together constitute the system of external forces (in a general sense since couples are also included in this system) acting on the body. Figure 1.11 shows the system of effective forces as well. The fundamental theorem of dynamics states that the system of effective forces is statically equivalent to the system of external forces. Consequently, it holds that

$$[\mathbf{a}\rho\, \mathrm{d}V] \overset{m}{=} [\mathbf{F}_1, \ldots, \mathbf{F}_\ell, \ldots, \mathbf{F}_K; \mathbf{M}_1, \ldots, \mathbf{M}_\ell, \ldots, \mathbf{M}_K],$$

where the letter m over the equality sign expresses the fact that the two force systems are statically equivalent (they have the same moment space).

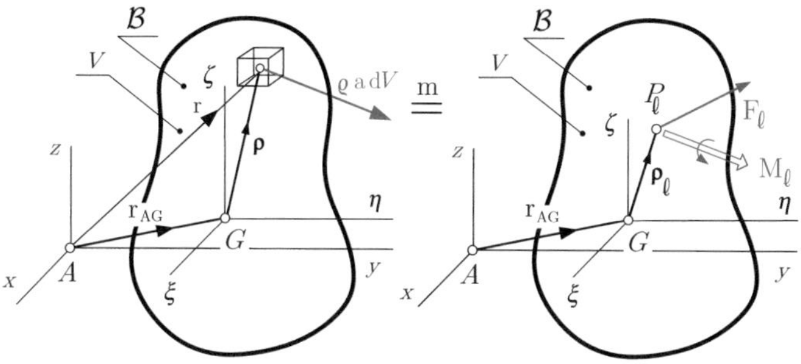

Fig. 1.11 Equivalence of the external and effective forces

We have already seen—recall Eq. (1.29)—that the equivalent force couple system $[\mathbf{R}, \mathbf{M}_G]$ for the external loads at the point G is given by

$$\mathbf{R} = \sum_{\ell=1}^{K} \mathbf{F}_\ell\,, \qquad \mathbf{M}_G = \sum_{\ell=1}^{K} \boldsymbol{\rho}_\ell \times \mathbf{F}_\ell + \sum_{\ell=1}^{K} \mathbf{M}_\ell\,. \tag{1.48a}$$

As regards the effective forces, the equivalent force couple system at G is of the form $[\mathcal{K}, \mathcal{D}_G]$. It follows from Eqs. (1.31b) and (1.37) that $\mathcal{K}$ and $\mathcal{D}_G$ can be calculated as

$$\underset{(3\times1)}{\underline{\boldsymbol{K}}} = m \underset{(3\times1)}{\underline{\mathbf{a}}_G}\,, \qquad \underset{(3\times1)}{\underline{\boldsymbol{D}}_G} = \underset{(3\times3)}{\underline{\mathbf{J}}_G} \underset{(3\times1)}{\underline{\boldsymbol{\alpha}}} + \underset{(3\times3)}{\underline{\boldsymbol{\omega}}_\times} \underset{(3\times1)}{\underline{\mathbf{H}}_G}\,. \tag{1.48b}$$

In tensorial notation, we can write

$$\mathcal{K} = m\,\mathbf{a}_G\,, \qquad \mathcal{D}_G = \mathbf{J}_G \cdot \boldsymbol{\alpha} + \boldsymbol{\omega} \times \mathbf{H}_G\,. \tag{1.48c}$$

For two force systems to be statically equivalent, it is necessary and sufficient that the resultants and moment resultants are equal:

$$\boxed{\mathcal{K} = m\,\mathbf{a}_G = \mathbf{R}\,, \qquad \mathcal{D}_G = \mathbf{M}_G\,.} \tag{1.49a}$$

These equations constitute the equations of motion for a rigid body. If the body rotates, say, about the fixed point A then the second equation takes the form

$$\mathcal{D}_A = \mathbf{M}_A\,. \tag{1.49b}$$

Since $\mathcal{K} = \dot{\mathbf{p}}$ (see Eq. (1.32)) and $\mathcal{D}_G = \dot{\mathbf{H}}_G$ (see Eqs. (1.39)) Eqs. (1.49a) can also be written in the form

$$\dot{\mathbf{p}} = \mathbf{R}\,, \qquad \dot{\mathbf{H}}_G = \mathbf{M}_G\,. \tag{1.49c}$$

Let t_1 and t_2 be two different points of time ($t_1 < t_2$). Integrate equations (1.49c) with respect to time. We obtain

$$\mathbf{p}(t_2) - \mathbf{p}(t_1) = \mathbf{p}_2 - \mathbf{p}_1 = \int_{t_1}^{t_2} \mathbf{R}\,\mathrm{d}t\,,$$

$$\mathbf{H}_G(t_2) - \mathbf{H}_G(t_1) = \mathbf{H}_{G2} - \mathbf{H}_{G1} = \int_{t_1}^{t_2} \mathbf{M}_G \mathrm{d}t$$

or

$$\boxed{\mathbf{p}_2 = \mathbf{p}_1 + \int_{t_1}^{t_2} \mathbf{R}\,\mathrm{d}t\,, \qquad \mathbf{H}_{G2} = \mathbf{H}_{G1} + \int_{t_1}^{t_2} \mathbf{M}_G \mathrm{d}t\,.} \tag{1.49d}$$

Equations (1.49d) are known as the principles of impulse and momentum for a rigid body [1].

1.5.5 Kinetic Energy of a Rigid Body. Power

The kinetic energy is denoted by $\mathcal{E}$. It is obvious from Fig. 1.12 that

$$\mathrm{d}\mathcal{E} = \frac{1}{2}\mathbf{v}^2\,\mathrm{d}m = \frac{1}{2}\mathbf{v}^2\,\rho \mathrm{d}V\,.$$

Thus

$$\mathcal{E} = \frac{1}{2}\int_V \mathbf{v}^2\,\rho \mathrm{d}V = \underset{(1.4)}{\uparrow} = \frac{1}{2}\int_V \mathbf{v}\cdot\left(\mathbf{v}_A + \boldsymbol{\omega}\times\hat{\boldsymbol{\rho}}\right)\rho \mathrm{d}V = \underset{(1.10)}{\uparrow}\;\underset{(1.24)}{\uparrow} =$$

$$= \frac{1}{2}\underbrace{\int_V \mathbf{v}\,\rho \mathrm{d}V}_{\mathbf{p}}\cdot\mathbf{v}_A + \frac{1}{2}\boldsymbol{\omega}\cdot\underbrace{\int_V \hat{\boldsymbol{\rho}}\times\mathbf{v}\,\rho \mathrm{d}V}_{\mathbf{H}_A}$$

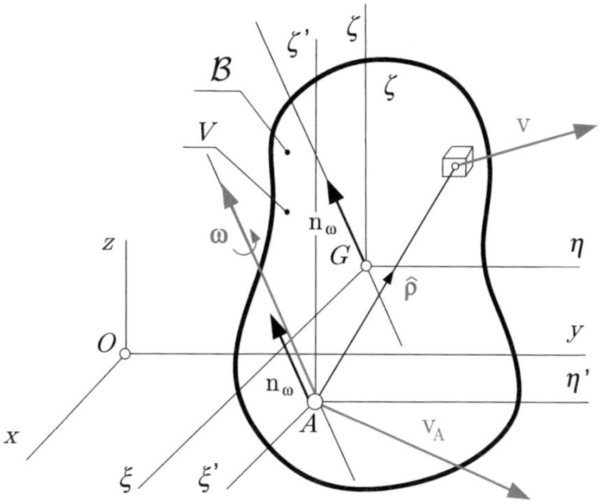

Fig. 1.12 Representation of the velocities for calculating the kinetic energy

or shortly

$$\boxed{\mathcal{E} = \frac{1}{2}\left(\mathbf{v}_A \cdot \mathbf{p} + \boldsymbol{\omega} \cdot \mathbf{H}_A\right) .} \tag{1.50}$$

Special cases:

- If $A = G$ then

$$\mathbf{v}_G \cdot \mathbf{p} \underset{(1.10)}{=\uparrow=} m\,\mathbf{v}_G^2 = m \underset{(1\times3)}{\underline{\mathbf{v}}_G^T}\ \underset{(3\times1)}{\underline{\mathbf{v}}_G} .$$

Let us give $\boldsymbol{\omega}$ in the form $\boldsymbol{\omega} = \omega\,\mathbf{n}_\omega$, $|\mathbf{n}_\omega| = 1$. Then we can write

$$\boldsymbol{\omega} \cdot \mathbf{H}_G \underset{(1.17a)}{=\uparrow} \underset{(1.17d)_1}{\uparrow=} \omega^2 \underbrace{\underset{(1\times3)}{\underline{\mathbf{n}}_\omega^T}\ \underset{(3\times3)}{\underline{\mathbf{J}}_G}\ \underset{(3\times1)}{\underline{\mathbf{n}}_\omega}}_{J_{G\omega}} ,$$

where $J_{G\omega}$ is the moment of inertia with respect to an axis parallel to $\boldsymbol{\omega}$ and passing through the mass center G. Consequently,

$$\boxed{\mathcal{E} = \frac{1}{2}\left(m \underset{(1\times3)}{\underline{\mathbf{v}}_G^T}\underset{(3\times1)}{\underline{\mathbf{v}}_G} + \omega^2 \underset{(1\times3)}{\underline{\mathbf{n}}_\omega^T}\ \underset{(3\times3)}{\underline{\mathbf{J}}_G}\ \underset{(3\times1)}{\underline{\mathbf{n}}_\omega}\right) = \frac{1}{2}\left(m\,\mathbf{v}_G^2 + J_{G\omega}\,\omega^2\right).} \tag{1.51a}$$

If $\omega = 0$ the above equation simplifies to

$$\mathcal{E} = \frac{1}{2} m\,\mathbf{v}_G^2 . \tag{1.51b}$$

- If $\mathbf{v}_A = \mathbf{0}$ then a similar line of thought yields

$$\boxed{\mathcal{E} = \frac{1}{2}\omega^2 \underbrace{\underset{(1\times3)}{\underline{\mathbf{n}}_\omega^T}\,\underset{(3\times3)}{\underline{\mathbf{J}}_A}\,\underset{(3\times1)}{\underline{\mathbf{n}}_\omega}}_{J_{A\omega}} = \frac{1}{2} J_{A\omega}\,\omega^2}, \tag{1.52}$$

where $J_{A\omega}$ is the moment of inertia with respect to an axis parallel to ω and passing through the point A.

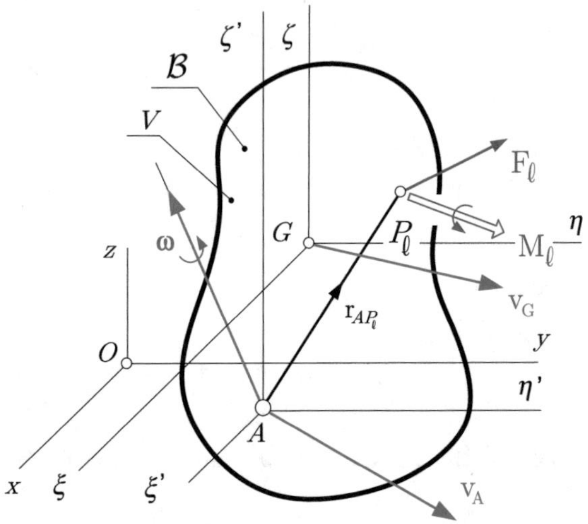

Fig. 1.13 Forces and velocities on a rigid body

Assume that the rigid body $\mathcal{B}$ is loaded at the points P_ℓ ($\ell = 1, \ldots, K$) by an external force $\mathbf{F}_\ell$ and an external couple $\mathbf{M}_\ell$—see Fig. 1.13 for details.

The equivalent force couple system at the point A is given by

$$\mathbf{R} = \mathbf{F} = \sum_{\ell=1}^{K} \mathbf{F}_\ell\,, \qquad \mathbf{M}_A = \sum_{\ell=1}^{K} \mathbf{M}_\ell + \sum_{\ell=1}^{K} \mathbf{r}_{AP_\ell} \times \mathbf{F}_\ell\,. \tag{1.53}$$

As is well known

$$\mathcal{P}_{\ell F} = \mathbf{F}_\ell \cdot \mathbf{v}_\ell \quad \text{and} \quad \mathcal{P}_{\ell M} = \mathbf{M}_\ell \cdot \boldsymbol{\omega} \tag{1.54}$$

are the powers of the force $\mathbf{F}_\ell$ and couple $\mathbf{M}_\ell$, where $\mathbf{v}_\ell$ is the velocity at P_ℓ and $\boldsymbol{\omega}$ is the angular velocity of the body. Hence, the power of this force system is the sum

$$\mathcal{P} = \sum_{\ell=1}^{K} (\mathcal{P}_{\ell F} + \mathcal{P}_{\ell M}) = \sum_{\ell=1}^{K} \mathbf{F}_\ell \cdot (\mathbf{v}_A + \boldsymbol{\omega} \times \mathbf{r}_{AP_\ell}) + \sum_{\ell=1}^{K} \mathbf{M}_\ell \cdot \boldsymbol{\omega} =$$

$$= \mathbf{v}_A \cdot \sum_{\ell=1}^{K} \mathbf{F}_\ell + \left[\sum_{\ell=1}^{K} \mathbf{M}_\ell + \sum_{\ell=1}^{K} \mathbf{r}_{AP_\ell} \times \mathbf{F}_\ell \right] \cdot \boldsymbol{\omega} = \mathbf{R} \cdot \mathbf{v}_A + \mathbf{M}_A \cdot \boldsymbol{\omega} . \tag{1.55a}$$

If $A = G$ Eq. (1.55a) is of the form

$$\mathcal{P} = \mathbf{R} \cdot \mathbf{v}_G + \mathbf{M}_G \cdot \boldsymbol{\omega} , \tag{1.55b}$$

in which $\mathbf{M}_G$ is the moment of the external forces about the mass center G.

1.5.6 Principle of Work and Energy

It can be shown that

$$\boxed{\frac{\mathrm{d}}{\mathrm{d}t}\mathcal{E} = \mathcal{P} .} \tag{1.56}$$

This equation is the differential form of the principle of work and energy.

For $A = G$, Eq. (1.50) assumes the form

$$\mathcal{E} = \frac{1}{2}(\mathbf{v}_G \cdot \mathbf{p} + \boldsymbol{\omega} \cdot \mathbf{H}_G) = \underbrace{\frac{1}{2}\mathbf{v}_G \cdot \mathbf{p}}_{\mathcal{E}_v} + \underbrace{\frac{1}{2}\boldsymbol{\omega} \cdot \mathbf{H}_G}_{\mathcal{E}_\omega}$$

As regards the time derivative of $\mathcal{E}_v$, we get

$$\dot{\mathcal{E}}_v = \frac{1}{2}\mathbf{a}_G \cdot \underbrace{\mathbf{p}}_{m\mathbf{v}_G} + \frac{1}{2}\mathbf{v}_G \cdot \dot{\mathbf{p}} = \frac{1}{2}\mathbf{v}_G \cdot \underbrace{m\mathbf{a}_G}_{\dot{\mathbf{p}}} + \frac{1}{2}\mathbf{v}_G \cdot \dot{\mathbf{p}} = \underset{(1.48\text{c})}{\uparrow} \underset{(1.49\text{a})_1}{\uparrow} = \mathbf{v}_G \cdot \underbrace{\dot{\mathbf{p}}}_{\mathbf{R}} = \mathbf{R} \cdot \mathbf{v}_G .$$

For the time derivative of $\mathcal{E}_\omega$, a similar line of thought yields

$$\dot{\mathcal{E}}_\omega = \frac{1}{2}\boldsymbol{\alpha} \cdot \mathbf{H}_G + \frac{1}{2}\boldsymbol{\omega} \cdot \dot{\mathbf{H}}_G = \underset{(1.17\text{a})}{\uparrow} \underset{(1.39)}{\uparrow} = \frac{1}{2}\boldsymbol{\alpha} \cdot \mathbf{J}_G \cdot \boldsymbol{\omega} + \frac{1}{2}\boldsymbol{\omega} \cdot \boldsymbol{\mathcal{D}}_G = \underset{(1.49\text{a})_2}{\uparrow} \underset{(1.48\text{c})_2}{\uparrow} =$$

$$= \frac{1}{2}\boldsymbol{\omega} \cdot \underbrace{\mathbf{J}_G \cdot \boldsymbol{\alpha}}_{=\boldsymbol{\mathcal{D}}_G - \boldsymbol{\omega} \times \mathbf{H}_G} + \frac{1}{2}\mathbf{M}_G \cdot \boldsymbol{\omega} = \frac{1}{2}\boldsymbol{\omega} \cdot \boldsymbol{\mathcal{D}}_G - \underbrace{\frac{1}{2}\boldsymbol{\omega} \cdot (\boldsymbol{\omega} \times \mathbf{H}_G)}_{=\mathbf{0}} + \frac{1}{2}\mathbf{M}_G \cdot \boldsymbol{\omega} = \mathbf{M}_G \cdot \boldsymbol{\omega} .$$

Consequently,

$$\frac{\mathrm{d}}{\mathrm{d}t}\mathcal{E} = \dot{\mathcal{E}} = \mathbf{R} \cdot \mathbf{v}_G + \mathbf{M}_G \cdot \boldsymbol{\omega} = \mathcal{P}, \tag{1.57}$$

which is a proof of Eq. (1.56).

Let t_1 and t_2 be two different points of time ($t_1 < t_2$). Integrate equation (1.56) with respect to time and take into account that the time integral of the power $\mathcal{P}$ is the work, which is denoted, in general, by W, done by the external forces in the time interval $[t_1, t_2]$. We get the integral form of the principle of work and energy:

$$\boxed{\mathcal{E}_2 - \mathcal{E}_1 = \int_{t_1}^{t_2} \mathcal{P}\mathrm{d}t = W_{12}\,.} \tag{1.58a}$$

Remark 1.5 If the external forces have a potential $\mathcal{U}$ then $W_{12} = \mathcal{U}_1 - \mathcal{U}_2$. Consequently

$$\boxed{\mathcal{E}_2 + \mathcal{U}_2 = \mathcal{E}_1 + \mathcal{U}_1\,.} \tag{1.58b}$$

This is the principle of the energy conservation.

Exercise 1.1 A cylinder with weight $m_{cyl}\,g$ (g is the gravitational acceleration) is lifted up by the crane shown in Fig. 1.14. Let J_e be the moment of inertia of the engine with respect to its axis. Further let $m_e\,g$ be the weight of the engine. The simply supported beam AB is a rigid one and the rope is inextensible. Using the notations of Fig. 1.14—M is the lifting moment, φ is the angle of rotation about the axis of the engine ($\varphi = 0$ if $z = h_o$)—determine the motion equation of the cylinder.

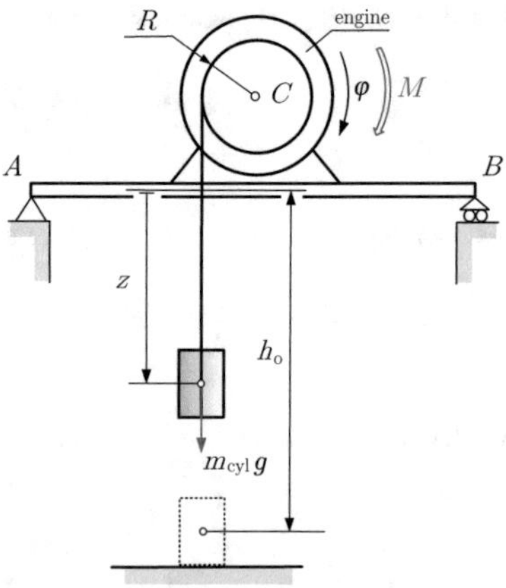

Fig. 1.14 A crane is lifting up a weight

It is clear from Fig. 1.14 that $z = h_o - R\varphi$ from where $\dot{z} = -R\dot{\varphi}$. The total kinetic energy is the sum of the kinetic energies the engine and the cylinder have. Using (1.51a), we get

$$\mathcal{E} = \frac{1}{2}\, m_{cyl}\dot{z}^2 + \frac{1}{2}\, J_e\dot{\varphi}^2 = \frac{1}{2}\, \underbrace{\left(m_{cyl} R^2 + J_e\right)}_{m_{gen}}\dot{\varphi}^2 = \frac{1}{2} m_{gen}\dot{\varphi}^2 \tag{1.59}$$

in which m_{gen} is referred to as generalized mass. The power of the external forces is due to the lifting moment M and the weight of the cylinder:

$$\mathcal{P} = M\dot{\varphi} + m_{cyl}\, g\, \dot{z} = \left(M - m_{cyl}\, gR\right)\dot{\varphi}. \tag{1.60}$$

Upon substitution of Eqs. (1.59) and (1.60) into the principle of work and energy (1.56), we obtain

$$m_{gen}\dot{\varphi}\,\ddot{\varphi} = \left(M - m_{cyl}\, gR\right)\dot{\varphi}$$

or

$$m_{gen}\ddot{\varphi} - M = -m_{cyl}\, gR\,. \tag{1.61}$$

If the lifting moment M is constant we have a simple linear differential equation (DE for short) with constant coefficients. It is worth mentioning that M, in general, depends on the angular velocity: $M = M(\dot{\varphi})$. Then, Eq. (1.61) is, in most cases, a non-linear differential equation.

Exercise 1.2 Assume that the beam and the rope in Exercise 1.1 are elastic. Further assume that their masses are negligible when compared to those of the engine and the cylinder. Determine the kinetic energy of the system (Fig. 1.15).

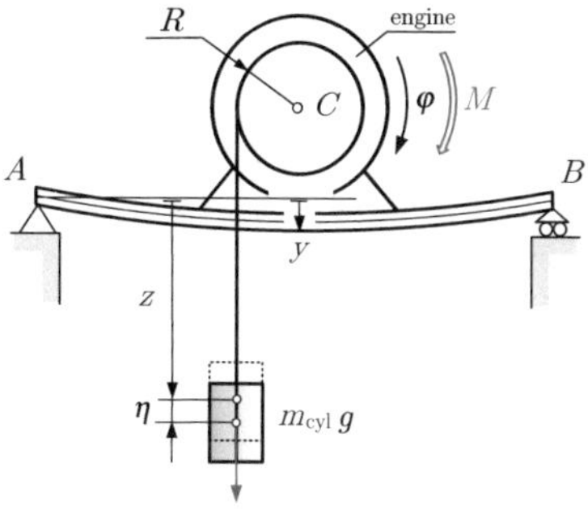

Fig. 1.15 A crane on an elastic beam is lifting up a weight

Let y be the vertical displacement at the middle of the beam. The elastic elongation of the rope is denoted by η. The total kinetic energy of the system is a sum: the kinetic energy of the engine plus that of the cylinder. Making use of (1.51a) and (1.52) we can write

$$\begin{aligned}
\mathcal{E} &= \frac{1}{2}\, m_{\rm cyl}\, (\dot{z} + \dot{\eta})^2 + \frac{1}{2}\, J_e\, \dot{\varphi}^2 + \frac{1}{2}\, m_e\, \dot{y}^2 = \\
&= \frac{1}{2}\, m_{\rm cyl}\, (-R\dot{\varphi} + \dot{\eta})^2 + \frac{1}{2}\, J_e\, \dot{\varphi}^2 + \frac{1}{2}\, m_e\, \dot{y}^2 = \\
&= \frac{1}{2}\, m_{\rm cyl}\, \left(R^2\dot{\varphi}^2 - 2R\dot{\varphi}\dot{\eta} + \dot{\eta}^2\right) + \frac{1}{2}\, J_e\, \dot{\varphi}^2 + \frac{1}{2}\, m_e\, \dot{y}^2
\end{aligned}$$

from where

$$\mathcal{E} = \frac{1}{2}\left[\dot{\varphi}\ \dot{y}\ \dot{\eta}\right] \underbrace{\begin{bmatrix} m_{\text{cyl}}R^2 + J_e & 0 & -m_{\text{cyl}}R \\ 0 & m_e & 0 \\ -m_{\text{cyl}}R & 0 & m_{\text{cyl}} \end{bmatrix}}_{\underset{(3\times 3)}{\underline{\mathbf{M}}}} \begin{bmatrix} \dot{\varphi} \\ \dot{y} \\ \dot{\eta} \end{bmatrix}$$

in which $\underline{\mathbf{M}} = \underline{\mathbf{M}}^T$ is the mass matrix.

Exercise 1.3 The mass m, the mass center G, i.e., r_G, and the geometric data are all known for the crankrod shown in Fig. 1.16. Perform measurements to determine the moment of inertia J_a of the crankrod with respect to the axis about which it rotates.

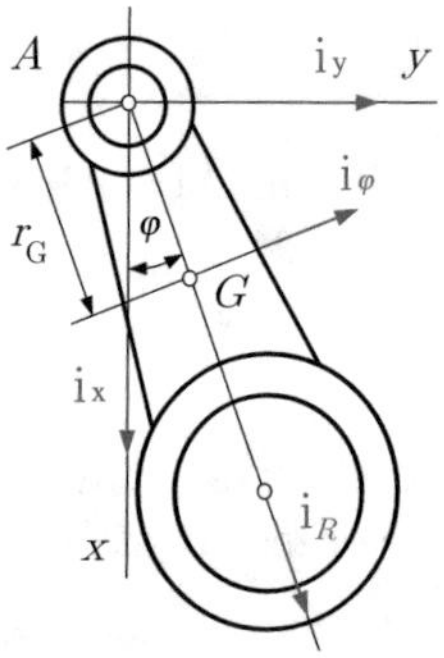

Fig. 1.16 A crankrod

Solution to the problem raised is based on the equation of motion that describes the rotation of the crankrod about A. We shall assume small rotations. Since

$$\mathcal{E} = \frac{1}{2}\,J_a\,\dot{\varphi}^2$$

is the kinetic energy of the crankrod and

$$\begin{aligned}\mathcal{P} = \mathbf{W}\cdot\mathbf{v}_G &= mg\,\mathbf{i}_x\cdot r_G\dot{\varphi}\,\mathbf{i}_\varphi = \\ &= mg\,\mathbf{i}_x\cdot r_G\dot{\varphi}\underbrace{\left(-\sin\varphi\,\mathbf{i}_x + \cos\varphi\,\mathbf{i}_y\right)}_{\mathbf{i}_\varphi} = \\ &= -mg\,r_G\dot{\varphi}\sin\varphi\end{aligned}$$

is the power of the weight $\mathbf{W} = mg\,\mathbf{i}_x$ of the crankrod it follows from the principle of work and energy (1.56) that

$$J_a\dot{\varphi}\,\ddot{\varphi} = -mg\,r_G\dot{\varphi}\sin\varphi$$

or

$$J_a\,\ddot{\varphi} + mg\,r_G\sin\varphi = 0\,.$$

Small rotation has been assumed. Then $|\varphi| \ll 1$ and $\sin\varphi = \varphi - \frac{\varphi^3}{3!} + \frac{\varphi^5}{5!} + \cdots \approx \varphi$. Consequently

$$\ddot{\varphi} + \underbrace{\frac{mg\,r_G}{J_a}}_{\omega_n^2}\varphi = \ddot{\varphi} + \omega_n^2\varphi = 0 \qquad (1.62)$$

is the linearized equation of motion. Its general solution is of the form

$$\varphi = C_1 \sin \omega_n t + C_2 \cos \omega_n t = \\ = \underbrace{A_o \cos \phi}_{C_1} \sin \omega_n t + \underbrace{A_o \sin \phi}_{C_2} \cos \omega_n t = A_o \sin(\omega_n t + \phi)$$

in which C_1 and C_2 are integration constants (they depend on the initial conditions), A_o is the amplitude, and ϕ is the so-called phase angle. The solution obtained shows that the crankrod performs harmonic vibration with an angular velocity ω_n, consequently the period of vibrations τ_n is

$$\tau_n = \frac{2\pi}{\omega_n} = 2\pi \sqrt{\frac{J_a}{mg\, r_G}} . \tag{1.63}$$

After we have measured the period of vibrations τ_n Eq. (1.63) can be used to calculate the moment of inertia sought:

$$J_a = \frac{1}{4\pi^2} mg\, r_G \tau_n^2 . \tag{1.64}$$

Exercise 1.4 Generalize the line of thought presented in the previous exercise for the case of an arbitrary rigid body.

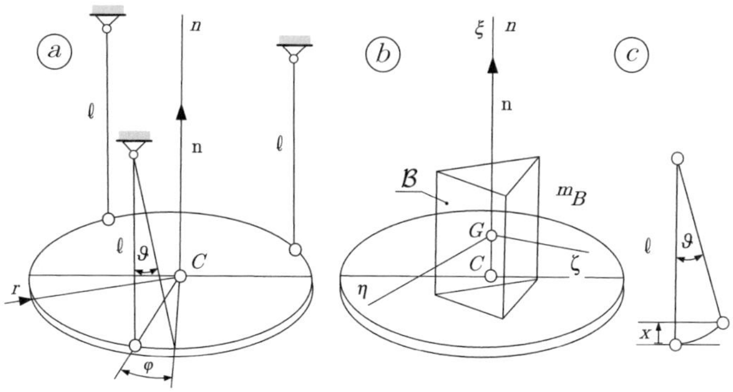

Fig. 1.17 A rigid circular tray

Figure 1.17a shows a rigid circular tray and the way it is hung up. Figure 1.17b shows the body $\mathcal{B}$ on the tray. Its mass center G is located on the vertical center line n ($|\mathbf{n}| = 1$) of the tray. The tray and the body can perform small torsional vibrations. The equation of motion established for the torsional vibrations makes it possible to determine the moment of inertia $J_n = J_\xi$ of body $\mathcal{B}$.

It is not too difficult to check that the following approximative kinematic relations hold:

$$\vartheta\ell = r\varphi \qquad \longrightarrow \qquad \vartheta = \frac{r}{\ell}\varphi\,,$$

and

$$x = \ell\,(1-\cos\vartheta) = \ell\left(1-\cos\frac{r}{\ell}\varphi\right)$$

from where we obtain

$$\dot{x} = r\dot{\varphi}\sin\frac{r}{\ell}\varphi \underset{\frac{r}{\ell}\varphi\ll 1}{=\uparrow\cong} \frac{r^2}{\ell}\dot{\varphi}\varphi\,.$$

Let m_{tray} be the mass of the tray. Let us denote the moment of inertia of the tray with respect to the axis n by J_{tray}. Then

$$\mathcal{E} = \frac{1}{2}\left(J_n + J_{\text{tray}}\right)\dot{\varphi}^2 + \frac{1}{2}\left(m_B + m_{\text{tray}}\right)\dot{x}^2$$

is the kinetic energy of the system. In practice $J_n = J_\xi \gg J_{\text{tray}}$ and the part of the kinetic energy due to the vertical motion can be neglected. Consequently, we may assume that

$$\mathcal{E} \cong \frac{1}{2}J_n\dot{\varphi}^2\,.$$

The power of the external forces is given by

$$\mathcal{P} = -\underbrace{\left(m_B + m_{\text{tray}}\right)}_{m}g\dot{x} = -mg\dot{x} = -mg\frac{r^2}{\ell}\dot{\varphi}\varphi\,.$$

With the knowledge of $\mathcal{E}$ and $\mathcal{P}$ we get the equation of motion from the principle of work and energy (1.56):

$$J_n\,\ddot{\varphi} + mg\,\frac{r^2}{\ell}\varphi = 0$$

or which is the same

$$\ddot{\varphi} + \underbrace{\frac{mg\,r^2}{\ell J_n}}_{\omega_n^2}\varphi = 0\,. \tag{1.65}$$

Since the torsional vibrations are harmonic it follows that

$$\tau_n = \frac{2\pi}{\omega_n}\,, \qquad \text{i.e.,} \qquad J_n = J_\xi = \frac{mg\,r^2}{4\pi^2\ell}\tau_n^2. \tag{1.66}$$

Assume now that the coordinate system $(\xi\eta\zeta)$ is rigidly attached to the mass center G of the body. Then, as we have just seen, $J_n = J_\xi$. Put the body $\mathcal{B}$ onto the tray in such a way that

$$
\begin{array}{llll}
\text{(a)} & \mathbf{n}=\mathbf{i}_\xi,\, \mathbf{i}_\eta,\, \mathbf{i}_\zeta & \text{then we get} & J_n = J_\xi,\ J_\eta,\ J_\zeta; \\
\text{(b)} & \mathbf{n}=\frac{1}{2}\left(\mathbf{i}_\xi + \mathbf{i}_\eta\right) & & J_n = \underset{(1\times3)}{\underline{\mathbf{n}}^T}\,\underset{(3\times3)}{\underline{\mathbf{J}}_G}\,\underset{(3\times1)}{\underline{\mathbf{n}}} = \frac{1}{2}\left(J_\xi + J_\eta - 2J_{\xi\eta}\right), \\
 & \mathbf{n}=\frac{1}{2}\left(\mathbf{i}_\eta + \mathbf{i}_\zeta\right) & \text{then we get} & J_n = \underset{(1\times3)}{\underline{\mathbf{n}}^T}\,\underset{(3\times3)}{\underline{\mathbf{J}}_G}\,\underset{(3\times1)}{\underline{\mathbf{n}}} = \frac{1}{2}\left(J_\eta + J_\zeta - 2J_{\eta\zeta}\right), \\
 & \mathbf{n}=\frac{1}{2}\left(\mathbf{i}_\zeta + \mathbf{i}_\xi\right) & & J_n = \underset{(1\times3)}{\underline{\mathbf{n}}^T}\,\underset{(3\times3)}{\underline{\mathbf{J}}_G}\,\underset{(3\times1)}{\underline{\mathbf{n}}} = \frac{1}{2}\left(J_\zeta + J_\xi - 2J_{\zeta\xi}\right).
\end{array}
$$

By solving the last three equations for $J_{\xi\eta}$, $J_{\eta\zeta}$, and $J_{\zeta\xi}$, we obtain the missing three elements of the matrix $\underline{\mathbf{J}}_G$.

1.6 Problems

Problems 1.1 Prove that J_n and J_{mn} are the moment of inertia and product of inertia.

Problems 1.2 Prove that the moment of momentum about the point A is given by Eq. (1.25a).

Problems 1.3 Prove the parallel axis theorem (1.28).

Problems 1.4 Show that Eq. (1.39) is true. (The proof is based on the Coriolis theorem. In this respect, the reader is referred to Appendix C.)

References

1. F.P. Beer, Jr. E.R. Johnston, D.F. Mazurek, P.J. Cornwell, E.R. Eisenberg, *Vector Mechanics for Engineers Statics and Dynamics* (McGraw-Hill, New York, 2010)
2. I. Sályi. *Engineering Mechanics I. Elements of Kinematics (in Hungarian)* (Tankönyvkiadó, Budapest, 1960)
3. G. Csernák. *Dynamics (in Hungarian)*. Akadémiai Kiadó (Publisher of the Hungarian Academy of Sciences, Budapest, 2018)

Chapter 2
Impact

2.1 What Is Meant by Impact

The collision between two bodies, which occurs in a very short time period and during which the two bodies exert relatively large impulsive forces on each other (if there are no constraints these forces are much greater than the forces exerted on the two bodies by other bodies—then the effects of the former forces can be neglected) is called impact. A typical example of impact is a bat striking a ball. Figure 2.1a, b shows bodies $\mathcal{B}_1$ and $\mathcal{B}_2$ just before collision and at the beginning of impact. The outward unit normal vectors of bodies $\mathcal{B}_1$ and $\mathcal{B}_2$ are denoted by $\mathbf{n}_1$ and $\mathbf{n}_2$. The common normal to the surfaces in contact is the *line of impact*, see Fig. 2.1b, c. If the mass centers G_1 and G_2 are located on this line then we speak about *central impact*, otherwise the impact is *eccentric*—the impact shown in Fig. 2.1 is central, the issue of eccentric impact will be considered later. The impact is called *direct* if $\mathbf{v}_1||\mathbf{v}_2||\mathbf{n}_1 = -\mathbf{n}_2$—see Fig. 2.1b—and is *oblique* if the velocity of one or both of the bodies is not parallel to the line of impact—see Fig. 2.1c. When clarifying how to solve impact problems we shall apply the following assumptions:

(a) Bodies $\mathcal{B}_1$ and $\mathcal{B}_2$ are rigid except a small neighborhood of the point $A_1 = A_2 = A$, i.e., the point at which they collide—$A_i = A$ is on body $\mathcal{B}_i$ $(i = 1, 2)$.
(b) The normals $\mathbf{n}_1 = -\mathbf{n}_2$ do not change during impact.
(c) If there are no constraints the contact forces $\mathbf{F}_{12} = -\mathbf{F}_{21}$—here $(\mathbf{F}_{12})[\mathbf{F}_{21}]$ is the force exerted by body $(\mathcal{B}_2)[\mathcal{B}_1]$ on body $(\mathcal{B}_1)[\mathcal{B}_2]$—are much greater than the other forces acting on bodies $\mathcal{B}_1$ and $\mathcal{B}_2$. Then the effects of the other forces can, therefore, be neglected.
(d) Bodies $\mathcal{B}_1$ and $\mathcal{B}_2$ are smooth, i.e., the lines of actions of the contact forces coincide with the common normal (with the line of impact).

G. Szeidl and L. P. Kiss, *Mechanical Vibrations*, Foundations of Engineering Mechanics,
https://doi.org/10.1007/978-3-030-45074-8_2

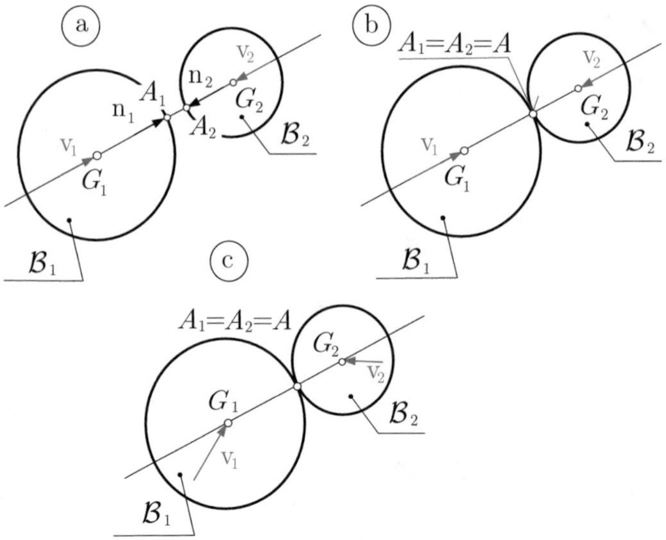

Fig. 2.1 Two bodies just before collision and at the beginning of the impact

Remark 2.1 If the impact is central the contact forces $\mathbf{F}_{12} = -\mathbf{F}_{21}$ have no moment about the mass centers G_1 and G_2. Consequently, the angular velocities $\boldsymbol{\omega}_1$ and $\boldsymbol{\omega}_2$ the bodies have when impact begins will remain unchanged during impact. For central impact the velocities at the mass centers before and after impact are denoted by $\mathbf{v}_1$, $\mathbf{v}_2$ and $\mathbf{v}_1'$, $\mathbf{v}_2'$.

2.2 Central Impact

Figure 2.2 shows the two bodies when the impact begins. If $\mathbf{v}_1 \cdot \mathbf{n}_1 > \mathbf{v}_2 \cdot \mathbf{n}_1$ then body $\mathcal{B}_1$ and $\mathcal{B}_2$ will deform. Let us denote the velocities during impact by $\mathbf{v}_{1c}$ and $\mathbf{v}_{2c}$—c is the first letter in the word collision. Deformation ends when

$$\mathbf{v}_{1c} \cdot \mathbf{n}_1 = \mathbf{v}_{2c} \cdot \mathbf{n}_1, \tag{2.1}$$

which means that the velocity components on the line of impact are equal. This quantity is denoted by c_n.

Hereupon a period of restitution takes place. By the end of the period of restitution (depending on the magnitude of the contact forces and the material properties of the two bodies) both bodies will have regained their original shape and form (the impact is then perfectly elastic) or will stay permanently deformed (the impact is then plastic). If there is no restitution at all the impact is called perfectly plastic.

We shall assume that $t = 0$ when the impact begins and $t = \tau$ when it ends. During impact time is denoted by $t_c \in [0, \tau]$.

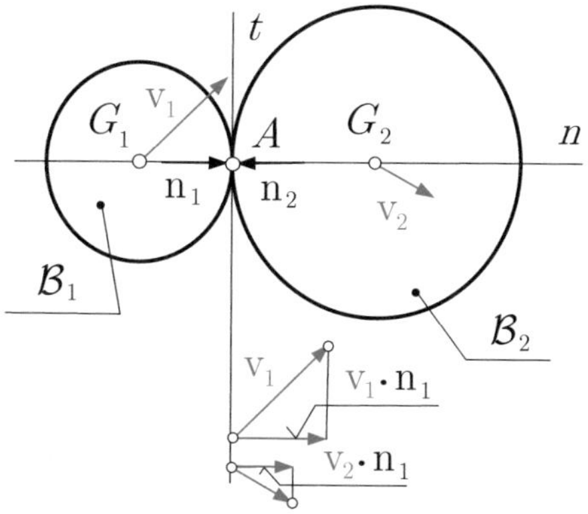

Fig. 2.2 Two bodies when impact begins

For a system of particles, the position vector of the mass center is given by Eq. (1.9b). If the system consists of two bodies only then

$$\mathbf{r}_{OG} = \mathbf{r}_{\mathcal{G}} = (m_1\mathbf{r}_1 + m_2\mathbf{r}_2)/(m_1 + m_2),$$

from where

$$\begin{aligned} m_1\mathbf{r}_1 + m_2\mathbf{r}_2 &= (m_1 + m_2)\mathbf{r}_{\mathcal{G}}\,, \\ m_1\mathbf{v}_1 + m_2\mathbf{v}_2 &= (m_1 + m_2)\mathbf{v}_{\mathcal{G}}\,, \\ m_1\mathbf{a}_1 + m_2\mathbf{a}_2 &= (m_1 + m_2)\mathbf{a}_{\mathcal{G}} = \mathcal{K} = \mathbf{R} = \mathbf{F}_{12} + \mathbf{F}_{21} = \mathbf{0}\,. \end{aligned} \tag{2.2}$$

Equation $(2.2)_3$ is a consequence of Eq. $(1.49\text{a})_2$ which is a necessary condition for the equivalence of the external forces and the effective forces. Equation $(2.2)_2$ yields

$$m_1\mathbf{v}_1 + m_2\mathbf{v}_2 = (m_1 + m_2)\mathbf{v}_{\mathcal{G}} = \mathbf{p}_1 + \mathbf{p}_2 = \mathbf{p} = m_1\mathbf{v}_{1c} + m_2\mathbf{v}_{2c} = \text{constant}. \tag{2.3}$$

This equation reflects that the momentum of the system is preserved during impact. For our later considerations, we rewrite the above equation into two different forms:

$$\begin{aligned} \mathbf{v}_{\mathcal{G}} &= \frac{m_1}{m_1 + m_2}\mathbf{v}_1 + \frac{m_2}{m_1 + m_2}\mathbf{v}_2 = \frac{m_1 + m_2}{m_1 + m_2}\mathbf{v}_1 + \frac{m_2}{m_1 + m_2}(\mathbf{v}_2 - \mathbf{v}_1) = \\ &= \mathbf{v}_1 + \underbrace{\frac{m_2}{m_1 + m_2}(\mathbf{v}_2 - \mathbf{v}_1)} = \mathbf{v}_2 + \underbrace{\frac{m_1}{m_1 + m_2}(\mathbf{v}_1 - \mathbf{v}_2)} \end{aligned} \tag{2.4a}$$

or

$$\begin{aligned} \mathbf{v}_{\mathcal{G}} &= \frac{m_1}{m_1 + m_2}\mathbf{v}_{1c} + \frac{m_2}{m_1 + m_2}\mathbf{v}_{2c} = \frac{m_1 + m_2}{m_1 + m_2}\mathbf{v}_{1c} + \frac{m_2}{m_1 + m_2}(\mathbf{v}_{2c} - \mathbf{v}_{1c}) = \\ &= \mathbf{v}_{1c} + \underbrace{\frac{m_2}{m_1 + m_2}(\mathbf{v}_{2c} - \mathbf{v}_{1c})} = \mathbf{v}_{2c} + \underbrace{\frac{m_1}{m_1 + m_2}(\mathbf{v}_{1c} - \mathbf{v}_{2c})}\,. \end{aligned} \tag{2.4b}$$

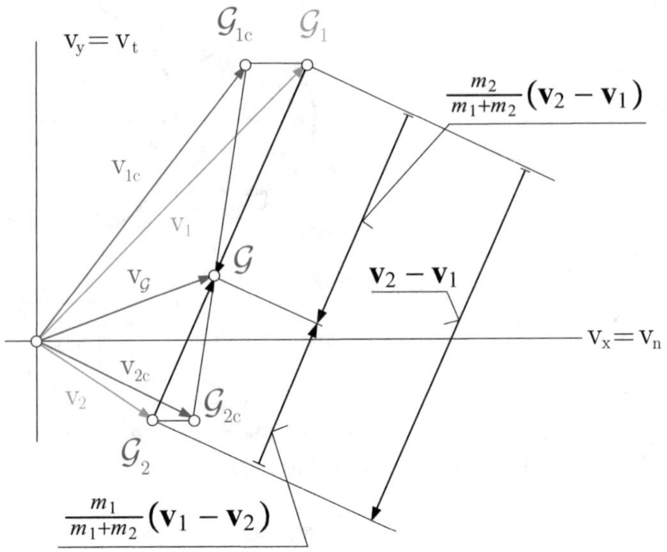

Fig. 2.3 Velocity diagram

Figure 2.3 represents the velocities $\mathbf{v}_1$, $\mathbf{v}_2$, $\mathbf{v}_{\mathcal{G}}$, and $\mathbf{v}_{1c}$, $\mathbf{v}_{2c}$ as well as the points $\mathcal{G}_1$, $\mathcal{G}_2$, $\mathcal{G}$, and $\mathcal{G}_{1c}$, $\mathcal{G}_{2c}$ in the coordinate system $v_x = v_n$, $v_y = v_t$. It follows from Eqs. (2.4)—observe what directions the terms that are underbraced have—that the points $\mathcal{G}_1$, $\mathcal{G}_2$, $\mathcal{G}$ constitute a straight line. The points $\mathcal{G}_{1c}$, $\mathcal{G}_{2c}$, and $\mathcal{G}$ are also on a straight line. The two straight lines intersect each other at the point $\mathcal{G}$.

It also holds that

$$\frac{\mathcal{G}\mathcal{G}_1}{\mathcal{G}\mathcal{G}_2} = \frac{\mathcal{G}\mathcal{G}_{1c}}{\mathcal{G}\mathcal{G}_{2c}} = \frac{m_2}{m_1}\,. \tag{2.5}$$

It is, however, an open issue what directions the line segments $\mathcal{G}_1\mathcal{G}_{1c}$ and $\mathcal{G}_2\mathcal{G}_{2c}$ have.

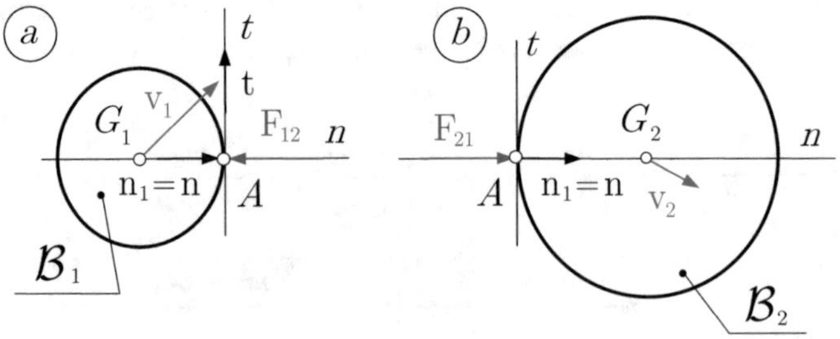

Fig. 2.4 The plane coordinate system $(n; t)$

Consider now the motion of body $\mathcal{B}_1$—see Fig. 2.4a. It is clear that

$$m_1\mathbf{a}_1 = m_1\dot{\mathbf{v}}_1 = \mathbf{F}_{12} = F_{12}\mathbf{n}_1\,, \qquad F_{12} < 0\,. \tag{2.6a}$$

Assume that $t = 0$ when the impact begins and let t_c be the time during impact. Integrating Eq. (2.6a) from $t = 0$ to t_c, we get

$$\underbrace{m_1 (\mathbf{v}_{1c} - \mathbf{v}_1)}_{\Delta \mathbf{p}_1} = \mathbf{n}_1 \int_{t=0}^{t_c} F_{12} \mathrm{d}t \,. \tag{2.6b}$$

Hence,

$$\Delta \mathbf{v}_1(t_c) = \mathbf{v}_{1c} - \mathbf{v}_1 \,||\, \mathbf{n}_1 \,. \tag{2.6c}$$

For body $\mathcal{B}_2$ a similar line of thought yields

$$m_2 \mathbf{a}_2 = m_2 \dot{\mathbf{v}}_2 = \mathbf{F}_{21} = F_{21} \mathbf{n}_1 \,, \qquad F_{21} > 0 \,, \tag{2.7a}$$

$$\underbrace{m_2 (\mathbf{v}_{2c} - \mathbf{v}_2)}_{\Delta \mathbf{p}_2} = \mathbf{n}_1 \int_{t=0}^{t_c} F_{21} \mathrm{d}t \,, \tag{2.7b}$$

$$\Delta \mathbf{v}_2(t_c) = \mathbf{v}_{2c} - \mathbf{v}_2 \,||\, \mathbf{n}_1 \,. \tag{2.7c}$$

For our later considerations, it is worth introducing the plane coordinate system (n, t); $\mathbf{n}_1 = \mathbf{n}$, $|\mathbf{t}| = 1$—see Fig. 2.4a.

As is mentioned in Remark 2.1 velocities at the end of restitution are denoted by primed letters: $\mathbf{v}_1'$, $\mathbf{v}_2'$. On the basis of all that has been said so far Fig. 2.5—referred to as Maxwell's diagram—shows how the velocities change during impact [1].

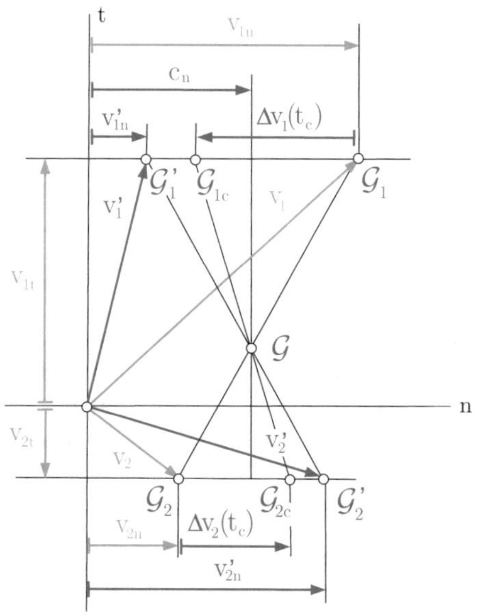

Fig. 2.5 Velocity diagram with velocity changes

Note that the velocity changes in Fig. 2.5 are, in accordance with Eqs. (2.6c) and (2.7c), parallel to **n**. However, we do not know the velocity changes, that is, the locations of the points $\mathcal{G}_1'$ and $\mathcal{G}_2'$ in Fig. 2.5.

In the sequel, we shall use the coordinate system (n, t) introduced above. In this coordinate system

$$\mathbf{v} = v_n \mathbf{n} + v_t \mathbf{t} \tag{2.8}$$

is, in general, the form of a velocity vector.

(a) It follows from Eqs. (2.6c) and (2.7c) that

$$\boxed{v_{1t}' = v_{1t}\,, \qquad v_{2t}' = v_{2t}\,,} \tag{2.9a}$$

which means that the velocity components perpendicular to the line of impact remain unchanged.

(b) The straight line perpendicular to the line of impact (to n) through the point $\mathcal{G}$—Fig. 2.5—determines c_n, i.e., the end of deformation since $v_{1cn} = v_{2cn} = c_n$. Recalling Eq. (2.3) we can write

$$\boxed{(m_1 + m_2)\, c_n = m_1 v_{1n} + m_2 v_{2n}} \tag{2.9b}$$

and

$$\boxed{m_1 v_{1n}' + m_2 v_{2n}' = m_1 v_{1n} + m_2 v_{2n}\,.} \tag{2.9c}$$

Equations (2.9a) (two equations) (2.9b) (one equation), (2.9c) (one equation) involve five unknowns:

$$v_{1n}'\,, \quad v_{1t}'\,, \quad v_{2n}'\,, \quad v_{2t}'\,, \quad c_n\,.$$

Since we have four equations for five unknowns a further equation is needed. It is an experimental experience that

$$\begin{aligned} c_n - v_{1n}' &= e\,(v_{1n} - c_n)\,, \\ c_n - v_{2n}' &= e\,(v_{2n} - c_n)\,, \end{aligned} \qquad e \in [0, 1], \tag{2.9d}$$

where e is the *coefficient of restitution*.

We remark that Eqs. (2.9d) uniquely determine the locations of points $\mathcal{G}_1'$ and $\mathcal{G}_2'$ in Fig. 2.5. It is worth now drawing a figure which may serve as a basis for the graphical solution of the impact problem. It is clear from Fig. 2.6, which is a summation of our results—recall Figs. 2.3, 2.5 and Eqs. (2.9d)—that the solution to the impact problem can be obtained by performing the following steps:

(a) Construct the straight line $\mathcal{G}_1\mathcal{G}_2$ and then locate the point $\mathcal{G}$ on that line.

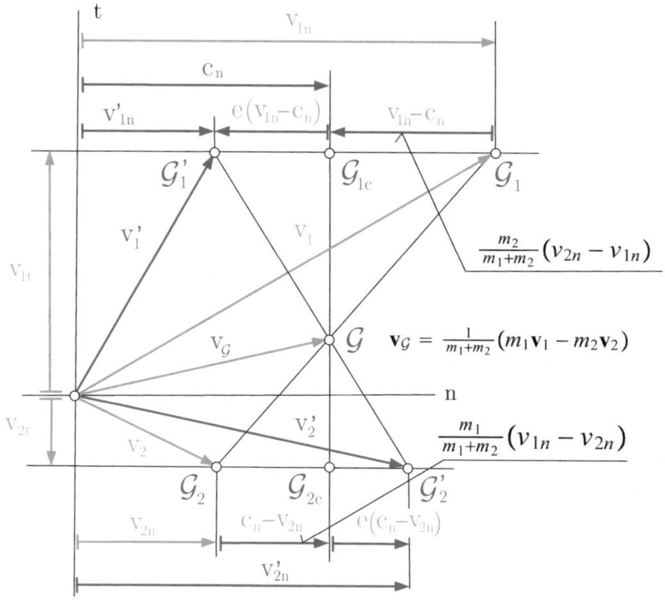

Fig. 2.6 Velocity diagram with the graphical solution

(b) Draw a straight line perpendicularly to n through the point $\mathcal{G}$. This makes possible to find the points $\mathcal{G}_{1c}$ and $\mathcal{G}_{2c}$ which lie on the horizontal lines that pass through the points $\mathcal{G}_1$ and $\mathcal{G}_2$.

(c) Mark off the distance ($|e\,(v_{1n} - c_n)\,|$) [$|e\,(v_{2n} - c_n)\,|$] starting at (and to the left of the point G_{1c} if $v_{1n} > v_{2_n}$ or to the right if $v_{1n} < v_{2_n}$) [and to the right of the point $\mathcal{G}_{2c}$ if $v_{1n} > v_{2_n}$ or to the left if $v_{1n} < v_{2_n}$]. Now you have the points $\mathcal{G}_1'$ and $\mathcal{G}_2'$ which determine the unknown velocities $\mathbf{v}_1'$ and $\mathbf{v}_2'$. If $e = 0$ this step is unnecessary: the unknown velocities are determined by the points $\mathcal{G}_{1c}$ and $\mathcal{G}_{2c}$.

By subtracting Eq. (2.9d)$_2$ from Eq. (2.9d)$_1$, we obtain

$$\boxed{v_{2n}' - v_{1n}' = e\,(v_{1n} - v_{2n})\,.} \tag{2.9e}$$

Equations (2.9a), (2.9c) and (2.9e)—four equations—involve four unknowns (v_{1n}', v_{1t}', v_{2n}', and v_{2t}'): we have as many equations as there are unknowns.

If $e = 1$ the impact is perfectly elastic—the two bodies regain their original shape.

If $e = 0$ the impact is perfectly plastic—there is no restitution at all.

If $e \in (0, 1)$ the impact is partially elastic.

With

$$c_n = \frac{m_1 v_{1n} + m_2 v_{2n}}{m_1 + m_2} \tag{2.10}$$

equation system (2.9c) and (2.9d) can be rewritten into the following form:

$$m_1 v'_{1n} + m_2 v'_{2n} = (m_1 + m_2)\, c_n \,, \qquad v'_{2n} - v'_{1n} = e\,(v_{1n} - v_{2n}) \,. \tag{2.11}$$

It is not too difficult to check that the solutions to this equation system are

$$\boxed{v'_{1n} = c_n - e\frac{m_2}{m_1 + m_2}(v_{1n} - v_{2n}) \,, \quad v'_{2n} = c_n + e\frac{m_1}{m_1 + m_2}(v_{1n} - v_{2n}) \,.} \tag{2.12}$$

We remark that these solutions can be read off Fig. 2.6 as well.

If $e \neq 1$ there is an energy loss which is given by the following equation:

$$\boxed{\mathcal{E}_{loss} = \frac{1}{2}\left(1 - e^2\right)\frac{m_1 m_2}{m_1 + m_2}(v_{1n} - v_{2n})^2 \,.} \tag{2.13}$$

The proof of this equation is left for Problem 2.1.

Exercise 2.1 Measure the coefficient of restitution e.

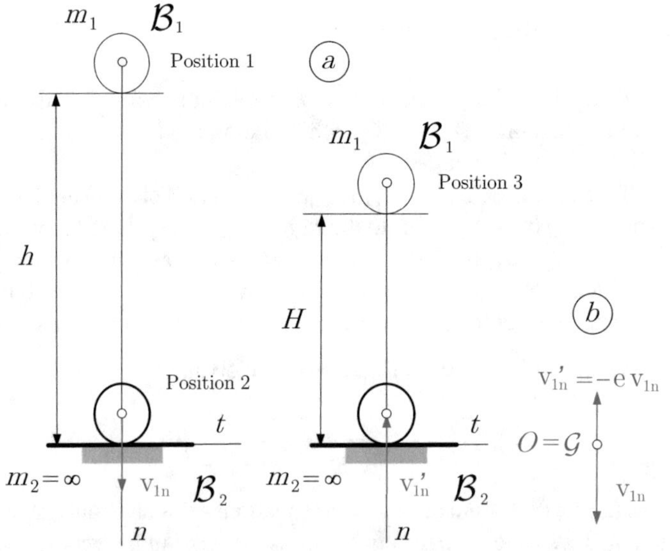

Fig. 2.7 A body bouncing off the ground

Let us drop body $\mathcal{B}_1$ (Particle 1) with mass m_1 from a height h onto body $\mathcal{B}_2$. It is assumed that the material properties are the same for the two bodies which collide—see Fig. 2.7a. Since body $\mathcal{B}_2$ is, in fact, a half-space it follows that $m_2 = \infty$. Then $v_{2n} = v'_{2n} = 0$ and Eq. (2.9d) yields

$$-v'_{1n} = e v_{1n}. \tag{2.14}$$

As regards the velocity v_{1n} the principle of work and energy (1.57) says

$$\mathcal{E}_2 - \mathcal{E}_1 = W_{12},$$

where

$$\mathcal{E}_1 = 0\,, \qquad \mathcal{E}_2 = \frac{1}{2} m_1 v_{1n}^2\,, \qquad W_{12} = m_1 g h\,.$$

Consequently,

$$\frac{1}{2} m_1 v_{1n}^2 = m_1 g h\,, \quad \Rightarrow \quad |v_{1n}| = \sqrt{2gh}. \tag{2.15a}$$

After impact body $\mathcal{B}_1$ bounces back with an initial velocity v'_{1n} till it reaches the height H ($H \leq h$). It is obvious that

$$|v'_{1n}| = \sqrt{2gH}\,. \tag{2.15b}$$

A comparison of Eqs. (2.14) and (2.15) gives the following result:

$$e = \frac{|v'_{1n}|}{|v_{1n}|} = \sqrt{\frac{H}{h}}. \tag{2.16}$$

This equation says that the measurement of e requires the measurement of distances.

Since $m_2 = \infty$ it follows from Eq. (2.4a) that

$$\mathbf{v}_{\mathcal{G}} = \mathbf{0}\,, \quad \Rightarrow \quad \mathcal{G} = O\,.$$

Under these conditions Maxwell's diagram simplifies to the diagram shown in Fig. 2.7b.

Exercise 2.2 A ball strikes a rigid wall. After impact the ball has a velocity $\mathbf{v}'_1 = -3\mathbf{i}_x - 4\mathbf{i}_y$ [m/s]. The coefficient of restitution $e = 0.5$. Determine the velocity the ball had before impact.

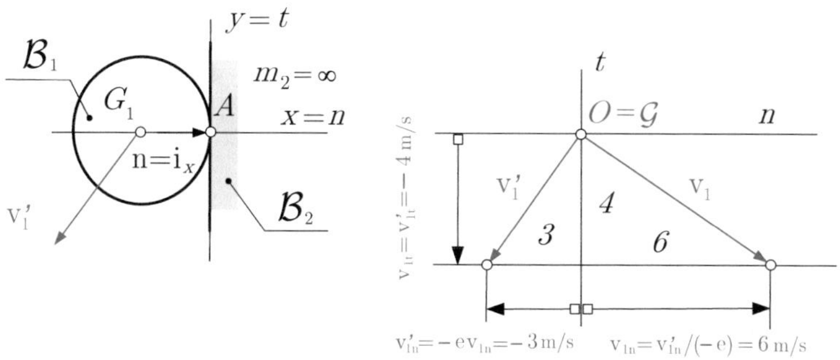

Fig. 2.8 Maxwell's diagram for the ball that strikes a rigid wall

The solution can be read off Fig. 2.8: $\mathbf{v}_1 = 6\mathbf{i}_x - 4\mathbf{i}_y$ [m/s].

Exercise 2.3 A sphere rebounds horizontally as shown in Fig. 2.9 after striking an inclined plane with a vertical velocity $\mathbf{v}_1$ of magnitude 0.5 m/s. Determine the velocity after impact and the coefficient of restitution if $\tan\alpha = 0.75$.

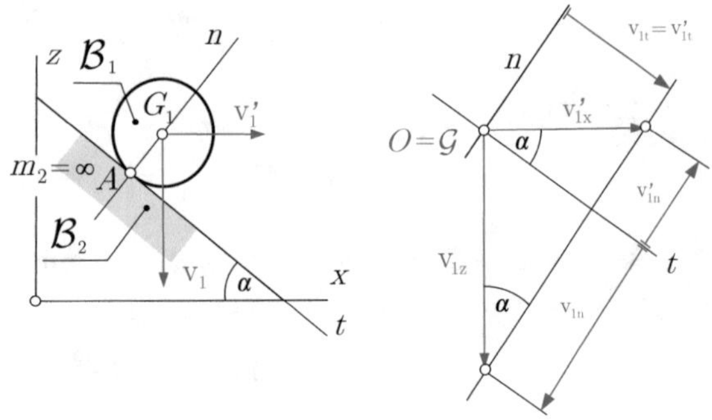

Fig. 2.9 Maxwell's diagram for a sphere rebounding horizontally

It follows from Maxwell's diagram that

$$\frac{|v_{1x}'|}{|v_{1z}|} = \tan\alpha\,, \qquad |v_{1x}'| = v_{1x}' = |v_{1z}|\tan\alpha = 0.5\cdot 0.75 = 0.375 \text{ m/s}\,.$$

On the other hand,

$$|v_{1n}| = |v_{1z}|\cos\alpha = |v_{1z}|\frac{1}{\sqrt{1+\tan^2\alpha}} = 0.5\times 0.8 = 0.4 \text{ m/s}\,,$$

$$|v_{1n}'| = |v_{1x}'|\sin\alpha = |v_{1x}'|\frac{\tan\alpha}{\sqrt{1+\tan^2\alpha}} = 0.375\times 0.6 = 0.225 \text{ m/s}\,.$$

Consequently, see Eq. (2.16)

$$e = \frac{|v_{1n}'|}{|v_{1n}|} = \frac{0.225}{0.4} = 0.5625.$$

Exercise 2.4 A ball is thrown with a velocity $\mathbf{v}_o = 4\mathbf{i}_x + 4\mathbf{i}_y$ [m/s]. What velocity will have had the ball after rebounding from the plane if the air resistance can be neglected and the coefficient of restitution $e = 0.75$?

If there is no air resistance

$$v_{1x} = v_{1t} = v_{ox} = 4 \text{ m/s}\,, \qquad -v_{1y} = v_{oy} = v_{1n} = 4 \text{ m/s}\,.$$

It follows from Maxwell's diagram—see Fig. 2.10 for details—that

$$v_{1n}' = -e v_{1n} = -0.75\times 4 = -3 \text{ m/s}\,; \qquad \mathbf{v}_1' = 4\mathbf{i}_x + 3\mathbf{i}_y \text{ [m/s]}\,.$$

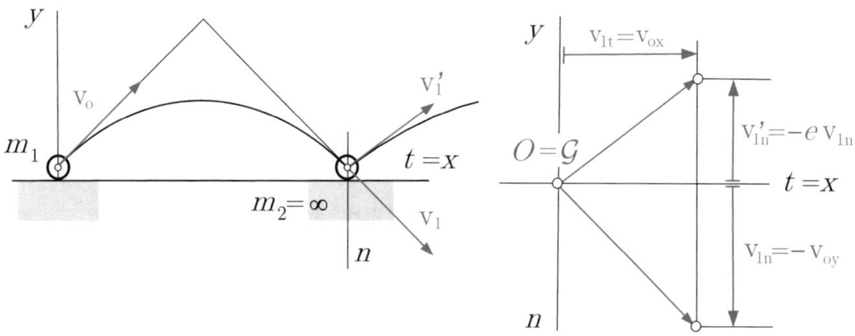

Fig. 2.10 Maxwell's diagram for a thrown ball

Exercise 2.5 Two balls with masses $m_1 = 2$ kg and $m_2 = 6$ kg collide—see Fig. 2.11 for details. The velocities $\mathbf{v}_1$ and $\mathbf{v}_2$ are known. Determine the velocities $\mathbf{v}_1'$ and $\mathbf{v}_2'$ if (a) $e = 0$ (perfectly plastic impact), (b) $e = 1$ (perfectly elastic impact), (c) $e = 0.6$ (partially elastic impact).

$$\mathbf{v}_1 = 4\mathbf{i}_x + 4\mathbf{i}_y \text{ [m/s]}, \qquad \mathbf{v}_2 = -4\mathbf{i}_y \text{ [m/s]}.$$

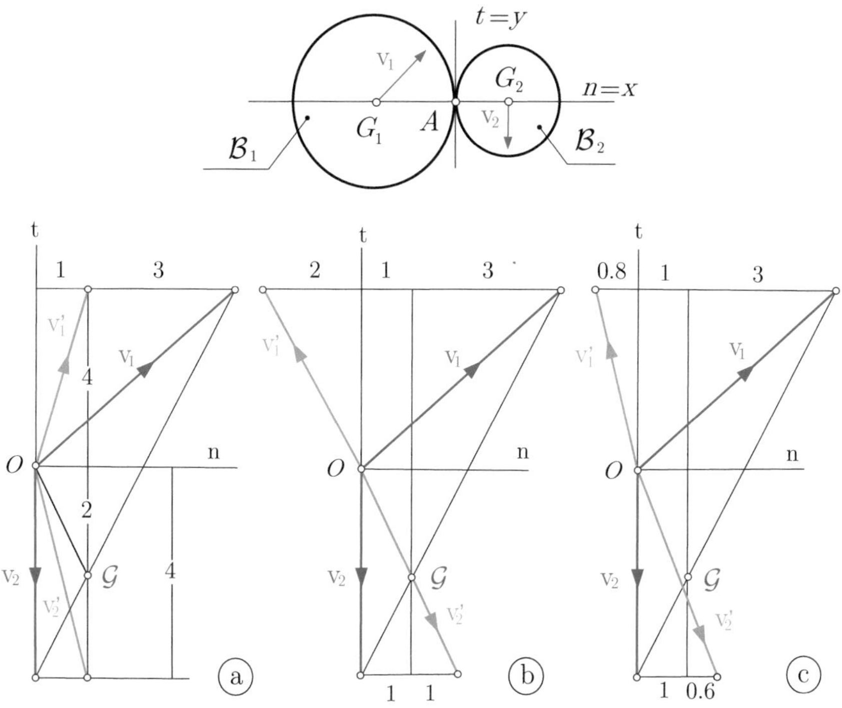

Fig. 2.11 Impact of two balls with graphical solution

The velocity $\mathbf{v}_G$ is obtained from Eq. (2.4):

$$\mathbf{v}_G = \frac{1}{m_1 + m_2}(m_1\mathbf{v}_1 + m_2\mathbf{v}_2) = \frac{1}{8}\left(8\mathbf{i}_x + 8\mathbf{i}_y - 24\mathbf{i}_y\right) = \mathbf{i}_x - 2\mathbf{i}_y \text{ [m/s]}.$$

After having determined $\mathbf{v}_G$ we can construct three Maxwell's diagrams:
The solutions in [m/s] can be read off from Fig. 2.11a–c.

$$\begin{array}{lll} \text{(a)} & \text{(b)} & \text{(c)} \\ \mathbf{v}_1' = \mathbf{i}_x + 4\mathbf{i}_y\,, & \mathbf{v}_1' = -2\mathbf{i}_x + 4\mathbf{i}_y\,, & \mathbf{v}_1' = -0.8\mathbf{i}_x + 4\mathbf{i}_y\,, \\ \mathbf{v}_2' = \mathbf{i}_x - 4\mathbf{i}_y\,, & \mathbf{v}_2' = 2\mathbf{i}_x - 4\mathbf{i}_y\,, & \mathbf{v}_2' = 1.6\mathbf{i}_x - 4\mathbf{i}_y\,. \end{array}$$

Exercise 2.6 Two balls with masses $m_1 = 2$ kg and $m_3 = 6$ kg collide—see Fig. 2.12 for details. The velocities $\mathbf{v}_1$ and $\mathbf{v}_2$ are known, $\mathbf{v}_1 = 6\mathbf{i}_x$ [m/s], $\mathbf{v}_2 = 2\mathbf{i}_x$ [m/s]. Determine the velocities $\mathbf{v}_1'$ and $\mathbf{v}_2'$ if (a) $e = 0$ (perfectly plastic impact), (b) $e = 1$ (perfectly elastic impact).

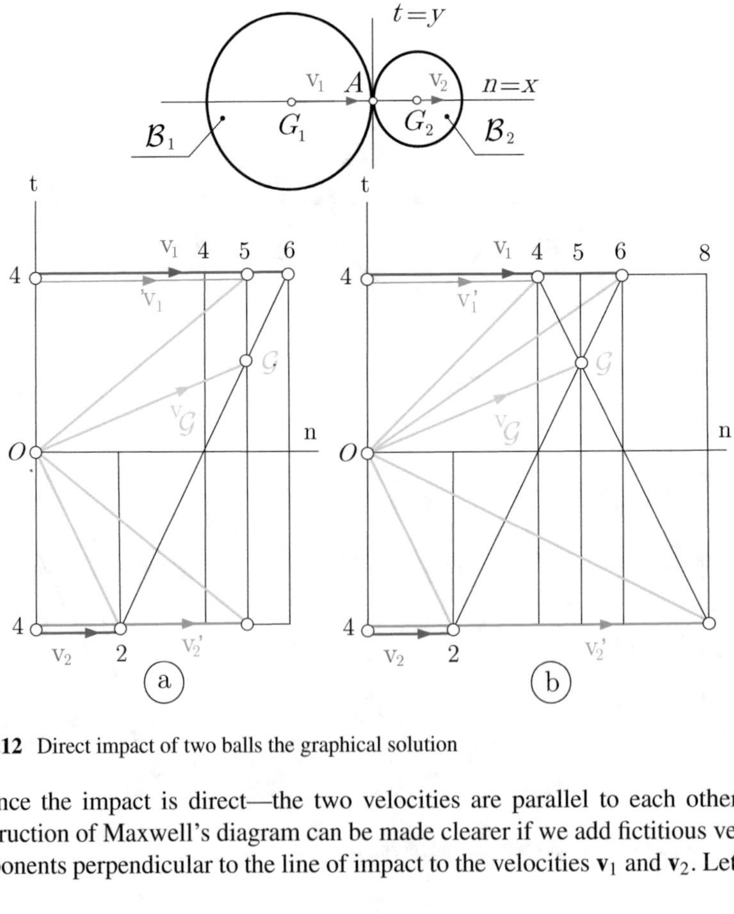

Fig. 2.12 Direct impact of two balls the graphical solution

Since the impact is direct—the two velocities are parallel to each other—the construction of Maxwell's diagram can be made clearer if we add fictitious velocity components perpendicular to the line of impact to the velocities $\mathbf{v}_1$ and $\mathbf{v}_2$. Let these velocity components be $4\mathbf{i}_y$ and $-4\mathbf{i}_y$. Then

$$\mathbf{v}_1 = 6\mathbf{i}_x + 4\mathbf{i}_y \text{ [m/s]}, \qquad \mathbf{v}_2 = 2\mathbf{i}_x - 4\mathbf{i}_y \text{ [m/s]}.$$

Therefore,

$$\mathbf{v}_{\mathcal{G}} = \frac{1}{m_1 + m_2}(m_1\mathbf{v}_1 + m_2\mathbf{v}_2) = \frac{1}{4}\left(18\mathbf{i}_x + 12\mathbf{i}_y + 2\mathbf{i}_x - 4\mathbf{i}_y\right) =$$
$$= 5\mathbf{i}_x + 2\mathbf{i}_y \text{ [m/s]}.$$

We can now construct two Maxwell's diagrams.

On the basis of the diagram the real solutions are as follows:

$$\begin{array}{ccc} \text{(a)} & \text{(b)} & \\ \mathbf{v}_1' = 5\mathbf{i}_x, & \mathbf{v}_1' = 4\mathbf{i}_x & \text{[m/s]}, \\ \mathbf{v}_2' = 5\mathbf{i}_x, & \mathbf{v}_2' = 8\mathbf{i}_x & \text{[m/s]}. \end{array}$$

It is worth mentioning that the fictitious velocity components can be selected arbitrarily.

2.3 Eccentric Impact

2.3.1 Impact of Two Bodies Which Perform Free Plane Motion

Figure 2.13 shows the two bodies when the impact begins. Let $\mathbf{v}_i$ and ω_i be the velocity at the mass center G_i and the angular velocity for body $\mathcal{B}_i$ $(i = 1, 2)$ when impact begins. We remark that primed letters denote these quantities right after the impact.

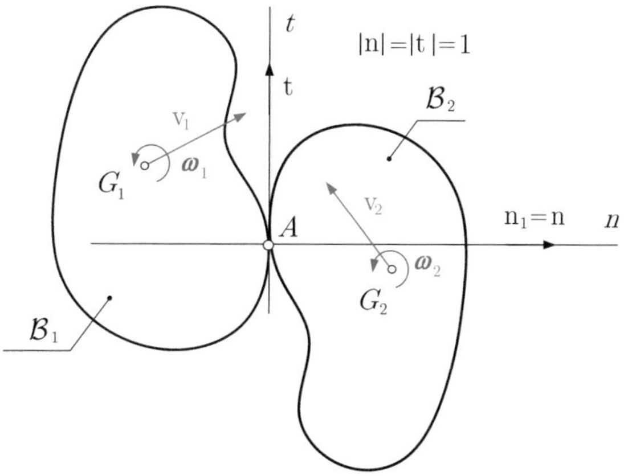

Fig. 2.13 Eccentric impact of two bodies in plane motion

The assumptions we make are the same as those on Sect. 2.2.

The state of velocity before and after impact is uniquely determined by

(a) $\mathbf{v}_1, \omega_1, \mathbf{v}_1', \omega_1'$ (body $\mathcal{B}_1$) and
(b) $\mathbf{v}_2, \omega_2, \mathbf{v}_2', \omega_2'$ (body $\mathcal{B}_2$).

It is worth, however, emphasizing that the angular velocities will also change since the contact forces have, now, moment about the mass centers.

Since the velocity states of the bodies before impact are known, the primed quantities are the unknowns: $\mathbf{v}_1'$ (two unknowns), ω_1' (one unknown—the angular velocity is perpendicular to the plane of motion), $\mathbf{v}_2'$ (two unknowns), ω_2' (one unknown).

It is our main objective to formulate the problem in such a way that the equations that provide the solutions be of the same structure as Eqs. (2.9a), (2.9c), (2.9d) valid for central impact. Then the problem can be solved by using Maxwell's diagram. To achieve this goal some manipulations are to be carried out.

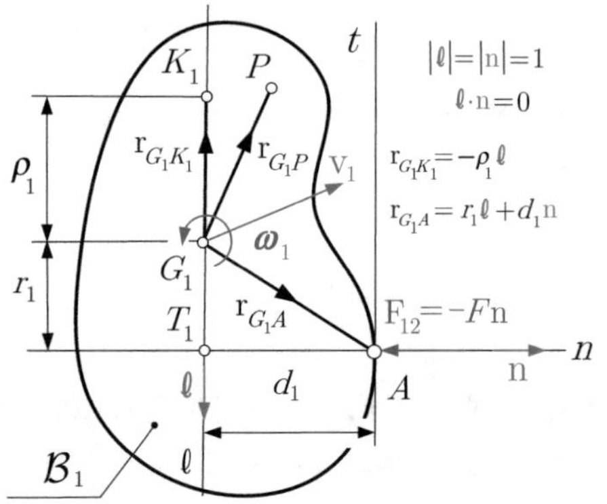

Fig. 2.14 Motion characteristics of body $\mathcal{B}_1$

First, we shall examine the motion of body $\mathcal{B}_1$ shown in Fig. 2.14. It follows from Eqs. (1.32), (1.31b) and (1.39), (1.46), (1.49a)$_2$ that

$$\dot{\mathbf{p}}_1 = m_1\,\mathbf{a}_1 = m_1\,\dot{\mathbf{v}}_1 = \mathbf{F}_{12} \tag{2.17a}$$

$$\dot{\mathbf{H}}_{G_1} = J_{G_1}\boldsymbol{\alpha}_1 = \mathbf{r}_{G_1A} \times \mathbf{F}_{12}, \tag{2.17b}$$

where $\mathbf{v}_{G_1}$ is the velocity at G_1 and $\boldsymbol{\alpha}_1$ is the angular acceleration if $t_c \in [0, \tau]$. Integrating these equations with respect to time, remember that $t = 0$ when the impact begins and $t = \tau$ when it ends, we have

$$m_1\left(\mathbf{v}_1' - \mathbf{v}_1\right) = -\mathbf{n}\int_0^\tau F\,\mathrm{d}t = \Delta\mathbf{p}_1, \tag{2.18a}$$

$$J_{G_1}\left(\omega_1' - \omega_1\right) = \mathbf{r}_{G_1A} \times \left(-\mathbf{n}\int_0^\tau F\,\mathrm{d}t\right) = \mathbf{r}_{G_1A} \times \Delta\mathbf{p}_1. \tag{2.18b}$$

Let P be an arbitrary point of $\mathcal{B}_1$. For our later considerations, we shall determine the velocity at P before and after impact. Making use of Eq. (1.4) we can write

$$\mathbf{v}_P = \mathbf{v}_1 + \boldsymbol{\omega}_1 \times \mathbf{r}_{G_1P}\,, \qquad \mathbf{v}'_P = \mathbf{v}'_1 + \boldsymbol{\omega}'_1 \times \mathbf{r}_{G_1P}. \tag{2.19}$$

Subtract now the first equation from the second one. The result is

$$\mathbf{v}'_P - \mathbf{v}_P = \mathbf{v}'_1 - \mathbf{v}_1 + \left(\boldsymbol{\omega}'_1 - \boldsymbol{\omega}_1\right) \times \mathbf{r}_{G_1P}\,. \tag{2.20}$$

If we multiply throughout by $m_1\, J_{G_1}$ and take into account (2.18b), we obtain

$$m_1 J_{G_1}\left(\mathbf{v}'_P - \mathbf{v}_P\right) = J_{G_1} \underbrace{m_1\left(\mathbf{v}'_1 - \mathbf{v}_1\right)}_{\Delta\mathbf{p}_1} + m_1 \underbrace{J_{G_1}\left(\boldsymbol{\omega}'_1 - \boldsymbol{\omega}_1\right)}_{\mathbf{r}_{G_1A}\times\Delta\mathbf{p}_1} \times \mathbf{r}_{G_1P}$$

in which

$$m_1\left(\mathbf{r}_{G_1A} \times \Delta\mathbf{p}_1\right) \times \mathbf{r}_{G_1P} = m_1\left[\left(\mathbf{r}_{G_1A}\cdot\mathbf{r}_{G_1P}\right)\Delta\mathbf{p}_1 - \mathbf{r}_{G_1A}\left(\mathbf{r}_{G_1P}\cdot\Delta\mathbf{p}_1\right)\right],$$

hence

$$m_1 J_{G_1}\left(\mathbf{v}'_P - \mathbf{v}_P\right) = \left[J_{G_1} + m_1\left(\mathbf{r}_{G_1A}\cdot\mathbf{r}_{G_1P}\right)\right]\Delta\mathbf{p}_1 - m_1\mathbf{r}_{G_1A}\left(\mathbf{r}_{G_1P}\cdot\Delta\mathbf{p}_1\right). \tag{2.21}$$

This equation expresses that the velocity difference at any point within the body is directly proportional to $\Delta\mathbf{p}_1$.

(a) If P is on the line ℓ then $\mathbf{r}_{G_1P}\cdot\Delta\mathbf{p}_1 = 0$ (the two vectors are perpendicular to each other) and Eq. (2.21) simplifies to

$$m_1 J_{G_1}\left(\mathbf{v}'_P - \mathbf{v}_P\right) = \left[J_{G_1} + m_1\left(\mathbf{r}_{G_1A}\cdot\mathbf{r}_{G_1P}\right)\right]\Delta\mathbf{p}_1. \tag{2.22}$$

(b) It is clear that $\mathbf{v}'_1 - \mathbf{v}_1$ is perpendicular to the line ℓ. In addition the cross product $\left(\boldsymbol{\omega}'_1 - \boldsymbol{\omega}_1\right) \times \mathbf{r}_{G_1P}$ is also perpendicular to the line ℓ if P is on the line ℓ. Consequently, there is a point on the line ℓ—this point is denoted by K_1—for which the velocity difference $\mathbf{v}'_{K_1} - \mathbf{v}_{K_1}$ vanishes

$$\mathbf{v}'_{K_1} - \mathbf{v}_{K_1} = \mathbf{v}'_1 - \mathbf{v}_1 + \left(\boldsymbol{\omega}'_1 - \boldsymbol{\omega}_1\right) \times \mathbf{r}_{G_1K_1} = \mathbf{0}\,.$$

Then Eq. (2.22) yields

$$\left[J_{G_1} + m_1\left(\mathbf{r}_{G_1A}\cdot\mathbf{r}_{G_1K_1}\right)\right]\underbrace{\Delta\mathbf{p}_1}_{\neq\mathbf{0}} = \mathbf{0},$$

from where—see Fig. 2.14—it follows that

$$J_{G_1} + m_1\left(r_1\boldsymbol{\ell} + d_1\mathbf{n}\right)\cdot\left(-\rho_1\boldsymbol{\ell}\right) = J_{G_1} - m_1 r_1\rho_1 = 0,$$

which means that

$$\boxed{\rho_1 = \frac{J_{G_1}}{m_1 r_1}} \tag{2.23}$$

is the vertical distance between the points G_1 and K_1. Observe that here we have used the notations of Fig. 2.14.

Point K_1 is the point of $\mathcal{B}_1$ with no velocity change. For this reason, it is referred to as the swaying center.

(c) If $P = T_1$ Eq. (2.22) yields

$$m_1 J_{G_1}\left(\mathbf{v}'_{T_1} - \mathbf{v}_{T_1}\right) = \left[J_{G_1} + m_1\left(\mathbf{r}_{G_1A}\cdot\mathbf{r}_{G_1T_1}\right)\right]\Delta\mathbf{p}_1 =$$
$$= \left[J_{G_1} + m_1\left(r_1\boldsymbol{\ell} + d_1\mathbf{n}\right)\cdot r_1\boldsymbol{\ell}\right]\Delta\mathbf{p}_1 = \underbrace{\left(J_{G_1} + m_1 r_1^2\right)}_{J_{T_1}}\Delta\mathbf{p}_1 = J_{T_1}\Delta\mathbf{p}_1$$

or

$$\underbrace{m_1\frac{J_{G_1}}{J_{T_1}}}_{\tilde{m}_1}\left(\mathbf{v}'_{T_1} - \mathbf{v}_{T_1}\right) = \Delta\mathbf{p}_1\,, \qquad \boxed{\tilde{m}_1 = m_1\frac{J_{G_1}}{J_{T_1}}\,.} \tag{2.24}$$

This means that the velocity difference at the point T_1 is parallel to $\Delta\mathbf{p}_1$. The physical content of this equation is the same as that of Eq. (2.6b) valid for central impact. $\tilde{m}_1$ is the mass reduced to point T_1 (Fig. 2.15).

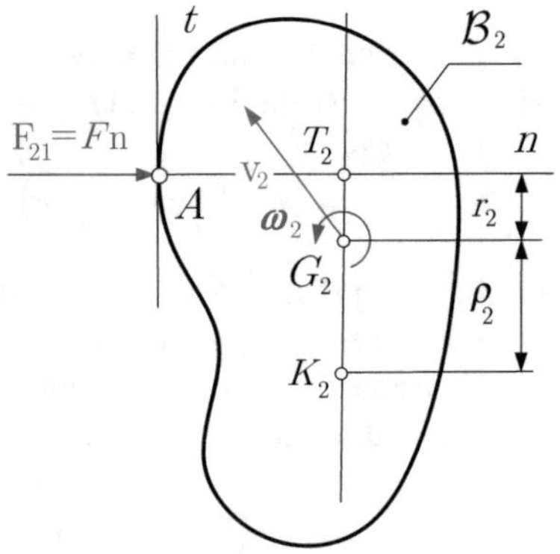

Fig. 2.15 Motion characteristics of body $\mathcal{B}_2$

As regards body $\mathcal{B}_2$ a similar line of thought results in

$$\boxed{\rho_2 = \frac{J_{G_2}}{m_2 r_2}} \tag{2.25}$$

and

$$\underbrace{m_2\frac{J_{G_2}}{J_{T_2}}}_{\tilde{m}_2}\left(\mathbf{v}'_{T_2} - \mathbf{v}_{T_2}\right) = \Delta\mathbf{p}_2 = -\Delta\mathbf{p}_1,$$

where

$$\boxed{\tilde{m}_2 = m_2 \frac{J_{G_2}}{J_{T_2}} \,.} \tag{2.26}$$

After adding Eqs. (2.24)$_1$–(2.26)$_1$, we have

$$\tilde{m}_1 \left(\mathbf{v}'_{T_1} - \mathbf{v}_{T_1} \right) + \tilde{m}_2 \left(\mathbf{v}'_{T_2} - \mathbf{v}_{T_2} \right) = \mathbf{0}$$

or

$$\boxed{\tilde{m}_1 \mathbf{v}_{T_1} + \tilde{m}_2 \mathbf{v}_{T_2} = \tilde{m}_1 \mathbf{v}'_{T_1} + \tilde{m}_2 \mathbf{v}'_{T_2} = \left(\tilde{m}_1 + \tilde{m}_2 \right) \mathbf{v}_G ,} \tag{2.27}$$

which expresses that the momenta that belong to the reduced masses are conserved. It also holds that

$$\boxed{\left(\mathbf{v}'_{T_2} - \mathbf{v}'_{T_1} \right) \cdot \mathbf{n} = e \left(\mathbf{v}_{T_1} - \mathbf{v}_{T_2} \right) \cdot \mathbf{n} \,.} \tag{2.28}$$

On the basis of Eqs. (2.27) and (2.28), one can construct Maxwell's diagram or can calculate the solution by using the following scalar equations:

$$\boxed{v'_{T_1 t} = v_{T_1 t} \,, \qquad v'_{T_2 t} = v_{T_2 t} \,;} \tag{2.29a}$$

$$\boxed{\tilde{m}_1 v_{T_1 n} + \tilde{m}_2 v_{T_2 n} = \tilde{m}_1 v'_{T_1 n} + \tilde{m}_2 v'_{T_2 n} \,;} \tag{2.29b}$$

$$\boxed{v'_{T_2 n} - v'_{T_1 n} = e \left(v_{T_1 n} - v_{T_2 n} \right) \,.} \tag{2.29c}$$

After having determined the solution for $\mathbf{v}'_{T_1}$ we can obtain the velocity $\mathbf{v}'_1$ at the mass center G. Comparison of Eqs. (2.26) and (2.18a) results in the equation

$$\tilde{m}_1 \left(\mathbf{v}'_{T_1} - \mathbf{v}_{T_1} \right) = \Delta \mathbf{p}_1 = m_1 \left(\mathbf{v}'_1 - \mathbf{v}_1 \right)$$

from where

$$\boxed{\mathbf{v}'_1 = \mathbf{v}_1 + \frac{\tilde{m}_1}{m_1} \left(\mathbf{v}'_{T_1} - \mathbf{v}_{T_1} \right) , \qquad \frac{\tilde{m}_1}{m_1} = \frac{J_{G_1}}{J_{T_1}} \,.} \tag{2.30}$$

As regards body $\mathcal{B}_2$ we should write 2 for 1 in Eq. (2.30).

The last issue is how to determine ω'_1. Substituting $\Delta \mathbf{p}_1$ from Eq. (2.26) into Eq. (2.18a), we get

$$J_{G_1} \left(\omega'_1 - \omega_1 \right) = \mathbf{r}_{G_1 A} \times \Delta \mathbf{p}_1 = \tilde{m}_1 \mathbf{r}_{G_1 A} \times \left(\mathbf{v}'_{T_1} - \mathbf{v}_{T_1} \right) ,$$

which can be solved for the unknown angular velocity:

$$\boxed{\omega'_1 = \omega_1 + \frac{\tilde{m}_1}{J_{G_1}} \mathbf{r}_{G_1 A} \times \left(\mathbf{v}'_{T_1} - \mathbf{v}_{T_1} \right) .} \tag{2.31}$$

Change 1 to 2 in Eq. (2.30) for body $\mathcal{B}_2$.

Exercise 2.7 The slender rod $\mathcal{B}_2$ with mass $m_2 = 3.5$ kg and length $\ell_2 = 1.5$ m is at rest when it is hit by the ball with mass $m_1 = 3$ kg. The velocity of the ball $\mathcal{B}_1$ is $\mathbf{v}_1 = 12\,\mathbf{i}_x$ [m/s]. Determine the state of velocity of the rod right after impact if $e = 0.8$ and $r_2 = \ell_2/3 = 0.5$ m.

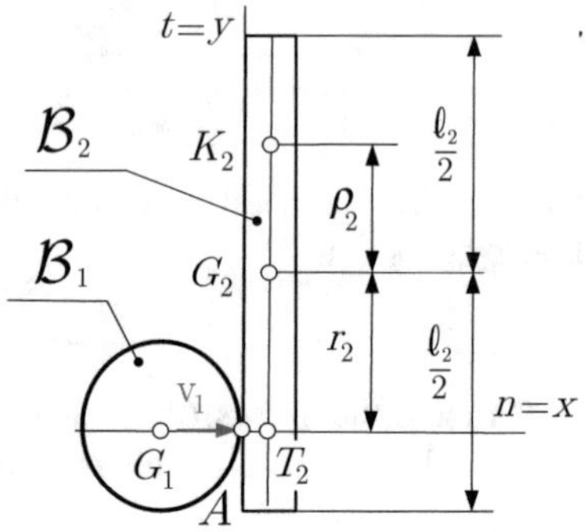

Fig. 2.16 Impact of slender rod and ball

It follows from Eq. (2.25) that

$$\rho_2 = \frac{J_{G_2}}{m_2 r_2} = \frac{\frac{1}{12} m_2 \ell_2^2}{m_2 \frac{\ell_2}{3}} = \frac{\ell_2}{4} = 0.375\,\text{m}\,.$$

Using the parallel axis theorem, we obtain

$$J_{T_2} = J_{G_2} + m_2\, r_2^2 = \frac{1}{12} m_2 \ell_2^2 + m_2 \frac{\ell_2^2}{9} = \frac{7}{36} m_2 \ell_2^2\,.$$

With this value Eq. (2.26) yields

$$\tilde{m}_2 = m_2 \frac{J_{G_2}}{J_{T_2}} = \frac{1}{12}\frac{36}{7} m_2 = \frac{3}{7} m_2 = 1.5\ \text{kg}\,.$$

It is obvious now that $\tilde{m}_1 = m_1 = 3$ kg, $\tilde{m}_2 = 1.5$ kg, $v_{T_1t} = v_{1t} = 0$, $v_{T_1n} = v_{1n} = 12$ m/s, $v_{T_2t} = 0$, $v_{T_2n} = 0$ and $e = 0.8$. With the knowledge of these values it is clear from Eqs. (2.29a) that $v'_{T_1t} = v'_{1t} = 0$, $v'_{T_2t} = 0$. Upon substitution of the known values into Eqs. (2.29b) and (2.29c), we have

$$3 \times 12 + 0 = 3v'_{T_1n} + 1.5v'_{T_2n}\,,$$
$$v'_{T_2n} - v'_{T_1n} = 0.8 \times (12 - 0) = 9.6$$

from where

$$v'_{T_2n} = 14.4\ \text{m/s} \quad \text{and} \quad v'_{T_1n} = 4.8\ \text{m/s}\,.$$

Since the point K_2 remains at rest during impact it is the instantaneous center of rotation when the impact ends. Consequently,

$$\omega_2' = \frac{v_{T_2 n}'}{r_2 + \rho_2} = \frac{14.4}{0.5 + 0.375} = 15.457 \ \frac{1}{\text{s}} .$$

2.3.2 *Impact of Two Bodies if One or Two Bodies Rotate About a Fixed Point*

There are some changes if a body, say body $\mathcal{B}_1$, rotates about a fixed axis. Figure 2.17 shows such a body, i.e., body $\mathcal{B}_1$ under the assumption that it rotates about the hinge at O. It is obvious that

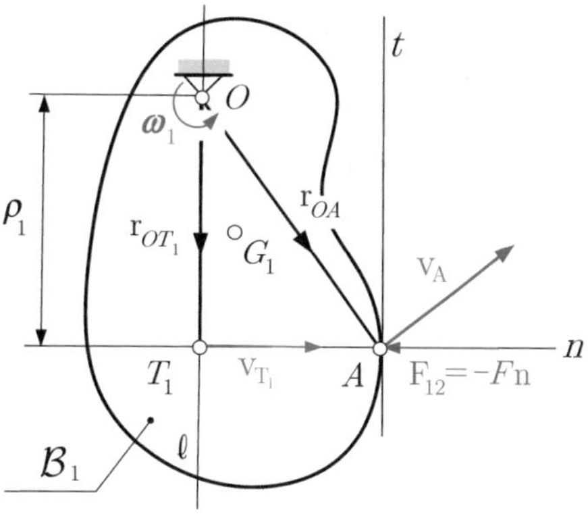

Fig. 2.17 Body rotating about the hinge at O

$$\mathbf{v}_A = \boldsymbol{\omega}_1 \times \mathbf{r}_{OA} , \qquad \text{and} \qquad \mathbf{v}_{T_1} = \boldsymbol{\omega}_1 \times \mathbf{r}_{OT_1}$$

from where

$$\mathbf{v}_{T_1} - \mathbf{v}_A = \boldsymbol{\omega}_1 \times \underbrace{\left(\mathbf{r}_{OT_1} - \mathbf{r}_{OA}\right)}_{||\mathbf{n}} \qquad | \cdot \mathbf{n},$$

which means that

$$\left(\mathbf{v}_{T_1} - \mathbf{v}_A\right) \cdot \mathbf{n} = 0 .$$

It is also obvious

$$J_O\, \boldsymbol{\alpha}_1 = \mathbf{r}_{OA} \times \mathbf{F}_{12}$$

from where by performing time integration, we get

$$J_O \left(\boldsymbol{\omega}_1' - \boldsymbol{\omega}_1\right) = \mathbf{r}_{OA} \times \left(-\mathbf{n} \int_0^{\tau} F \, \mathrm{d}t\right) = \mathbf{r}_{OA} \times (-\Delta \mathbf{p}_2) .$$

If we now cross multiply the above equation from right by $\mathbf{r}_{OT_1}$, we obtain

$$J_O\left(\boldsymbol{\omega}_1' \times \mathbf{r}_{OT_1} - \boldsymbol{\omega}_1 \times \mathbf{r}_{OT_1}\right) = -\left(\mathbf{r}_{OA} \times \Delta\mathbf{p}_2\right) \times \mathbf{r}_{OT_1} =$$
$$= -\underbrace{\left(\mathbf{r}_{OA} \cdot \mathbf{r}_{OT_1}\right)}_{\rho_1^2}\Delta\mathbf{p}_2 + \underbrace{\left(\mathbf{r}_{OT_1} \cdot \Delta\mathbf{p}_2\right)}_{=0}\mathbf{r}_{OA}$$

or

$$-\Delta\mathbf{p}_2 = \frac{J_O}{\rho_1^2}\left(\mathbf{v}_{T_1}' - \mathbf{v}_{T_1}\right) = \tilde{m}_1\left(\mathbf{v}_{T_1}' - \mathbf{v}_{T_1}\right), \qquad \tilde{m}_1 = \frac{J_O}{\rho_1^2}. \tag{2.32}$$

After adding this equation to Eq. (2.26)$_2$ we arrive at (2.27). Equation (2.28) is still valid. Consequently, solutions of impact problems with bodies rotating about a fixed point can be calculated in the same way as shown in Sect. 2.3.1.

Exercise 2.8 A bullet with mass m_2 is fired with a horizontal velocity $\mathbf{v}_2 = -v_2\mathbf{n}$ into the side of a panel suspended from a hinge at B. Knowing that the panel is initially at rest determine the angular velocity of the panel immediately after the bullet is embedded and the impulsive reaction $\mathbf{R}_B$ at the hinge B assuming that the bullet is embedded in τ sec. ($\tau \ll 1$).

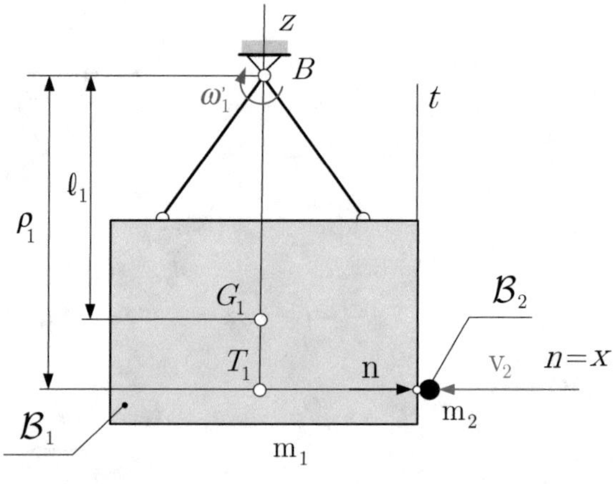

Fig. 2.18 A panel hit by a bullet

We shall assume that the necessary data of the panel—m_1, J_B, ℓ_1 and ρ_1—are all known. According to Eq. (2.32)

$$\tilde{m}_1 = J_B/\rho_1^2 .$$

Since the bullet is embedded it follows that

$$v_2' = v_{2n}' = v_{T_1n}' ; \qquad e = 0 .$$

Equation (2.29b) can now be utilized to find the unknowns $v_2' = v_{2n}' = v_{T_1n}'$. We can write

$$\underbrace{\tilde{m}_1 v_{T_1n}}_{=0} + m_2 v_n = \tilde{m}_1 v_{T_1n}' + m_2 v_2' ,$$

or

$$m_2 v_n = (\tilde{m}_1 + m_2)\, v_{T_1n}'$$

from where

$$v_2' = v_{T_1n}' = \frac{m_2}{\tilde{m}_1 + m_2}\, v_2 .$$

It can be checked with ease that

$$\omega_1' = \frac{v_{T_1n}'}{\rho_1} = \frac{v_2'}{\rho_1} .$$

We remark that ω_1' is shown in Fig. 2.18.

Let $\mathbf{p}$ and $\mathbf{p}'$ be the momenta of the whole system before and after impact. Using the principle of impulse and momentum $(1.49\text{d})_1$, we can write

$$\mathbf{p}' = \mathbf{p} + \int_0^{\tau} \mathbf{R}_B \,\mathrm{d}t \cong (R_{Bx}\mathbf{i}_x + R_{Bz}\mathbf{i}_z)\underbrace{\Delta t}_{\tau} , \qquad (2.33)$$

where

$$\mathbf{p}' \cong -m_1 v_1' \mathbf{i}_x = -m_1 \omega_1' \ell_1 \mathbf{i}_x = -m_1 v_2' \frac{\ell_1}{\rho_1} \mathbf{i}_x \qquad (m_1 \gg m_2!)$$

and

$$\mathbf{p} = -m_2 v_2 \mathbf{i}_x .$$

Hence,

$$R_{Bz} = 0$$

and

$$R_{Bx} = \frac{1}{\tau}\left(m_2 v_2 - m_1 v_2' \frac{\ell_1}{\rho_1}\right) = \frac{1}{\tau}\left(1 - \frac{m_1}{\tilde{m}_1 + m_2}\frac{\ell_1}{\rho_1}\right) m_2 v_2$$

are the components of the impulsive reaction at B.

Exercise 2.9 The panel in the previous exercise will sway to the left till it reaches an extreme position. Determine what angle φ belongs to this position.

Using the principle of work an energy (1.58a), we can write

$$\mathcal{E}_2 - \mathcal{E}_1 = W_{12} = U_{12} = U_1 - U_2,$$

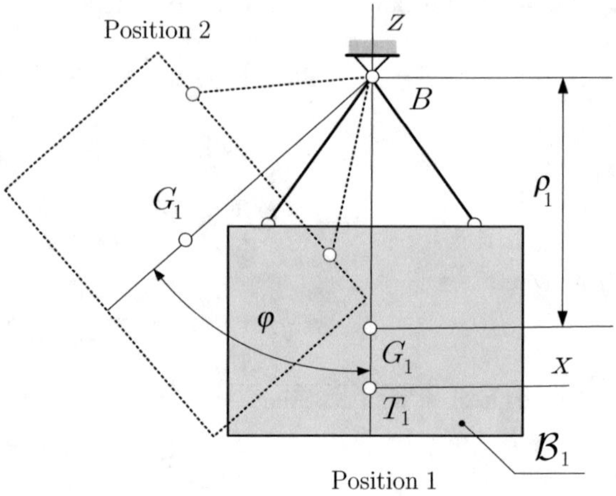

Fig. 2.19 The swaying panel

where

$$\mathcal{E}_1 = \frac{1}{2}m_2\left(v_2'\right)^2 + \frac{1}{2}J_B\frac{\left(v_2'\right)^2}{\rho_1^2} = \frac{1}{2}\frac{(\tilde{m}_1+m_2)\,m_2}{(\tilde{m}_1+m_2)^2}m_2v_2^2 = \frac{m_2}{\tilde{m}_1+m_2}\underbrace{\frac{1}{2}m_2v_2^2}_{\mathcal{E}_o}$$

and

$$\mathcal{E}_2 = 0.$$

On the other hand,

$$U_1 = 0 \quad \text{and} \quad U_2 = mg\rho_1\left(1-\cos\varphi\right),$$

where

$$m = m_1 + m_2 \approxeq m_1\,, \quad m_1 \gg m_2\,.$$

Thus,

$$\frac{m_2}{\tilde{m}_1+m_2}\,\frac{1}{2}m_2v_2^2 = m_1g\rho_1\underbrace{\left(1-\cos\varphi\right)}_{2\sin^2\frac{\varphi}{2}}$$

from where

$$v_2 = 2\sqrt{\frac{m_1g\rho_1\left(\tilde{m}_1+m_2\right)}{m_1^2}}\,\sin\frac{\varphi}{2}$$

is the equation for φ. Conversely, if we measure φ we can determine the velocity v_2 of the bullet.

2.4 Problems

Problem 2.1 Prove formula (2.13) which provides the energy loss during impact.

Problem 2.2 Solve Exercise 2.5 by using Eqs. (2.9a), (2.9c), and (2.9e).

Problem 2.3 Two balls with masses $m_1 = 2\,\text{kg}$ and $m_2 = 6\,\text{kg}$ collide—see Fig. 2.20 for details. The velocities $\mathbf{v}_1'$ and $\mathbf{v}_2'$ the balls have after impact are known. Determine the velocities $\mathbf{v}_1$ and $\mathbf{v}_2$ before impact if $e = 1$ (perfectly elastic impact).

$$\mathbf{v}_1' = -2\mathbf{i}_x + 4\mathbf{i}_y \text{ [m/s]}\,, \quad \mathbf{v}_2' = 2\mathbf{i}_x - 4\mathbf{i}_y \text{ [m/s]}\,.$$

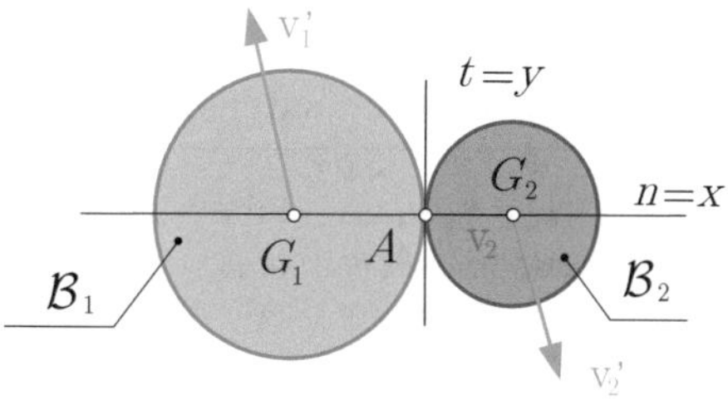

Fig. 2.20 Impact of two balls

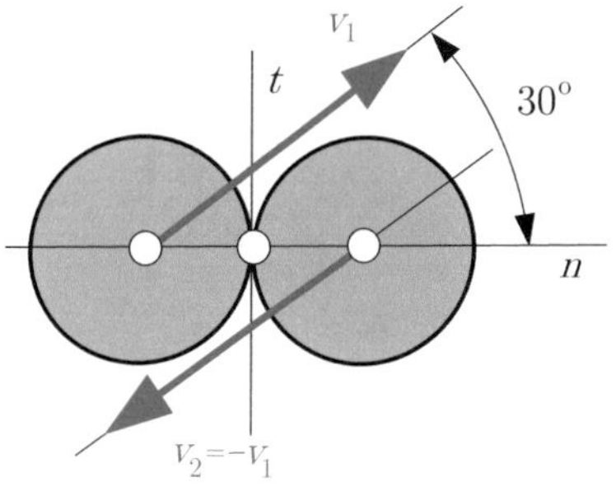

Fig. 2.21 Impact of two hockey pucks

Problem 2.4 Two identical hockey pucks are moving on a hockey rink at the same speed of 4 m/s in parallel but opposite directions till they strike each other. Determine the velocity of each puck after impact if the coefficient of restitution $e = 1$.

Problem 2.5 Figure 2.22 shows two pendulums. Each pendulum has the same mass and length. Pendulum 1 is released from a horizontal position with no initial velocity. It swings from the horizontal position till it hits pendulum 2 which is at rest in a vertical position. Find the velocities after impact if $e = 1$. What happens to the system after impact?

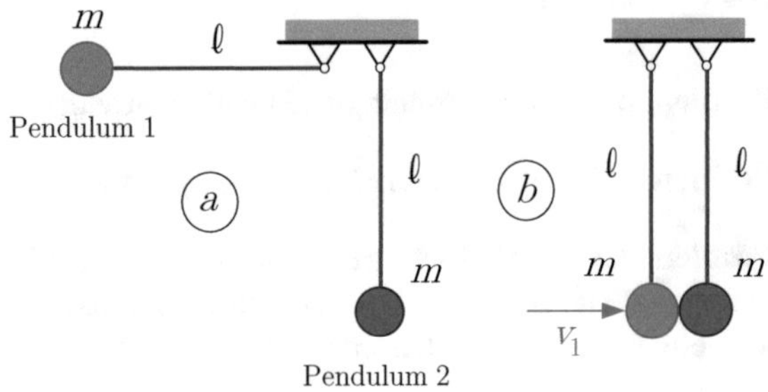

Fig. 2.22 Impact of two pendulums

Problem 2.6 Assume that the pendulums in the previous problem have different masses, i.e., $m_1 \neq m_2$—the lengths of the pendulums are the same. Assume further that $e = 0.9$. Determine the velocity of Pendulum 2 after impact. What equations can be used to determine the angle ϑ Pendulum 2 swings through after impact?

Problem 2.7 A slender rod of length ℓ and mass m is dropped onto the rigid supports B and D. Since the support B is slightly higher than support D the rod strikes support B first. Then the velocity of the rod is $\mathbf{v}_0 = -v_o\mathbf{i}_y$, $v_o > 0$. Assume that the impact is perfectly elastic both at B and D. Determine the angular velocity of the rod and the velocity at its mass center right after the rod (i) strikes support B, (ii) strikes support D, and (iii) strikes support B again.

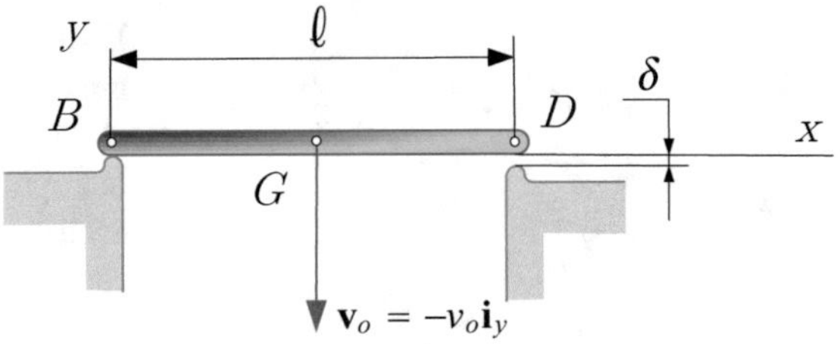

Fig. 2.23 Slender rod dropped onto rigid supports

Reference

1. I. Sályi, *Kinetics (in Hungarian)* (Tankönyvkiadó, Budapest, 1961)

Chapter 3
Some Vibration Problems

3.1 Single Degree of Freedom Systems

3.1.1 Introductory Remarks

Let us denote the coordinate that describes the motion by $q(t)$. Vibration is a mechanical phenomenon whereby oscillations occur about an equilibrium point. More precisely the motion $q(t)$ is said to be vibration if

- $q(t)$ is limited and describes backward and forward movement hence
- the derivative $\mathrm{d}q(t)/\mathrm{d}t$ constantly changes its sign.

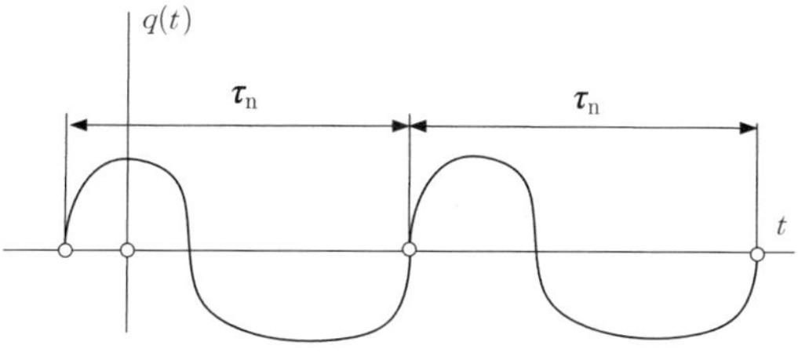

Fig. 3.1 Periodic function

The vibration is *periodic* if

$$q(t) = q(t + \tau_n)\,,$$

in which τ_n is the *period* (Fig. 3.1), and *random* if $\mathrm{d}q(t)/\mathrm{d}t$ changes its sign a lot of times (e.g., movement of a tire on a gravel road).

For periodic vibrations, the number of cycles per unit time defines the *frequency* and the maximum of $|q(t)|$ is called *amplitude* of the vibration.

G. Szeidl and L. P. Kiss, *Mechanical Vibrations*, Foundations of Engineering Mechanics,
https://doi.org/10.1007/978-3-030-45074-8_3

Mechanical vibration takes place when a system is displaced from a position of stable equilibrium. Then the system tends to the equilibrium position under the action of restoring forces (spring, gravitational pendulum). When reaching this position the system has a velocity which carries it beyond that position. The process can be repeated infinitely, i.e., the system keeps moving back and forth across the equilibrium position.

When the motion is maintained by the restoring force only the vibration is said to be *free vibration*; if a periodic external force is applied to the system then the resulting motion is *forced vibration*. When the effect of friction can be neglected the vibrations are said to be undamped, otherwise we speak about *damped vibrations*. (Each real system is actually damped to some degree.)

3.1.2 Classification

3.1.2.1 Undamped Free Vibrations

Figure 3.2a shows a *spring–mass system* which, in accordance with its name, consists of a spring and a mass (a particle). The horizontal displacement of the mass center G is denoted by x. If $x = 0$ the system is at rest, i.e., there is no force in the spring. It is obvious that $(x > 0)[x < 0]$ is the (elongation) [shortening] of the spring. Figure 3.2b represents graphically how the spring force F_k (the force exerted on the spring) may depend on x. If the function $F_k(x)$ is linear it can be given in the form

$$F_k = k\,x = \frac{1}{f}x \tag{3.1}$$

in which k [N/m] is the *spring constant* (or spring stiffness) and f is referred to as *spring flexibility*. The spring characteristic $F_k(x)$ is (degressive)[progressive] if the derivative $\mathrm{d}F_k/\mathrm{d}x$ is (decreasing)[increasing]. We shall, in general, assume that the spring characteristic is linear.

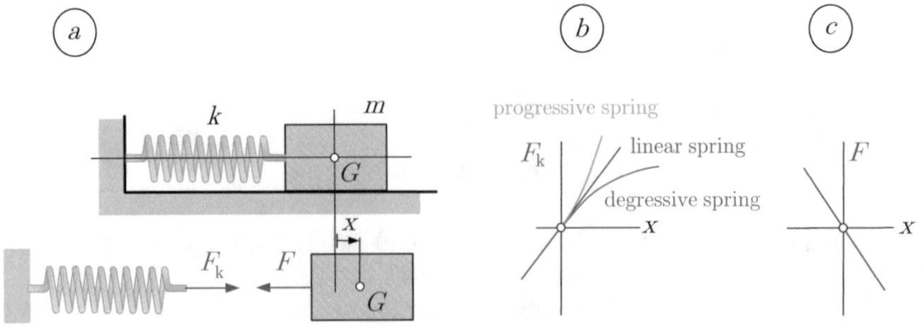

Fig. 3.2 Spring mass system with spring characteristic

If we displace the particle from its equilibrium position—see Fig. 3.2a—then it moves under the action of the restoring force $F = -F_k$. Our aim is to find the motion law (the function) $x(t)$. Using Newton's second law, we get the equation of motion

$$ma = m\ddot{x} = F = -kx \ .$$

This equation is a homogeneous ordinary differential equation of order two:

$$\boxed{\ddot{x} + \omega_n^2 x = 0\,, \qquad \omega_n^2 = \frac{k}{m}\ .} \tag{3.2}$$

Here ω_n [1/s] is the *natural circular (or angular) frequency* (or eigenfrequency for short). The general solution of the differential equation (3.2) takes the form

$$x = A\cos\omega_n t + B\sin\omega_n t, \tag{3.3}$$

where A and B are undetermined integration constants. Let x_o and v_o be the initial displacement and velocity of the particle m. The undetermined constants of integration A and B can be determined from the initial conditions:

$$\begin{aligned} t = t_o = 0 \qquad & x(t_o) = x(0) = x_o = \left(A\cos\omega_n t + B\sin\omega_n t\right)\big|_{t=0} = A\,, \\ t = t_o = 0 \qquad & \dot{x}(t_o) = v(0) = v_o = \omega_n\left(-A\sin\omega_n t + B\cos\omega_n t\right)\big|_{t=0} = \omega_n B\,. \end{aligned} \tag{3.4}$$

With the integration constants $A = x_o$ and $B = v_o/\omega_n$

$$x = \underbrace{x_o}_{D\sin\phi}\cos\omega_n t + \underbrace{\frac{v_o}{\omega_n}}_{D\cos\phi}\sin\omega_n t = D\sin\left(\omega_n t + \phi\right) \tag{3.5a}$$

is the solution in which

$$D = \sqrt{x_o^2 + \left(\frac{v_o}{\omega_n}\right)^2}\,, \qquad \phi = \arctan\frac{\omega_n x_o}{v_o} \tag{3.5b}$$

are the amplitude and the phase angle. Equation

$$\tau_n = \tau = 2\pi/\omega_n \tag{3.6a}$$

gives the period of the spring–mass system for which the unit is, in general, 1/s. This definition is the same as that given by equation (1.63). Its reciprocal is denoted by f_n and is called natural frequency:

$$f_n = 1/\tau_n = \omega_n/2\pi\ . \tag{3.6b}$$

The unit of the natural frequency is 1/s. It is called hertz and is abbreviated as Hz. Figure 3.3 shows the displacement time curve.

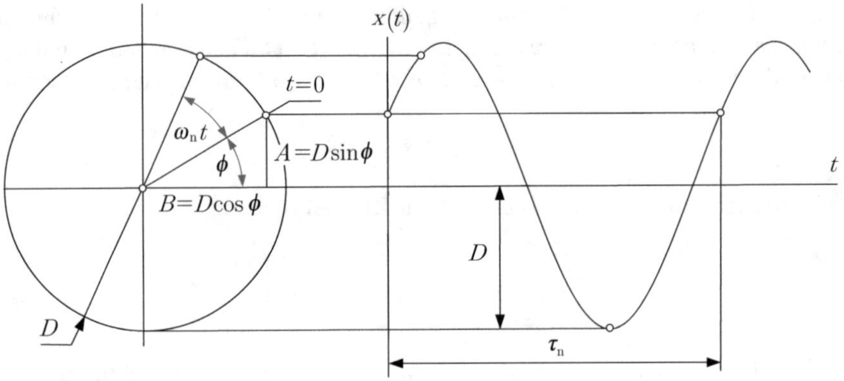

Fig. 3.3 Displacement time curve

Remark 3.1 The kinetic energy of the particle and the potential energy stored in the spring are given by the following relations:

$$\mathcal{E} = \frac{1}{2}m\dot{x}^2, \qquad U = \frac{1}{2}kx^2, \tag{3.7}$$

where

$$x = D\sin(\omega_n t + \phi), \qquad \dot{x} = D\omega_n \cos(\omega_n t + \phi). \tag{3.8}$$

It is obvious that

$$\mathcal{E}_{\max} = \frac{1}{2}\omega_n^2 D^2 m, \qquad U_{\max} = \frac{1}{2}kD^2. \tag{3.9}$$

are maximum of $\mathcal{E}$ and U. Since $\mathcal{E} = 0$ if $U = U_{\max}$ and $U = 0$ if $\mathcal{E} = \mathcal{E}_{\max}$ it follows from the principle of energy conservation (1.58b) that

$$\mathcal{E}_{\max} = U_{\max}. \tag{3.10}$$

Hence

$$\boxed{\omega_n^2 = \frac{k}{m} = \frac{U_{\max}}{\mathcal{E}_{\max}}.} \tag{3.11}$$

Exercise 3.1 Liquid vibrates in the U-pipe shown in Fig. 3.4. Find the eigenfrequency of the oscillating liquid column.

Fig. 3.4 Vibrating liquid in a U pipe

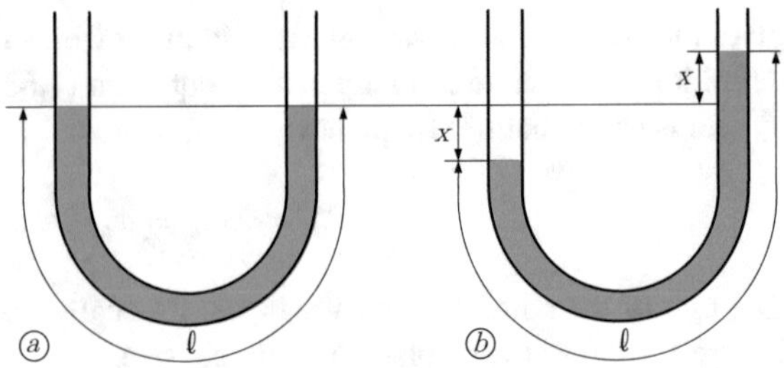

Figure 3.4a, b shows (the equilibrium position of the liquid) [the liquid in motion]. Let ρ and A be the density of the liquid and the cross-sectional area of the pipe. Further let x be the displacement of the liquid column in motion. It can be seen that the change in potential energy consists of two parts: (a) the potential energy of the raised liquid column plus (b) the potential energy of the depressed liquid column. Hence

$$\mathcal{U} = \frac{x}{2}\rho A g x + \frac{x}{2}\rho A g x = \rho A g x^2 .$$

The kinetic energy of the moving liquid is given by the following equation:

$$\mathcal{E} = \frac{1}{2}\rho A \ell \, \dot{x}^2 .$$

Let us assume that the motion of the liquid is harmonic. Then

$$x = D \cos \omega_n t$$

and

$$U = \underbrace{\rho A g D^2}_{U_{\max}} \cos^2 \omega_n t , \qquad \mathcal{E} = \underbrace{\frac{1}{2}\rho A \ell \, \omega_n^2 D^2}_{\mathcal{E}_{\max}} \sin^2 \omega_n t .$$

With $U_{\max}$ and $\mathcal{E}_{\max}$ Eq. (3.11) yields the sought eigenfrequency:

$$\omega_n^2 = \frac{k}{m} = \frac{U_{\max}}{\mathcal{E}_{\max}} = \frac{2g}{\ell} .$$

Remark 3.2 For our later considerations we shall introduce some notational conventions. The beam shown in Fig. 3.5 is subjected to vertical unit forces at the points P_i and P_j. The vertical displacement at P_j due to the unit force at P_i is denoted by f_{ji}—here the first subscript identifies the point where we consider the displacement, whereas the second subscript identifies the point where the unit force is acting. In accordance with this rule f_{ii} is the displacements at P_i due to the unit force exerted on the beam at the same point and f_{ij} is the displacement at P_i under the action of the unit force at P_j, etc. The displacements f_{ji}, f_{ij} f_{ii}, and f_{jj} are called flexibility influence coefficients (or simply flexibilities).

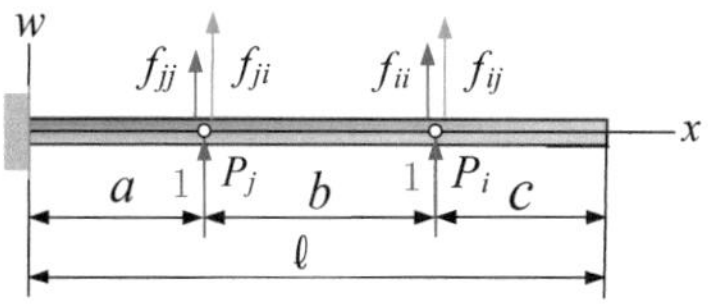

Fig. 3.5 Cantilever beam

Table 3.1 Eigenfrequencies of beams with one concentrated mass

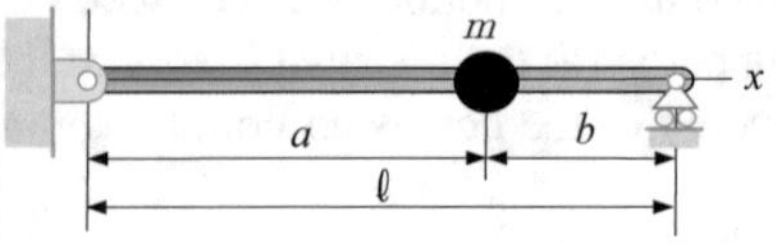	$\omega_n^2 = \dfrac{3IE\ell}{ma^2b^2}$
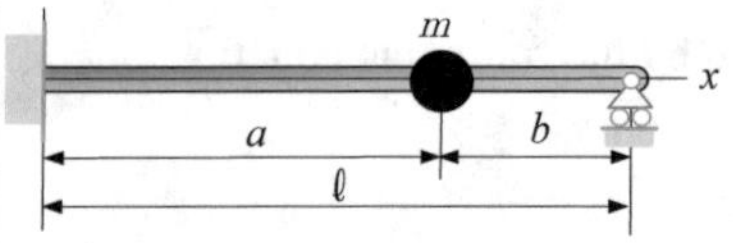	$\omega_n^2 = \dfrac{12IE\ell^3}{ma^3b^2(3\ell + b)}$
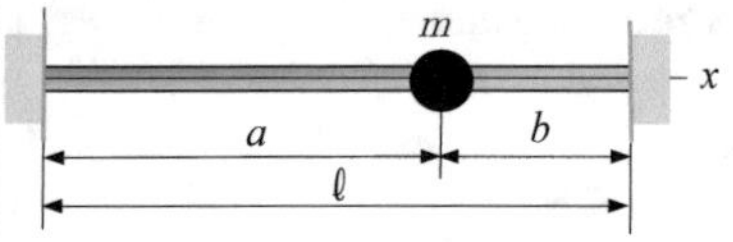	$\omega_n^2 = \dfrac{3IE\ell^3}{ma^3b^3}$
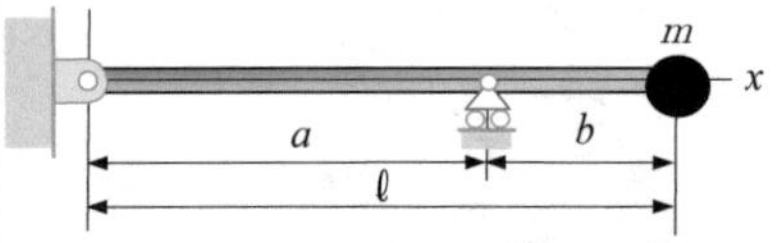	$\omega_n^2 = \dfrac{3IE}{mb^2\ell}$
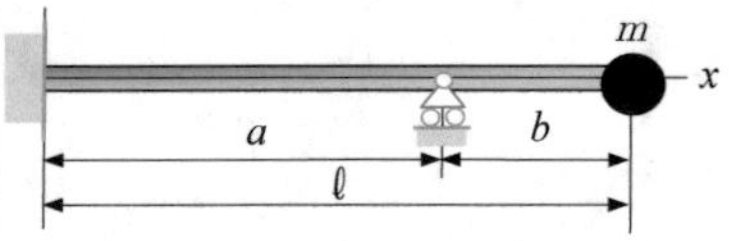	$\omega_n^2 = \dfrac{3IE}{mb^2\ell(1 - a/4\ell)}$

Exercise 3.2 The cantilever beam shown in Fig. 3.6 carries a mass attached to its right end. We shall assume that m is much greater than the mass of the beam. We shall denote the flexibility influence coefficient at the right end of the beam by f_{11}. It is known that

$$f_{11} = \frac{\ell^3}{3IE}, \tag{3.12}$$

where ℓ is the length of the beam, I is the moment of inertia, and E is Young's modulus. Find the eigenfrequency of the oscillating beam.

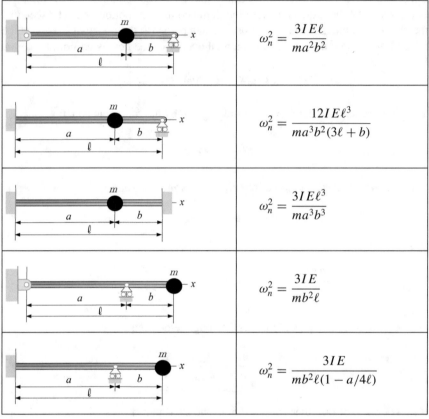

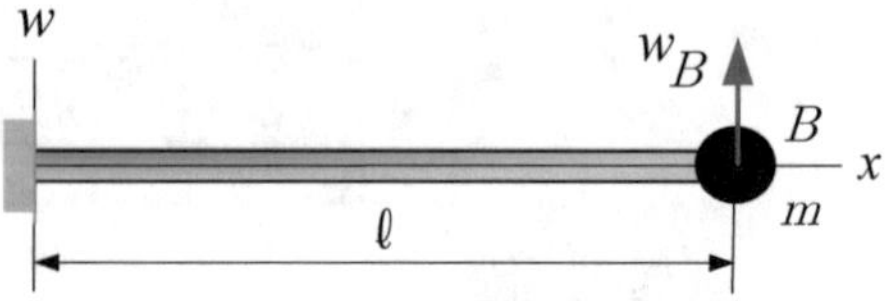

Fig. 3.6 Cantilever beam with a mass at its right end

If the beam oscillates the displacement w_B at the right end of the beam is due to the force of inertia $-m\ddot{w}_B$. We can, therefore, write

$$w_B = -f_{11}\, m\ddot{w}_B\,. \tag{3.13}$$

After rearranging this equation, we get

$$m\ddot{w}_B + \frac{1}{f_{11}} w_B = m\ddot{w}_B + k_{11} w_B = 0\,, \qquad k_{11} = \frac{1}{f_{11}} = \frac{3IE}{\ell^3}\,. \tag{3.14}$$

This is the equation of motion of an undamped free system with one degree of freedom in which $k_{11} = 1/f_{11}$ is the spring constant (or spring stiffness). Thus, it follows from Eq. $(3.2)_2$ that

$$\omega_n^2 = \frac{k_{11}}{m} = \frac{3IE}{m\ell^3}\,. \tag{3.15}$$

Based on the solution presented in Exercise 3.2, Table 3.1 contains the square of the eigenfrequencies for some beam supports.

3.1.2.2 Damped Free Vibrations

We have mentioned that the vibrations are always damped to some degree. By damping we mean an influence within or upon an oscillatory system that has the effect of reducing, restricting, or preventing its oscillations. Damping is caused, in general, by various friction forces such as dry friction or fluid friction—if a rigid body moves in a fluid then the forces exerted on it slacken the speed of motion. It is worth mentioning that damping is always associated with energy dissipation. We shall assume viscous damping which is caused by fluid friction. If the velocity of the particle is under a certain and not too high limit the damping force exerted on the body is directly proportional and opposite in direction to the velocity. Then the damping force F_c can be given in the form

$$F_c = -c\,\dot{x}, \tag{3.16}$$

where c [Ns/m] is called *damping coefficient*.

Fig. 3.7 Viscous damper

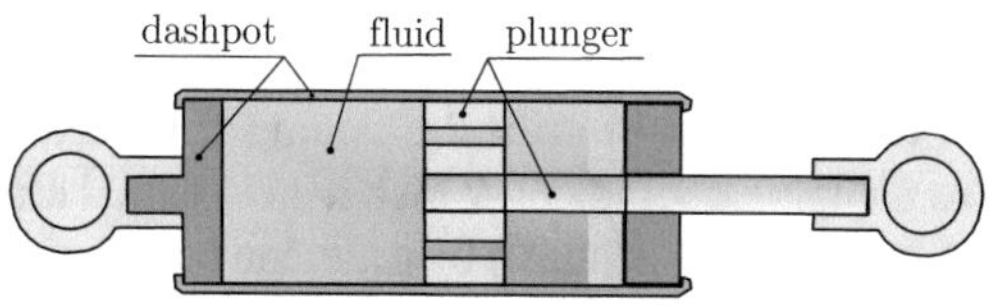

Figure 3.7 shows the sketch of a vibration damper and its main parts.

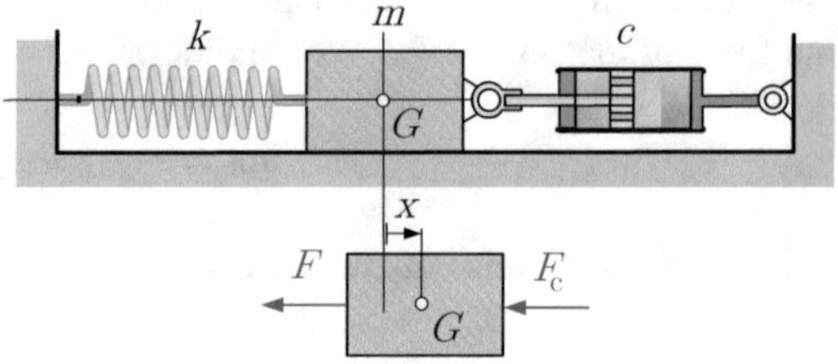

Fig. 3.8 Damped free spring-mass system

Figure 3.8 is that of a damped free system with one degree of freedom if $\dot{x} > 0$ which means that $F_c < 0$.

The equation of motion takes the form

$$m\,\ddot{x} + c\,\dot{x} + k\,x = 0, \tag{3.17a}$$

which can be rewritten as

$$\boxed{\ddot{x} + 2\beta\dot{x} + \omega_n^2\,x = 0\,, \qquad 2\beta = \frac{c}{m}\,, \quad \omega_n^2 = \frac{k}{m}\,.} \tag{3.17b}$$

Let us assume that $x = e^{\lambda t}$ is the solution to the differential equation (3.17b). Here λ is the characteristic value. Upon substitution of the solution into Eq. (3.17b) we get the characteristic equation

$$\lambda^2 + 2\beta\lambda + \omega_n^2 = 0 \tag{3.18a}$$

from where

$$\lambda_{1,2} = -\beta \pm \sqrt{\beta^2 - \omega_n^2} = -\beta \pm \mu\,, \qquad \mu = \sqrt{\beta^2 - \omega_n^2}\,. \tag{3.18b}$$

Assume that the two roots are different. Then

$$x = Ae^{\lambda_1 t} + Be^{\lambda_2 t} = Ae^{-(\beta-\mu)\,t} + Be^{-(\beta+\mu)t} \tag{3.19}$$

is the general solution. The behavior of the system depends on the amount of damping.

The undetermined constants of integration A and B can be determined by utilizing the initial conditions.

We speak about critical damping if we have double roots, that is, $\mu = 0$. Then the damping coefficient, which is now called critical damping coefficient, cannot be arbitrary. Let us denote it by c_n. It can be calculated if we utilize the condition $\mu = 0$ and Eq. (3.17b)$_{2,3}$:

$$\beta^2 = \omega_n^2 \quad \rightarrow \quad \left(\frac{c_n}{2m}\right)^2 = \frac{k}{m}$$

from where

$$\boxed{c_n = 2m\sqrt{\frac{k}{m}} = 2m\omega_n.} \tag{3.20}$$

Depending on what sign the expression $\beta^2 - \omega_n^2$ under the square root has, we distinguish three cases:

(a) If $\beta > \omega_n$ ($c > c_n$) then μ is real. Since $\beta > \mu > 0$ the roots $\lambda_1 = -\beta + \mu$ and $\lambda_2 = -\beta - \mu$ are negative real numbers. Consequently, the solution (3.19) is non-vibratory and tends to zero as t tends to infinity. In accordance with (3.4) let x_o and v_o be the displacement and velocity at $t = 0$. It can be checked with paper and pencil calculations that

$$\begin{aligned} x =& \frac{1}{\lambda_2 - \lambda_1}\left[-\left(v_o - \lambda_2 x_o\right) e^{\lambda_1 t} + \left(v_o - \lambda_1 x_o\right) e^{\lambda_2 t}\right] = \\ =& \frac{1}{2\mu}\left[\left(v_o + (\beta + \mu) x_o\right) e^{-(\beta-\mu)t} - \left(v_o + (\beta - \mu) x_o\right) e^{-(\beta+\mu)t}\right] \end{aligned} \tag{3.21}$$

is the solution in terms of the initial values. Figure 3.9 shows the quotient x/x_o for a given system in the time interval $t \in [0, 2]$ if (i) $v_o > 0$, (ii) $v_o = 0$, and finally (iii) if $v_o < 0$. It should be mentioned that there exist such negative initial velocities for which the solution does not change sign. Since solution (3.21) is not oscillatory *the system is overdamped*. In other words, we are faced with the case of heavy damping.

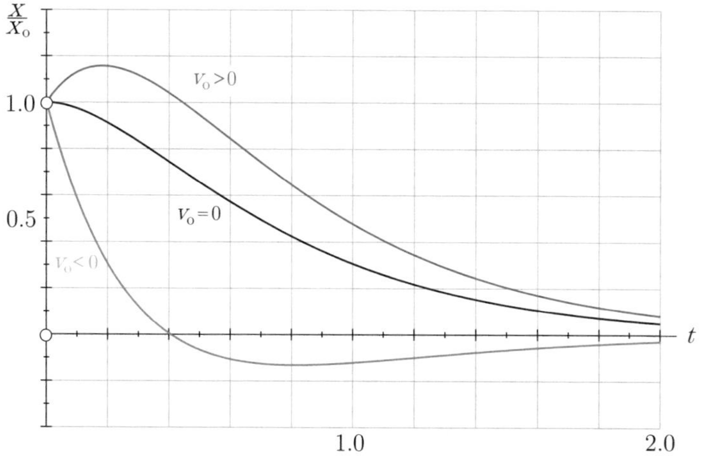

Fig. 3.9 Dependence of x/x_0 on time $-v_0$ is the parameter

(b) If $\beta = \omega_n$ ($c = c_n$) we have double roots and the solution assumes the form

$$x = (A + Bt)\, e^{-\beta t} = [x_o + (\beta x_o + v_o)\, t]\, e^{-\beta t}. \tag{3.22}$$

This solution is also non-vibratory. Its graphical representation is more or less the same as that shown in Fig. 3.9. The damping is called critical since the condition $c = c_n$ separates the vibratory and non-vibratory solutions.

(c) If $\beta < \omega_n$ ($c < c_n$) we speak about light damping: the system is underdamped. The solutions for λ are complex and conjugate:

$$\lambda_{1,2} = -\beta \pm i\sqrt{\omega_n^2 - \beta^2} = -\beta \pm i\omega_d \tag{3.23a}$$

in which

$$\omega_d^2 = \omega_n^2 - \beta^2 = \frac{k}{m} - \left(\frac{c}{2m}\right)^2 = \frac{k}{m}\left(1 - \frac{c}{\underbrace{4m^2\frac{k}{m}}_{c_n^2}}\right) = \omega_n^2\left(1 - \frac{c^2}{c_n^2}\right)$$

hence

$$\boxed{\omega_d = \omega_n\sqrt{1 - \frac{c^2}{c_n^2}} \underset{\zeta=\frac{c}{c_n}}{=\uparrow=} \omega_n\sqrt{1-\zeta^2}\,.} \tag{3.23b}$$

Here ω_d is the natural circular frequency (eigenfrequency) of the damped vibrations and

$$\boxed{\zeta = \frac{c}{c_n}} \tag{3.23c}$$

is the damping factor. The real solution for x is of the form

$$\boxed{x = e^{-\beta t}\left(A\cos\omega_d t + B\sin\omega_d t\right) = e^{-\beta t}\left(x_o\cos\omega_d t + \frac{v_o + \beta x_o}{\omega_d}\sin\omega_d t\right)} \tag{3.24a}$$

or

$$\boxed{x = e^{-\beta t}\left(\underbrace{x_o}_{D\sin\phi}\cos\omega_d t + \underbrace{\frac{v_o + \beta x_o}{\omega_d}}_{D\cos\phi}\sin\omega_d t\right) = e^{-\beta t} D\sin\left(\omega_d t + \phi\right),} \tag{3.24b}$$

where

$$D = \underbrace{x_o\sqrt{1 + \left(\frac{v_o + \beta x_o}{\omega_d x_o}\right)^2}}_{\hat{D}} \quad \text{and} \quad \phi = \arctan\frac{\omega_d x_o}{v_o + \beta x_o}\,. \tag{3.25}$$

Solution (3.24) describes a vibratory motion with diminishing amplitude and the time interval

$$\tau_d = \frac{2\pi}{\omega_d} = \underbrace{\frac{2\pi}{\omega_n}}_{\tau_n} \frac{1}{\sqrt{1-\zeta^2}} = \tau_n \frac{1}{\sqrt{1-\zeta^2}} > \tau_n, \tag{3.26}$$

which separates two successive points where the solution touches one of the limiting curves—see Fig. 3.10 for details—is called period of the damped vibration.

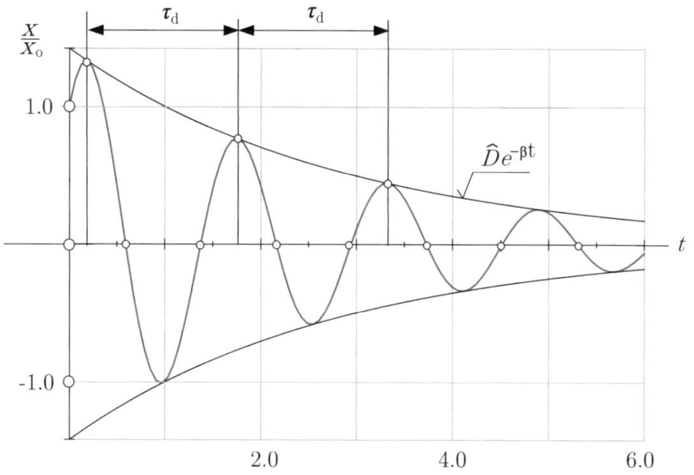

Fig. 3.10 Dependence of x/x_0 on time

The logarithmic decrement is defined by the equation

$$\delta = \ln \frac{x(t)}{x(t+\tau_d)} = \ln \frac{e^{-\beta t} D \sin(\omega_d t + \phi)}{e^{-\beta(t+\tau_d)} D \sin[\omega_d (t+\tau_d) + \phi]} \underset{\omega_d \tau_d = 2\pi}{=\uparrow=}$$

$$= \ln \frac{e^{-\beta t} D \sin(\omega_d t + \phi)}{e^{-\beta(t+\tau_d)} D \sin(\omega_d t + \phi + 2\pi)} = \ln e^{-\beta t} e^{\beta(t+\tau_d)} = \ln e^{\frac{c}{2m}\tau_d} = \frac{c}{2m}\tau_d. \tag{3.27}$$

With δ, τ_d (these two quantities are measurable), and m, we can determine the damping coefficient.

3.1.2.3 Undamped Forced Vibrations

Excitations play an important role in various engineering applications since they are the sources of the forced vibrations of a system. These vibrations are often caused by a periodic force (by a force of excitation) exerted on the system. It may also occur that the system considered is connected elastically to such machine part which performs an alternating motion. Then we speak about displacement excitation which results in, again, forced vibrations. We shall assume that the force of excitation

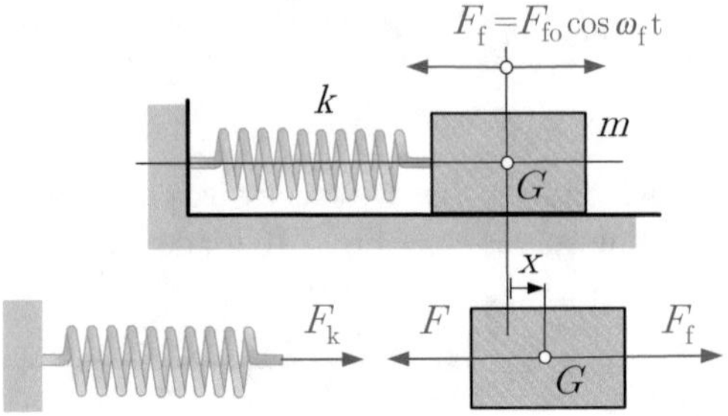

Fig. 3.11 Excitation on a spring mass system

is a harmonic one. Consider now the spring–mass system shown in Fig. 3.11. The system is subjected to a harmonic external force $F_f = F_{fo} \cos \omega_f t$ where F_{fo} is the amplitude of the excitation force while ω_f is referred to as forced circular frequency.

Remark 3.3 Here and in the sequel a harmonic vectorial quantity (F_f in the present case) is depicted as is shown in Fig. 3.11 since the two extreme values of the vector differ from each other in sign.

Using Newton's second law, we get the equation of motion in the form

$$m\ddot{x} = -F + F_f = -kx + F_{fo} \cos \omega_f t \tag{3.28a}$$

or after dividing throughout by m it is

$$\ddot{x} + \omega_n^2 x = \frac{F_{fo}}{m} \cos \omega_f t\,, \qquad \omega_n^2 = \frac{k}{m}, \tag{3.28b}$$

where

$$\frac{F_{fo}}{m} = \underbrace{\frac{F_{fo}}{k}}_{\delta_f} \frac{k}{m} = \omega_n^2 \delta_f\,. \tag{3.28c}$$

Here δ_f is the elongation due to the amplitude F_{fo} of the excitation force. Using this notation, we can rewrite Eq. (3.28b) in the following form:

$$\boxed{\ddot{x} + \omega_n^2 x = \omega_n^2 \delta_f \cos \omega_f t\,, \qquad \omega_n^2 = \frac{k}{m}, \qquad \delta_f = \frac{F_{fo}}{k}\,.} \tag{3.29}$$

The above equation is an inhomogeneous differential equation with constant coefficients. As is well known its general solution can be expressed in the form

$$x = x_{\text{hom}}(t) + x_{\text{part}}(t), \tag{3.30a}$$

where

$$x_{\text{hom}}(t) = A \cos \omega_n t + B \sin \omega_n t \tag{3.30b}$$

is the solution of the homogeneous differential equation

$$\ddot{x} + \omega_n^2 x = 0$$

and

$$x_{\text{part}} = x_m \cos \omega_f t \tag{3.30c}$$

is the particular solution that satisfies the inhomogeneous equation—it never satisfies the homogeneous equation. The unknown amplitude x_m can be obtained from the condition that the particular solution should satisfy the inhomogeneous equation. Upon substitution of the particular solution (3.30c) into $(3.29)_1$, we have

$$-\omega_f^2 x_m \cos \omega_f t + \omega_n^2 x_m \cos \omega_f t = \omega_n^2 \delta_f \cos \omega_f t$$

from where the unknown amplitude is

$$\boxed{x_m = \frac{\omega_n^2}{\omega_n^2 - \omega_f^2} \delta_f = \frac{1}{1 - \frac{\omega_f^2}{\omega_n^2}} \delta_f = \frac{1}{1 - \eta^2} \delta_f \,, \quad \eta^2 = \frac{\omega_f^2}{\omega_n^2}.} \tag{3.31}$$

With x_m the general solution takes the form

$$\boxed{x(t) = \underbrace{A \cos \omega_n t + B \sin \omega_n t}_{\text{transient function}} + \underbrace{\overbrace{\frac{\delta_f}{1 - \eta^2}}^{x_m} \cos \omega_f t}_{\text{steady-state vibration}}.} \tag{3.32}$$

The transient vibration is a free vibration of the system which soon disappears due to the damping inherent in the system. The second term represents the steady-state vibrations caused and maintained by the force of excitation. Its frequency coincides with that of the excitation force. The amplitude x_m depends on the frequency ratio η. The magnitude of the quotient x_m/δ_f is called *magnification factor*:

$$\boxed{m_f = \left| \frac{x_m}{\delta_f} \right| = \frac{1}{|1 - \eta^2|} \,.} \tag{3.33}$$

The quotient x_m/δ_f has the following properties:

(a) If $\eta = \omega_f/\omega_n \in [0, 1)$ then x_m/δ_f is a positive quantity and so is the amplitude x_m of the steady-state vibrations. Then we say that the vibration is in phase with the excitation.

(b) If $\eta = \omega_f/\omega_n \to 1$ then for $(\eta < 1)$ $[\eta > 1]$ the quotient x_m/δ_f tends to $(+\infty)$ $[-\infty]$ and so does the amplitude x_m of the steady-state vibrations. This phenomenon is known as *resonance* which is to be avoided since it may result in harmful vibrations.

(c) If $\eta = \omega_f/\omega_n > 1$ then the quotient x_m/δ_f is a negative quantity and so is the amplitude x_m of the steady-state vibrations. Then we say that the vibration is 180° out of phase with respect to the excitation.

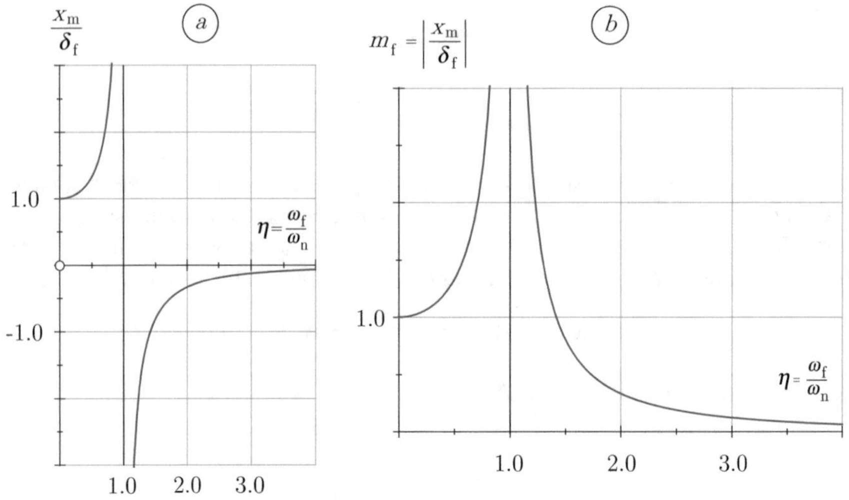

Fig. 3.12 Magnification factor (Resonance curve)

Figure 3.12 represents the magnification factor and its magnitude against the dimensionless parameter η.

3.1.2.4 Forced Vibrations with Damping

If the damped single degree of freedom system shown in Fig. 3.8 is subjected to a harmonic excitation force $F_f = F_{fo} \cos \omega_f t$ we speak about forced-damped vibrations.

Fig. 3.13 Excitation on a damped spring-mass system

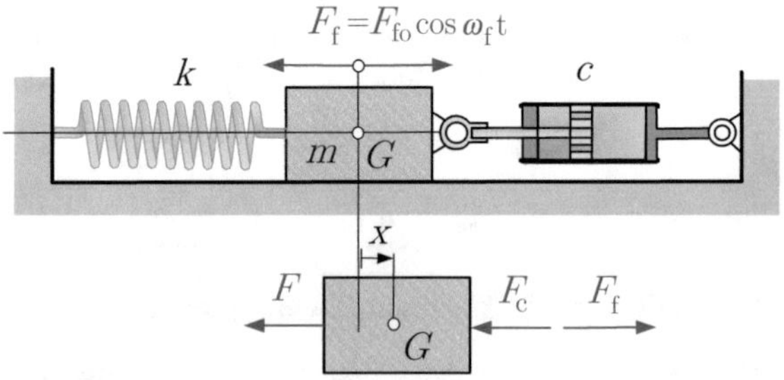

The equation of motion for forced vibrations with damping can easily be established if we make use of the notations from Fig. 3.13 which depicts a harmonically excited and damped single degree of freedom system:

$$m\,\ddot{x} + c\,\dot{x} + kx = F_{fo} \cos \omega_f t. \tag{3.34a}$$

After dividing throughout by m and utilizing then the notations $(3.17\text{b})_{1,2}$ and Eq. $(3.28\text{c})_2$, we get

$$\boxed{\ddot{x} + 2\beta\,\dot{x} + \omega_n^2 x = \omega_n^2 \delta_f \cos\omega_f t.} \tag{3.34b}$$

Assume that the system is underdamped. The general solution is again the sum of the general solution to the homogeneous equation, which is given by Eq. (3.24b), and a harmonic particular solution:

$$x(t) = \underbrace{e^{-\beta t}\left(A\cos\omega_d t + B\sin\omega_d t\right)}_{\text{transient function}} + \underbrace{x_m \cos\left(\omega_f t + \phi\right)}_{\text{steady-state vibration}}.$$

Since the transient function is usually negligible it is our aim to determine both the amplitude x_m of the steady-state vibrations and the phase angle ϕ. Substituting the particular solution

$$x_{\text{part}} = x_m \cos\left(\omega_f t + \phi\right) \tag{3.35}$$

into the equation of motion (3.34b), we obtain

$$-\omega_f^2 x_m \cos\left(\omega_f t + \phi\right) - 2\beta\omega_f x_m \sin\left(\omega_f t + \phi\right) + \omega_n^2 x_m \cos\left(\omega_f t + \phi\right) = \\ = \omega_n^2 \delta_f \cos\omega_f t.$$

Making $\omega_f t + \phi$ equal to 0 ($\omega_f t = -\phi$) and $\pi/2$ ($\omega_f t = \pi/2 - \phi$), we can write

$$\left(\omega_n^2 - \omega_f^2\right) x_m = \omega_n^2 \delta_f \cos\phi\,, \tag{3.36a}$$

$$-2\beta\omega_f x_m = \omega_n^2 \delta_f \cos\left(\pi/2 - \phi\right) = \omega_n^2 \delta_f \sin\phi\,. \tag{3.36b}$$

Squaring equations (3.36) and adding them, we have

$$x_m^2\left[\left(\omega_n^2 - \omega_f^2\right)^2 + 4\beta^2\omega_f^2\right] = \omega_n^4 \delta_f^2$$

from where

$$m_f = \left|\frac{x_m}{\delta_f}\right| = \frac{1}{\sqrt{\left(1 - \dfrac{\omega_f^2}{\omega_n^2}\right)^2 + \left(\dfrac{2\beta}{\omega_n}\dfrac{\omega_f}{\omega_n}\right)^2}} \tag{3.37}$$

is the magnification factor in which

$$\frac{\omega_f}{\omega_n} \underset{(3.31)_2}{=\uparrow=} \eta\,, \quad \frac{2\beta}{\omega_n} \underset{(3.17\text{b})_2}{=\uparrow=} \frac{c}{m}\frac{1}{\omega_n} = 2c\frac{1}{2m\omega_n} \underset{(3.20)}{=\uparrow=} \frac{2c}{c_n} \underset{(3.23\text{c})}{=\uparrow=} 2\zeta\,. \tag{3.38}$$

Thus

$$\boxed{m_f = \frac{1}{\sqrt{\left(1 - \eta^2\right)^2 + (2\zeta\eta)^2}}\,.} \tag{3.39}$$

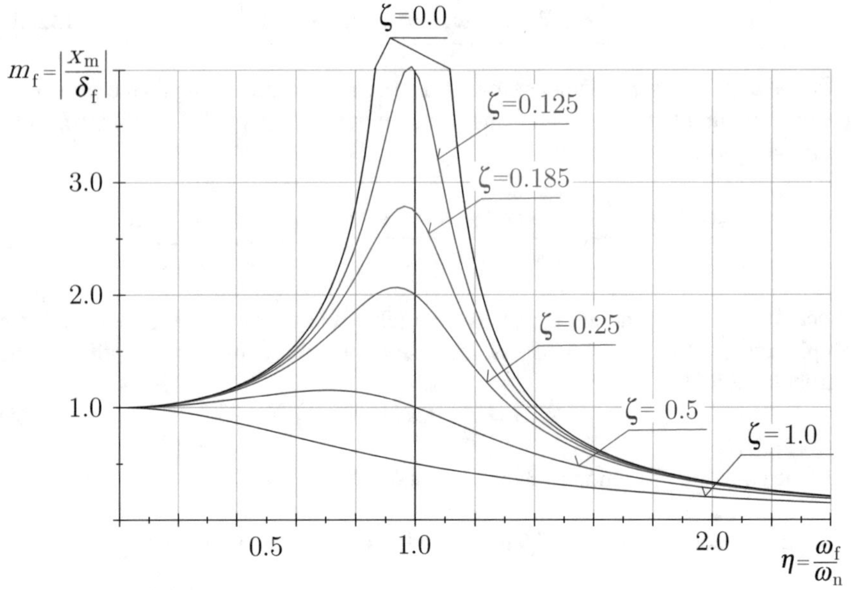

Fig. 3.14 Magnification factor for a damped system

Figure 3.14 shows the magnification factor $m_f = |x_m/\delta_f|$ for different values of the damping factor ζ.

It is clear from Fig. 3.14 that the amplitude x_m of the forced vibrations has a local maximum if $\eta \approx 1$ which is the place of resonance for the undamped system. The greater the difference $\eta - 1.0$ for $\eta > 1$ the smaller the amplitude x_m. If we increase the damping factor ζ we can also reduce the amplitude x_m.

To get the phase angle we have to divide the left and right sides of Eq. (3.36b) by the left and right sides of Eq. (3.36a). We obtain

$$\tan\phi = -\frac{2\beta\omega_f}{\omega_n^2 - \omega_f^2} \underset{\eta=\frac{\omega_f}{\omega_n}}{=\uparrow=} -\frac{2\beta\omega_f}{\omega_n^2\left(1-\eta^2\right)} =$$

$$\underset{(3.20)\to 2\beta=\frac{c}{m}=\frac{c}{c_n/2\omega_n}=2\zeta\omega_n}{=\uparrow=} -\frac{2\zeta\eta}{1-\eta^2} \tag{3.40a}$$

from where

$$\boxed{\phi = \arctan\frac{2\zeta\eta}{\eta^2-1}.} \tag{3.40b}$$

Exercise 3.3 A rigid block with mass $m = 100$ kg moves horizontally as shown in Fig. 3.15. The block is moved $x_o = 30$ mm to the right from its equilibrium position and then released. Knowing that $k_1 = 5$ kN/m, $k_2 = 3.1$ kN/m determine the period of the vibration, the maximum velocity, and acceleration of the block for each spring arrangement.

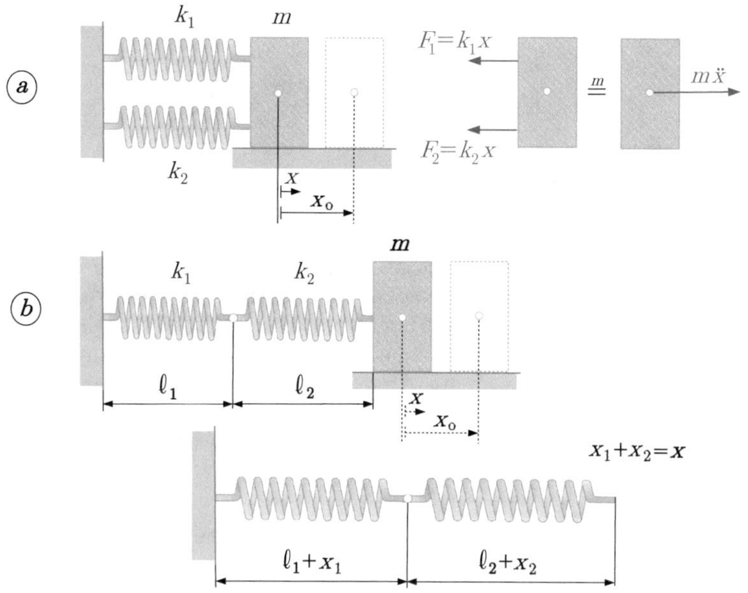

Fig. 3.15 Springs attached in parallel and series

The two springs can be replaced by a single one for each spring arrangement.

(a) The two springs are attached in parallel. The force exerted by the two springs on the block should be the same as the force exerted on the block by the single spring that replaces the original two. Let us denote the spring constant of the equivalent spring by k. It is obvious that the following equation should be satisfied for any x during the motion:

$$k_1 x + k_2 x = (k_1 + k_2)\, x \, .$$

This means that

$$\boxed{k = k_1 + k_2 \, , \qquad \frac{1}{f} = \frac{1}{f_1} + \frac{1}{f_2} ,} \tag{3.41}$$

where we have utilized the definition (3.1) of the spring flexibility. In other words, if two springs are attached in parallel the stiffness of the equivalent spring is the sum of the spring constants the two springs have. Since the initial velocity v_o is zero it follows from (3.5a), (3.2), and (3.6a) that

$$x = x_o \cos \omega_n t \, , \quad \omega_n = \sqrt{\frac{k}{m}} = \sqrt{\frac{k_1 + k_2}{m}} = \sqrt{\frac{8.1 \cdot 10^3 \text{kg m/m s}^2}{100 \, \text{kg}}} = 9 \, \frac{1}{\text{s}} ,$$

$$\tau_n = \frac{2\pi}{\omega_n} = \frac{2\pi}{9} \, \text{s} = 0.698 \, \text{s} \, .$$

Making use of the solution, we get

$$v = \dot{x} = -x_o\omega_n \sin\omega_n t \quad \text{and} \quad a = \dot{v} = -x_o\omega_n^2 \cos\omega_n t$$

hence

$$v_{\max} = x_o\omega_n = 30 \cdot 9 \frac{\text{mm}}{\text{s}} = 0.27 \frac{\text{m}}{\text{s}}$$

and

$$a_{\max} = x_o\omega_n^2 = 30 \cdot 81 \frac{\text{mm}}{\text{s}^2} = 2.43 \frac{\text{m}}{\text{s}^2}$$

are the maximum velocity and acceleration.

(b) The two springs are attached in series. The equivalent spring stiffness can easily be obtained from the fact that the total elongation x is the sum of the elongations the two springs have under the action of the spring force F. Making use of (3.1), we can write

$$x_1 + x_2 = \frac{F}{k_1} + \frac{F}{k_2} = F\left(\frac{1}{k_1} + \frac{1}{k_2}\right) = F\frac{1}{k}.$$

Consequently

$$\boxed{\frac{1}{k} = \frac{1}{k_1} + \frac{1}{k_2}, \qquad f = f_1 + f_2.} \tag{3.42}$$

In other words, the reciprocal of the equivalent spring stiffness is the sum of the reciprocals of the two original spring constants. Utilizing the data given it is not too difficult to check that

$$k = \frac{1}{\frac{1}{k_1} + \frac{1}{k_2}} = \frac{k_1 k_2}{k_1 + k_2} = \frac{5 \cdot 3.1}{5 + 3.1} = 1.914 \frac{\text{kN}}{\text{m}},$$

$$\omega_n = \sqrt{\frac{k}{m}} = \sqrt{\frac{1.914 \cdot 10^3}{100}} = 4.374 \frac{1}{\text{s}}, \quad \tau_n = \frac{2\pi}{\omega_n} = \frac{2\pi}{4.374} = 1.436\,\text{s},$$

$$v_{\max} = x_o\omega_n = 30 \cdot 4.374 \frac{\text{mm}}{\text{s}} = 0.131 \frac{\text{m}}{\text{s}}$$

and

$$a_{\max} = x_o\omega_n^2 = 30 \cdot (4.374)^2 \frac{\text{mm}}{\text{s}} = 0.573 \frac{\text{m}}{\text{s}^2}.$$

Exercise 3.4 Figure 3.16 shows a uniform rod of length ℓ and mass m. The rod is supported by a pin at A and a spring of constant k at D and is connected at B to a dashpot of damping coefficient c. Knowing m, ℓ, k, and c determine, for small oscillations, (a) the equation of motion regarding the polar angle φ as an unknown, (b) the circular frequency ω_n for the undamped system, and (c) the critical damping coefficient c_n.

The moment of inertia of the rod with respect to the axes passing through the point A perpendicularly to the plane of motion, the forces F_k and F_c exerted on the rod by the spring are all given by the following equations:

$$J_A = \frac{1}{3}m\ell^2, \qquad F_k = k\frac{\ell\varphi}{2}, \qquad F_c = c\ell\dot{\varphi}.$$

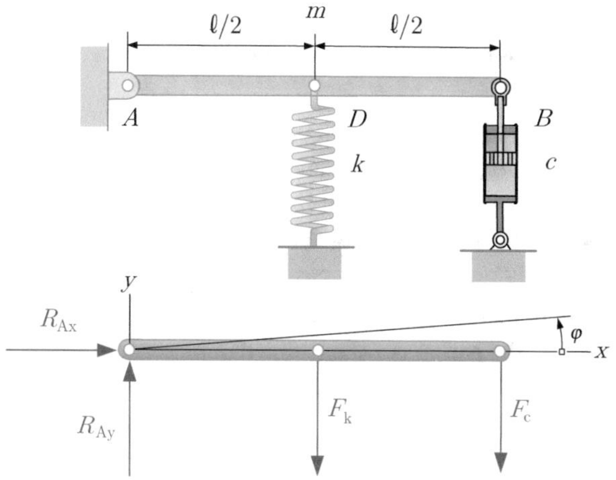

Fig. 3.16 Vibrating uniform rod

Since the moment of the effective forces about the point A should be the same as that of the external forces, we can write utilizing Eqs. (1.47) and (1.49b) that

$$J_A\ddot{\varphi} = -F_c\ell - F_k\frac{\ell}{2}$$

from where by dividing throughout by J_A it follows that

$$\underbrace{\frac{1}{3}m\ell^2}_{m_g}\ddot{\varphi} + \underbrace{c\ell^2}_{c_g}\dot{\varphi} + \underbrace{\frac{k\ell^2}{4}}_{k_g}\varphi = 0 \quad \text{or} \quad \ddot{\varphi} + \frac{3c}{m}\dot{\varphi} + \frac{3k}{4m}\varphi = 0,$$

which is the equation of motion in two formally different (but equivalent) forms. Recalling $(3.2)_2$ and (3.20) we obtain the circular frequency of the undamped system and the critical damping coefficient:

$$\omega_n = \sqrt{\frac{k_g}{m_g}} = \sqrt{\frac{3k}{4m}}, \quad c_{gn} = 2\omega_n m_g = \sqrt{\frac{3k}{m}}\frac{1}{3}m\ell^2 = \ell^2\sqrt{\frac{km}{3}},$$

$$c_n = \frac{c_{gn}}{\ell^2} = \sqrt{\frac{km}{3}}.$$

3.2 Machine Foundations and Vibration Isolation

3.2.1 *Introductory Remarks*

Machines (machine tools) are put, in general, onto a foundation in order

(a) to reduce the harmful effects of those vibrations which are transmitted from the machine (the source of vibration) to the environment or/and

(b) to reduce the harmful vibrations that are transmitted from the environment (neighboring machines—a forging hammer, for instance) to the machine considered.

As regards their effects, vibrations can be either harmful (e.g., pneumatic hammer, chain saw, ventilator, forging machine) or useful (e.g., riddle machine, bolting machine, compression machine, etc.). Vibration reduction can be achieved in various ways. Some of the most important techniques are listed below:

- Mass balancing by changing the mass distribution.
- Active vibration reduction by changing the parameters of the vibrating system appropriately (e.g., the spring constants, the damping coefficients, etc.). This can be achieved by measuring a time signal of the vibration which is then processed and used to control an actuator, i.e., the device that changes the system parameters.
- Passive vibration reduction by means of preventing vibration transmission from the neighborhood by applying

(a) some vibration isolation devices and/or
(b) damping elements which are capable of absorbing vibration energy, etc.

As regards the issue of machine foundations it is worth emphasizing that there are a number of books devoted to this important engineering problem: we cite here only two [1, 2].

3.2.2 Minimal Model with One Degree of Freedom

The machine foundation can be modeled, in many cases, as a rigid body which has six degrees of freedom. However, a model with one degree of freedom (block foundation) is often sufficient for clarifying some basic questions of vibration isolation as regards the effect of periodic excitation. Figure 3.17 depicts a simple one degree of freedom model for machine foundation. The block foundation is a rigid body which can move

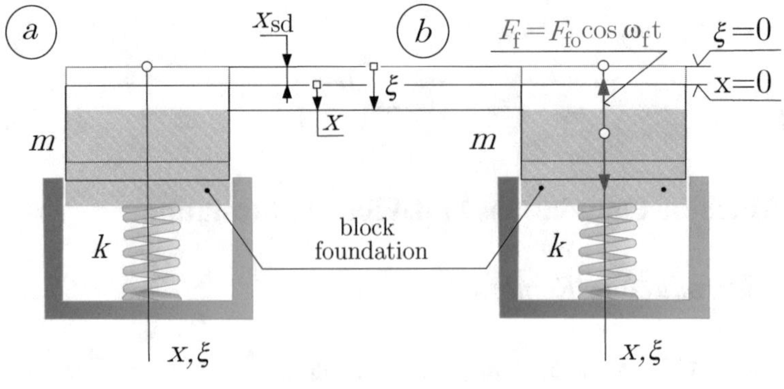

Fig. 3.17 Single degree of freedom model for machine foundation

only vertically. Let x_{sd} be the static deflection of the foundation caused by the weight of the machine that is put onto the foundation (Fig. 3.17 does not show the machine) and the weight of the foundation itself. Let m be the total mass of the foundation and the machine. If the foundation and the machine vibrates together vertically, the displacement that belongs to this motion is denoted by x. We shall assume that the ground is an elastic body: k stands for the ground rigidity.

The total vertical displacement is given by the following equation:

$$\xi = x_{sd} + x = \frac{mg}{k} + x\,. \tag{3.43}$$

(a) If there is no excitation—see Fig. 3.17a—the equation of motion is as follows:

$$m\ddot{\xi} + k\xi = mg, \tag{3.44a}$$

where $\ddot{\xi} = \ddot{x}_{sd} + \ddot{x}$, $\ddot{x}_{sd} = 0$ and $k\xi = kx_{sd} + kx = mg + kx$. Hence

$$\ddot{x} + \omega_n^2 x = 0\,, \qquad \omega_n^2 = \frac{k}{m} \tag{3.44b}$$

is the final form of the equation of motion. Observe that this equation coincides with Eq. (3.2). With x_{sd}—this quantity can be measured easily—the ground rigidity is given by

$$k = \frac{mg}{x_{sd}}\,. \tag{3.45}$$

Hence

$$\boxed{\omega_n = \sqrt{\frac{k}{m}} = \sqrt{\frac{g}{x_{sd}}}\,.} \tag{3.46a}$$

This means that the static deflection determines the natural circular frequency of the free vibrations for the system constituted by the foundation and the machine. As regards the natural frequency of the free vibrations, we get

$$\boxed{f_n = \frac{1}{\tau_n} = \frac{\omega_n}{2\pi} = \frac{\sqrt{g}}{2\pi}\frac{1}{\sqrt{x_{sd}}} = \frac{\sqrt{981}}{2\pi}\frac{1}{\sqrt{x_{sd}}} \simeq \frac{5}{\sqrt{x_{sd}}}} \tag{3.46b}$$

provided that x_{sd} is given in cm.

(b) If there is an excitation—see Fig. 3.17b—then the equation of motion is the same as Eq. (3.29) which, for the sake of completeness, is repeated and supplemented here

$$\boxed{\ddot{x} + \omega_n^2 x = \omega_n^2 \delta_f \cos\omega_f t\,, \quad \omega_n^2 = \frac{k}{m} = \frac{g}{x_{sd}}\,, \quad \delta_f = \frac{F_{fo}}{k} = \frac{F_{fo}x_{sd}}{mg}\,.} \tag{3.47}$$

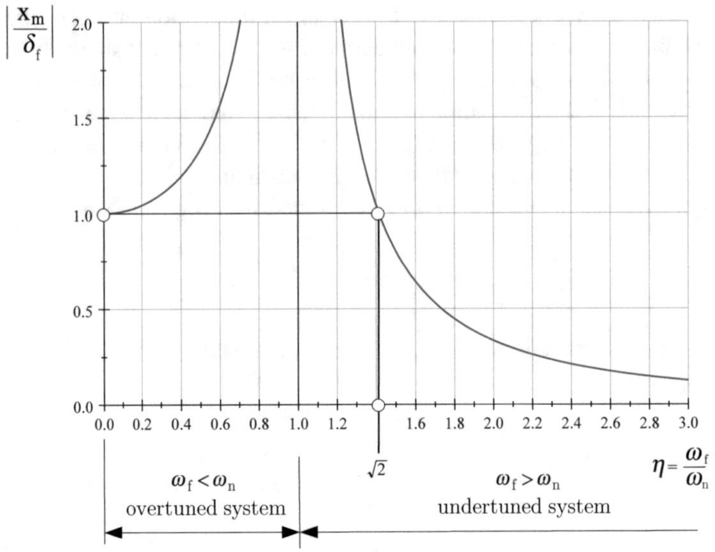

Fig. 3.18 Resonance curve

It follows from the resonance curve—see Fig. 3.18—that the amplitude x_m of the steady-state vibrations satisfies the inequality $|x_m/\delta_f| < 1$ if $\omega_f > \sqrt{2}\,\omega_n$ ($\eta > \sqrt{2}$).

For given m and ω_f, the condition $\omega_f > \sqrt{2}\,\omega_n = \sqrt{2k/m}$ can only be satisfied by changing (if necessary) the ground rigidity k.

The system is [overtuned]{undertuned} if $[\omega_f < \omega_n]\{\omega_f > \omega_n\}$.

The force exerted on the ground is given by

$$F_{gr} = kx = kx_m \cos\omega_f t = F_{gr\,o} \cos\omega_f t \tag{3.48a}$$

from where

$$F_{gr\,o} = kx_m = \underbrace{k\,\delta_f}_{F_{fo}} \frac{1}{\left|1 - \left(\dfrac{\omega_f}{\omega_n}\right)^2\right|} = F_{fo}\frac{1}{|1-\eta^2|} \tag{3.48b}$$

is the amplitude of the force F_{gr} which is less than F_{fo} if $\omega_f > \sqrt{2}\omega_n$.

(c) Figure 3.19 is a sketch of the machine foundation if damping is taken into account. The equation of motion is the same as Eq. (3.34b) which is supplemented here by two formulae for ω_n and δ_f each taken form (3.47). Thus, we have

$$\boxed{\begin{aligned} &\ddot{x} + 2\beta\,\dot{x} + \omega_n^2 x = \underbrace{\omega_n^2\delta_f}_{x_m} \cos\omega_f t\,, \\ &\omega_n^2 = \frac{g}{x_{sd}}\,, \quad \delta_f = \frac{F_{fo}\,x_{sd}}{mg}\,. \end{aligned}} \tag{3.49}$$

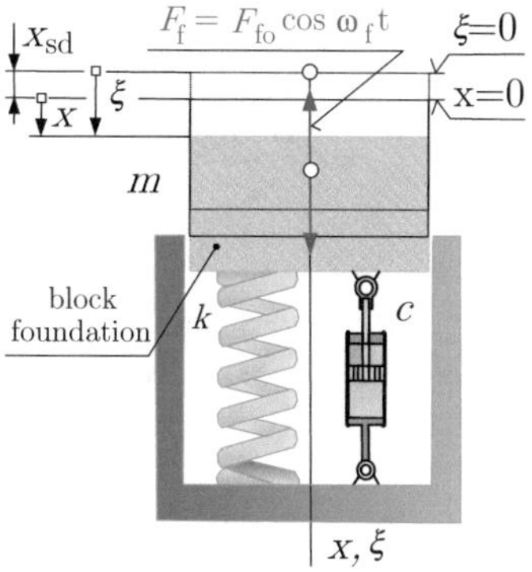

Fig. 3.19 Model for the machine foundation with damping

The steady-state solution—see Eq. (3.35)—is given by

$$x_{\text{part}} = x_m \cos\left(\omega_f t + \phi\right) \tag{3.50}$$

in which—it is worth reminding the reader of Eqs. (3.38), (3.39), and (3.40) here—

$$x_m = \delta_f \frac{1}{\sqrt{\left(1-\eta^2\right)^2 + (2\zeta\eta)^2}} , \quad \eta = \frac{\omega_f}{\omega_n} , \quad \zeta = \frac{c}{c_n} \tag{3.51a}$$

and

$$\tan\phi = -\frac{2\zeta\eta}{1-\eta^2} . \tag{3-42b}$$

The force exerted on the ground reflects the effects of the spring and the damping:

$$\begin{aligned} F_{gr} = c\dot{x} + kx = kx_m \cos\left(\omega_f t + \phi\right) - c\omega_f x_m \sin\left(\omega_f t + \phi\right) = \\ = x_m k \left[1 \cos\left(\omega_f t + \phi\right) - \frac{c}{k}\omega_f \sin\left(\omega_f t + \phi\right)\right] . \end{aligned} \tag{3.52}$$

Here it will be assumed that

$$1 = A_{gr} \cos\gamma , \qquad \frac{c}{k}\omega_f = A_{gr} \sin\gamma \tag{3.53}$$

from where

$$1 + \left(\frac{c}{k}\omega_f\right)^2 = A_{gr}^2 \tag{3.54a}$$

in which

$$\frac{c}{k}\omega_f = \frac{c}{c_n}\frac{c_n}{k}\omega_f \underset{(3.23c),\,(3.20)}{=\!\!\uparrow\!\!=}$$

$$= \zeta\frac{2m\omega_n}{k}\omega_f \underset{(3.2)_2}{=\!\!\uparrow\!\!=} 2\zeta\frac{\omega_f}{\omega_n} = 2\zeta\eta. \quad (3.54b)$$

Hence

$$\boxed{A_{gr} = \sqrt{1+\left(\frac{c}{k}\omega_f\right)^2} = \sqrt{1+4\zeta^2\eta^2}} \quad (3.55a)$$

and

$$\boxed{\tan\gamma = \frac{c}{k}\omega_f = 2\zeta\eta\,.} \quad (3.55b)$$

A comparison of Eqs. (3.52), (3.53), and (3.55a) yields

$$F_{gr} = \underbrace{x_m k A_{gr}}_{F_{gr\,o}} \cos\left(\omega_f t + \phi + \gamma\right) =$$

$$\underset{(3.51a),(3.55a)}{=\!\!\uparrow\!\!=} \delta_f\, k\frac{\sqrt{1+4\zeta^2\eta^2}}{\sqrt{\left(1-\eta^2\right)^2 + (2\zeta\eta)^2}}\cos\left(\omega_f t + \phi + \gamma\right). \quad (3.56)$$

Let us introduce the following notations:

$$V_1(\eta,\zeta) = \frac{1}{\sqrt{\left(1-\eta^2\right)^2 + (2\zeta\eta)^2}}\,,$$
$$V_2(\eta,\zeta) = \frac{\sqrt{1+4\zeta^2\eta^2}}{\sqrt{\left(1-\eta^2\right)^2 + (2\zeta\eta)^2}}\,. \quad (3.57)$$

Making use of the notations introduced we get, in simple forms, the amplitudes x_m and $F_{gr\,o}$ of the vertical displacement x and the force F_{gr} exerted on the ground by the machine and its foundation:

$$\boxed{x_m = \delta_f V_1(\eta,\zeta)\,, \qquad F_{gr\,o} = \underbrace{\delta_f k}_{F_{fo}}\, V_2(\eta,\zeta)\,.} \quad (3.58)$$

Figures 3.20 and 3.21 depict the functions $V_1(\eta,\zeta)$ and $V_2(\eta,\zeta)$. Observe that Fig. 3.20 is actually the same as Fig. 3.14.

Remark 3.4 If the system is overtuned, i.e., $\omega_n > \omega_f$ $(\eta < 1)$ then the amplitude of vibrations x_m is greater than the deflection δ_f caused by the amplitude F_{fo} of the excitation force.

Remark 3.5 If the system is undertuned, i.e., $\omega_n < \omega_f$ $(\eta > 1)$ then the amplitude x_m of the vibrations can be made smaller than the deflection δ_f caused by the amplitude F_{fo} of the excitation force.

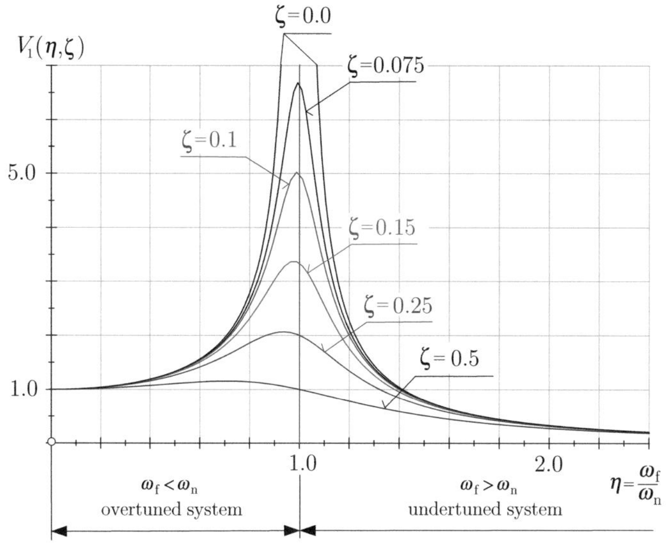

Fig. 3.20 $V_1(\eta;\zeta)$ against η (ζ is a parameter)

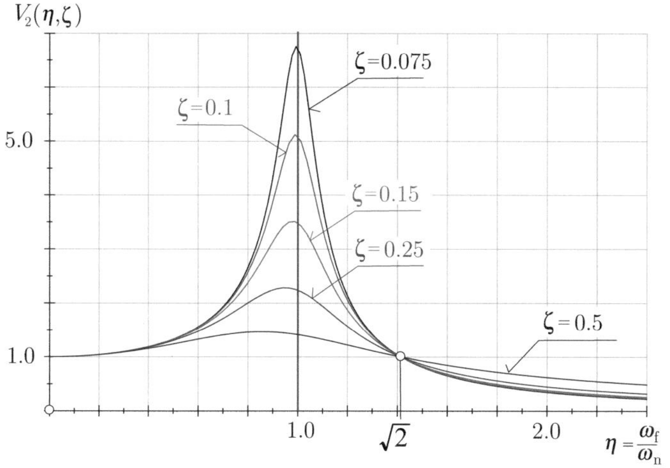

Fig. 3.21 $V_2(\eta;\zeta)$ against η (ζ is a parameter)

Exercise 3.5 A sawing machine[1] (see Fig. 3.22) is excited (due to the motion of the sawing tool, i.e., the saw and its frame): $F_f = F_{fo} \cos \omega_f t$, $F_{fo} = 16 \cdot 10^4$ N, $f_f = 600$/min ($\omega_f = 2\pi f_f = 62.83$ 1/s). Determine the mass of the machine and foundation if (a) the force exerted on the ground cannot be greater than $0.05F_{fo}$ and if (b) the inequality $x_m < 1$ mm should also be satisfied.

[1]Photo courtesy of Zhejiang Guanbao Industrial Co., China.

Fig. 3.22 Sawing machine

Under the assumption of an undamped system ($\zeta = 0$) we can use the model shown in Fig. 3.17. If we neglect the transient function the solution for x follows from (3.32), $(3.58)_1$, $(3.57)_1$, and $(3.47)_3$:

$$x = x_m \cos\left(\omega_f t + \phi\right) , \qquad x_m = \delta_f V_1(\eta, 0) = \delta_f \frac{1}{\left|1 - \eta^2\right|} , \qquad \delta_f = \frac{F_{fo}}{k} .$$

The force exerted on the ground is obtained from $(3.56)_2$, $(3.58)_2$, and $(3.57)_2$:

$$F_{gr} = F_{gr\,o} \cos\left(\omega_f t + \phi + \gamma\right) ,$$

where

$$F_{gr\,o} = \underbrace{\delta_f k}_{F_{fo}} \; V_2\left(\eta, 0\right) = F_{fo} V_1\left(\eta, 0\right) = F_{fo} \frac{1}{\left|1 - \eta^2\right|} .$$

Since $F_{gr\,o} \leqq 0.05 F_{fo}$ we have to choose an undertuned system: $\omega_f > \omega_n$, $\eta = \omega_f / \omega_n > 1$. Then

$$F_{gr\,o} = 0.05 F_{fo} = F_{fo} \frac{1}{\eta^2 - 1}$$

from where

$$\eta^2 = 1 + \frac{1}{0.05} = 21 , \qquad \eta = \sqrt{21} = 4.583$$

and

$$\omega_n = \frac{\omega_f}{\eta} = \frac{62.83}{4.583} = 13.71 \, \frac{1}{\mathrm{s}} .$$

Since $x_m = 1$ we can write

$$x_m = 1 = \delta_f \frac{1}{\eta^2 - 1} = \frac{F_{fo}}{k} \frac{1}{\eta^2 - 1}$$

and it follows that

$$k = \frac{F_{fo}}{\eta^2 - 1} = \frac{16.0 \cdot 10^4}{20.0} = 0.8 \cdot 10^4 \, \frac{\mathrm{N}}{\mathrm{mm}} \underset{1\mathrm{N} = 1000 \frac{\mathrm{kg\,mm}}{\mathrm{s}^2}}{\overset{\uparrow}{=}} = 0.8 \cdot 10^7 \, \frac{\mathrm{kg}}{\mathrm{s}^2} .$$

Using now $(3.28b)_2$ we find the total mass

$$m = \frac{k}{\omega_n^2} = \frac{F_{fo}}{\eta^2 - 1} = \frac{0.8 \cdot 10^7}{13.71^2} \cong 42558 \text{ kg} .$$

Equation $(3.49)_2$ yields the static deflection

$$x_{sd} = \frac{g}{\omega_n^2} = \frac{9.81 \cdot 10^3}{13.71^2} \approx 52 \text{ mm} .$$

If the fulfillment of the more rigorous inequality $F_{gr} \leq 0.005 F_{fo}$ is the requirement in the manner of the previous solution, we obtain

$$\eta = \frac{\omega_f}{\omega_n} = \sqrt{2001} = 44.73 , \qquad \omega_n = \frac{\omega_f}{\eta} = \frac{62.83}{44.73} = 1.405 \ \frac{1}{\text{s}} ,$$

$$m = \frac{k}{\omega_n^2} = \frac{0.8 \cdot 10^7}{1.405^2} \cong 4.055112 \times 10^6 \text{ kg} .$$

This mass is very large, and therefore it is not allowable.

3.2.3 *Machine Foundation Under the Action of a Rotating Unbalance*

Figure 3.23 shows a block foundation of mass M. The foundation carries a rotating machine—its total mass is denoted by m—which is unbalanced since the mass center does not coincide with the center of rotation. The effect of the ground on the foundation is modeled by a spring of stiffness k and a viscous vibration damper with damping coefficient c. It will be assumed that the unbalance can be represented by a mass m_u which rotates on a circle of radius e (sometimes called the eccentricity) with the angular velocity ω_f of the rotating machine. We assume further that the rotating

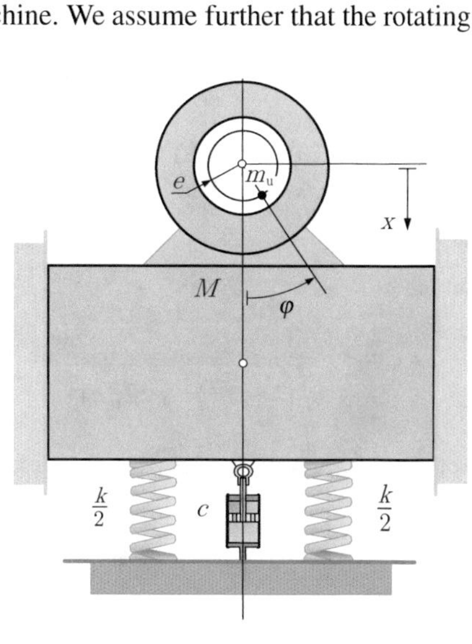

Fig. 3.23 Rotating machine on block foundation

machine and its foundation is constrained to move vertically. We have, therefore, a single degree of freedom system. Let x be the vertical displacement. When the system is at rest $x = 0$, it is clear that

$$\varphi = \omega_f t\,, \qquad v_{ux} = \dot{x} - e\omega_f \sin\omega_f t$$

are the polar angle and the vertical velocity component of the unbalanced mass m_u. The equation of motion is as follows:

$$(M + m - m_u)\,\ddot{x} + m_u \frac{\mathrm{d}}{\mathrm{d}t} v_{ux} = -c\dot{x} - kx$$

or

$$(M + m)\,\ddot{x} + c\dot{x} + kx = \underbrace{m_u e\omega_f^2}_{F_{fo}} \cos\omega_f t \tag{3.59a}$$

in which F_{fo} is the amplitude of the excitation force. If the rotating machine is balanced $e = 0$, then there is no excitation. Otherwise, the amplitude of the excitation force F_{fo} depends on m_u and its distance e from the axis of rotation as well as on the angular velocity ω_f. For small e, however, the effect of the excitation force can be neglected.

If we divide throughout by $\mathcal{M} = M + m$ we can manipulate Eq. (3.59a) into a form similar to that of Eq. (3.34b)

$$\ddot{x} + \underbrace{\frac{c}{\mathcal{M}}}_{2\beta}\dot{x} + \underbrace{\frac{k}{\mathcal{M}}}_{\omega_n^2} x = \underbrace{\frac{k}{\mathcal{M}}}_{\omega_n^2}\underbrace{\frac{m_u e\omega_f^2}{k}}_{\delta_f} \cos\omega_f t \tag{3.59b}$$

or

$$\ddot{x} + 2\beta\dot{x} + \omega_n^2 x = \omega_n^2 \delta_f \cos\omega_f t \tag{3.59c}$$

in which

$$2\beta = \frac{c}{\mathcal{M}}\,, \qquad \omega_n^2 = \frac{k}{\mathcal{M}}\,, \qquad \delta_f = \frac{F_{fo}}{k} = \frac{m_u e\omega_f^2}{k}\,. \tag{3.60}$$

Recalling (3.50) and (3.51), we have the steady-state motion in the form

$$x_{\text{part}} = x_m \cos\left(\omega_f t + \phi\right),$$

where

$$x_m = \delta_f \frac{1}{\sqrt{\left(1 - \eta^2\right)^2 + (2\zeta\eta)^2}} \underset{(3.58)}{=\uparrow=} \delta_f V_1(\eta, \zeta)\,, \quad \eta = \frac{\omega_f}{\omega_n}\,, \quad \zeta = \frac{c}{c_n}$$

is the amplitude of the steady-state motion. By utilizing the transformation

$$\delta_f = \frac{m_u e\omega_f^2}{k} = \frac{m_u}{\mathcal{M}} \frac{1}{\frac{k}{\mathcal{M}}}\omega_f^2 = \frac{m_u e}{\mathcal{M}}\frac{\omega_f^2}{\omega_n^2} = \frac{m_u e}{\mathcal{M}}\eta^2 \tag{3.61}$$

we can rewrite the amplitude x_m into the form

$$x_m = \frac{m_u e}{\mathcal{M}} \eta^2 V_1(\eta, \zeta) = \frac{m_u e}{\mathcal{M}} V_3(\eta, \zeta) ,$$

$$V_3(\eta, \zeta) = \eta^2 V_1(\eta, \zeta) = \frac{\eta^2}{\sqrt{\left(1-\eta^2\right)^2 + (2\zeta\eta)^2}} .$$

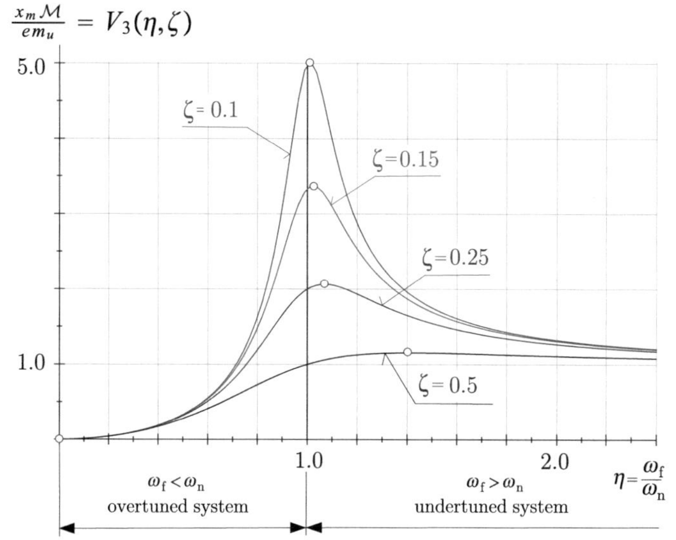

Fig. 3.24 Dimensionless amplitude of the rotating machine

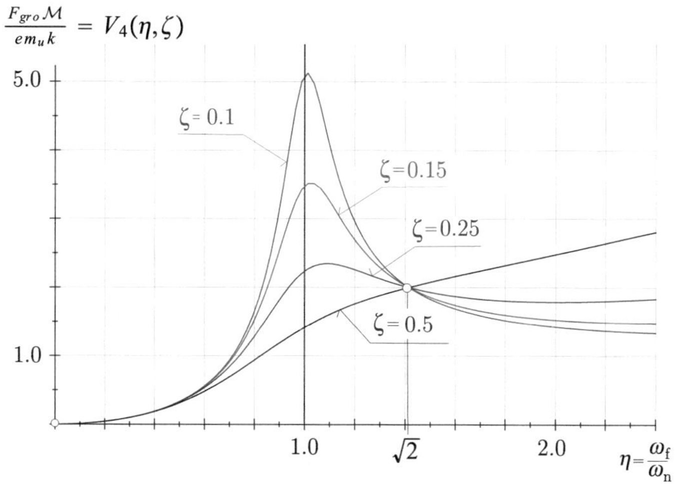

Fig. 3.25 Dimensionless force exerted on the ground by the rotating machine

Figure 3.24 is obtained by plotting the dimensionless quotient

$$\frac{x_m \mathcal{M}}{e\, m_u} = V_3\,(\eta, \zeta)$$

against η. For small η (for small ω_f), the amplitude x_m of the motion is nearly 0: then the system is overtuned. For relatively large η, the amplitude x_m becomes a constant of value $e\, m_u/\mathcal{M}$ independently of the amount of damping and the angular velocity ω_f.

Condition

$$\frac{\mathrm{d}V_3\,(\eta, \zeta)}{\mathrm{d}\eta} = 0 \qquad \longrightarrow \qquad 1 - \eta^2\left(1 - 2\zeta^2\right) = 0$$

yields that the amplitude x_m has a finite maximum if $0 < \zeta < 1/\sqrt{2} = 0.707$ and

$$\eta^2 = \frac{1}{1 - 2\zeta^2}\,.$$

The greater $\zeta \in (0, 1/\sqrt{2})$ is the smaller the maximum amplitude is.

Remark 3.6 If we want to reduce x_m it is worthwhile to overtune the system. For internal combustion engines (e.g., four-stroke engines), the frequency spectrum contains subharmonic frequencies. Overtuning is, however, possible for one (say a subharmonic) frequency only.

If the wheels of your car are unbalanced you have to speed up to increase ω_f when your car begins to shake. This is the way to avoid shaking of your car.

Remark 3.7 The rotating machine can be balanced if we place a balancing mass (denoted by m_b) onto the machine diametrically opposite to m_u. Let h be the distance of the balancing mass m_b from the axis of rotation. The rotating machine is balanced if $h\, m_b = e\, m_u$ since then the mass center of the two masses is located on the axis of rotation.

The amplitude of the force exerted on the ground by the foundation is obtained from $(3.58)_2$ and $(3.57)_2$:

$$F_{gr\,o} = \delta_f k\, V_2(\eta, \zeta) = \delta_f k \frac{\sqrt{1 + 4\zeta^2\eta^2}}{\sqrt{\left(1 - \eta^2\right)^2 + (2\zeta\,\eta)^2}}\,, \tag{3.62}$$

where utilizing (3.61), we get

$$\delta_f k = \frac{m_u e k}{\mathcal{M}}\eta^2\,.$$

Hence

$$F_{gr\,o} = \frac{m_u e k}{\mathcal{M}} \eta^2 V_2(\eta, \zeta) = \frac{m_u e k}{\mathcal{M}} \underbrace{\frac{\eta^2 \sqrt{1 + 4\zeta^2 \eta^2}}{\sqrt{\left(1 - \eta^2\right)^2 + (2\zeta \eta)^2}}}_{V_4(\eta,\zeta)} = \frac{m_u e k}{\mathcal{M}} V_4(\eta, \zeta). \tag{3.63}$$

Figure 3.25 depicts the dimensionless quotient

$$\frac{F_{gr\,o}\,\mathcal{M}}{e\,m_u k} = V_4(\eta, \zeta)$$

against η.

Remark 3.8 Since $\lim_{\eta \to \infty} V_4(\eta, \zeta) = \infty$ it follows that the force exerted by the foundation on the ground can be reduced effectively only if we design an overtuned system.

3.2.4 Displacement Excitation on a Machine Foundation

Excitation can be caused not only by a force but by giving a prescribed motion to an elastic element of the system. If we apply again a single degree of freedom model for describing the behavior of a block foundation the displacement excitation means that the motion of the ground on which the block lies is prescribed due to an external effect. An analysis which clarifies how such an excitation influences the

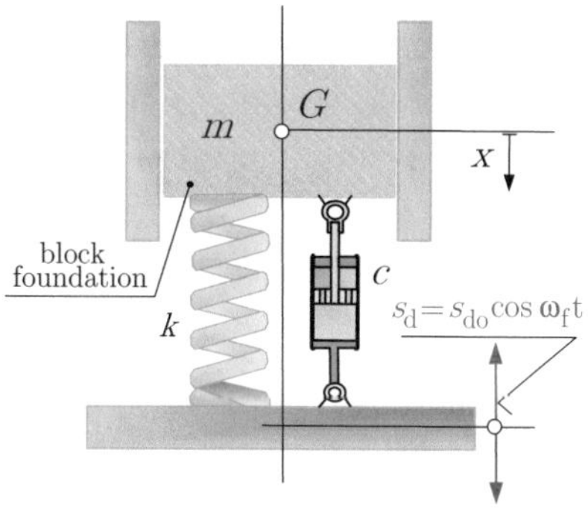

Fig. 3.26 Displacement excitation

behavior of the foundation could help us to reduce the harmful vibrations that are transmitted from the environment. Figure 3.26 shows the model we apply to examine this phenomenon. It is not too difficult to check that the change in length of the spring is $x - s_d$. Consequently, the velocity difference is given by $\dot{x} - \dot{s}_d$. In terms of these quantities, the equation of motion takes the form

$$m\ddot{x} + c\,(\dot{x} - \dot{s}_d) + k\,(x - s_d) = 0$$

or

$$m\ddot{x} + c\dot{x} + kx = c\dot{s}_d + ks_d\,.$$

If we divide throughout by m and apply the notations $(3.17\text{b})_{2,3}$, we get

$$\ddot{x} + 2\beta\dot{x} + \omega_n^2 x = s_{do}\left(-2\beta\omega_f \sin\omega_f t + \omega_n^2 \cos\omega_f t\right). \tag{3.64}$$

Let us manipulate the right side of this equation into a cos function. The transformation is based on the equation

$$-\underbrace{2\beta\omega_f s_{do}}_{\mathcal{A}\sin\gamma} \sin\omega_f t + \underbrace{\omega_n^2 s_{do}}_{\mathcal{A}\cos\gamma} \cos\omega_f t = \omega_n^2\,\delta_f\,\cos\left(\omega_f t + \gamma\right) \tag{3.65}$$

in which A and γ are the unknowns. If we square both sides of the formulas

$$\mathcal{A}\sin\gamma = 2\beta\omega_f s_{do}\,, \qquad \mathcal{A}\cos\gamma = \omega_n^2 s_{do}$$

and then add them to each other we get

$$\mathcal{A} = s_{do}\sqrt{\left(\omega_n^2\right)^2 + \left(2\beta\omega_f\right)^2} = s_{do}\omega_n^2\sqrt{1 + \left(\frac{2\beta}{\omega_n}\frac{\omega_f}{\omega_n}\right)^2}$$

in which according to (3.38)

$$\frac{\omega_f}{\omega_n} = \eta \quad \text{and} \quad \frac{2\beta}{\omega_n} = 2\zeta\,. \tag{3.66}$$

With this result we get on the base of (3.65) that

$$\mathcal{A} = s_{do}\omega_n^2\sqrt{1 + (2\eta\zeta)^2} = \omega_n^2\delta_f \tag{3.67a}$$

from where

$$\delta_f = s_{do}\sqrt{1 + (2\eta\zeta)^2}\,. \tag{3.67b}$$

It is obvious that

$$\tan\gamma = 2\beta\omega_f/\omega_n^2 = 2\eta\zeta\,. \tag{3.68}$$

Upon substitution of (3.65) and (3.67a), the equation of motion (3.64) can be rewritten in the form

$$\ddot{x} + 2\beta\dot{x} + \omega_n^2 x = \omega_n^2\delta_f \cos\left(\omega_f t + \gamma\right) = \omega_n^2 s_{do}\sqrt{1 + (2\eta\zeta)^2}\,\cos\left(\omega_f t + \gamma\right). \tag{3.69}$$

The amplitude of the steady-state solution

$$x_{\text{part}} = x_m \cos\left(\omega_f t + \gamma + \phi\right) \tag{3.70}$$

(see Eqs. (3.35) and (3.34b)—$(\omega_f t + \gamma + \phi)$ corresponds to $(\omega_f t + \phi)$ in (3.35)) follows from (3.39) by utilizing (3.67b) and (3.57)

$$x_m = \frac{\delta_f}{\sqrt{\left(1-\eta^2\right)^2 + (2\zeta\eta)^2}} \underset{(3.67b)}{=\uparrow=} \frac{s_{do}\sqrt{1+(2\eta\zeta)^2}}{\sqrt{\left(1-\eta^2\right)^2 + (2\zeta\eta)^2}} \underset{(3.57)}{=\uparrow=} s_{do} V_2(\eta, \zeta)\,. \tag{3.71}$$

Remark 3.9 For passive vibration reduction, it is worth designing an undertuned system for which ω_n is much smaller than ω_f, i.e., the mass of foundation is relatively large and the spring stiffness is small (Fig. 3.27).

Exercise 3.6 The switch box of a machine tool is to be protected against the vibrations that are transmitted from the environment. The machine and the switch box are on the second floor. The vibrations of the floor have an amplitude of $s_{do} = 20\,\mu\text{m} = 20 \times 10^{-6}$ m. The natural frequency is $f_f = 960\,\text{rpm} = 16/\text{s}$. The amplitude of the switch box vibrations cannot be more than $x_m = 2\,\mu\text{m} = 2 \cdot 10^{-6}$ m. The mass of the switch box is 300 kg. It is put onto four identical springs. Determine the spring constants if the damping is negligible.

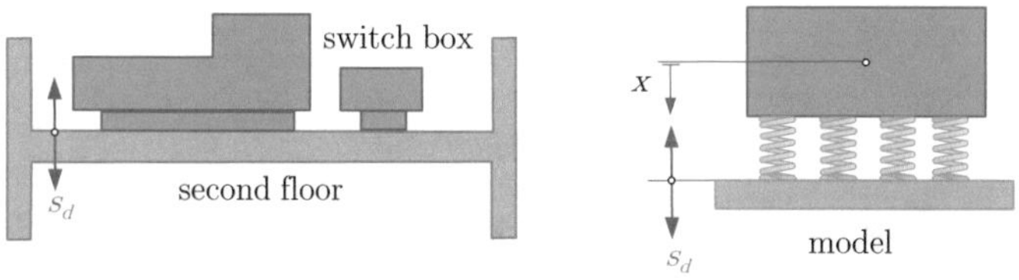

Fig. 3.27 Switch box of a machine tool

We have a displacement excitation of the form $s_d = s_{do} \cos \omega_f t$. If there is damping the solution is given by (3.70). Since now we have no damping $\zeta = 0$ and it follows from (3.71) that

$$x_m = s_{do} V_2(\eta, \zeta)\,|_{\zeta=0} = s_{do} \frac{1}{|1-\eta^2|} \underset{x_m < s_d}{=\uparrow=} s_{do} \frac{1}{\left(\frac{\omega_f}{\omega_n}\right)^2 - 1}$$

in which

$$\omega_f = 2\pi f_f = 2\pi \times 16 = 100.53\ \frac{\text{rad}}{\text{s}}.$$

Hence

$$\frac{\omega_f^2}{\omega_n^2} = 1 + \frac{s_{do}}{x_m} = 1 + \frac{20 \times 10^{-6}}{2 \times 10^{-6}} = 11$$

from where

$$\omega_n = \frac{\omega_f}{\sqrt{11}} = \frac{100.53}{\sqrt{11}} = 30.31\ \frac{\text{rad}}{\text{s}}\,.$$

We have four identical springs. Consequently,

$$\omega_n^2 = \frac{4k}{m} \quad \rightarrow \quad k = \frac{m\omega_n^2}{4} = \frac{300 \times (30.31)^2}{4} = 68.91 \times 10^3 \; \frac{\mathrm{N}}{\mathrm{m}} \, .$$

As regards the static deflection Eq. (3.45) yields

$$x_{sd} = \frac{mg}{4k} = \frac{mg}{m\omega_n^2} = \frac{g}{\omega_n^2} = \frac{9.81}{30.31^2} = 1.0677 \times 10^{-2} \, \mathrm{m} \, .$$

3.3 Forced Vibrations of Undamped Systems Under Special Conditions

3.3.1 First We Shall Assume That $\omega_n = \omega_f$

The particular solution to the equation of motion

$$\ddot{x} + \omega_n^2 x = \omega_n^2 \delta_f \cos \omega_n t \, , \quad \omega_n^2 = \frac{k}{m} \, , \quad \delta_f = \frac{F_{fo}}{k} \, , \quad F_f = F_{fo} \cos \omega_n t \tag{3.72}$$

is of the form

$$x_{\text{part}} = K t \sin \omega_n t \tag{3.73}$$

in which K is an unknown parameter. We shall assume that the transient function in the complete solution

$$x(t) = \underbrace{A \cos \omega_n t + B \sin \omega_n t}_{\text{transient function}} + \underbrace{K t \sin \omega_n t}_{\text{particular solution}} \tag{3.74}$$

can be neglected. Making use of the derivatives

$$\begin{aligned} \dot{x} &= K \sin \omega_n t + K t \omega_n \cos \omega_n t \, , \\ \ddot{x} &= 2K \, \omega_n \cos \omega_n t - K t \omega_n^2 \sin \omega_n t \end{aligned} \tag{3.75}$$

we obtain from the differential equation that

$$2K \, \omega_n \cos \omega_n t - K t \omega_n^2 \sin \omega_n t + K t \omega_n^2 \sin \omega_n t = \omega_n^2 \delta_f \cos \omega_n t \, .$$

Thus

$$K = \frac{\omega_n \delta_f}{2} = \frac{F_{fo}}{2\omega_n m} \tag{3.76a}$$

and

$$x \approx x_{\text{part}} = \frac{\omega_n \delta_f}{2} t \sin \omega_n t = \frac{F_{fo}}{2\omega_n m} t \sin \omega_n t \, . \tag{3.76b}$$

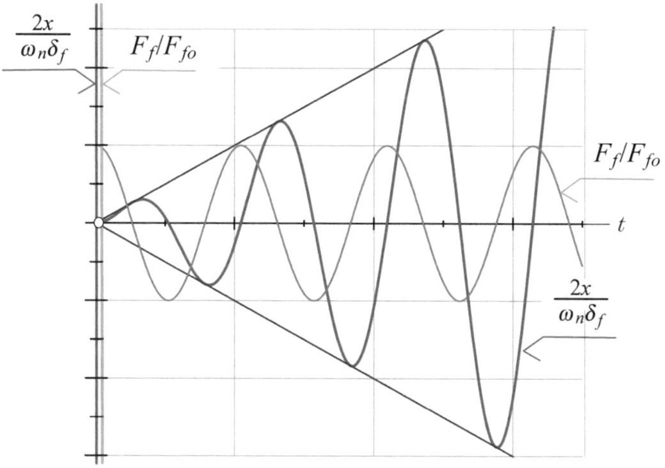

Fig. 3.28 Dimensionless displacement and force against time if $\omega_f = \omega_n$

The displacement function $2x/(\omega_n\delta_f)$ and the quotient $F_f/F_{fo} = \cos\omega_n t$ are plotted against time in Fig. 3.28. The displacement function tends linearly to infinity as t tends to infinity. Observe further that there is a phase difference of an angle $\pi/2$ between the force of excitation and the displacement function.

3.3.2 Second We Shall Assume That $|\omega_f - \omega_n| = \varepsilon \ll 1$

Let now $F_f = F_{fo}\sin\omega_f t$ be the force of excitation. Consider the differential equation

$$\ddot{x} + \omega_n^2 x = \omega_n^2\delta_f \sin\omega_f t\,, \qquad \omega_n^2\delta_f = \frac{k}{m}\frac{F_{fo}}{k} \tag{3.77}$$

which describes the behavior of an undamped spring mass system subjected to the harmonic force of excitation $F_f = F_{fo}\sin\omega_f t$ and its total solution is given by

$$x = \underbrace{A\cos\omega_n t + B\sin\omega_n t}_{\text{complementary function}} + \underbrace{\delta_f\frac{1}{1-\eta^2}\sin\omega_f t}_{\text{steady-state solution}}\,. \tag{3.78}$$

Making use of the initial conditions $x(t)|_{t=0} = x_o$ and $v(t)|_{t=0} = v_o$ we can determine the integration constants A and B. The first initial condition yields

$$A = x_o\,. \tag{3.79a}$$

As regards the second initial condition, we can write

$$v_o = \left[-A\omega_n\sin\omega_n t + B\omega_n\cos\omega_n t + \delta_f\frac{1}{1-\eta^2}\omega_f\cos\omega_f t\right]\Bigg|_{t=0} = B\omega_n + \frac{\delta_f\omega_f}{1-\eta^2}$$

from where

$$B=\frac{v_o}{\omega_n}-\delta_f\frac{\eta}{1-\eta^2}\,. \tag{3.79b}$$

If $x_o = v_o = 0$ then

$$A=0\,,\qquad B=-\delta_f\frac{\eta}{1-\eta^2} \tag{3.80}$$

and solution (3.78) takes the form

$$x=\delta_f\frac{\eta}{1-\eta^2}\left(\sin\omega_f t-\sin\omega_n t\right)=\delta_f\frac{\eta}{1-\eta^2}\left(\sin\eta\,\omega_n t-\sin\omega_n t\right)\,. \tag{3.81}$$

Let $f(x)$ and $g(x)$ are such functions for which it holds that $\lim_{x\to x_o} f(x)=0$ and $\lim_{x\to x_o} g(x)=0$. Then, as is well known, the limit of the quotient $f(x)/g(x)$ as $x\to x_o$ is given by the L'Hospital rule:

$$\lim_{x\to x_o}\frac{f(x)}{g(x)}=\lim_{x\to x_o}\frac{\frac{\mathrm{d}f(x)}{\mathrm{d}x}}{\frac{\mathrm{d}g(x)}{\mathrm{d}x}}\,. \tag{3.82}$$

Let us now take the limit of solution (3.81) as $\eta\to 1$. By applying the L'Hospital rule, we get

$$\lim_{\eta\to 1}x=\delta_f\lim_{\eta\to 1}\frac{\frac{\mathrm{d}}{\mathrm{d}\eta}(\sin\eta\,\omega_n t-\eta\sin\omega_n t)}{\frac{\mathrm{d}}{\mathrm{d}\eta}(1-\eta^2)}=\delta_f\lim_{\eta\to 1}\frac{\omega_n t\cos\eta\,\omega_n t-\sin\omega_n t}{-2\eta}$$

from where

$$\lim_{\eta\to 1}x=\frac{\delta_f}{2}\left(\sin\omega_n t-\omega_n t\cos\omega_n t\right) \tag{3.83}$$

is the limit in question.

Figure 3.29 graphically represents the dimensionless function $\sin\omega_n t - \omega_n t\cos\omega_n t$ and its parts.

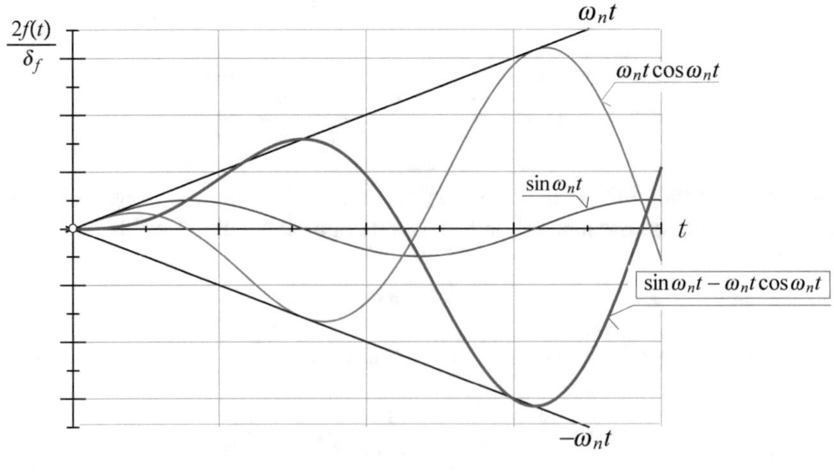

Fig. 3.29 Dimensionless displacement components against time if $\omega_f = \omega_n$

Remark 3.10 The previous line of thought, recall formula (3.83), resulted in the system behavior again if $\eta = 1$ ($\omega_n = \omega_f$) though now the excitation is given, in contrast to the right side of Eq. (3.72), by the harmonic function $F_f = F_{fo} \sin \omega_f t$. However, it is still an open question what happens if $\omega_n \approx \omega_f$ but $|\omega_f - \omega_n| \neq 0$.

If $\omega_n \approx \omega_f$ we may assume that $\eta = 1$ in the numerator of solution (3.81). Then

$$x \cong \delta_f \frac{1}{1-\eta^2} \left(\sin \omega_f t - \sin \omega_n t \right), \tag{3.84}$$

where

$$\begin{aligned} \sin \omega_f t - \sin \omega_n t &= 2 \sin \frac{\omega_f - \omega_n}{2} t \cos \frac{\omega_f + \omega_n}{2} t \\ &\approx 2 \sin \frac{\omega_f - \omega_n}{2} t \cos \omega_f t . \end{aligned} \tag{3.85}$$

Thus

$$x \cong \underbrace{\frac{\delta_f}{1-\eta^2} 2 \sin \frac{\omega_f - \omega_n}{2} t}_{\text{amplitude}} \cos \omega_f t . \tag{3.86}$$

The above result should be proved. If we utilize the relationships

$$\begin{aligned} \sin \frac{\omega_f - \omega_n}{2} t &= \sin \frac{\omega_f}{2} t \cos \frac{\omega_n}{2} t - \cos \frac{\omega_f}{2} t \sin \frac{\omega_n}{2} t, \\ \cos \frac{\omega_f + \omega_n}{2} t &= \cos \frac{\omega_f}{2} t \cos \frac{\omega_n}{2} t - \sin \frac{\omega_f}{2} t \sin \frac{\omega_n}{2} t, \end{aligned}$$

we can write

$$\begin{aligned} & 2 \sin \frac{\omega_f - \omega_n}{2} t \cos \frac{\omega_f + \omega_n}{2} t = \\ & = \Big[\sin \frac{\omega_f}{2} t \cos \frac{\omega_f}{2} t \cos^2 \frac{\omega_n}{2} t + \sin \frac{\omega_f}{2} t \cos \frac{\omega_f}{2} t \sin^2 \frac{\omega_n}{2} t - \\ & - \sin \frac{\omega_n}{2} t \cos \frac{\omega_n}{2} t \cos^2 \frac{\omega_f}{2} t - \sin \frac{\omega_n}{2} t \cos \frac{\omega_n}{2} t \sin^2 \frac{\omega_f}{2} t \Big] = \\ & \qquad = \sin \omega_f t - \sin \omega_n t \end{aligned}$$

by the use of which we can easily check the rightfulness of Eq. (3.86).

Figure 3.30 represents the solution. Observe that the function $\cos \omega_f t$ will go through several cycles while the amplitude function $2 \sin \frac{\omega_f - \omega_n}{2} t$ goes through a single cycle. This phenomenon is called *beating*.

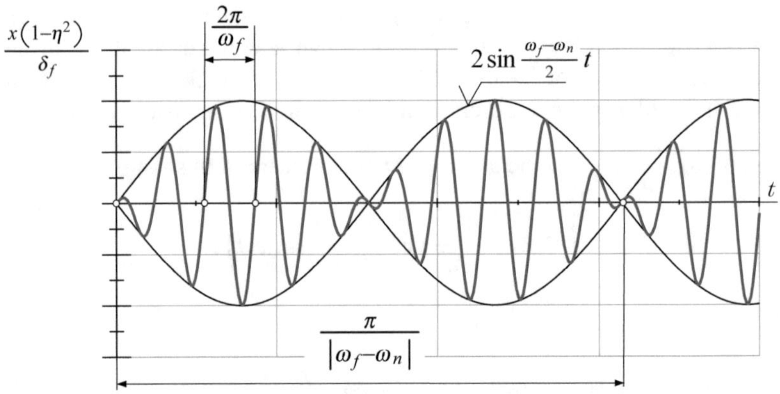

Fig. 3.30 Displacement against time—beating vibration phenomenon

3.3.3 *Third We Shall Assume That the System Is Subjected to a Constant Force*

If this is the case the differential equation to be solved is

$$\ddot{x} + \omega_n^2 x = \omega_n^2 \delta_f = \frac{F}{m}, \qquad F = \text{constant}. \tag{3.87}$$

The general solution takes the form

$$x = \underbrace{A\cos\omega_n t + B\sin\omega_n t}_{\text{complementary function}} + \underbrace{\delta_f}_{\substack{\text{particular}\\\text{solution}}} . \tag{3.88}$$

It follows from the initial conditions $x_o = v_o = 0$ that

$$\begin{aligned} x_o &= x(t=0) = A + \delta_f && \rightarrow \quad A = -\delta_f \\ v_o &= \dot{x}(t=0) = [-\omega_n A\sin\omega_n t + B\omega_n t\sin\omega_n t]\,|_{t=0} = 0 && \rightarrow \quad B = 0 . \end{aligned}$$

Consequently,

$$x = \delta_f\,(1 - \cos\omega_n t) . \tag{3.89}$$

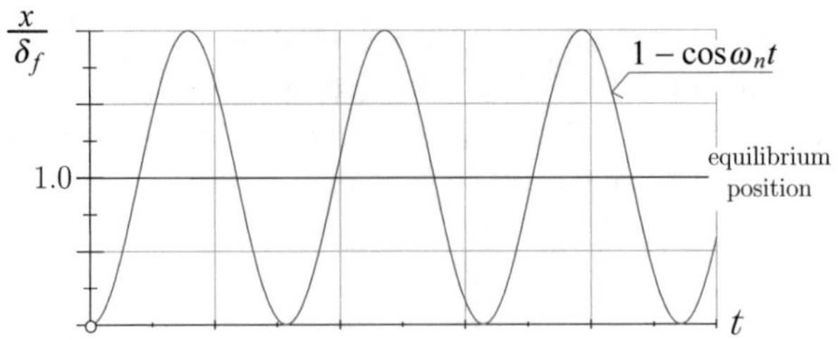

Fig. 3.31 Motion about equilibrium position

It is obvious from Fig. 3.31 that the motion is a harmonic vibration about the equilibrium position δ_f.

Exercise 3.7 Block 1 of mass m_1 is dropped from a height h onto block 2 of mass m_2. Block 2 is supported by a spring of stiffness k. Assuming that (a) block 2 is at rest before impact and that (b) there is no rebound (the impact is then perfectly plastic) determine the dynamic displacement x_d.

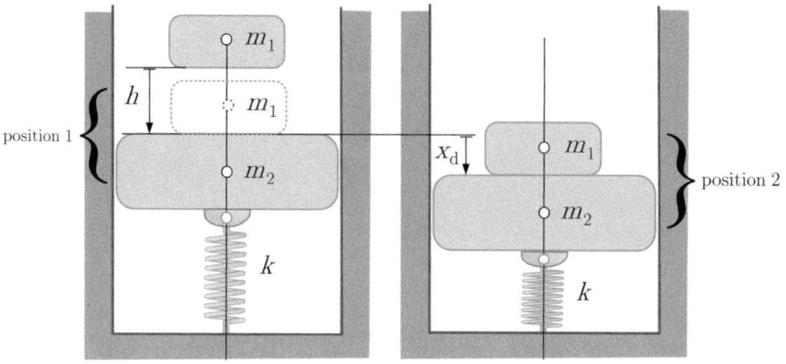

Fig. 3.32 Block m_1 dropped onto block m_2

When impact begins the velocities are given by

$$v_1 = v_{1n} = \sqrt{2gh}\,, \quad \text{and} \quad v_{2n} = 0\,.$$

Since the impact is perfectly plastic, Eq. (2.9b) yields the common velocity of the two bodies right after impact (Fig. 3.32):

$$c_n = \frac{m_1 v_1}{m_1 + m_2}\,.$$

To find x_d, we have to apply the principle of work an energy (1.58a):

$$\mathcal{E}_2 - \mathcal{E}_1 = W_{12},$$

where

$$\mathcal{E}_1 = \frac{1}{2}(m_1 + m_2)\,c_n^2 = \frac{1}{2}\frac{m_1^2 v_1^2}{m_1 + m_2} = \frac{m_1^2}{m_1 + m_2}gh\,, \quad \mathcal{E}_2 = 0$$

and

$$W_{12} = \underbrace{(m_1 + m_2)g\,x_d}_{\substack{\text{work done by the}\\ \text{weight force}}} - \underbrace{\int_0^{x_d} F_k\,\mathrm{d}x}_{\substack{\text{work done by the}\\ \text{spring force } F_k}}\,.$$

Therefore, it holds

$$-\frac{m_1^2}{m_1+m_2}gh = (m_1+m_2)g\,x_d - \int_0^{x_d} F_k\,\mathrm{d}x \tag{3.90}$$

in which

$$F_k = m_2 g + kx \qquad \rightarrow \qquad \int_0^{x_d} F_k\,\mathrm{d}x = m_2 g x_d + \frac{1}{2}kx_d^2. \tag{3.91}$$

After substituting (3.91) into (3.90), we have

$$x_d^2 - 2\underbrace{\frac{m_1 g}{k}}_{\delta_f}x_d - 2\underbrace{\frac{m_1 g}{k}}_{\delta_f}\frac{1}{1+\frac{m_2}{m_1}}h = 0$$

or

$$x_d^2 - 2\delta_f x_d - 2\delta_f\frac{1}{1+\frac{m_2}{m_1}}h = 0$$

from where

$$x_d = \delta_f + \sqrt{\delta_f^2 + 2\delta_f\frac{h}{1+\frac{m_2}{m_1}}}\,.$$

The final form of the solution obtained is given by

$$x_d = \delta_f\underbrace{\left(1+\sqrt{1+2\frac{h}{\delta_f}\frac{1}{1+\frac{m_2}{m_1}}}\right)}_{\nu} = \nu\delta_f\,. \tag{3.92}$$

Here ν is the dynamic factor. Observe that

$$\nu \qquad \begin{bmatrix} \text{increases if } h & \text{increases,} \\ \text{decreases if } k \text{ and } m_2 & \text{increase} \end{bmatrix}.$$

Exercise 3.8 What changes if damping is taken into account in Eq. (3.87)? If damping is present and the solution is vibratory, i.e., $\zeta < 1$

$$\ddot{x} + 2\beta\dot{x} + \omega_n^2 x = \omega_n^2\delta_f = \frac{F}{m}, \qquad F = \text{constant}$$

is the equation of motion with solution

$$x = e^{-\beta t}\left(A\cos\omega_d t + B\sin\omega_d t\right) + \delta_f\,, \qquad \omega_d = \omega_n\sqrt{1-\zeta^2}.$$

If $x(t)|_{t=0} = x_o$ and $\dot{x}(t)|_{t=0} = v_o$ are the initial conditions, the solution is of the form

$$x(t) = e^{-\beta t}\left[(x_o - \delta_f)\cos\omega_d t + \frac{v_o + \beta x_o - \delta_f}{\omega_d}\sin\omega_d t\right].$$

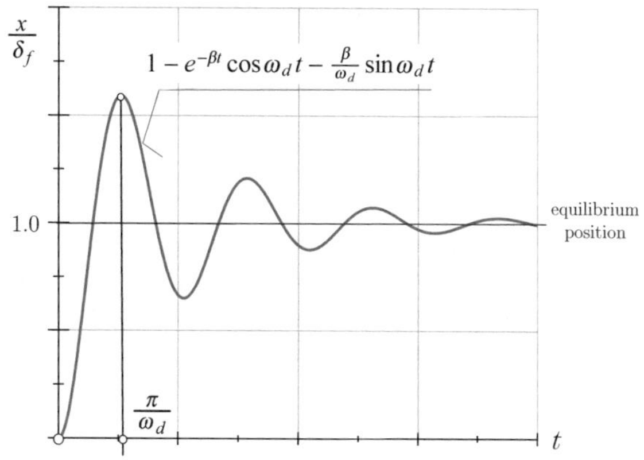

Fig. 3.33 Damped motion about equilibrium position

Assume, in accordance with Sect. 3.3.3, that $x_o = 0$ and $v_o = 0$ are the initial conditions. Then the solution simplifies to

$$x(t) = \delta_f \left[1 - e^{-\beta t} \cos \omega_d t - \frac{\beta}{\omega_d} \sin \omega_d t \right] .$$

Figure 3.33 shows that the motion is, now, a diminishing vibration about the static equilibrium position δ_f.

3.4 Problems

Problem 3.1 A particle moves in a simple harmonic motion with a maximum acceleration of 50 m/s^2. Knowing that the frequency is 30 Hz determine the amplitude and maximum velocity of the particle.

Problem 3.2 Assume that we consider the free vibrations of a single degree of freedom system with damping. Assume further that $c > c_n$ which means that we have heavy damping. Prove that the body never passes (again) through its position of equilibrium (a) if it is released with $v_o = 0$ initial velocity from an arbitrary position x_o or (b) if it is started from its equilibrium position (for which naturally $x_o = 0$) with an arbitrary initial velocity v_o. (Hint: the proofs should be based on Eq. (3.21).)

Problem 3.3 Show that the logarithmic decrement δ—see Eq. (3.27)—can be given in the form

$$\delta = \frac{2\pi c/c_n}{\sqrt{1 - (c/c_n)^2}} .$$

Problem 3.4 Show that the logarithmic decrement can also be given as

$$\delta = \frac{1}{k} \ln \frac{x(t)}{x(t + k\tau_d)},$$

where k is a natural number and $k \geq 2$. For $k = 1$, we get back definition (3.27).

Problem 3.5 Figure 3.34 shows two support arrangements for a slender uniform rod of length ℓ and mass m. The rod is subjected to a harmonic force of excitation at its middle point. Examine which support arrangement results in a steady-state response function with a smaller amplitude.

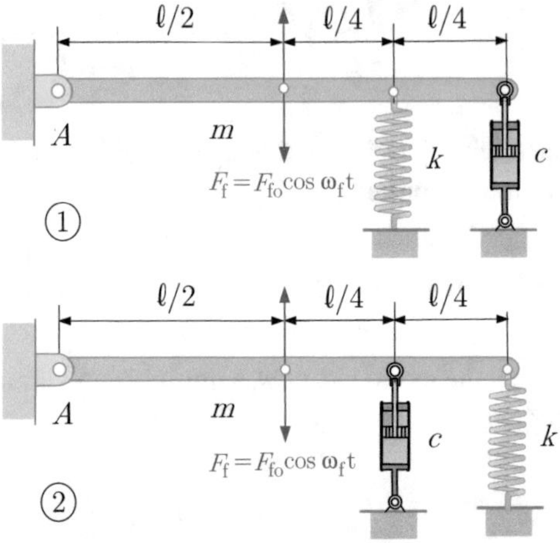

Fig. 3.34 Models of vibrating slender rods

Problem 3.6 A block with mass $m_1 = 20$ kg is dropped from a height of 2.5 m onto the $m_2 = 10$ kg pan of a spring scale. Determine the maximum deflection of the pan if the impact is perfectly plastic. The spring constant is $k = 20$ kN/m.

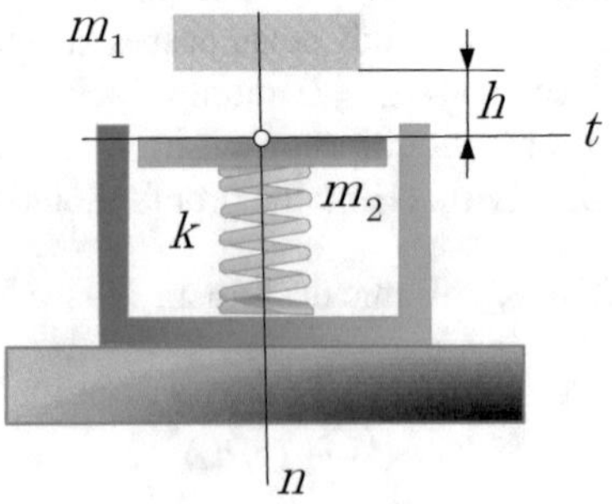

Fig. 3.35 Block m_1 dropped onto block m_2

References

1. F.E. Richart, J.R. Hall, R.D. Woods, *Vibrations of Soils and Foundations* (Prentice Hall, Inc., Upper Saddle River, 1970)
2. H. Dreisig, F. Holzweißig, *Dynamics of Machinery* (Springer, Berlin, 2010)

Chapter 4
Introduction to Multidegree of Freedom Systems

4.1 Lagrange's Equations of Motion of the Second Kind

4.1.1 Generalized Coordinates

If a single particle moves freely in space it has three degrees of freedom which means that three Cartesian coordinates are needed to give its position in space. If a rigid body moves freely in the three-dimensional space we need six independent quantities to give the position of the body uniquely. Consider this issue in a bit more detail. The rigid body $\mathcal{B}$ shown in Fig. 4.1 moves freely in space. The Cartesian coordinate system (xyz) is the absolute coordinate system. The Greek Cartesian coordinate system $(\xi\eta\zeta)$ is attached to the mass center G and moves together with the body in such a way that the orientation of the coordinate axes remains unchanged, i.e., $\xi||x$,

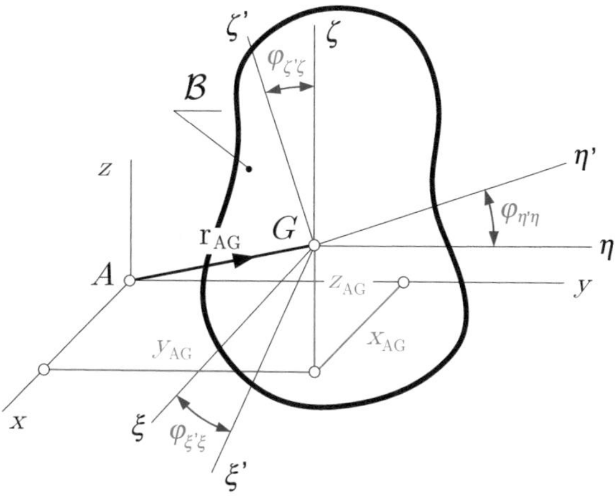

Fig. 4.1 Coordinate systems for a body moving freely in space

G. Szeidl and L. P. Kiss, *Mechanical Vibrations*, Foundations of Engineering Mechanics,
https://doi.org/10.1007/978-3-030-45074-8_4

$\eta || y$, and $\zeta || z$ during the motion. The third coordinate system $(\xi'\eta'\zeta')$ is also attached to the body at the mass center G but, in contrast to the coordinate system $(\xi\eta\zeta)$, it rotates together with the body. When the motion begins it will be assumed that the two Greek coordinate systems coincide. During the motion the angles formed by the axes ξ and ξ', η and η', ζ and ζ' will be denoted by $\varphi_{\xi'\xi}$, $\varphi_{\eta'\eta}$, $\varphi_{\zeta'\zeta}$, respectively. It is obvious that the six time functions $x_{AG}(t)$, $y_{AG}(t)$, $z_{AG}(t)$, $\varphi_{\xi'\xi}(t)$, $\varphi_{\eta'\eta}(t)$, $\varphi_{\zeta'\zeta}(t)$ uniquely determine the position of the rigid body in space: which shows that the degree of freedom is really six for the free motion of a rigid body. It follows from the mentioned two examples that the degree of freedom is, in fact, the number of independent coordinates we need to describe the motion.

Observe that the quantities $\varphi_{\xi'\xi}(t)$, $\varphi_{\eta'\eta}(t)$, $\varphi_{\zeta'\zeta}(t)$ are not Cartesian coordinates; they are, in fact, angle coordinates. For the sake of generality, the quantities needed to describe the motion uniquely will also be referred to as generalized coordinates which are, in general, denoted by the letter q. We can now say that we need the six generalized coordinates

$$q_1 = x_{AG}, \quad q_2 = y_{AG}, \quad q_3 = z_{AG} \quad \text{and} \quad q_4 = \varphi_{\xi'\xi}, \quad q_5 = \varphi_{\eta'\eta}, \quad q_6 = \varphi_{\zeta'\zeta} \tag{4.1}$$

to describe the free motion of the rigid body considered.

Motions of particles and bodies are not always free motions since in many cases there are various restrictions, called constrains, which force the bodies to move in a predetermined manner. If there are constraints the motion is a constrained motion.

Exercise 4.1 Examine what coordinates could be used to describe the motion of the spherical pendulum (a mass m on an inextensible cord of length ℓ_r) shown in Fig. 4.2.

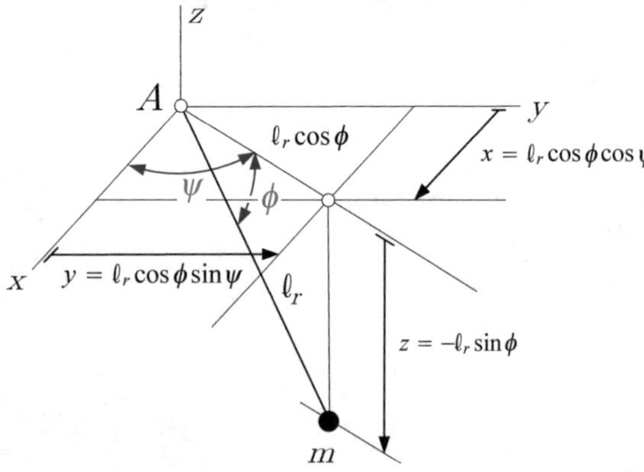

Fig. 4.2 Spherical pendulum

Observe that the set of coordinates we can use to describe the motion is not unique. The position of the mass m at any point of time may be given in Cartesian coordinates x, y, and z—they are, however, not independent on each other—or in terms of the spherical coordinates $r = \ell_r$, ψ, and ϕ:

$$x = \ell_r \cos\phi \cos\psi, \quad y = \ell_r \cos\phi \sin\psi, \quad z = -\ell_r \sin\phi. \tag{4.2}$$

If we use spherical coordinates only two, i.e., ϕ and ψ are independent since the length ℓ_r is constant in time. If we use Cartesian coordinates, we can choose, for instance, x and y as independent coordinates while z is the solution of the constraint equation

$$f_c(x, y, z, t) = f_c(x, y, z) = \ell_r - \sqrt{x^2 + y^2 + z^2} = 0, \tag{4.3}$$

which expresses that ℓ_r is constant. For this choice z is a superfluous coordinate which can be eliminated by utilizing the constraint equation (4.3). Since we have two independent coordinates $q_1 = \phi$ and $q_2 = \psi$ (or $q_1 = x$ and $q_2 = y$) the spherical pendulum represents a system of two degrees of freedom. Equation (4.3) is a geometric constraint equation.

The constraint equation of the form

$$f_c(x, y, z, t) = 0 \tag{4.4}$$

is called holonomic.

Exercise 4.2 Particles m_1 and m_2 are connected with a weightless rod of constant length ℓ. Assume that these two mass systems move in such a manner in the coordinate plane xy that the velocity at the midpoint C is parallel to the rod. What are the constraint equations?

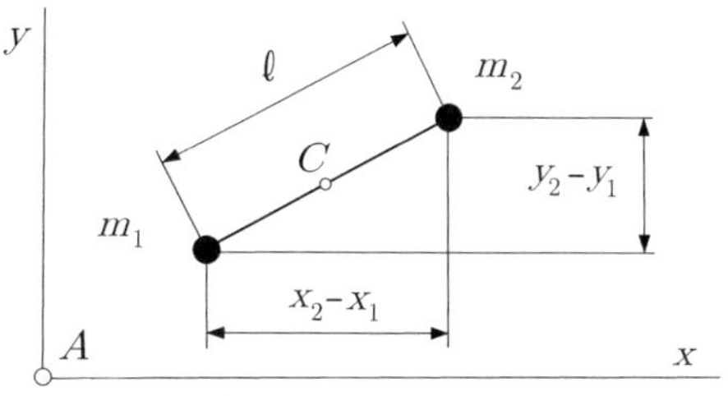

Fig. 4.3 Weightless rod with two masses at the endpoints

Since the system moves in the coordinate plane xy and the length of the rod is constant the following equations should be satisfied:

$$\begin{aligned} f_{c1}(x_1, y_1, z_1, x_2, y_2, z_2, t) &= z_1 = 0, \\ f_{c2}(x_1, y_1, z_1, x_2, y_2, z_2, t) &= z_2 = 0, \\ f_{c3}(x_1, y_1, z_1, x_2, y_2, z_2, t) &= \ell - \sqrt{(x_2 - x_1)^2 + (y_2 - y_1)^2} = 0\,. \end{aligned} \tag{4.5a}$$

Constraints (4.5a) are all holonomic.

It is obvious on the base of Eq. (1.4) that the velocities linearly depend on the distances. Consequently,

$$v_{Cx} = \frac{1}{2}(\dot{x}_1 + \dot{x}_2)\,, \quad v_{Cy} = \frac{1}{2}(\dot{y}_1 + \dot{y}_2)$$

are the two velocity components at the midpoint C. If $\mathbf{v}_C$ is parallel to the rod then

$$\frac{v_{Cy}}{v_{Cx}} = \frac{y_2 - y_1}{x_2 - x_1}$$

from where it follows that

$$f_{c4}(x_1, y_1, z_1, x_2, y_2, z_2, \dot{x}_1, \dot{y}_1, \dot{z}_1, \dot{x}_2, \dot{y}_2, \dot{z}_2, t) = \frac{\dot{y}_1 + \dot{y}_2}{y_2 - y_1} - \frac{\dot{x}_1 + \dot{x}_2}{x_2 - x_1} = 0\,. \quad (4.5b)$$

The constraint equations that involve velocities are called nonholonomic constraint equations provided that the velocity components, which are various time derivatives, cannot be removed form the constraint equation by integration. Equation (4.5b) is a nonholonomic constraint equation. It can also be called kinematic constraint equation since it expresses that the velocity at C is parallel to the rod.

4.1.2 *Principle of Virtual Work*

A virtual displacement (or displacement variation) is an infinitesimal change in the corresponding coordinate which may be chosen arbitrarily irrespective of the time provided that it is consistent with the constraints, i.e., it does not violate the constraint equations of the system.

Exercise 4.3 Determine the virtual displacements for the spherical pendulum shown in Fig. 4.2.

Since ψ and ϕ may change arbitrarily (independently of the constraint equation) it follows from Eq. (4.2) that

$$\begin{aligned}
\delta x &= \frac{\partial x}{\partial \phi}\delta\phi + \frac{\partial x}{\partial \psi}\delta\psi = -\ell_r\,(\sin\phi\,\cos\psi\,\delta\phi + \cos\phi\,\sin\psi\,\delta\psi)\,,\\
\delta y &= \frac{\partial y}{\partial \phi}\delta\phi + \frac{\partial y}{\partial \psi}\delta\psi = \ell_r\,(-\sin\phi\,\sin\psi\delta\phi + \cos\phi\,\cos\psi\,\delta\psi)\,,\\
\delta z &= \frac{\partial z}{\partial \phi}\delta\phi + \frac{\partial z}{\partial \psi}\delta\psi = -\ell_r\cos\phi\,\delta\phi
\end{aligned} \quad (4.6)$$

are the three virtual displacements. We remark that the constraint condition

$$\ell_r - \sqrt{(x+\delta x)^2 + (y+\delta y)^2 + (z+\delta z)^2} = 0$$

should also be satisfied by the virtual displacements.

Consider a system of N ($N \geq 2$) particles in motion—see Fig. 4.4. We assume that the system has n degrees of freedom $2 \leq n < 3N$. Here and in the sequel we

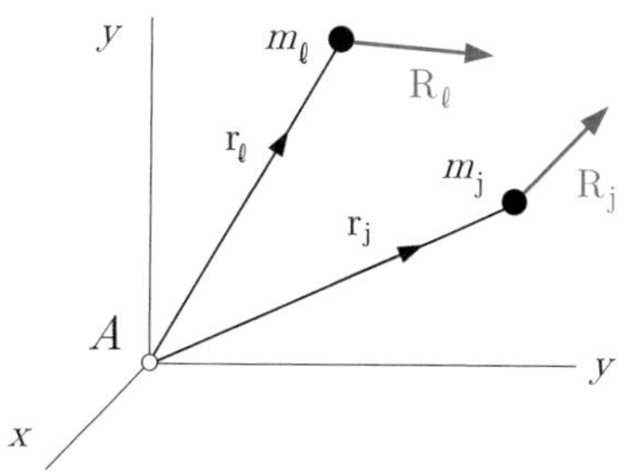

Fig. 4.4 A system of particles in motion

will also assume that the constraint equations are all holonomic, i.e., they can be given in the form:

$$f_{ci}(x_1, y_1, z_1, x_2, y_2, z_2, \ldots, x_N, y_N, z_N, t) = 0, \quad i = 1, 2, \ldots, 3N - n\,. \tag{4.7}$$

Let $\mathbf{R}_j$ be the resultant of the forces acting on particle m_j. It can be resolved into two parts

$$\mathbf{R}_j = \mathbf{F}_j + \mathbf{f}_{jc}, \tag{4.8}$$

where $\mathbf{f}_{jc}$ is the constraint force exerted on the jth particle. It is a further assumption that the constraint forces do no work at all (for example, if the constraints are smooth contacting surfaces then the constraint forces (contact forces) are perpendicular to the velocity, and therefore they have no power and do no work). The constraints are called ideal ones if the constraint forces do no work. According to Newton's second law, the motion of the jth particle is described by the dynamic equilibrium equation

$$\mathbf{F}_j + \mathbf{f}_{jc} - m_j\ddot{\mathbf{r}}_j = \mathbf{0}\,, \qquad j = 1, \ldots, N\,. \tag{4.9}$$

Let $\delta\mathbf{r}_j$ be the virtual displacement of the jth particle. If we dot multiply Eq. (4.8) by $\delta\mathbf{r}_j$ and add the dot products obtained to each other, we get

$$\sum_{j=1}^{N}\left(\mathbf{F}_j + \mathbf{f}_{jc} - m_j\ddot{\mathbf{r}}_j\right)\cdot\delta\mathbf{r}_j = 0 \tag{4.10a}$$

in which $\mathbf{f}_{jc}\cdot\delta\mathbf{r}_j = 0$—the constraint forces do no work—hence

$$\sum_{j=1}^{N}\left(\mathbf{F}_j - m_j\ddot{\mathbf{r}}_j\right)\cdot\delta\mathbf{r}_j = 0, \tag{4.10b}$$

where

$$\delta W = \sum_{j=1}^{N}\mathbf{F}_j\cdot\delta\mathbf{r}_j \quad \text{and} \quad \delta W_{\text{inertia}} = -\sum_{j=1}^{N} m_j\ddot{\mathbf{r}}_j\cdot\delta\mathbf{r}_j \tag{4.11}$$

are the virtual works done by the forces $\mathbf{F}_j$ and the forces of inertia $m_j\ddot{\mathbf{r}}_j$ throughout the virtual displacements $\delta\mathbf{r}_j$.

Equation (4.10b) is known as the principle of virtual work (or the weak form of the dynamic equilibrium equations) which says that the sum of the two virtual works is equal to zero

$$\delta W + \delta W_{\text{inertia}} = 0\,. \tag{4.12}$$

4.1.3 Derivation of Lagrange's Equation of Motion of the Second Kind

It is worth mentioning that these equations were derived by Joseph Louis Lagrange [1, 2]. Since the system has n degrees of freedom the position vector of the jth particle can be given in terms of the generalized coordinates $q_1, q_2, \ldots, q_n$ in the form

$$\mathbf{r}_j = \mathbf{r}_j\,(q_1, q_2, \ldots, q_n; t)\;. \tag{4.13}$$

The virtual displacements of particle j can now be calculated in the same manner as in Exercise 4.3:

$$\delta\mathbf{r}_j = \frac{\partial\mathbf{r}_j}{\partial q_1}\,\delta q_1 + \frac{\partial\mathbf{r}_j}{\partial q_2}\,\delta q_2 + \cdots + \frac{\partial\mathbf{r}_j}{\partial q_n}\,\delta q_n = \sum_{\ell=1}^{n}\frac{\partial\mathbf{r}_j}{\partial q_\ell}\,\delta q_\ell\,. \tag{4.14}$$

Making use of the previous equation one can rewrite the virtual work δW in such a manner which makes possible to introduce the concept of generalized forces:

$$\delta W = \sum_{j=1}^{N}\mathbf{F}_j\cdot\delta\mathbf{r}_j = \sum_{j=1}^{N}\mathbf{F}_j\cdot\left(\sum_{\ell=1}^{n}\frac{\partial\mathbf{r}_j}{\partial q_\ell}\,\delta q_\ell\right) = \sum_{\ell=1}^{n}\underbrace{\left(\sum_{j=1}^{N}\mathbf{F}_j\cdot\frac{\partial\mathbf{r}_j}{\partial q_\ell}\right)}_{Q_\ell}\delta q_\ell\,. \tag{4.15}$$

Here

$$\boxed{Q_\ell = \sum_{j=1}^{N}\mathbf{F}_j\cdot\frac{\partial\mathbf{r}_j}{\partial q_\ell}} \tag{4.16}$$

is the generalized force which belongs to the generalized coordinate q_ℓ. (We have as many generalized forces as there are generalized coordinates.)

The work δW can now be given in terms of the generalized forces:

$$\delta W = \sum_{\ell=1}^{n} Q_\ell\,\delta q_\ell\,. \tag{4.17}$$

Consider now, for the sake of our further consideration, the velocity of particle j. Recalling (4.13), we can write

$$\mathbf{v}_j = \frac{\mathrm{d}\mathbf{r}_j}{\mathrm{d}t} = \frac{\partial \mathbf{r}_j}{\partial q_1}\dot{q}_1 + \frac{\partial \mathbf{r}_j}{\partial q_2}\dot{q}_2 + \cdots + \frac{\partial \mathbf{r}_j}{\partial q_n}\dot{q}_n + \frac{\partial \mathbf{r}_j}{\partial t} = \sum_{\ell=1}^{n} \frac{\partial \mathbf{r}_j}{\partial q_\ell}\dot{q}_\ell + \frac{\partial \mathbf{r}_j}{\partial t}. \quad (4.18)$$

Hence

$$\boxed{\frac{\partial \mathbf{v}_j}{\partial \dot{q}_\ell} = \frac{\partial \dot{\mathbf{r}}_j}{\partial \dot{q}_\ell} = \frac{\partial \mathbf{r}_j}{\partial q_\ell},} \quad (4.19)$$

which shows that the derivative

$$\frac{\partial \mathbf{v}_j}{\partial \dot{q}_\ell}$$

is the velocity that belongs to a unit generalized coordinate velocity. Let us now substitute (4.19) into (4.16). We get

$$\boxed{Q_\ell = \sum_{j=1}^{N} \underbrace{\mathbf{F}_j \cdot \frac{\partial \mathbf{v}_j}{\partial \dot{q}_\ell}}_{\mathcal{P}_{j\ell}}\ .} \quad (4.20)$$

In other words, the generalized force is the sum of the powers that are calculated as the product of a force and a velocity which belongs to a unit coordinate velocity.

In what follows, our aim is to manipulate $\delta W_{\text{inertia}}$ in Eq. (4.11) into a more suitable form. To this end, substituting (4.14) into the expression $m_j\ddot{\mathbf{r}}_j \cdot \delta\mathbf{r}_j$, we have

$$m_j\ddot{\mathbf{r}}_j \cdot \delta\mathbf{r}_j = \sum_{\ell=1}^{n} m_j\ddot{\mathbf{r}}_j \cdot \frac{\partial \mathbf{r}_j}{\partial q_\ell}\delta q_\ell, \quad (4.21)$$

where

$$m_j\ddot{\mathbf{r}}_j \cdot \frac{\partial \mathbf{r}_j}{\partial q_\ell} = \frac{\mathrm{d}}{\mathrm{d}t}\left(m_j\dot{\mathbf{r}}_j \cdot \frac{\partial \mathbf{r}_j}{\partial q_\ell}\right) - m_j\dot{\mathbf{r}}_j \frac{\mathrm{d}}{\mathrm{d}t}\frac{\partial \mathbf{r}_j}{\partial q_\ell} \underset{\frac{\partial \mathbf{r}_j}{\partial q_\ell} = \frac{\partial \dot{\mathbf{r}}_j}{\partial \dot{q}_\ell}}{=\uparrow=}$$

$$= \frac{\mathrm{d}}{\mathrm{d}t}\left(m_j\dot{\mathbf{r}}_j \cdot \frac{\partial \dot{\mathbf{r}}_j}{\partial \dot{q}_\ell}\right) - m_j\dot{\mathbf{r}}_j \frac{\partial \dot{\mathbf{r}}_j}{\partial q_\ell} = \left(\frac{\mathrm{d}}{\mathrm{d}t}\frac{\partial}{\partial \dot{q}_\ell} - \frac{\partial}{\partial q_\ell}\right) \underbrace{\frac{1}{2}m_j\dot{\mathbf{r}}_j \cdot \dot{\mathbf{r}}_j}_{\mathcal{E}_j} =$$

$$= \left(\frac{\mathrm{d}}{\mathrm{d}t}\frac{\partial}{\partial \dot{q}_\ell} - \frac{\partial}{\partial q_\ell}\right)\mathcal{E}_j$$

in which $\mathcal{E}_j$ is the kinetic energy of particle j. If we insert this result into (4.21), we get

$$m_j\ddot{\mathbf{r}}_j \cdot \delta\mathbf{r}_j = \sum_{\ell=1}^{n}\left(\frac{\mathrm{d}}{\mathrm{d}t}\frac{\partial}{\partial\dot{q}_\ell} - \frac{\partial}{\partial q_\ell}\right)\mathcal{E}_j\,\delta q_\ell\,.$$

Let us now sum over the N particles. We obtain

$$\delta W_{\text{inertia}} = -\sum_{j=1}^{N} m_j\ddot{\mathbf{r}}_j \cdot \delta\mathbf{r}_j = -\sum_{\ell=1}^{n}\left(\frac{\mathrm{d}}{\mathrm{d}t}\frac{\partial}{\partial\dot{q}_\ell} - \frac{\partial}{\partial q_\ell}\right)\underbrace{\left(\sum_{j=1}^{N}\mathcal{E}_j\right)}_{\mathcal{E}}\delta q_\ell \tag{4.22}$$

in which $\mathcal{E}$ is the total kinetic energy of the system. To complete the derivation, we shall add Eqs. (4.17) and (4.22). If we take (4.12) into account, we get

$$\delta W + \delta W_{\text{inertia}} = \sum_{\ell=1}^{n}\left(Q_\ell - \frac{\mathrm{d}}{\mathrm{d}t}\frac{\partial\mathcal{E}}{\partial\dot{q}_\ell} + \frac{\partial\mathcal{E}}{\partial q_\ell}\right)\delta q_\ell = 0$$

from where with regard to the arbitrariness of δq_ℓ it follows

$$\boxed{\frac{\mathrm{d}}{\mathrm{d}t}\frac{\partial\mathcal{E}}{\partial\dot{q}_\ell} - \frac{\partial\mathcal{E}}{\partial q_\ell} = Q_\ell\,, \qquad \ell = 1, 2, \ldots, n\,.} \tag{4.23}$$

This equation is known as Lagrange's equation of motion of the second kind (or Lagrange's equations of the second kind).

Generalizations:

1. Assume that the system with n degrees of freedom consists of N rigid bodies. It is worthy of mentioning that Lagrange's equations of motion of the second kind are valid for such systems as well.
2. Assume that the external forces exerted on the jth rigid body are replaced by the equivalent force couple system $[\mathbf{F}_j, \mathbf{M}_j]$ at the point P_j of the body. Let $\mathbf{v}_j$ be the velocity of the point P_j. Further let $\boldsymbol{\omega}_j$ be the angular velocity of the jth body. With the knowledge of these quantities we can calculate the generalized forces Q_ℓ by using the following equation:

$$Q_\ell = \sum_{j=1}^{N}\mathbf{F}_j \cdot \frac{\partial\mathbf{v}_j}{\partial\dot{q}_\ell} + \sum_{j=1}^{N}\mathbf{M}_j \cdot \frac{\partial\boldsymbol{\omega}_j}{\partial\dot{q}_\ell}\,, \qquad \ell = 1, 2, \ldots, n\,. \tag{4.24}$$

Assume that the system considered is a conservative one. Then there exists a potential function $U(q_1, q_2, \ldots, q_n)$ such that the work done by the forces acing on the system is equal to the negative of the potential function:

$$W(q_1, q_2, \ldots, q_n) = -U(q_1, q_2, \ldots, q_n), \tag{4.25}$$

while the virtual work is given by

$$\delta W(q_1, q_2, \dots, q_n) = -\sum_{\ell=1}^{n} \frac{\partial U(q_1, q_2, \dots, q_n)}{\partial q_\ell} \delta q_\ell \, . \tag{4.26}$$

Comparison of Eqs. (4.17) and (4.25) yields formula

$$Q_\ell = -\frac{\partial U(q_1, q_2, \dots, q_n)}{\partial q_\ell} \tag{4.27}$$

for the generalized forces. Thus, we can write $-\partial U/\partial q_\ell$ for Q_ℓ in Eq. (4.23) obtaining in this manner a further form of Lagrange's equations:

$$\boxed{\frac{\mathrm{d}}{\mathrm{d}t}\frac{\partial \mathcal{E}}{\partial \dot{q}_\ell} - \frac{\partial \mathcal{E}}{\partial q_\ell} + \frac{\partial U}{\partial q_\ell} = 0\,, \qquad \ell = 1, 2, \dots, n\,.} \tag{4.28}$$

The Lagrangian can then be defined by the relation

$$\mathcal{L} = \mathcal{E} - U\,. \tag{4.29}$$

The dimension of the Lagrangian is, of course, energy. Since U is independent of the generalized coordinate velocities $\dot{q}_\ell$ Lagrange's equations (4.28) can be rewritten into the form

$$\boxed{\frac{\mathrm{d}}{\mathrm{d}t}\frac{\partial \mathcal{L}}{\partial \dot{q}_\ell} - \frac{\partial \mathcal{L}}{\partial q_\ell} = 0\,, \qquad \ell = 1, 2, \dots, n\,.} \tag{4.30}$$

4.2 Applications of Lagrange's Equations

4.2.1 Calculations of Generalized Forces

Spring forces. Figure 4.5a shows the rod AB which is supported by a spring at the right end. Let q be the displacement at the point B in the direction of the spring. It is obvious that the strain energy stored in the spring and the force exerted by the spring on the rod are given by

$$U = \frac{1}{2}kq^2\,, \qquad \mathbf{F}_k = -kq\mathbf{n}\,. \tag{4.31}$$

Utilizing the power $\mathcal{P} = \mathbf{F}_k \cdot \mathbf{v}_B$ of the spring force we obtain that the work done by the force in the time interval $[0, t]$ is

Fig. 4.5 Rods supported by springs

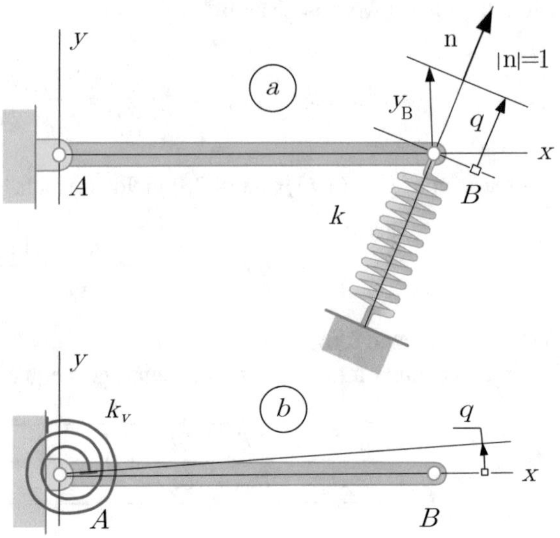

$$W=\int_0^t \mathcal{P}\,\mathrm{d}t=\int_0^t \mathbf{F}_k\cdot\mathbf{v}_B\,\mathrm{d}t=\underset{\mathbf{v}_B=\frac{\partial \mathbf{r}_B}{\partial q}\dot{q}}{\uparrow}=\int_0^t \underbrace{\mathbf{F}_k\cdot\frac{\partial \mathbf{r}_B}{\partial q}}_{Q_k}\underbrace{\dot{q}\,\mathrm{d}t}_{\mathrm{d}q}=\int_0^q Q_k\,\mathrm{d}q \tag{4.32a}$$

or

$$W=\int_0^t \mathbf{F}_k\cdot\mathbf{v}_B\,\mathrm{d}t=\int_0^t -kq\mathbf{n}\cdot\mathbf{v}_B\,\mathrm{d}t=\underset{\mathbf{n}\cdot\mathbf{v}_B=\dot{q}}{\uparrow}=-\int_0^q kq\,\mathrm{d}q=-\frac{1}{2}kq^2=-U\,. \tag{4.32b}$$

Comparison of Eqs. (4.32a) and (4.32b) yields the generalized spring force:

$$Q_k=-\frac{\partial U}{\partial q}\,. \tag{4.33}$$

Note that the above line of thought is, in fact, a proof of equation (4.27) for this simple case.

Rod AB in Fig. 4.5b is supported by a volute spring of spring constant k_v at the left end. The angle of rotation is denoted by q. If we rotate the rod about the axis at A

$$U=\frac{1}{2}k_v q^2\,,\qquad \mathbf{M}_A=-k_v q\mathbf{i}_z \tag{4.34}$$

are the strain energy stored in the volute spring and the restoring moment exerted by the volute spring on the rod. The angular velocity of the rod is $\boldsymbol{\omega}=\dot{q}\mathbf{i}_z$. Since the power of the restoring moment is given by the relation $\mathcal{P}=\mathbf{M}_A\cdot\boldsymbol{\omega}=-k_v q\dot{q}$ it follows that the work done by the restoring force in the time interval $[0,t]$ can be calculated as

$$W = \int_0^t \mathbf{M}_A \cdot \boldsymbol{\omega}\, \mathrm{d}t = \int_0^t -k_v q \underbrace{\dot{q}\mathrm{d}t}_{\mathrm{d}q} = -\int_0^q k_v q\, \mathrm{d}q = -\frac{1}{2}k_v q^2 = -U\,. \quad (4.35)$$

On the other hand,

$$W = \int_0^q Q_k\, \mathrm{d}q\,. \quad (4.36)$$

Comparison of Eqs. (4.35) and (4.36) results in relation (4.33) again.

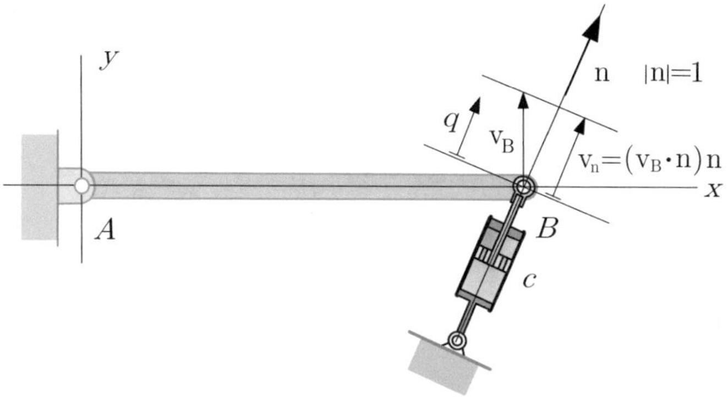

Fig. 4.6 Rod supported by a vibration damper

Damping forces. Figure 4.6 shows the rod AB which is now supported by a vibration damper at the right end—viscous damping is assumed. The unit vector $\mathbf{n}$ lies on the axis of the vibration damper. On the base of Eq. (4.20) we can write for the generalized damping force that

$$Q_c = \mathbf{F}_c \cdot \frac{\partial \mathbf{v}_B}{\partial \dot{q}}, \quad (4.37)$$

where $\mathbf{F}_c$ is the damping force, $\mathbf{v}_B$ is the velocity of the point B, and q is the generalized coordinate (the displacement component at B parallel to $\mathbf{n}$). The damping force is given by the equation

$$\mathbf{F}_c = -c\mathbf{v}_n \quad (4.38)$$

in which c is the damping factor and $\mathbf{v}_n$ is that velocity component of $\mathbf{v}_B$ which is parallel to $\mathbf{n}$. It is not too difficult to check if we utilize Eq. (4.19) that

$$\mathbf{v}_n = (\mathbf{v}_B \cdot \mathbf{n})\,\mathbf{n} = \Big[\underbrace{\frac{\partial \mathbf{r}_B}{\partial q}\dot{q}}_{\mathbf{v}_B} \cdot \mathbf{n}\Big]\mathbf{n} = \underset{\frac{\partial \mathbf{r}_B}{\partial q} = \frac{\partial \mathbf{v}_B}{\partial \dot{q}}}{\uparrow} = \Big[\frac{\partial \mathbf{v}_B}{\partial \dot{q}}\dot{q} \cdot \mathbf{n}\Big]\mathbf{n}\,. \quad (4.39)$$

Substituting first (4.38) and then (4.39) into (4.37), we get

$$Q_c = -c\,\mathbf{v}_n \cdot \frac{\partial \mathbf{v}_B}{\partial \dot{q}} = -c\left[\frac{\partial \mathbf{v}_B}{\partial \dot{q}}\dot{q}\cdot\mathbf{n}\right]\frac{\partial \mathbf{v}_B}{\partial \dot{q}}\cdot\mathbf{n} = \underbrace{-c\left[\frac{\partial \mathbf{v}_B}{\partial \dot{q}}\cdot\mathbf{n}\right]^2}_{c_g}\dot{q} = -c_g\dot{q}, \tag{4.40}$$

where

$$c_g = c\left[\frac{\partial \mathbf{v}_B}{\partial \dot{q}}\cdot\mathbf{n}\right]^2 \tag{4.41}$$

is the generalized damping coefficient.

Generalized excitation forces. For a single degree of freedom system subjected to only one excitation force

$$Q_f = \mathbf{F}_f \cdot \frac{\partial \mathbf{v}}{\partial \dot{q}} \tag{4.42a}$$

is the formula for the generalized excitation force in which $\mathbf{F}_f$ is the excitation force and $\mathbf{v}$ is the velocity of its point of application.

For a multidegree of freedom system

$$Q_{f\ell} = \sum_{j=1}^{N}\mathbf{F}_{fj}\cdot\frac{\partial \mathbf{v}_j}{\partial \dot{q}_\ell}, \qquad \ell = 1,\ldots,n \tag{4.42b}$$

is the ℓth generalized excitation force where $\mathbf{F}_{fj}$ is the force of excitation exerted on the jth particle (or body), $\mathbf{v}_j$ is the velocity of the point where the force $\mathbf{F}_{fj}$ is applied.

Remark 4.1 For a single degree of freedom system

$$\frac{\mathrm{d}}{\mathrm{d}t}\frac{\partial \mathcal{E}}{\partial \dot{q}} - \frac{\partial \mathcal{E}}{\partial q} = Q_k + Q_c + Q_f\,, \qquad Q_k = -\frac{\partial U}{\partial q} \tag{4.43}$$

is the equation of motion where $\mathcal{E}$ is the total kinetic energy, Q_k, Q_c, and Q_f are the generalized spring force, damping force, and excitation force. Since the spring force has, in general, a potential U it can be given as the negative of the derivative $\partial U/\partial q$.

Exercise 4.4 Figure 4.7 shows a single degree of freedom system. The solid disk of mass m and radius R is hinged at A. The vibration damper with a damping coefficient c is attached to the mass center. The motion of the right end of the spring (the spring constant is k), which is attached to the top of the disk, is prescribed. Derive the equation of motion by choosing ϑ for the generalized coordinate q and determine the circular frequency of the undamped free vibrations. Assume that ϑ is small.

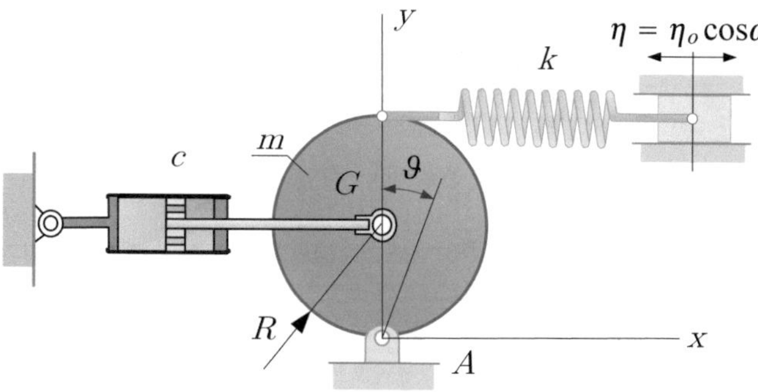

Fig. 4.7 Displacement excitation on a spring mass system

The kinetic energy of the disk is given by

$$\mathcal{E} = \frac{1}{2} J_A \dot{\vartheta}^2 , \qquad J_A = \frac{3}{2} m R^2 . \tag{4.44}$$

Using the length change $\eta_o \cos\omega_f t - 2R\vartheta$ of the spring we get the potential energy in the form

$$U = \frac{1}{2} k \left(\eta_o \cos\omega_f t - 2R\vartheta \right)^2 . \tag{4.45}$$

The corresponding generalized force is

$$-\frac{\partial U}{\partial \vartheta} = -k \left(\eta_o \cos\omega_f t - 2R\vartheta \right) (-2R) = \underbrace{2Rk\eta_o}_{Q_{fo}} \cos\omega_f t - \underbrace{4R^2 k}_{k_g} \vartheta =$$

$$= \underbrace{Q_{fo} \cos\omega_f t}_{Q_f} + \underbrace{k_g \vartheta}_{Q_k} , \tag{4.46}$$

where Q_{fo} and k_g are the amplitudes of the generalized excitation force and the generalized spring constant while Q_f and Q_k are the generalized excitation and spring forces. The generalized damping force is obtained from (4.40) by taking into account that $\mathbf{v}_G = \dot{\vartheta} R \mathbf{i}_x$, $\mathbf{n} = \mathbf{i}_x$. We get

$$Q_c = -c \left[\frac{\partial \mathbf{v}_G}{\partial \dot{\vartheta}} \cdot \mathbf{i}_x \right]^2 \dot{\vartheta} = -\underbrace{cR^2}_{c_g} \dot{\vartheta} = -c_g \dot{q} \tag{4.47}$$

in which c_g is the generalized damping coefficient. Upon substitution of (4.44), (4.46), and (4.47) into (4.43) the equation of motion is obtained

$$\underbrace{J_A}_{m_g}\ \ddot{\vartheta} - 0 = \underbrace{2Rk\eta_o}_{Q_{fo}} \cos\omega_f t - \underbrace{4R^2k}_{k_g}\,\vartheta - \underbrace{cR^2}_{c_g}\,\dot{\vartheta} \tag{4.48a}$$

or

$$m_g\,\ddot{\vartheta} + c_g\dot{\vartheta} + k_g\,\vartheta = Q_{fo}\,\cos\omega_f t, \tag{4.48b}$$

where m_g is the generalized mass. For the circular frequency of the undamped free vibrations, Eqs. (3.2)$_2$ and (4.48b) yield

$$\omega_n = \sqrt{\frac{k_g}{m_g}} = \sqrt{\frac{4R^2k}{\frac{3}{2}mR^2}} = \sqrt{\frac{8k}{3m}}\,. \tag{4.49}$$

Exercise 4.5 Figure 4.8 depicts a simple model of a gear box (a geared torsional system). When investigating the torsional vibrations of the shafts we apply the following assumptions: (a) shafts 1 and 2 are elastic. They behave if they were linear volute springs for which

$$k_{1v} = \frac{I_{p1}G_1}{\ell_1}, \qquad k_{2v} = \frac{I_{p2}G_2}{\ell_2} \tag{4.50}$$

are the spring constants where I_{pi}, G_i, and ℓ_i $(i = 1, 2)$ are the polar moments of inertia, the shear moduli, and the lengths of the shafts. It will be assumed the masses of the shafts are much less than those of the gears. Therefore, their effect on the kinetic energy will be neglected. (b) Gears 1, 2′, 2″, and 3 are rigid, their moments of inertia with respect their axes of symmetry are denoted by J_1, J_2', J_2'' and J_3. Let $\vartheta_1, \vartheta_1', \vartheta_1''$, and ϑ_3 be the angles of rotation of the gears. Find the equations of motion provided that a harmonic torsional moment $M_f = M_{fo}\cos\omega_f t$ is acting on gear 3.

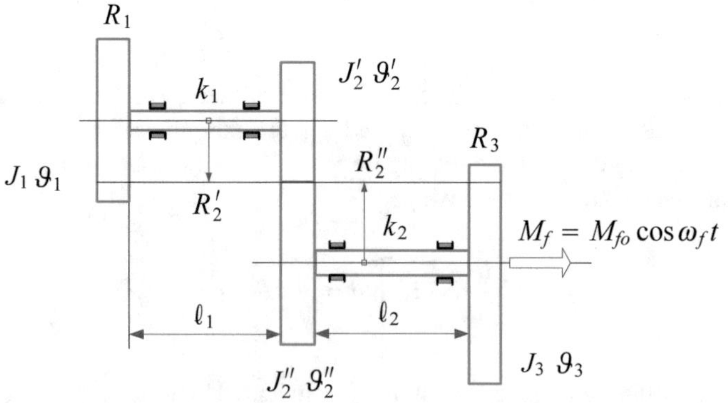

Fig. 4.8 A simple model for a gear box

The generalized coordinates q_1, q_2, and q_3 are defined by the following relations:

$$q_1 = R_2'\vartheta_1\,, \qquad q_2 = R_2'\vartheta_2' = R_2''\vartheta_2''\,, \qquad q_3 = R_2''\vartheta_3\,. \tag{4.51}$$

Making use of the angular velocities

$$\dot{\vartheta}_1 = \frac{\dot{q}_1}{R_2'}\,, \qquad \dot{\vartheta}_2' = \frac{\dot{q}_2}{R_2'}\,, \qquad \dot{\vartheta}_2'' = \frac{\dot{q}_2}{R_2''}\,, \qquad \dot{\vartheta}_3 = \frac{\dot{q}_3}{R_2''} \tag{4.52}$$

it is easy to calculate the total kinetic energy of the system:

$$\begin{aligned}\mathcal{E} &= \frac{1}{2}J_1\left(\dot{\vartheta}_1\right)^2 + \frac{1}{2}J_2'\left(\dot{\vartheta}_2'\right)^2 + \frac{1}{2}J_2''\left(\dot{\vartheta}_2''\right)^2 + \frac{1}{2}J_3\left(\dot{\vartheta}_3\right)^2 = \\ &= \frac{1}{2}\Bigg[\underbrace{\frac{J_1}{\left(R_2'\right)^2}}_{m_1}\dot{q}_1^2 + \underbrace{\left(\frac{J_2'}{\left(R_2'\right)^2} + \frac{J_2''}{\left(R_2''\right)^2}\right)}_{m_2}\dot{q}_2^2 + \underbrace{\frac{J_3}{\left(R_2''\right)^2}}_{m_3}\dot{q}_3^2\Bigg]\end{aligned} \tag{4.53a}$$

in which

$$m_1 = \frac{J_1}{\left(R_2'\right)^2}\,, \qquad m_2 = \frac{J_2'}{\left(R_2'\right)^2} + \frac{J_2''}{\left(R_2''\right)^2}\,, \qquad m_3 = \frac{J_3}{\left(R_2''\right)^2} \tag{4.53b}$$

are the generalized masses. Hence,

$$\mathcal{E} = \frac{1}{2}\left(m_1\,\dot{q}_1^2 + m_2\,\dot{q}_2^2 + m_3\,\dot{q}_3^2\right) \tag{4.53c}$$

is the total kinetic energy.

The potential energy (strain energy) stored in the shafts is given by

$$\begin{aligned}U &= \frac{1}{2}k_{1v}\left(\vartheta_2' - \vartheta_1\right)^2 + \frac{1}{2}k_{2v}\left(\vartheta_3 - \vartheta_2''\right)^2 = \\ &= \frac{1}{2}\underbrace{\frac{k_{1v}}{\left(R_2'\right)^2}}_{k_1}\left(q_2 - q_1\right)^2 + \frac{1}{2}\underbrace{\frac{k_{2v}}{\left(R_2''\right)^2}}_{k_2}\left(q_3 - q_2\right)^2\,,\end{aligned} \tag{4.54a}$$

where

$$k_1 = \frac{k_{1v}}{\left(R_2'\right)^2} \quad \text{and} \quad k_2 = \frac{k_{2v}}{\left(R_2''\right)^2} \tag{4.54b}$$

are generalized spring constants. Consequently,

$$U = \frac{1}{2}k_1\left(q_2 - q_1\right)^2 + \frac{1}{2}k_2\left(q_3 - q_2\right)^2\,. \tag{4.54c}$$

If we utilize Eq. (4.24) we get the generalized excitation force

$$
\begin{aligned}
Q_{f3}=Q_f=M_f\frac{\partial\omega_3}{\partial\dot{q}_3}=M_f\frac{\partial\dot{\vartheta}_3}{\partial\dot{q}_3}&=\frac{M_f}{R_2''}\frac{\partial\dot{q}_3}{\partial\dot{q}_3}=\\
&=\underbrace{\frac{M_{fo}}{R_2''}}_{Q_{fo}}\cos\omega_f t=Q_{fo}\cos\omega_f t\,.
\end{aligned}
\tag{4.55}
$$

Substituting now Eqs. (4.53c), (4.54c), and (4.55) into (4.23) by taking relation (4.27) into account, we get

$$
\begin{aligned}
&\frac{\mathrm{d}}{\mathrm{d}t}\frac{\partial\mathcal{E}}{\partial\dot{q}_1}-\frac{\partial\mathcal{E}}{\partial q_1}=-\frac{\partial U}{\partial q_1} &&\rightarrow m_1\ddot{q}_1=k_1\,(q_2-q_1)\,,\\
&\frac{\mathrm{d}}{\mathrm{d}t}\frac{\partial\mathcal{E}}{\partial\dot{q}_2}-\frac{\partial\mathcal{E}}{\partial q_2}=-\frac{\partial U}{\partial q_2} &&\rightarrow m_2\ddot{q}_2=-k_1\,(q_2-q_1)+k_2\,(q_3-q_2)\,,\\
&\frac{\mathrm{d}}{\mathrm{d}t}\frac{\partial\mathcal{E}}{\partial\dot{q}_3}-\frac{\partial\mathcal{E}}{\partial q_3}=-\frac{\partial U}{\partial q_3}+Q_f &&\rightarrow m_3\ddot{q}_3=-k_2\,(q_3-q_2)+Q_{fo}\cos\omega_f t
\end{aligned}
\tag{4.56a}
$$

or

$$
\begin{aligned}
m_1\ddot{q}_1+k_1\,(q_1-q_2)&=0\,,\\
m_2\ddot{q}_2-k_1\,(q_1-q_2)+k_2\,(q_2-q_3)&=0\,,\\
m_3\ddot{q}_3-k_2\,(q_2-q_3)&=Q_{fo}\cos\omega_f t\,.
\end{aligned}
\tag{4.56b}
$$

Equations (4.56b) can be rewritten in matrix form

$$
\begin{aligned}
&\underbrace{\begin{bmatrix} m_1 & 0 & 0\\ 0 & m_2 & 0\\ 0 & 0 & m_3\end{bmatrix}}_{\underset{(3\times3)}{\underline{\mathbf{M}}}}
\underbrace{\begin{bmatrix}\ddot{q}_1\\ \ddot{q}_2\\ \ddot{q}_3\end{bmatrix}}_{\underset{(3\times1)}{\underline{\ddot{\mathbf{q}}}}}
+\underbrace{\begin{bmatrix} k_1 & -k_1 & 0\\ -k_1 & k_1+k_2 & -k_2\\ 0 & -k_2 & k_2\end{bmatrix}}_{\underset{(3\times3)}{\underline{\mathbf{K}}}}
\underbrace{\begin{bmatrix} q_1\\ q_2\\ q_3\end{bmatrix}}_{\underset{(3\times1)}{\underline{\mathbf{q}}}}=\\
&\qquad=\underbrace{\begin{bmatrix} 0\\ 0\\ Q_{fo}\cos\omega_f t\end{bmatrix}}_{\underset{(3\times1)}{\underline{\mathbf{Q}}}}
\end{aligned}
\tag{4.56c}
$$

in which

$$
\underset{(3\times3)}{\underline{\mathbf{M}}},\quad \underset{(3\times3)}{\underline{\mathbf{K}}},\quad \underset{(3\times1)}{\underline{\mathbf{q}}}\quad \text{and}\quad \underset{(3\times1)}{\underline{\mathbf{Q}}}
$$

are the mass and stiffness matrices, the generalized displacement matrix, and the loading matrix.

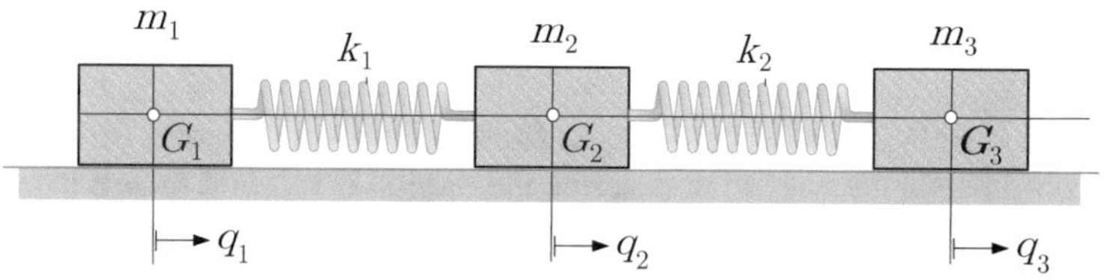

Fig. 4.9 Three degree of freedom spring-mass system

It is not too difficult to check that the equations of motion (4.56) are those of a spring–mass system shown in Fig. 4.9.

4.3 Multidegree of Freedom Systems with Solutions

4.3.1 Examples for Multidegree of Freedom Systems

Figure 4.10 shows a few multidegree of freedom systems: (a) a coupled pendulum, (b) a branched geared system, (c) a belt drive (pulley *A* drives pulley *B* via a belt), and (d) an unbranched geared system (in this respect we remind the reader of Exercise 4.5). The examples shown in Fig 4.10 represent that the dynamical behavior of some machine parts can be understood only if the model established for the investigation has more than one degree of freedom: it is not too difficult to check that (a) the coupled pendulum has two, (b) the branched geared system has four, (c) the belt drive has again two, and finally the unbranched geared system has three degrees of freedom. We shall consider systems with two degrees of freedom first.

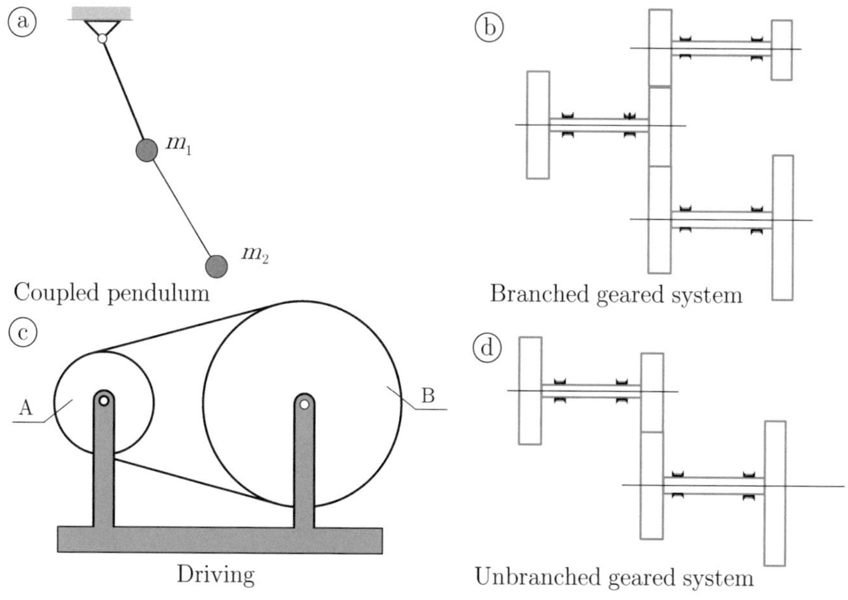

Fig. 4.10 Multidegree of freedom systems

4.3.2 Two Degree of Freedom Systems

Coordinate coupling. A vibrating system for which two coordinates (or two generalized coordinates) are needed to describe the motion is referred to as a two degree of freedom system. The equations of motion for two degree of freedom system are, in general, coupled. This means that both generalized coordinates appear in each equation. In the most general case, the equations of motion for an undamped free system assume the following form:

$$\underbrace{\begin{bmatrix} m_{11} & m_{12} \\ m_{21} & m_{22} \end{bmatrix}}_{\underset{(2\times 2)}{\underline{\mathbf{M}}}} \underbrace{\begin{bmatrix} \ddot{q}_1 \\ \ddot{q}_2 \end{bmatrix}}_{\underset{(2\times 1)}{\underline{\ddot{\mathbf{q}}}}} + \underbrace{\begin{bmatrix} k_{11} & k_{12} \\ k_{21} & k_{22} \end{bmatrix}}_{\underset{(2\times 2)}{\underline{\mathbf{K}}}} \underbrace{\begin{bmatrix} q_1 \\ q_2 \end{bmatrix}}_{\underset{(2\times 1)}{\underline{\mathbf{q}}}} = \begin{bmatrix} 0 \\ 0 \end{bmatrix}, \tag{4.57}$$

where $\underline{\mathbf{M}}$ is the mass matrix and $\underline{\mathbf{K}}$ is the stiffness matrix. These matrices are symmetric, i.e., $\underline{\mathbf{M}} = \underline{\mathbf{M}}^T$ and $\underline{\mathbf{K}} = \underline{\mathbf{K}}^T$. We speak about (a) mass or dynamical coupling if the mass matrix is nondiagonal and (b) stiffness or static coupling if the stiffness matrix is nondiagonal. For undamped systems, it is always possible to find such generalized coordinates for which the mass and stiffness matrices are both diagonal. Then the equations of motion are decoupled and can be solved independently of each other. The coordinates in this coordinate system are called principal or normal coordinates.

Exercise 4.6 Figure 4.11 shows a lathe machine[1] on its foundation which is modeled as a rigid bar with a mass center not coinciding with the geometrical center of the bar. Investigate how to choose the coordinates so that the equations of motion are statically coupled, dynamically coupled, and both statically and dynamically coupled.

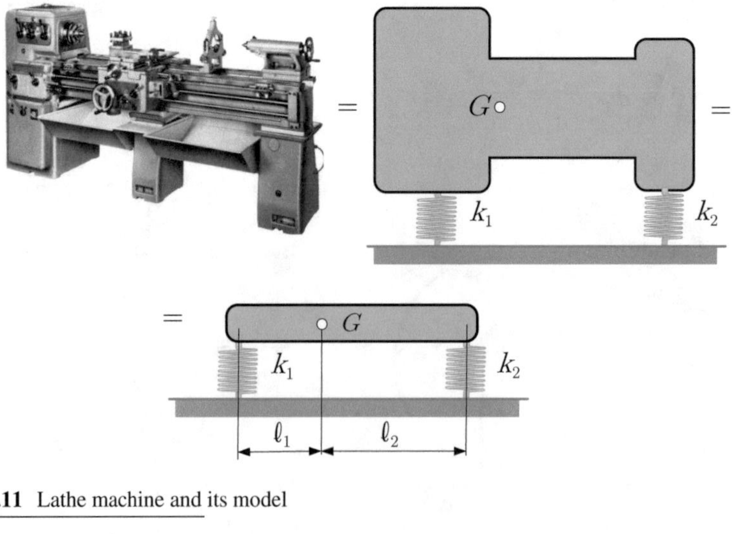

Fig. 4.11 Lathe machine and its model

[1]Photo courtesy of Tony Griffiths of lathes.co.uk

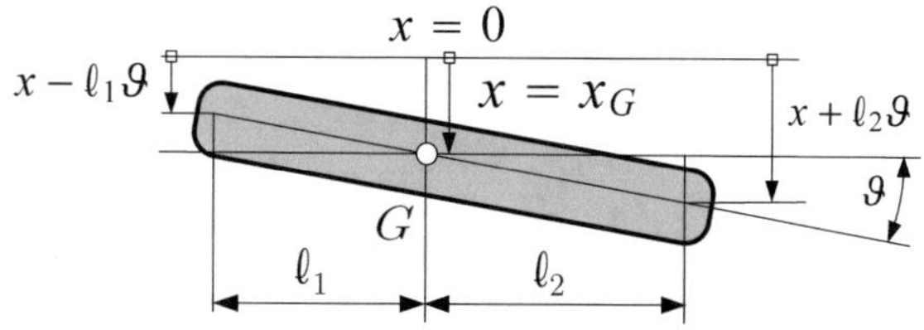

Fig. 4.12 Data for static coupling

Static coupling. Let x and ϑ be the two generalized coordinates. Making use of Eq. (1.51a) and the data of Fig. 4.12 we get the kinetic energy of the bar and the strain energy stored in the two springs:

$$\mathcal{E} = \frac{1}{2} m \underbrace{v_G^2}_{\dot{x}_G = \dot{x}} + \frac{1}{2} J_G \dot{\vartheta}^2 , \qquad U = \frac{1}{2} k_1 (x - \ell_1 \vartheta)^2 + \frac{1}{2} k_2 (x + \ell_2 \vartheta)^2 ,$$

where m and J_G are the mass of the bar and the moment of inertia with respect to the mass center. Substituting these quantities into Lagrange's equation (4.28), we obtain

$$\frac{\mathrm{d}}{\mathrm{d}t} \frac{\partial \mathcal{E}}{\partial \dot{x}} - \frac{\partial \mathcal{E}}{\partial x} + \frac{\partial U}{\partial x} = m \ddot{x} + k_1 (x - \ell_1 \vartheta) + k_2 (x + \ell_2 \vartheta) = 0 ,$$

$$\frac{\mathrm{d}}{\mathrm{d}t} \frac{\partial \mathcal{E}}{\partial \dot{\vartheta}} - \frac{\partial \mathcal{E}}{\partial \vartheta} + \frac{\partial U}{\partial \vartheta} = J_G \ddot{\vartheta} - \ell_1 k_1 (x - \ell_1 \vartheta) + \ell_2 k_2 (x + \ell_2 \vartheta) = 0 .$$

These equations can be rewritten into a matrix form

$$\begin{bmatrix} m & 0 \\ 0 & J_G \end{bmatrix} \begin{bmatrix} \ddot{x} \\ \ddot{\vartheta} \end{bmatrix} + \begin{bmatrix} k_1 + k_2 & \ell_2 k_2 - \ell_1 k_1 \\ \ell_2 k_2 - \ell_1 k_1 & \ell_1^2 k_1 + \ell_2^2 k_2 \end{bmatrix} \begin{bmatrix} x \\ \vartheta \end{bmatrix} = \begin{bmatrix} 0 \\ 0 \end{bmatrix}$$

from where it is clear that the mass matrix is a diagonal matrix. The system is, therefore, dynamically uncoupled but statically coupled.

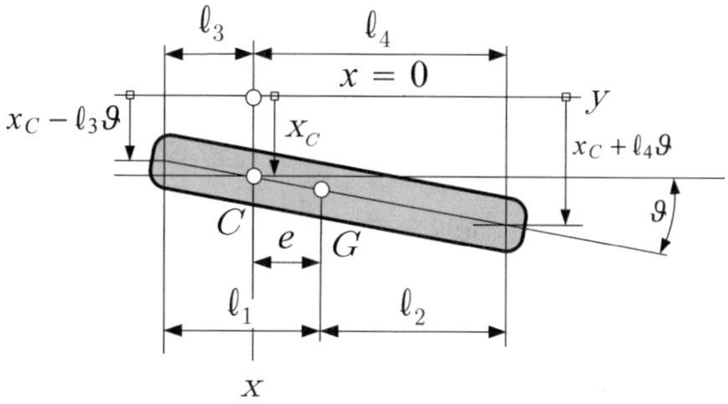

Fig. 4.13 Data for dynamic coupling

Dynamic coupling. There exists such a point C on the geometric centerline of the bar (Fig. 4.13) where a vertical force exerted on the bar yields pure vertical translation, the bar does not rotate. Then the two spring forces acting on the bar have no moment about the point C. This condition results in the following equation:

$$\underbrace{k_1 x_C}_{\text{spring force}}\ \ell_3 = \underbrace{k_2 x_C}_{\text{spring force}}\ \ell_4 \,.$$

Consequently,

$$\ell_3 \, k_1 = \ell_4 \, k_2 .$$

As regards the kinetic energy of the bar it is worth recalling Eq. (1.51a) again in which

$$v_G = \dot{x}_C + e\dot{\vartheta}$$

hence

$$\begin{aligned}
\mathcal{E}\frac{1}{2} m\, v_G^2 + \frac{1}{2} J_G\, \dot{\vartheta}^2 &= \frac{1}{2} m\left(\dot{x}_C + e\dot{\vartheta}\right)^2 + \frac{1}{2} J_G\, \dot{\vartheta}^2 = \\
&= \frac{1}{2}\left(m\dot{x}_C^2 + 2me\dot{x}_C\dot{\vartheta}\right) + \frac{1}{2}\underbrace{\left(J_G + me^2\right)}_{J_C}\dot{\vartheta}^2 = \\
&= \frac{1}{2}\left(m\dot{x}_C^2 + 2me\dot{x}_C\dot{\vartheta}\right) + \frac{1}{2} J_C\, \dot{\vartheta}^2 .
\end{aligned}$$

The strain energy stored in the springs is

$$U = \frac{1}{2} k_1 \left(x_C - \ell_3 \vartheta\right)^2 + \frac{1}{2} k_2 \left(x_C + \ell_4 \vartheta\right)^2 .$$

With $\mathcal{E}$ and U, Lagrange's equations (4.28) yield

$$\begin{aligned}
&\frac{\mathrm{d}}{\mathrm{d}t}\frac{\partial \mathcal{E}}{\partial \dot{x}_C} - \frac{\partial \mathcal{E}}{\partial x_C} + \frac{\partial U}{\partial x_C} = \\
&\qquad = m\,\ddot{x}_C + me\ddot{\vartheta} + k_1 x_C - \underline{\ell_3 k_1 \vartheta} + k_2 x_C + \underline{\ell_4 k_2 \vartheta} = 0\,, \\
&\frac{\mathrm{d}}{\mathrm{d}t}\frac{\partial \mathcal{E}}{\partial \dot{\vartheta}} - \frac{\partial \mathcal{E}}{\partial \vartheta} + \frac{\partial U}{\partial \vartheta} = \\
&\qquad = J_C\ddot{\vartheta} + me\ddot{x}_C - \underline{\ell_3 k_1 x_C} + \ell_3^2 k_1 \vartheta + \underline{\ell_4 k_2 x_C} + \ell_2^2 k_2 \vartheta = 0\,.
\end{aligned}$$

By taking into account that the terms underlined cancel each other, we get the matrix equation

$$\begin{bmatrix} m & me \\ me & J_C \end{bmatrix}\begin{bmatrix} \ddot{x}_C \\ \ddot{\vartheta} \end{bmatrix} + \begin{bmatrix} k_1 + k_2 & 0 \\ 0 & \ell_3^2 k_1 + \ell_4^2 k_2 \end{bmatrix}\begin{bmatrix} x_C \\ \vartheta \end{bmatrix} = \begin{bmatrix} 0 \\ 0 \end{bmatrix},$$

which shows that the system is now dynamically coupled but statically uncoupled.

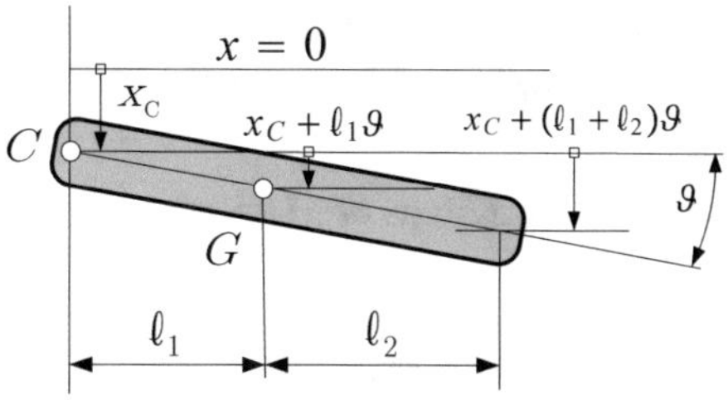

Fig. 4.14 Data for static and dynamic coupling

Static and dynamic coupling. Let now the point C be located at the left end of the bar. It is not too difficult to check using the data of Fig. 4.14 that

$$\begin{aligned}\mathcal{E} &= \frac{1}{2}m\,v_G^2 + \frac{1}{2}J_G\,\dot{\vartheta}^2 = \frac{1}{2}m\left(\dot{x}_C + \ell_1\dot{\vartheta}\right)^2 + \frac{1}{2}J_G\,\dot{\vartheta}^2 = \\ &= \frac{1}{2}\left(m\dot{x}_C^2 + 2m\ell_1\dot{x}_C\dot{\vartheta}\right) + \frac{1}{2}\underbrace{\left(J_G + m\ell_1^2\right)}_{J_C}\dot{\vartheta}^2 = \\ &= \frac{1}{2}\left(m\dot{x}_C^2 + 2m\ell_1\dot{x}_C\dot{\vartheta}\right) + \frac{1}{2}J_C\,\dot{\vartheta}^2\end{aligned}$$

and

$$U = \frac{1}{2}k_1x_C^2 + \frac{1}{2}k_2\left(x_C + \ell\vartheta\right)^2\,, \qquad \ell = \ell_1 + \ell_2$$

by the use of which, we get

$$\begin{aligned}&\frac{\mathrm{d}}{\mathrm{d}t}\frac{\partial\mathcal{E}}{\partial\dot{x}_C} - \frac{\partial\mathcal{E}}{\partial x_C} + \frac{\partial U}{\partial x_C} = \\ &\qquad = m\,\ddot{x}_C + m\ell_1\ddot{\vartheta} + k_1\left(x_C + \ell_1\vartheta\right) + k_2\left(x_C + \ell\vartheta\right) = 0\,, \\ &\frac{\mathrm{d}}{\mathrm{d}t}\frac{\partial\mathcal{E}}{\partial\dot{\vartheta}} - \frac{\partial\mathcal{E}}{\partial\vartheta} + \frac{\partial U}{\partial\vartheta} = \\ &\qquad = J_C\ddot{\vartheta} + m\ell_1\ddot{x}_C + \ell_1k_1\left(x_C + \ell_1\vartheta\right) + \ell k_2\left(x_C + \ell\vartheta\right) = 0\end{aligned}$$

or in matrix form

$$\begin{bmatrix} m & m\ell_1 \\ m\ell_1 & J_C \end{bmatrix}\begin{bmatrix} \ddot{x}_C \\ \ddot{\vartheta} \end{bmatrix} + \begin{bmatrix} k_1 + k_2 & \ell_1k_1 + \ell k_2 \\ \ell_1k_1 + \ell k_2 & \ell_1^2k_1 + \ell^2k_2 \end{bmatrix}\begin{bmatrix} x_C \\ \vartheta \end{bmatrix} = \begin{bmatrix} 0 \\ 0 \end{bmatrix}.$$

This system is a dynamically and statically coupled one.

4.3.3 *Free Vibrations of Two Degree of Freedom Spring–Mass Systems*

4.3.3.1 Equations of Motion

Figure 4.15 shows the typical spring–mass systems with two degrees of freedom. System A is free at its ends, System B is fixed at the left end while System C is fixed at both ends. One can check with ease by using Lagrange's equations that the equations of motion are

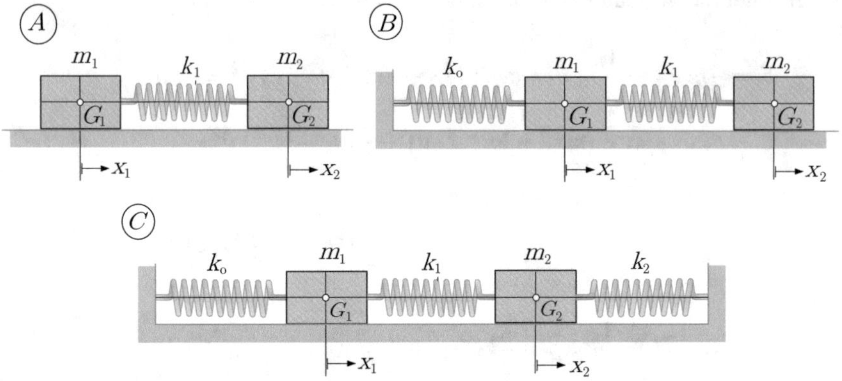

Fig. 4.15 Spring-mass systems of two degrees of freedom

A.

$$\begin{bmatrix} m_1 & 0 \\ 0 & m_2 \end{bmatrix} \begin{bmatrix} \ddot{x}_1 \\ \ddot{x}_2 \end{bmatrix} + \begin{bmatrix} k_1 & -k_1 \\ -k_1 & k_1 \end{bmatrix} \begin{bmatrix} x_1 \\ x_2 \end{bmatrix} = \begin{bmatrix} 0 \\ 0 \end{bmatrix}, \tag{4.58a}$$

B

$$\begin{bmatrix} m_1 & 0 \\ 0 & m_2 \end{bmatrix} \begin{bmatrix} \ddot{x}_1 \\ \ddot{x}_2 \end{bmatrix} + \begin{bmatrix} k_o + k_1 & -k_1 \\ -k_1 & k_1 \end{bmatrix} \begin{bmatrix} x_1 \\ x_2 \end{bmatrix} = \begin{bmatrix} 0 \\ 0 \end{bmatrix}, \tag{4.58b}$$

C.

$$\underbrace{\begin{bmatrix} m_1 & 0 \\ 0 & m_2 \end{bmatrix}}_{\underset{(2\times 2)}{\underline{\mathbf{M}}}} \underbrace{\begin{bmatrix} \ddot{x}_1 \\ \ddot{x}_2 \end{bmatrix}}_{\underset{(2\times 1)}{\ddot{\underline{\mathbf{q}}}}} + \underbrace{\begin{bmatrix} k_o + k_1 & -k_1 \\ -k_1 & k_1 + k_2 \end{bmatrix}}_{\underset{(2\times 2)}{\underline{\mathbf{K}}}} \underbrace{\begin{bmatrix} x_1 \\ x_2 \end{bmatrix}}_{\underset{(2\times 1)}{\underline{\mathbf{q}}}} = \underbrace{\begin{bmatrix} 0 \\ 0 \end{bmatrix}}_{\underset{(2\times 1)}{\underline{\mathbf{0}}}}, \tag{4.58c}$$

which can be given in terms of the mass matrix $\underline{\mathbf{M}}$, stiffness matrix $\underline{\mathbf{K}}$, and the matrix $\underline{\mathbf{q}}$ of generalized displacements:

$$\underset{(2\times2)}{\underline{\mathbf{M}}}\ \underset{(2\times1)}{\ddot{\underline{\mathbf{q}}}} + \underset{(2\times2)}{\underline{\mathbf{K}}}\ \underset{(2\times1)}{\underline{\mathbf{q}}} = \underset{(2\times1)}{\underline{\mathbf{0}}}\,. \tag{4.59}$$

4.3.3.2 General Solution

Solution of differential equation (4.59) is sought in the form

$$\underset{(2\times1)}{\underline{\mathbf{q}}} = \underset{(2\times1)}{\underline{\mathbf{A}}} \cos\omega_n t, \tag{4.60}$$

where the matrix $\underline{\mathbf{A}}^T = [A_1|A_2]$ contains the amplitudes for harmonic motion of the two masses. This motion is called normal mode oscillation. After substituting solution (4.60) into the differential equation (4.59) and taking into account that

$$\dot{\underline{\mathbf{q}}} = -\omega_n \underline{\mathbf{A}} \sin\omega_n t\,, \qquad \ddot{\underline{\mathbf{q}}} = -\omega_n^2 \underline{\mathbf{A}} \cos\omega_n t = -\omega_n^2 \underline{\mathbf{q}}\,, \tag{4.61a}$$

we get

$$\left(\underline{\mathbf{K}} - \omega_n^2 \underline{\mathbf{M}}\right) \underline{\mathbf{A}} \cos\omega_n t = \underline{\mathbf{0}} \quad \text{or} \quad \Big(\underset{(2\times2)}{\underline{\mathbf{K}}} - \omega_n^2 \underset{(2\times2)}{\underline{\mathbf{M}}} \Big) \underset{(2\times1)}{\underline{\mathbf{A}}} = \underset{(2\times1)}{\underline{\mathbf{0}}}\,, \tag{4.61b}$$

which is a homogeneous linear equation system with the unknowns A_1 and A_2. Solution for $\underline{\mathbf{A}}$ which is different from the trivial one exists if and only if the determinant of the coefficient matrix vanishes

$$\det\left(\underline{\mathbf{K}} - \omega_n^2 \underline{\mathbf{M}}\right) = \underline{0}\,. \tag{4.62}$$

Introduce the notation $\omega_n^2 = \lambda$ and then expand the determinant. In this manner, three characteristic equations are obtained:

A.

$$\begin{vmatrix} k_1 - m_1\lambda & -k_1 \\ -k_1 & k_1 - m_2\lambda \end{vmatrix} = \lambda\left[m_1 m_2 \lambda - k_1\left(m_1 + m_2\right)\right] = 0 \tag{4.63a}$$

or

$$m_1 \frac{1}{k_1} m_2 \lambda - \left(m_1 + m_2\right) = 0\,, \tag{4.63b}$$

B.

$$\begin{vmatrix} k_o + k_1 - m_1\lambda & -k_1 \\ -k_1 & k_1 - m_2\lambda \end{vmatrix} =$$

$$= m_1 m_2 \lambda^2 - \left[k_1\left(m_1 + m_2\right) + k_o m_2\right] + k_o k_1 = 0 \tag{4.64a}$$

or

$$\frac{1}{k_o} m_1 \frac{1}{k_1} m_2 \lambda^2 - \left[k_1\left(m_1 + m_2\right) + k_o m_2\right] \lambda + 1 = 0\,, \tag{4.64b}$$

C.

$$\begin{vmatrix} k_o + k_1 - m_1\lambda & -k_1 \\ -k_1 & k_1 + k_2 - m_2\lambda \end{vmatrix} =$$
$$= m_1 m_2 \lambda^2 - [(k_o + k_1)\, m_2 + (k_1 + k_2)\, m_1] + (k_o + k_1)\,(k_1 + k_2) - k_1^2 = 0\,. \tag{4.65}$$

4.3.3.3 Solutions for System *A*

Characteristic equation (4.63a) has the following solutions for λ:

$$\lambda_o = 0\,, \qquad \lambda_1 = \frac{m_1 + m_2}{m_1 \frac{1}{k_1} m_2}\,. \tag{4.66a}$$

If $\lambda = \lambda_o = 0$ then $\omega_n = 0$. Consequently,

$$\ddot{\underline{\mathbf{q}}} = -\omega_n^2 \underline{\mathbf{q}} = \underline{\mathbf{0}} \tag{4.66b}$$

from where we have

$$\underset{(2\times1)}{\underline{\mathbf{q}}} = \underset{(2\times1)}{\underline{\mathbf{C}}_1} + \underset{(2\times1)}{\underline{\mathbf{C}}_2}\, t \tag{4.66c}$$

in which $\underline{\mathbf{C}}_1$ and $\underline{\mathbf{C}}_2$ are undetermined integration constant matrices. The motion described by Eq. (4.66c) is the well-known rigid body motion. Since neither the left end nor the right end of system *A* is fixed it can really perform a translational motion (a non-vibratory motion).

If $\lambda = \lambda_1$ equation system (4.61b) takes the form

$$\begin{bmatrix} k_1 - \lambda_1 m_1 & -k_1 \\ -k_1 & k_1 - \lambda_1 m_2 \end{bmatrix} \begin{bmatrix} A_{11} \\ A_{21} \end{bmatrix} = \begin{bmatrix} 0 \\ 0 \end{bmatrix} \tag{4.67a}$$

or

$$\begin{aligned} \left(k_1 - k_1 \frac{m_1 + m_2}{m_2}\right) A_{11} - k_1 A_{21} = 0, \\ -k_1 A_{11} + \left(k_1 - k_1 \frac{m_1 + m_2}{m_1}\right) A_{21} = 0, \end{aligned} \tag{4.67b}$$

where the second subscript, which is here 1, shows that $\underline{\mathbf{A}} = \underline{\mathbf{A}}_1$ belongs to λ_1. If we multiply Eq. $(4.67b)_1$ by m_2 and Eq. $(4.67b)_2$ by m_1 we arrive at the same result:

$$-k_1 m_1 A_{11} - k_1 m_2 A_{21} = 0$$

from where it follows that

$$\frac{A_{21}}{A_{11}} = -\frac{m_1}{m_2} \quad \text{or} \quad A_{21}m_2 + A_{11}m_1 = 0\,. \tag{4.68}$$

Consequently, Eq. (4.67b) is not independent. Comparison of (4.60) and (4.68) yields the solution

$$\underline{\mathbf{q}} = \begin{bmatrix} A_{11} \\ A_{21} \end{bmatrix} \cos \omega_{n1} t = \mathcal{A} \begin{bmatrix} m_2 \\ -m_1 \end{bmatrix} \cos \omega_{n1} t\,, \quad \omega_{n1} = \sqrt{\lambda_1} \tag{4.69}$$

in which $\mathcal{A}$ is an arbitrary constant.

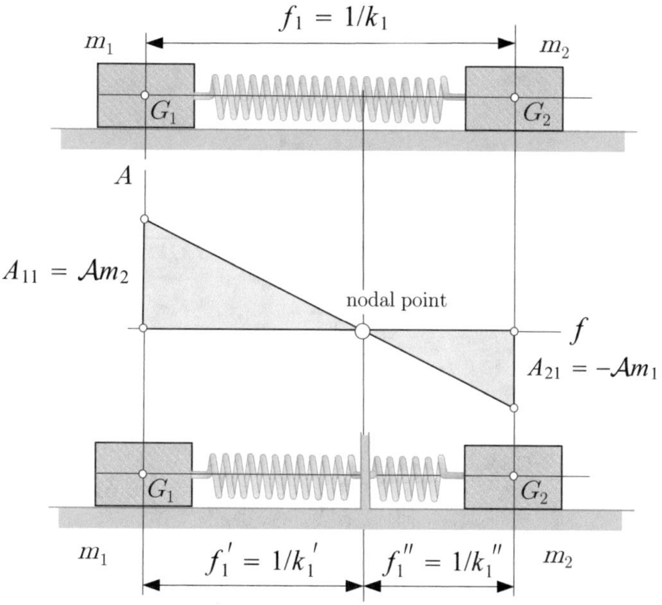

Fig. 4.16 Single degree of freedom systems equivalent to a two degree of freedom system

Assume that the distance between the two mass centers G_1 and G_2 is the flexibility $f_1 = 1/k_1$—see Fig. 4.16 for details. Since the masses m_1 and m_2 move in opposite directions during motion there exists such a point—called nodal point—between the two masses which remains at rest. Figure 4.16 shows how the amplitude of the motion (as a function of the flexibility f) changes between the two masses. The nodal point divides the two degree of freedom system into two single degree of freedom systems which have the same natural circular frequency. The bottom part of Fig. 4.16 shows the two single degree of freedom systems. Recalling that $\omega_n^2 = k/m$ for a single degree of freedom system we may write

$$\omega_{n1}^2 = \frac{1}{m_1 f_1'} = \frac{k_1'}{m_1} = \frac{m_1 + m_2}{m_1 \frac{1}{k_1} m_2} = \frac{k_1''}{m_2} = \frac{1}{m_2 f_1''} \quad \text{and} \quad f_1' + f_1'' = f_1 \tag{4.70}$$

from where it follows that

$$f_1' = \frac{1}{k_1'} = \frac{m_2}{m_1 + m_2}\frac{1}{k_1}\,, \qquad f_1'' = \frac{1}{k_1''} = \frac{m_1}{m_1 + m_2}\frac{1}{k_1}\,. \tag{4.71}$$

Exercise 4.7 Determine the natural frequencies and the solution $\underline{\mathbf{q}}$ for system B if

$$m_1 = m_2 = m\,, \qquad \text{and} \qquad k_o = k_1 = k. \tag{4.72}$$

Under these conditions Eq. (4.64a) yields

$$\begin{vmatrix} 2k - \lambda m & -k \\ -k & k - \lambda m \end{vmatrix} = m^2\lambda^2 - 3mk\lambda + k^2 = 0 \tag{4.73}$$

or

$$\lambda^2 - 3\frac{k}{m}\lambda + \frac{k^2}{m^2} = 0\,.$$

Introduce a new variable defined by the relation $\hat{\omega} = \sqrt{k/m}$. Using this quantity we can rewrite the above equation into the form

$$\lambda^2 - 3\,\hat{\omega}^2\lambda + \hat{\omega}^4 = 0\,,$$

hence

$$\lambda_{1,2} = \hat{\omega}^2\frac{3 \mp \sqrt{9-4}}{2} = \frac{k}{m}\begin{cases}\left(3 - \sqrt{5}\right)/2 = 0.3820 \\ \left(3 + \sqrt{5}\right)/2 = 2.6180\end{cases}\,. \tag{4.74}$$

Since determinant (4.61b) is that of the coefficient matrix in Eq. (4.61b) the amplitudes A_{ji} $(i, j = 1, 2)$ should satisfy the following linear equation system:

$$\begin{bmatrix} 2k - \lambda_i m & -k \\ -k & k - \lambda_i m \end{bmatrix}\begin{bmatrix} A_{1i} \\ A_{2i}\end{bmatrix} = \begin{bmatrix} 0 \\ 0 \end{bmatrix}. \tag{4.75}$$

If $i = 1$ we have to solve the equations

$$\begin{aligned}(2k - \lambda_1 m)\,A_{11} - kA_{21} &= 0\,, \\ -kA_{11} + (k - \lambda_1 m)\,A_{21} &= 0,\end{aligned}$$

which are, however, not independent equations. Therefore, it is sufficient to consider the first equation. If we divide throughout by m, we get

$$\left(2\frac{k}{m} - \frac{k}{m}\frac{3-\sqrt{5}}{2}m\right)A_{11} - \frac{k}{m}A_{21} = 0$$

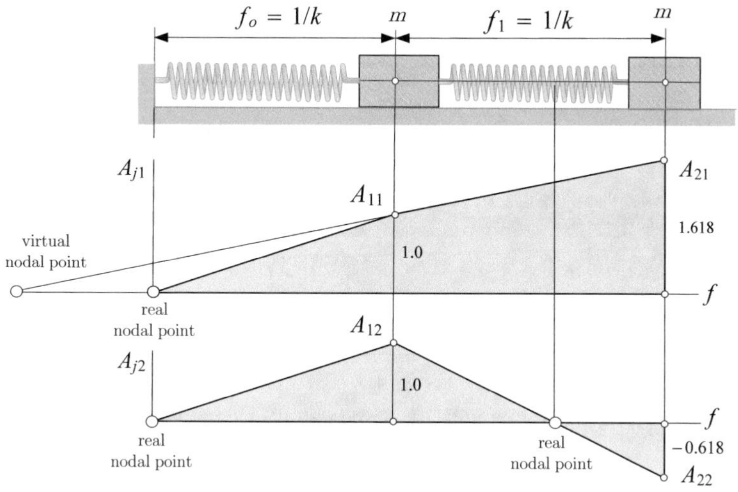

Fig. 4.17 Amplitude functions

from where

$$\frac{A_{21}}{A_{11}} = \frac{1+\sqrt{5}}{2} = 1.6180\,.$$

If $i = 2$ the first equation of (4.75) is of the form

$$(2k - \lambda_2 m)\,A_{12} - kA_{22} = 0\,.$$

Dividing throughout again by m, we obtain the equation

$$\left(2\frac{k}{m} - \frac{k}{m}\frac{3+\sqrt{5}}{2}\right)A_{12} - \frac{k}{m}A_{22} = 0,$$

which yields

$$\frac{A_{22}}{A_{12}} = \frac{1-\sqrt{5}}{2} = -0.6180\,.$$

Assume that $A_{11} = A_{12} = 1$. Under this condition Fig. 4.17 shows the amplitudes as functions of f, i.e., the two nodal modes. When the system vibrates in its first mode there exists only one real nodal point. The straight line obtained by joining the points $A_{11} = 1.0$ and $A_{12} = 1.618$ intersects the axis f outside the interval $[0, 2f_o]$. This nodal point is referred to as virtual nodal point.

When the system vibrates in its second mode there are two real nodal points. One of them is located in the interval $[f_o, 2f_o]$. The system can, therefore, be replaced by two single degree of freedom systems as shown in Fig. 4.18.

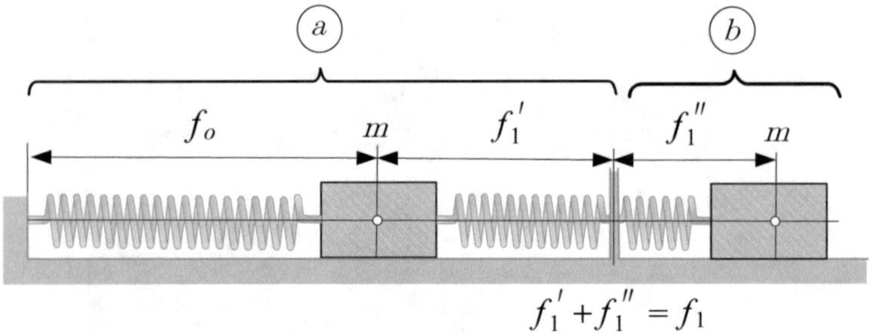

Fig. 4.18 Equivalent single degree of freedom systems

Systems (a) and (b) vibrate with the same natural frequency

$$\omega_{n2} = \sqrt{\frac{3+\sqrt{5}}{2}}\,\hat{\omega}\,, \qquad \hat{\omega} = \sqrt{k/m} = \sqrt{\frac{1}{f_o m}}\,.$$

As regards system (a) we can write using Eq. (3.2), in which $k + k_1'$ should be substituted for k on the base of Eq. (3.41) that

$$\omega_{n2}^2 = \frac{3+\sqrt{5}}{2}\hat{\omega}^2 = \frac{k+k_1'}{m} = \frac{1}{m}\left(\frac{1}{f_o} + \frac{1}{f_1'}\right).$$

Hence,

$$\frac{k}{m}\frac{3+\sqrt{5}}{2} = \frac{1}{mf_o} + \frac{1}{mf_1'} \qquad \longrightarrow \qquad \frac{1}{f_1'} = k_1' = \frac{3+\sqrt{5}}{2}\frac{1}{f_o}.$$

We obtain in the same manner for system (b) that

$$\omega_{n2}^2 = \frac{3+\sqrt{5}}{2}\frac{1}{mf_1} = \frac{1}{mf_1''} \qquad \longrightarrow \qquad \frac{1}{f_1''} = k_1'' = \frac{3+\sqrt{5}}{2}\frac{1}{f_1}.$$

It can easily be checked that

$$f_1' + f_1'' = f_1 = f_o.$$

As regards system C in Fig. 4.15 the methods that should be applied to finding solutions are basically the same as those we have utilized in connection with systems A and B.

4.3.4 Forced Vibrations of Two Degree of Freedom Systems

It will be assumed that the spring–mass system considered is subjected to a harmonic excitation force

$$\underset{(2\times 1)}{\underline{\mathbf{F}}_f} = \begin{bmatrix} F_1 \\ 0 \end{bmatrix} \cos\omega_f t = \underset{(2\times 1)}{\underline{\mathbf{F}}_{fo}} \cos\omega_f t\,. \tag{4.76}$$

It is also assumed that the motion can be described by the matrix equation

$$\begin{bmatrix} m_1 & 0 \\ 0 & m_2 \end{bmatrix}\begin{bmatrix} \ddot{x}_1 \\ \ddot{x}_2 \end{bmatrix} + \begin{bmatrix} k_{11} & k_{12} \\ k_{21} & k_{22} \end{bmatrix}\begin{bmatrix} x_1 \\ x_2 \end{bmatrix} = \begin{bmatrix} F_1 \\ 0 \end{bmatrix} \cos\omega_f t\,. \tag{4.77}$$

Depending on how the elements of the symmetric stiffness matrix $\underline{\mathbf{K}}$ are selected, this equation is the equation of motion for one of the systems shown in Fig. 4.15. The excitation force is acting on mass m_1. We seek the steady-state solution in the form

$$\begin{bmatrix} x_1 \\ x_2 \end{bmatrix} = \begin{bmatrix} A_{f1} \\ A_{f2} \end{bmatrix} \cos\omega_f t = \underset{(2\times 1)}{\underline{\mathbf{A}}_f} \cos\omega_f t\,. \tag{4.78}$$

Substituting it into Eq. (4.77), we obtain

$$\begin{bmatrix} k_{11} - m_1\omega_f^2 & k_{12} \\ k_{21} & k_{22} - m_2\omega_f^2 \end{bmatrix}\begin{bmatrix} A_{f1} \\ A_{f2} \end{bmatrix} = \begin{bmatrix} F_1 \\ 0 \end{bmatrix} \tag{4.79a}$$

or

$$\Big(\underset{(2\times 2)}{\underline{\mathbf{K}}} - \omega_f^2 \underset{(2\times 2)}{\underline{\mathbf{M}}}\Big) \underset{(2\times 1)}{\underline{\mathbf{A}}_f} = \underset{(2\times 1)}{\underline{\mathbf{F}}_{fo}}\,. \tag{4.79b}$$

We should notice that

$$\Big|\underset{(2\times 2)}{\underline{\mathbf{K}}} - \omega_f^2 \underset{(2\times 2)}{\underline{\mathbf{M}}}\Big| = \left(k_{11} - \omega_f^2 m_1\right)\left(k_{22} - \omega_f^2 m_2\right) - k_{12}^2 =$$

$$= m_1 m_2 \left(\omega_f^2 - \omega_{n1}^2\right)\left(\omega_f^2 - \omega_{n2}^2\right), \tag{4.80}$$

where $\omega_{n1}^2 = \lambda_1$ and $\omega_{n2}^2 = \lambda_2$ $(\lambda_1 \le \lambda_2)$ are the roots of the characteristic equation

$$\Big|\underset{(2\times 2)}{\underline{\mathbf{K}}} - \lambda \underset{(2\times 2)}{\underline{\mathbf{M}}}\Big| = 0\,,$$

i.e., they are the squares of normal mode frequencies (or the natural frequencies) of the free system. Consequently,

$$\underset{(2\times 1)}{\underline{\mathbf{A}}_f} = \frac{\operatorname{adj}\Big|\underset{(2\times 2)}{\underline{\mathbf{K}}} - \omega_f^2 \underset{(2\times 2)}{\underline{\mathbf{M}}}\Big|}{\Big|\underset{(2\times 2)}{\underline{\mathbf{K}}} - \omega_f^2 \underset{(2\times 2)}{\underline{\mathbf{M}}}\Big|} \underset{(2\times 1)}{\underline{\mathbf{F}}_{fo}} = \frac{\begin{bmatrix} k_{22} - m_2\omega_f^2 & -k_{12} \\ -k_{21} & k_{11} - m_1\omega_f^2 \end{bmatrix}\begin{bmatrix} F_1 \\ 0 \end{bmatrix}}{m_1 m_2 \left(\omega_f^2 - \omega_{n1}^2\right)\left(\omega_f^2 - \omega_{n2}^2\right)} \tag{4.81}$$

or

$$A_{f1} = \frac{k_{22} - m_2\omega_f^2}{m_1 m_2 \left(\omega_f^2 - \omega_{n1}^2\right)\left(\omega_f^2 - \omega_{n2}^2\right)} F_1\,, \tag{4.82a}$$

$$A_{f2} = \frac{-k_{21}}{m_1 m_2 \left(\omega_f^2 - \omega_{n1}^2\right)\left(\omega_f^2 - \omega_{n2}^2\right)} F_1\,. \tag{4.82b}$$

Exercise 4.8 Examine the steady-state vibrations of the system shown in Fig. 4.19. Plot its frequency response curve.

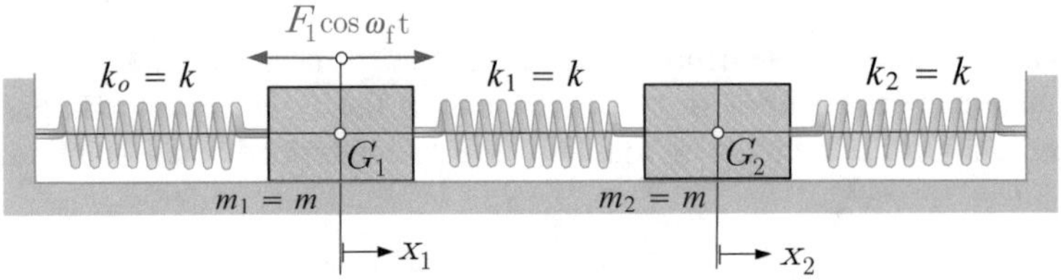

Fig. 4.19 Excitation on a two degree of freedom spring-mass system

Comparison of Eqs. (4.58c) and (4.77) yields the equations of motion in the form

$$\begin{bmatrix} m & 0 \\ 0 & m \end{bmatrix}\begin{bmatrix} \ddot{x}_1 \\ \ddot{x}_2 \end{bmatrix} + \begin{bmatrix} 2k & -k \\ -k & 2k \end{bmatrix}\begin{bmatrix} x_1 \\ x_2 \end{bmatrix} = \begin{bmatrix} F_1 \\ 0 \end{bmatrix} \cos \omega_f t\,, \tag{4.83}$$

where it is taken into account that

$$k_{11} = k_{22} = 2k\,, \qquad k_{12} = k_{21} = -k\,.$$

Consequently,

$$\left| \underset{(2\times2)}{\mathbf{K}} - \omega_f^2 \underset{(2\times2)}{\mathbf{M}} \right| = \begin{vmatrix} 2k - m\,\omega_f^2 & -k \\ -k & 2k - m\,\omega_f^2 \end{vmatrix} = m^2\omega_f^4 - 4k\omega_f^2 + 3k^2 =$$

$$= m^2 \left(\omega_f^2 - \frac{k}{m}\right)\left(\omega_f^2 - \frac{3k}{m}\right) = 0 \tag{4.84a}$$

is the characteristic equation and its product form since

$$\omega_{n1}^2 = \frac{k}{m} \quad \text{and} \quad \omega_{n2}^2 = \frac{3k}{m} \tag{4.84b}$$

are the roots. Equation (4.82a), therefore, becomes

$$A_{f1} = \frac{2k - m\,\omega_f^2}{m^2\left(\omega_{n1}^2 - \omega_f^2\right)\left(\omega_{n2}^2 - \omega_f^2\right)} F_1 = \frac{C_1}{\omega_{n1}^2 - \omega_f^2} + \frac{C_2}{\omega_{n2}^2 - \omega_f^2}, \tag{4.85}$$

where[2]

$$C_1 = \frac{\left(2k - m\,\omega_{n1}^2\right) F_1}{m^2\left(\omega_{n2}^2 - \omega_{n1}^2\right)} = \frac{2k - k}{m\,(3k - k)} F_1 = \frac{F_1}{2m}$$

and

$$C_2 = \frac{\left(2k - m\,\omega_{n2}^2\right) F_1}{m^2\left(\omega_{n1}^2 - \omega_{n2}^2\right)} = \frac{F_1}{2m},$$

hence,

$$A_{f1} = \frac{F_1}{2m}\left[\frac{1}{\omega_{n1}^2 - \omega_f^2} + \frac{1}{\omega_{n2}^2 - \omega_f^2}\right] = \frac{F_1}{2k}\left[\frac{1}{1 - \left(\omega_f/\omega_{n1}\right)^2} + \frac{1}{3 - \left(\omega_f/\omega_{n1}\right)^2}\right]. \tag{4.86a}$$

Treating A_2 in the same manner, we have

$$A_{f2} = \frac{F_1}{2k}\left[\frac{1}{1 - \left(\omega_f/\omega_{n1}\right)^2} - \frac{1}{3 - \left(\omega_f/\omega_{n1}\right)^2}\right]. \tag{4.86b}$$

Figure 4.20 shows the frequency response curve (the dimensionless quotient $A_{fi}/(F_1/2k)$ $(i = 1, 2)$ against ω_f/ω_{n1}). Resonance occurs if $\omega_f = \omega_{n1}$ or $\omega_f = \omega_{n2} = \sqrt{3}\,\omega_{n1}$.

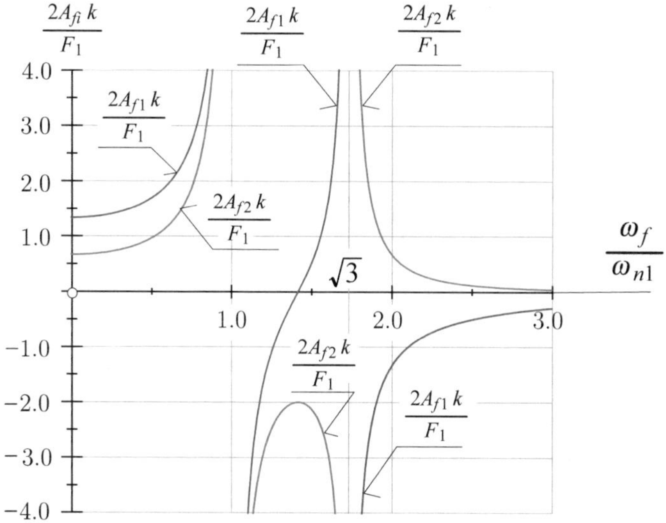

Fig. 4.20 Frequency response curve

[2]To solve for $[C_1]$ $\{C_2\}$ multiply (4.85) by $[\omega_{n1}^2 - \omega_f^2]$ $\{\omega_{n2}^2 - \omega_f^2\}$ and set ω_f to $[\omega_{n1}]\{\omega_{n2}\}$.

4.3.5 *Vibration Absorbers*

4.3.5.1 The Two Degree of Freedom Tuned Mass Damper

Consider a single degree of freedom system with mass m_1 and spring constant k_o (Fig. 4.21). We shall assume that the system is subjected to a harmonic force of excitation $F_f = F_{fo} \cos \omega_f t$. It is our aim to find a way for reducing the amplitude of the steady-state vibrations. To this end, a second spring–mass system is attached to the main mass m_1 for which we prescribe that its natural circular frequency coincides with the circular frequency of the excitation force, i.e., it holds that

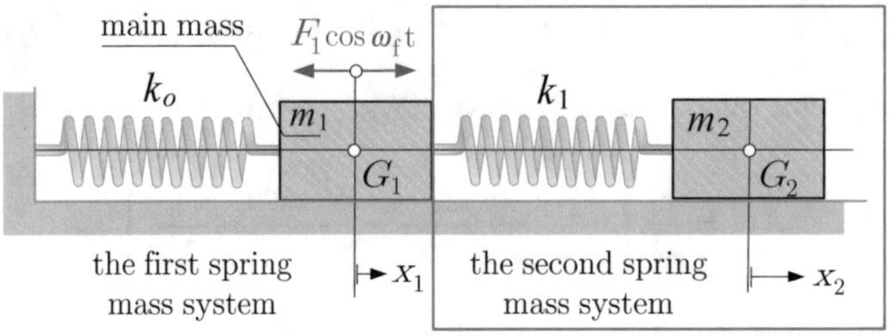

Fig. 4.21 Two degree of freedom tuned mass damper

$$\omega_{12}^2 = \frac{k_1}{m_2} = \omega_f^2 \,. \tag{4.87}$$

The equation of motion

$$\underbrace{\begin{bmatrix} m_1 & 0 \\ 0 & m_2 \end{bmatrix}}_{\substack{\mathbf{K} \\ (2\times2)}} \underbrace{\begin{bmatrix} \ddot{x}_1 \\ \ddot{x}_2 \end{bmatrix}}_{\substack{\ddot{\mathbf{q}} \\ (2\times1)}} + \underbrace{\begin{bmatrix} k_o + k_1 & -k_1 \\ -k_1 & k_1 \end{bmatrix}}_{\substack{\mathbf{K} \\ (2\times2)}} \underbrace{\begin{bmatrix} x_1 \\ x_2 \end{bmatrix}}_{\substack{\mathbf{q} \\ (2\times1)}} = \underbrace{\begin{bmatrix} F_1 \\ 0 \end{bmatrix}}_{\substack{\mathbf{F}_{fo} \\ (2\times1)}} \cos \omega_f t \tag{4.88}$$

can be obtained from Eq. (4.77) if in the former we set

$$k_{11} = k_o + k_1 \,, \quad k_{12} = k_{21} = -k_1 \,, \quad k_{22} = k_1 \,. \tag{4.89}$$

Consequently, we can use Eq. (4.82) to determine amplitude A_1. After substituting relations (4.89) into Eq. (4.82), we get

$$\underset{(2\times1)}{\mathbf{A}} = \frac{\operatorname{adj}\left| \underset{(2\times2)}{\mathbf{K}} - \omega_f^2 \underset{(2\times2)}{\mathbf{M}} \right|}{\left| \underset{(2\times2)}{\mathbf{K}} - \omega_f^2 \underset{(2\times2)}{\mathbf{M}} \right|} = \frac{\begin{bmatrix} k_1 - m_2\omega_f^2 & k_1 \\ k_1 & k_o + k_1 - m_1\omega_f^2 \end{bmatrix}}{\begin{vmatrix} k_o + k_1 - m_1\omega_f^2 & -k_1 \\ -k_1 & k_1 - m_2\omega_f^2 \end{vmatrix}} \begin{bmatrix} F_1 \\ 0 \end{bmatrix}$$

from where

$$A_1 = \frac{k_1 - m_2\omega_f^2}{\left(k_o + k_1 - m_1\omega_f^2\right)\left(k_1 - m_2\omega_f^2\right) - k_1^2} F_1 \underset{\frac{k_o}{m_1}=\omega_{11}^2}{=} \uparrow \underset{\frac{k_1}{m_2}=\omega_{12}^2}{\uparrow} =$$

$$= \frac{k_1\left(1 - \frac{\omega_f^2}{\omega_{12}^2}\right)}{k_o k_1\left\{\left(1 + \frac{k_1}{k_o} - \frac{\omega_f^2}{\omega_{11}^2}\right)\left(1 - \frac{\omega_f^2}{\omega_{12}^2}\right) - \frac{k_1}{k_o}\right\}} F_1$$

or

$$\frac{A_1 k_o}{F_1} = \frac{1 - \frac{\omega_f^2}{\omega_{12}^2}}{\left(1 + \frac{k_1}{k_o} - \frac{\omega_f^2}{\omega_{11}^2}\right)\left(1 - \frac{\omega_f^2}{\omega_{12}^2}\right) - \frac{k_1}{k_o}} \tag{4.90}$$

in which ω_{11} is the natural circular frequency of the original spring–mass system, and ω_{12} is that of the second spring–mass system—see Eq. (4.87). The natural circular frequencies of the two degree of freedom system obtained by adding a single degree of freedom system to the original single degree of freedom system are denoted in the usual way by ω_{n1} and ω_{n2}.

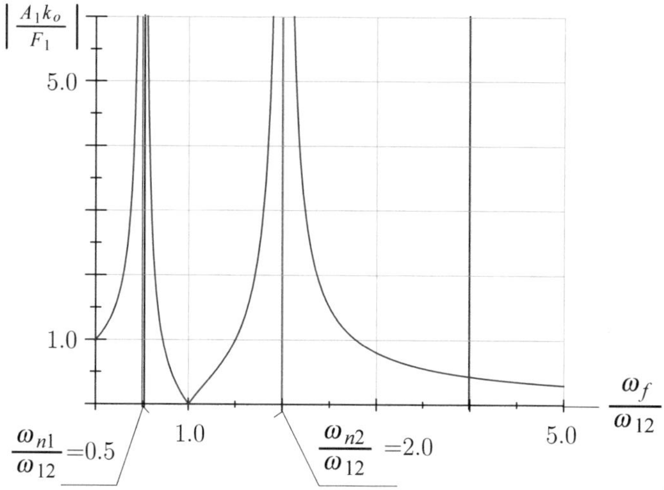

Fig. 4.22 Dimensionless amplitude as a function of ω_f

Figure 4.22 shows the quotient $|A_1 k_o/F_1|$ against the quotient ω_f/ω_{12} for the case $\omega_{11} = \omega_{12}$, $k_1/k_o = 0.5$. It follows form Eq. (4.90) and it can also be seen in Fig. 4.22 that the amplitude of the steady-state vibrations A_1 is zero when $\omega_{12} = \omega_f$. It is an important question how to select mass m_2. Let A_2 be the amplitude of mass m_2. Further let F_{k1} be the amplitude of the force exerted by spring k_1 on mass m_2.

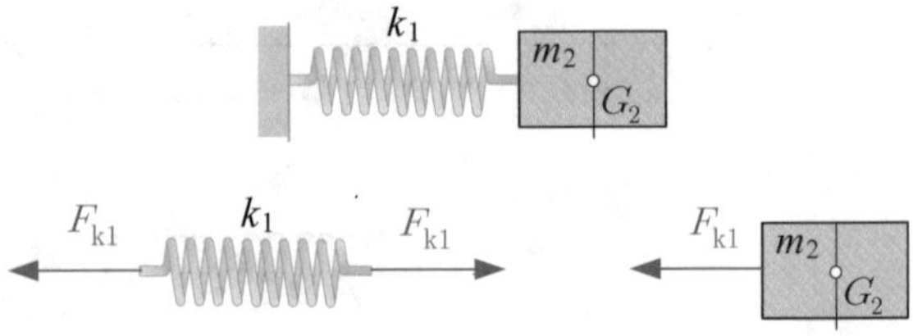

Fig. 4.23 Inner forces

Figure 4.23 shows the forces acting on the second spring and the absorber mass m_2. Since the left end of the spring k_1 is at rest the equation of motion for the absorber mass is of the form

$$m_2\ddot{x}_2 + k_1 x_2 = 0 \qquad (\ddot{x}_2 = -\omega_{12}^2 x_2 = -\omega_f^2 x_2)\ .$$

This equation is always valid. If $x_2 = A_2$ we can write

$$\underbrace{m_2\ddot{x}_2}_{-\omega_f^2 A_2 m_2} + \underbrace{k_1 x_2}_{k_1 A_2 = F_{k1}} = \boxed{-\underbrace{\omega_f^2}_{\frac{k_1}{m_2}} A_2 m_2 + F_{k1} = -k_1 A_2 + F_{k1} = 0} \tag{4.91}$$

from where

$$A_2 = \frac{F_{k1}}{k_1}\ .$$

These results show that the spring constant k_1 and the mass m_2 depend on the allowable values of A_2 and F_{k1}.

4.3.5.2 Untuned Viscous Damper

An untuned viscous damper consists of a disk (pulley) which can rotate freely because it is either mounted on bearings inside the housing of the damper or for the simple reason that the clearance between the housing and the disk (pulley) is filled with silicone fluid. The latter is preferred to oil for two reasons: (a) because of its high viscosity index and (b) the fact that its viscosity is not sensitive to the temperature changes. These dampers are used to limit the amplitudes of torsional vibrations in crankshafts (or shafts) for which the excitation frequency is, in general, not constant. They have the advantage in contrast to the tuned two degree of freedom mass damper that they reduce the amplitude of steady-state vibrations in a wide interval of the

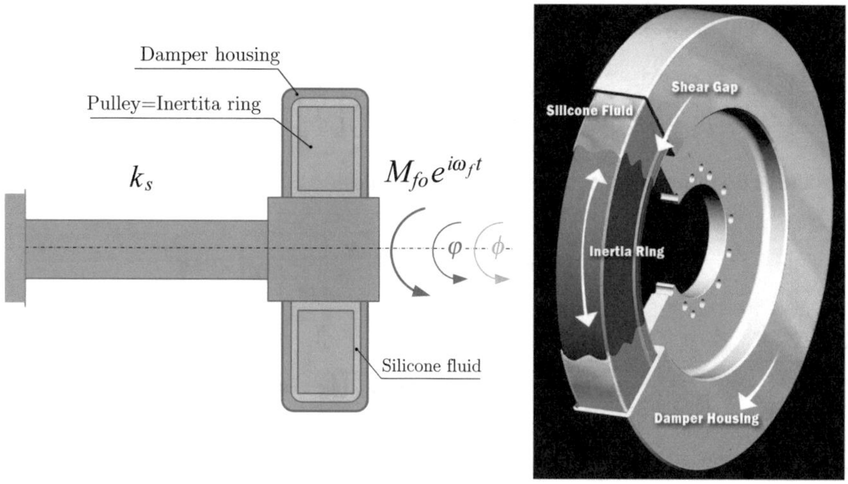

Fig. 4.24 Untuned viscous damper

excitation frequency ω_f. Figure 4.24 shows both the mechanical model of the damper and shaft and the picture of a real damper.[3] In the later case, the shaft is not shown. To derive the equations of motion we shall introduce the following notations: let J_p be the moment of inertia of the disk (pulley) about the axis of rotation. Further let J_s be the moment of inertia of the shaft about the same axis. The angle of rotation of the disk (pulley) and the housing are denoted by ϕ and φ, respectively. The torsional stiffness of the shaft is k and the harmonic excitation is the torque $M_f\, e^{i\omega_f t}$ where i is the imaginary unit.

The damper torque is proportional to the relative angular velocity $\dot{\varphi} - \dot{\phi}$ of the housing with respect to the disk and it is given by

$$M_c = c\left(\dot{\varphi} - \dot{\phi}\right), \tag{4.92}$$

where c is the damping coefficient. Making use of the notations introduced it is not too difficult to check that the equations of motion for the housing and the disk are as follows:

$$J_s\ddot{\varphi} + c\left(\dot{\varphi} - \dot{\phi}\right) + k\varphi = M_{fo}\, e^{i\omega_f t}, \tag{4.93a}$$

$$J_p\ddot{\phi} - c\left(\dot{\varphi} - \dot{\phi}\right) = 0. \tag{4.93b}$$

[3]Photo courtesy of Vibratech TVD 180 Zoar Valley Rd. Springville, New York USA.

These equations can be rewritten in matrix form:

$$\begin{bmatrix} J_s & 0 \\ 0 & J_p \end{bmatrix}\begin{bmatrix} \ddot{\varphi} \\ \ddot{\phi} \end{bmatrix}+\begin{bmatrix} c & -c \\ -c & c \end{bmatrix}\begin{bmatrix} \dot{\varphi} \\ \dot{\phi} \end{bmatrix}+\begin{bmatrix} k & 0 \\ 0 & 0 \end{bmatrix}\begin{bmatrix} \varphi \\ \phi \end{bmatrix}=\begin{bmatrix} M_{fo}e^{i\omega_f t} \\ 0 \end{bmatrix}. \tag{4.94}$$

If there is no damping we get the equation of motion of the free vibrations:

$$\begin{bmatrix} J_s & 0 \\ 0 & J_p \end{bmatrix}\begin{bmatrix} \ddot{\varphi} \\ \ddot{\phi} \end{bmatrix}+\begin{bmatrix} k & 0 \\ 0 & 0 \end{bmatrix}\begin{bmatrix} \varphi \\ \phi \end{bmatrix}=\begin{bmatrix} 0 \\ 0 \end{bmatrix}.$$

Hence,

$$\begin{vmatrix} k-\omega_n^2 J_s & 0 \\ 0 & -\omega_n^2 J_p \end{vmatrix} = -\omega_n^2 J_p\left(k-\omega_n^2 J_s\right)=0$$

is the characteristic equation and

$$\omega_{n1}=0 \quad \text{and} \quad \omega_{n2}=\omega_n=\sqrt{k/J_s}\neq 0 \tag{4.95}$$

are the two natural frequencies. Let us assume that

$$\begin{bmatrix} \varphi \\ \phi \end{bmatrix}=\begin{bmatrix} \varphi_o \\ \phi_o \end{bmatrix}e^{i\omega_f t} \tag{4.96}$$

is the steady-state solution in which φ_o and ϕ_o are the two unknown amplitudes. Upon substitution of the steady-state solution into the equations of motion (4.94) we arrive at the following system of linear equations for the unknown amplitudes φ_o and ϕ_o:

$$\begin{bmatrix} ic\omega_f-\omega_f^2 J_s+k & -ic\omega_f \\ -ic\omega_f & ic\omega_f-\omega_f^2 J_p \end{bmatrix}\begin{bmatrix} \varphi_o \\ \phi_o \end{bmatrix}=\begin{bmatrix} M_{fo} \\ 0 \end{bmatrix}. \tag{4.97}$$

The solution is

$$\begin{aligned}\begin{bmatrix} \varphi_o \\ \phi_o \end{bmatrix} &= \frac{\operatorname{adj}\begin{bmatrix} ic\omega_f-\omega_f^2 J_s+k & -ic\omega_f \\ -ic\omega_f & ic\omega_f-\omega_f^2 J_p \end{bmatrix}}{\begin{vmatrix} ic\omega_f-\omega_f^2 J_s+k & -ic\omega_f \\ -ic\omega_f & ic\omega_f-\omega_f^2 J_p \end{vmatrix}}\begin{bmatrix} M_{fo} \\ 0 \end{bmatrix} = \\ &= \frac{\begin{bmatrix} ic\omega_f-\omega_f^2 J_p & ic\omega_f \\ ic\omega_f & ic\omega_f-\omega_f^2 J_s+k \end{bmatrix}}{\begin{vmatrix} ic\omega_f-\omega_f^2 J_s+k & -ic\omega_f \\ -ic\omega_f & ic\omega_f-\omega_f^2 J_p \end{vmatrix}}\begin{bmatrix} M_{fo} \\ 0 \end{bmatrix},\end{aligned} \tag{4.98}$$

where the numerator is the adjugate of the coefficient matrix in (4.97) and

$$d = \begin{vmatrix} ic\omega_f - \omega_f^2 J_s + k & -ic\omega_f \\ -ic\omega_f & ic\omega_f - \omega_f^2 J_p \end{vmatrix} =$$
$$= -\omega_f^2 J_p \left(k - \omega_f^2 J_s\right) - ic\omega_f \left(\omega_f^2 J_p - \left(k - \omega_f^2 J_s\right)\right) \tag{4.99}$$

is the denominator. Recalling (3.20)—J_s corresponds to m—we get the critical damping coefficient:

$$c_n = 2J_s\omega_n \,.$$

Hence

$$c = \frac{c}{c_n} c_n = \zeta c_n = 2\zeta J_s \omega_n \tag{4.100}$$

is the damping coefficient in terms of the damping factor ζ. Let us introduce the dimensionless quantities

$$\mu = \frac{J_p}{J_s} \qquad \text{and} \qquad \eta = \frac{\omega_f}{\omega_n}$$

by the use of which we can manipulate the denominator d into a more suitable form:

$$\begin{aligned} d = &- \omega_f^2 J_p \left(k - \omega_f^2 J_s\right) - ic\omega \left(\omega_f^2 J_p - \left(k - \omega_f^2 J_s\right)\right) = \\ &= -J_s k \omega_f^2 \frac{J_p}{J_s} \left(1 - \frac{\omega_f^2}{\frac{k}{J_s}}\right) - i2\zeta J_s k \omega_n \omega_f \left(\omega_f^2 \frac{J_p}{J_s \frac{k}{J_s}} - \left(1 - \frac{\omega_f^2}{\frac{k}{J_s}}\right)\right) = \\ &= -J_s k \omega_n \omega_f \left[\eta\mu\left(1-\eta^2\right) + i2\zeta\left(\mu\eta^2 - \left(1-\eta^2\right)\right)\right] . \end{aligned} \tag{4.101a}$$

For the first element of the adjugate in (4.98), we can write in a similar manner that

$$ic\omega_f - \omega_f^2 J_p = -J_s \omega_n \omega_f \left(\frac{\omega_f}{\omega_n}\frac{J_p}{J_s} - i2\zeta\right) . \tag{4.101b}$$

Making use of Eq. (4.101) we obtain from solution (4.98) that

$$\frac{k\varphi_o}{M_{fo}} = \frac{\mu\eta - i2\zeta}{\eta\mu\left(1-\eta^2\right) + i2\zeta\left[\mu\eta^2 - \left(1-\eta^2\right)\right]} \tag{4.102a}$$

or

$$\boxed{\left|\frac{k\varphi_o}{M_{fo}}\right| = \sqrt{\frac{(\mu\eta)^2 + 4\zeta^2}{(\eta\mu)^2\left(1-\eta^2\right)^2 + 4\zeta^2\left(\mu\eta^2 - \left(1-\eta^2\right)\right)^2}} \,.} \tag{4.102b}$$

Figure 4.25 shows the graph $\left|k\varphi_o/M_{fo}\right|$ for $\mu = 1$ and different values of ζ—keep in mind that the relation $\zeta \in [0, 1]$ should be satisfied. It is a further issue how to reduce the peak values of the graph. To find an optimum we have to clarify the effect of the optimum conditions

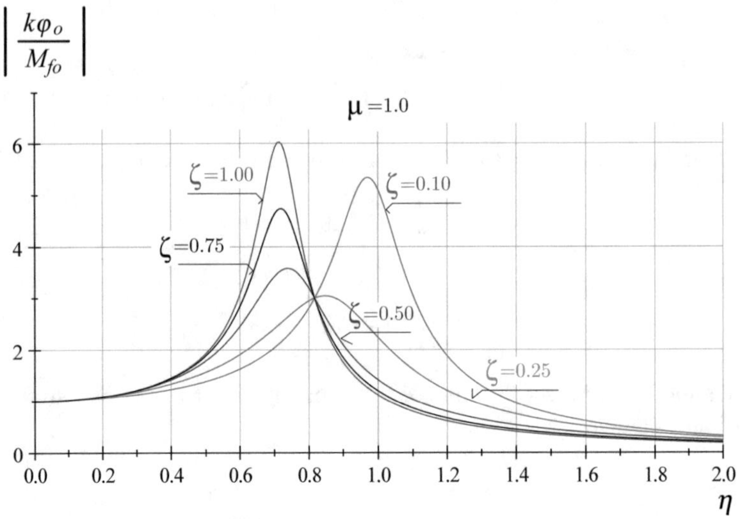

Fig. 4.25 The amplitude $|k\varphi_o/M_{fo}|$ against η

$$\frac{\mathrm{d}}{\mathrm{d}\left(\zeta^2\right)}\frac{\eta^2+4\frac{\zeta^2}{\mu^2}}{\eta^2\left(1-\eta^2\right)^2+4\frac{\zeta^2}{\mu^2}\left(1-(1+\mu)\,\eta^2\right)^2}=0 \tag{4.103a}$$

and

$$\frac{\mathrm{d}}{\mathrm{d}\left(\eta^2\right)}\frac{\eta^2+4\frac{\zeta^2}{\mu^2}}{\eta^2\left(1-\eta^2\right)^2+4\frac{\zeta^2}{\mu^2}\left(1-(1+\mu)\,\eta^2\right)^2}=0 \tag{4.103b}$$

on the system parameters. Let us introduce the following notations for the numerator and denominator in (4.103):

$$u=\eta^2+4\frac{\zeta^2}{\mu^2} \quad \text{and} \quad v=\eta^2\left(1-\eta^2\right)^2+4\frac{\zeta^2}{\mu^2}\left(1-(1+\mu)\,\eta^2\right)^2. \tag{4.104}$$

Using these notations, the optimum conditions (4.103) can be rewritten in the following forms:

$$v\frac{\mathrm{d}u}{\mathrm{d}\left(\zeta^2\right)}=u\frac{\mathrm{d}v}{\mathrm{d}\left(\zeta^2\right)} \quad \text{and} \quad v\frac{\mathrm{d}u}{\mathrm{d}\left(\eta^2\right)}=u\frac{\mathrm{d}v}{\mathrm{d}\left(\eta^2\right)}. \tag{4.105}$$

Consider now optimum condition $(4.105)_1$ in which

$$\frac{\mathrm{d}u}{\mathrm{d}\left(\zeta^2\right)}=\frac{4}{\mu^2}, \qquad \frac{\mathrm{d}v}{\mathrm{d}\left(\zeta^2\right)}=\frac{4}{\mu^2}\left(\mu\eta^2-1+\eta^2\right)^2.$$

Hence,

$$\left(\eta^2\left(1-\eta^2\right)^2+4\frac{\zeta^2}{\mu^2}\left(1-(1+\mu)\,\eta^2\right)^2\right)-\left(\eta^2+4\frac{\zeta^2}{\mu^2}\right)\left(\mu\eta^2-1+\eta^2\right)^2=0$$

which yields after performing some manipulations that

$$-\eta^4\mu\left(\mu\eta^2-2+2\eta^2\right)=0$$

is the first optimum condition. Consequently,

$$\boxed{\eta_{\text{opt}}=\sqrt{\frac{2}{2+\mu}}} \tag{4.106}$$

is the optimal value of η. Consider now optimum condition $(4.105)_2$ by taking the relations

$$\frac{\mathrm{d}u}{\mathrm{d}\left(\eta^2\right)}=\frac{\mathrm{d}}{\mathrm{d}\left(\eta^2\right)}\left(\eta^2+4\frac{\zeta^2}{\mu^2}\right)=1$$

and

$$\begin{aligned}\frac{\mathrm{d}v}{\mathrm{d}\left(\eta^2\right)}&=\frac{\mathrm{d}}{\mathrm{d}\left(\eta^2\right)}\left(\eta^2\left(1-\eta^2\right)^2+4\frac{\zeta^2}{\mu^2}\left(1-(1+\mu)\,\eta^2\right)^2\right)=\\&=\left(1-\eta^2\right)^2-2\left(1-\eta^2\right)\eta^2-8\frac{\zeta^2}{\mu^2}\left(1-(1+\mu)\,\eta^2\right)(1+\mu)\end{aligned}$$

into account. We get in this manner optimum condition $(4.105)_2$ in the form

$$\begin{aligned}&\eta^2\left(1-\eta^2\right)^2+4\frac{\zeta^2}{\mu^2}\left(1-(1+\mu)\,\eta^2\right)^2-\\&\qquad\left(\eta^2+4\frac{\zeta^2}{\mu^2}\right)\left(\left(1-\eta^2\right)^2-2\left(1-\eta^2\right)\eta^2-\right.\\&\qquad\qquad\left.-8\frac{\zeta^2}{\mu^2}\left(1-(1+\mu)\,\eta^2\right)\left(1+\eta^2\right)\right)=0\,.\end{aligned} \tag{4.107}$$

If we substitute here η_{opt} from Eq. (4.105) we find by shaking the second optimum condition down that

$$\left(\mu^2+4\zeta^2+2\zeta^2\mu\right)\left(-\mu^2+4\zeta^2+6\zeta^2\mu+2\zeta^2\mu^2\right)=0$$

from where we get

$$\boxed{\zeta = \frac{\mu}{\sqrt{2\,(1+\mu)\,(2+\mu)}}} \tag{4.108}$$

since ζ^2 cannot be negative.

4.4 Problems

Problem 4.1 Prove Eq. (4.24).

Problem 4.2 A cart of mass m_1 and a rod of mass m_2 and length $2L$ are connected to each other by a pin joint. The cart is subjected to a force of excitation $F_f = F_1 \cos \omega_f t$. Find the equations of motion in terms of q and ϑ by using Lagrange's equations. Linearize the equations of motion and give them in matrix form.

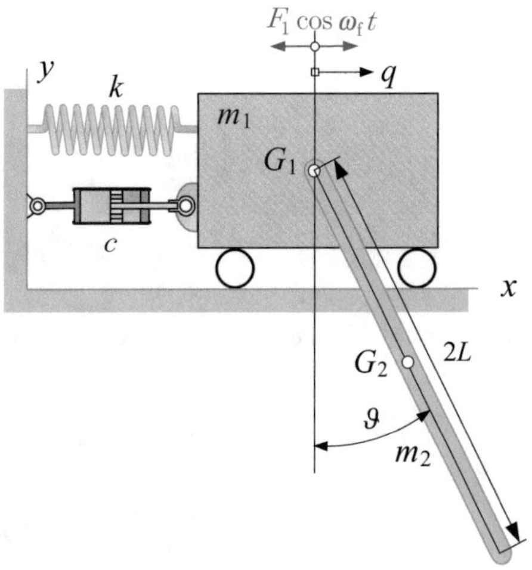

Fig. 4.26 Vibrating system constituted by a cart and rod

Problem 4.3 Determine the natural frequencies and the mode shapes for system C if $m_1 = m_2 = m$ and $k_o = k_1 = k_2 = k$—see Fig. 4.15. (Hint: Make use of Eq. (4.65).)

Problem 4.4 Solve Problem 4.3 if $k_o = k_2 = k$ and $k_1 = \mu k$, where μ is a positive integer.

References

1. J.L. Lagrange, *Mécanique Analytique*, 2nd edn. (Cambridge University Press, Cambridge, 2009). The first edition was published in Paris (1788)
2. J.L. Lagrange, *Analytical Mechanics*. Translated from the Mécanique analytique, novelle edition of 1811 and 1815, eds. by R.S. Cohen. Translated by A. Boissonnade, V.N. Vagliente, vol. 191. Boston Study of Philosophy and Science (Springer Netherlands, Dordrecht, 1997). https://doi.org/10.1007/978-94-015-8903-1

Chapter 5
Some Problems of Multidegree of Freedom Systems

5.1 Equations of Motion

5.1.1 Spring–Mass Systems

Figure 5.1 depicts (a) a free spring–mass system with five degrees of freedom, (b) a four degree of freedom spring–mass system fixed at the left end and free at the right end, and finally (c) a three degree of freedom system fixed at both ends. These

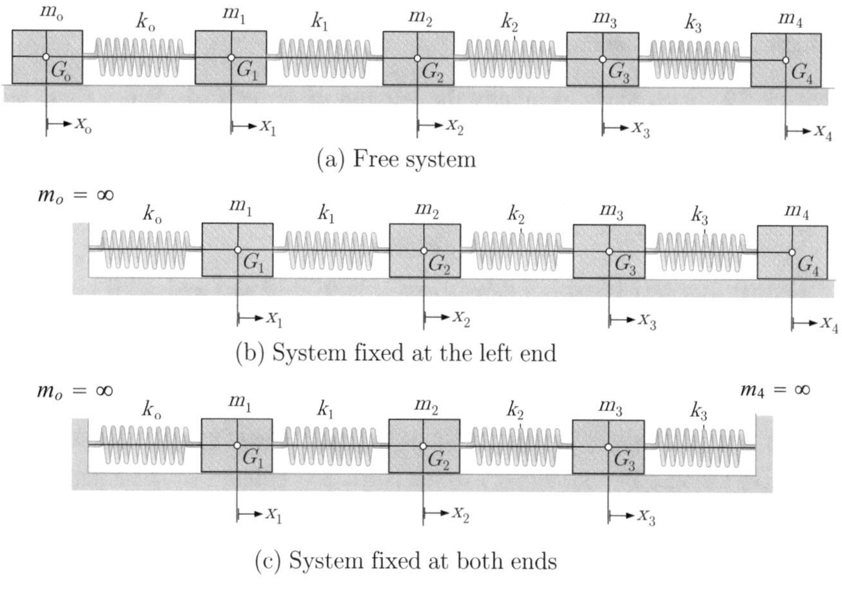

Fig. 5.1 Spring-mass systems

G. Szeidl and L. P. Kiss, *Mechanical Vibrations*, Foundations of Engineering Mechanics,
https://doi.org/10.1007/978-3-030-45074-8_5

systems represent the three possible categories for the spring–mass systems. The equation of motion for the free spring–mass system is given by the equation

$$\begin{bmatrix} m_o & 0 & 0 & 0 & 0 \\ 0 & m_1 & 0 & 0 & 0 \\ 0 & 0 & m_2 & 0 & 0 \\ 0 & 0 & 0 & m_3 & 0 \\ 0 & 0 & 0 & 0 & m_4 \end{bmatrix} \begin{bmatrix} \ddot{x}_o \\ \ddot{x}_1 \\ \ddot{x}_2 \\ \ddot{x}_3 \\ \ddot{x}_4 \end{bmatrix} + \begin{bmatrix} k_o & -k_o & 0 & 0 & 0 \\ -k_o & k_o+k_1 & -k_1 & 0 & 0 \\ 0 & -k_1 & k_1+k_2 & -k_2 & 0 \\ 0 & 0 & -k_2 & k_2+k_3 & -k_3 \\ 0 & 0 & 0 & -k_3 & k_3 \end{bmatrix} \begin{bmatrix} x_o \\ x_1 \\ x_2 \\ x_3 \\ x_4 \end{bmatrix} = \begin{bmatrix} 0 \\ 0 \\ 0 \\ 0 \\ 0 \end{bmatrix} \tag{5.1}$$

which can be rewritten in matrix form as

$$\underset{(5\times5)}{\underline{\mathbf{M}}}\ \underset{(5\times1)}{\underline{\ddot{\mathbf{q}}}} + \underset{(5\times5)}{\underline{\mathbf{K}}}\ \underset{(5\times1)}{\underline{\mathbf{q}}} = \underset{(5\times1)}{\underline{\mathbf{0}}}\,, \tag{5.2}$$

where $\underline{\mathbf{M}}$ is the mass matrix, $\underline{\mathbf{K}}$ is the stiffness matrix, and $\underline{\mathbf{q}}$ is the matrix of the generalized displacements. Comparison of Figs. 4.15, and 5.1 shows that the equations of the three spring–mass systems in Fig. 5.1 have the same structure as those in Fig. 4.15.

Remark 5.1 For the free system, the sum of the elements in a row or column in the stiffness matrix is equal to zero.

Remark 5.2 If m_o tends to infinity the left end of spring k_o becomes fixed and system (a) changes into system (b). The equations of motion are obtained from the equations of motion (5.1) by canceling the first row in each matrix and the first column in the mass and stiffness matrices. The result is presented in frames in Eq. (5.1). For completeness, it is repeated here:

$$\begin{bmatrix} m_1 & 0 & 0 & 0 \\ 0 & m_2 & 0 & 0 \\ 0 & 0 & m_3 & 0 \\ 0 & 0 & 0 & m_4 \end{bmatrix} \begin{bmatrix} \ddot{x}_1 \\ \ddot{x}_2 \\ \ddot{x}_3 \\ \ddot{x}_4 \end{bmatrix} + \begin{bmatrix} k_o+k_1 & -k_1 & 0 & 0 \\ -k_1 & k_1+k_2 & -k_2 & 0 \\ 0 & -k_2 & k_2+k_3 & -k_3 \\ 0 & 0 & -k_3 & k_3 \end{bmatrix} \begin{bmatrix} x_1 \\ x_2 \\ x_3 \\ x_4 \end{bmatrix} = \begin{bmatrix} 0 \\ 0 \\ 0 \\ 0 \end{bmatrix}. \tag{5.3}$$

Remark 5.3 If m_4 tends to infinity the right end of spring k_4 becomes fixed and system (b) changes into system (c) which is fixed at both ends. The equations of motion are obtained from the equations of motion (5.1) by canceling the first and last row in each matrix as well as the first and last column in the mass and stiffness matrices. The result is presented again in frames drawn by dashed lines in Eq. (5.1). For completeness, we give it here:

$$\begin{bmatrix} m_1 & 0 & 0 \\ 0 & m_2 & 0 \\ 0 & 0 & m_3 \end{bmatrix} \begin{bmatrix} \ddot{x}_1 \\ \ddot{x}_2 \\ \ddot{x}_3 \end{bmatrix} + \begin{bmatrix} k_o+k_1 & -k_1 & 0 \\ -k_1 & k_1+k_2 & -k_2 \\ 0 & -k_2 & k_2+k_3 \end{bmatrix} \begin{bmatrix} x_1 \\ x_2 \\ x_3 \end{bmatrix} = \begin{bmatrix} 0 \\ 0 \\ 0 \end{bmatrix}. \tag{5.4}$$

5.1.2 The General Form of the Equation of Motion

For a free n degree of freedom system ($n \geq 2$) with no damping the equation of motion has the following general form:

$$\underbrace{\begin{bmatrix} m_{11} & m_{12} & \dots & m_{1n} \\ m_{21} & m_{22} & \dots & m_{2n} \\ \vdots & \vdots & \ddots & \vdots \\ m_{n1} & m_{n2} & \dots & m_{nn} \end{bmatrix}}_{\underset{(n\times n)}{\underline{\mathbf{M}}}} \underbrace{\begin{bmatrix} \ddot{q}_1 \\ \ddot{q}_2 \\ \vdots \\ \ddot{q}_n \end{bmatrix}}_{\underset{(n\times 1)}{\underline{\ddot{\mathbf{q}}}}} + \underbrace{\begin{bmatrix} k_{11} & k_{12} & \dots & k_{1n} \\ k_{21} & k_{22} & \dots & k_{2n} \\ \vdots & \vdots & \ddots & \vdots \\ k_{n1} & k_{n2} & \dots & k_{nn} \end{bmatrix}}_{\underset{(n\times n)}{\underline{\mathbf{K}}}} \underbrace{\begin{bmatrix} q_1 \\ q_2 \\ \vdots \\ q_n \end{bmatrix}}_{\underset{(n\times 1)}{\underline{\mathbf{q}}}} = \underbrace{\begin{bmatrix} 0 \\ 0 \\ \vdots \\ 0 \end{bmatrix}}_{\underset{(n\times 1)}{\underline{\mathbf{0}}}}. \tag{5.5}$$

Remark 5.4 In contrast to the equations that describe the motions of the considered spring–mass systems—see, for instance, (5.1)–(5.4)—the elements of the matrix $\underline{\mathbf{q}}$ are now denoted by q_ℓ ($\ell = 1, \dots, n$) since there is no guarantee that the general displacements q_ℓ are all horizontal vector components, i.e., parallel to each other.

Let $\underline{\mathbf{v}}$, ($|\underline{\mathbf{v}}| \neq 0$) be an ($n \times 1$) matrix. Further let $\underline{\mathbf{A}}$ be an ($n \times n$) symmetric matrix.

$$\underline{\mathbf{A}} \text{ is said to be } \left\{ \begin{array}{l} \text{positive definite} \\ \text{positive semidefinite} \\ \text{negative semidefinite} \\ \text{negative definite} \end{array} \right\} \text{ if } \left\{ \begin{array}{l} \underline{\mathbf{v}}^T \underline{\mathbf{A}}\, \underline{\mathbf{v}} > 0 \\ \underline{\mathbf{v}}^T \underline{\mathbf{A}}\, \underline{\mathbf{v}} \geq 0 \\ \underline{\mathbf{v}}^T \underline{\mathbf{A}}\, \underline{\mathbf{v}} \leq 0 \\ \underline{\mathbf{v}}^T \underline{\mathbf{A}}\, \underline{\mathbf{v}} < 0 \end{array} \right\} \text{ for any } \underline{\mathbf{v}}.$$

If none of the above relations is satisfied the matrix $\underline{\mathbf{A}}$ is indefinite.

Remark 5.5 The mass matrix is a symmetric and positive definite matrix:

$$\underline{\mathbf{M}} = \underline{\mathbf{M}}^T, \qquad \mathcal{E} = \frac{1}{2}\underline{\dot{\mathbf{q}}}^T \underline{\mathbf{M}}\, \underline{\dot{\mathbf{q}}} > 0. \tag{5.6}$$

Equation $(5.6)_2$ shows that the positive definiteness is a consequence of the fact that the kinetic energy $\mathcal{E}$ is a positive quantity.

Remark 5.6 Let us assume that the kinetic energy can be given in the form $(5.6)_2$ but the mass matrix is not symmetric. We define the following two matrices:

$$\underline{\mathbf{M}}_s = \frac{1}{2}\left(\underline{\mathbf{M}} + \underline{\mathbf{M}}^T\right), \qquad \underline{\mathbf{M}}_a = \frac{1}{2}\left(\underline{\mathbf{M}} - \underline{\mathbf{M}}^T\right)$$

which obviously satisfy the relations

$$\underline{\mathbf{M}} = \underline{\mathbf{M}}_s + \underline{\mathbf{M}}_a, \qquad \underline{\mathbf{M}}_s = \underline{\mathbf{M}}_s^T, \qquad \text{and} \qquad \underline{\mathbf{M}}_a = -\underline{\mathbf{M}}_a^T.$$

Because of these properties $\underline{\mathbf{M}}_s$ and $\underline{\mathbf{M}}_a$ are called the symmetric and antisymmetric (or skew) parts of $\underline{\mathbf{M}}$. As is well known it holds for any $\underline{\mathbf{v}}$ that

$$\underline{\mathbf{v}}^T \underline{\mathbf{M}}_a \underline{\mathbf{v}} = 0.$$

Consequently,

$$\mathcal{E} = \frac{1}{2}\underline{\dot{\mathbf{q}}}^T \underline{\mathbf{M}}\, \underline{\dot{\mathbf{q}}} = \frac{1}{2}\underline{\dot{\mathbf{q}}}^T \underline{\mathbf{M}}_s \underline{\dot{\mathbf{q}}} + \underbrace{\frac{1}{2}\underline{\dot{\mathbf{q}}}^T \underline{\mathbf{M}}_a \underline{\dot{\mathbf{q}}}}_{=0} = \frac{1}{2}\underline{\dot{\mathbf{q}}}^T \underline{\mathbf{M}}_s \underline{\dot{\mathbf{q}}}$$

which shows that the antisymmetric matrix $\underline{\mathbf{M}}_a$ has no effect on the kinetic energy. On the basis of this fact, we shall assume that the mass matrix is always symmetric.

Remark 5.7 The stiffness matrix is also symmetric but positive semidefinite matrix:

$$\underline{\mathbf{K}} = \underline{\mathbf{K}}^T, \qquad \mathcal{U} = \frac{1}{2}\underline{\mathbf{q}}^T \underline{\mathbf{K}}\, \underline{\mathbf{q}} \geq 0. \tag{5.7}$$

Equation $(5.7)_2$ shows that the positive semidefiniteness is a consequence of the fact that the strain energy $\mathcal{U}$ stored in the elastic elements of the system (in the springs) is a positive quantity except the case when the body (the elastic elements) has a rigid body motion.

Exercise 5.1 The cantilever beam of length $\ell = 3a$ is subjected to the vertical forces Q_1, Q_2, and Q_3. Find the vertical displacements q_1, q_2, and q_3 at the points P_1, P_2, and P_3, i.e., at the points where the forces are applied. Use the results obtained to prove that the stiffness matrix is symmetric (Fig.5.2).

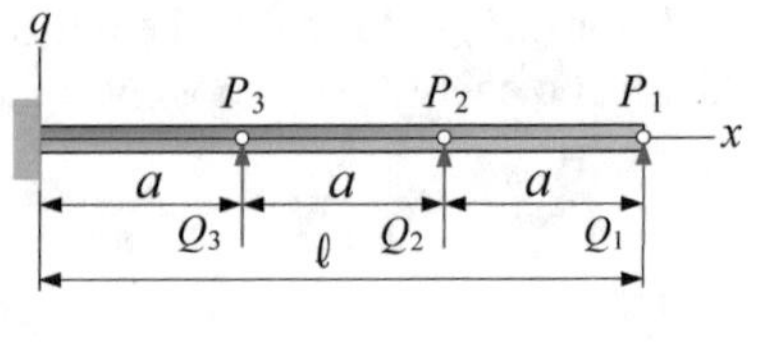

Fig. 5.2 Cantilever beam with the forces acting on it

In accordance with Remark 3.2 the vertical displacement at the point P_i due to a vertical unit force applied at the point P_j $(i, j = 1, 2, 3)$ is denoted by f_{ij} and is called flexibility. Since the bending problem of an elastic beam is governed by linear equations we can apply the principle of superposition to determine the displacements in question:

$$\begin{aligned} q_1 &= f_{11}Q_1 + f_{12}Q_2 + f_{13}Q_3\,, \\ q_2 &= f_{21}Q_1 + f_{22}Q_2 + f_{23}Q_3\,, \\ q_3 &= f_{31}Q_1 + f_{32}Q_2 + f_{33}Q_3\,. \end{aligned} \tag{5.8a}$$

These equations can be rewritten in matrix form

$$\underbrace{\begin{bmatrix} q_1 \\ q_2 \\ q_3 \end{bmatrix}}_{\underset{(3\times1)}{\underline{\mathbf{q}}}} = \underbrace{\begin{bmatrix} f_{11} & f_{12} & f_{13} \\ f_{21} & f_{22} & f_{23} \\ f_{31} & f_{32} & f_{33} \end{bmatrix}}_{\underset{(3\times3)}{\underline{\mathbf{f}}}} \underbrace{\begin{bmatrix} Q_1 \\ Q_2 \\ Q_3 \end{bmatrix}}_{\underset{(3\times1)}{\underline{\mathbf{Q}}}}, \tag{5.8b}$$

or concisely as

$$\underset{(3\times1)}{\underline{\mathbf{q}}} = \underset{(3\times3)}{\underline{\mathbf{f}}}\ \underset{(3\times1)}{\underline{\mathbf{Q}}}\,, \tag{5.8c}$$

where $\underline{\mathbf{f}}$ is the flexibility matrix. Multiply throughout by the inverse of the flexibility matrix. The result is

$$\underset{(3\times3)}{\underline{\mathbf{f}}}^{-1}\ \underset{(3\times1)}{\underline{\mathbf{q}}} = \underset{(3\times3)}{\underline{\mathbf{K}}}\ \underset{(3\times1)}{\underline{\mathbf{q}}} = \underset{(3\times1)}{\underline{\mathbf{Q}}} \qquad \underset{(3\times3)}{\underline{\mathbf{f}}}^{-1} = \underset{(3\times3)}{\underline{\mathbf{K}}}^{-1} \tag{5.9a}$$

or

$$\begin{bmatrix} Q_1 \\ Q_2 \\ Q_3 \end{bmatrix} = \begin{bmatrix} k_{11} & k_{12} & k_{13} \\ k_{21} & k_{22} & k_{23} \\ k_{31} & k_{32} & k_{33} \end{bmatrix} \begin{bmatrix} q_1 \\ q_2 \\ q_3 \end{bmatrix}. \tag{5.9b}$$

It follows from the structure of Eqs. (5.9) that $\underline{\mathbf{K}}$ is the stiffness matrix.

Maxwell's reciprocity theorem states that the flexibility matrix is symmetric

$$\underline{\mathbf{f}} = \underline{\mathbf{f}}^T \qquad \longrightarrow \qquad f_{ij} = f_{ji} \quad (i, j = 1, 2, 3)\,. \tag{5.10}$$

For the proof of this theorem we shall assume that the cantilever beam is subjected to two forces Q_I and Q_J only where the capital letters mean that the subscripts $I, J \in (1, 2, 3)$ are fixed. Maxwell's theorem is based on the fact that the work done by the two forces is independent of the loading order. Let W_{II} and W_{JJ} be the works done by the forces when we apply them one by one independently of each other. First let I, J be the loading order. Then W_{IJ} denotes the work done by the force Q_I throughout the displacements caused by the force Q_J. The total work for this loading order is given by

$$W = W_{II} + W_{IJ} + W_{JJ} = \frac{1}{2} f_{II}\, Q_I^2 + Q_I f_{IJ} Q_J + \frac{1}{2} f_{JJ}\, Q_J^2 \,. \tag{5.11a}$$

For the reverse loading order W_{JI} is the work done by the force Q_J throughout the displacements caused by the force Q_I. Consequently,

$$W = W_{JJ} + W_{JI} + W_{II} = \frac{1}{2} f_{II}\, Q_I^2 + Q_I f_{JI} Q_J + \frac{1}{2} f_{JJ}\, Q_J^2 \tag{5.11b}$$

is the total work for the second loading order. Comparison of Eqs. (5.11) yields the symmetry condition to be proved:

$$f_{IJ} = f_{JI} \,. \tag{5.12}$$

Conclusion: If the flexibility matrix is symmetric then so is the stiffness matrix.

5.2 Eigenvalue Problem of Symmetric Matrices

5.2.1 Eigenvalues and Eigenvectors

Let us introduce the notations

$$\underline{\mathbf{q}}^T \underline{\mathbf{M}}\, \underline{\mathbf{q}} = \left(\underline{\mathbf{q}}, \underline{\mathbf{q}}\right)_M \,, \qquad \underline{\mathbf{q}}^T \underline{\mathbf{K}}\, \underline{\mathbf{q}} = \left(\underline{\mathbf{q}}, \underline{\mathbf{q}}\right)_K \tag{5.13}$$

which reflect that the products are performed on the matrices $\underline{\mathbf{M}}$ and $\underline{\mathbf{K}}$. Since the mass matrix is positive definite and the stiffness matrix is positive semidefinite it holds that

$$\left(\underline{\mathbf{q}}, \underline{\mathbf{q}}\right)_M > 0 \,, \qquad \left(\underline{\mathbf{q}}, \underline{\mathbf{q}}\right)_K \geq 0 \,. \tag{5.14}$$

It is well known from the theory of differential equations that the equations of motion

$$\underset{(n\times n)}{\underline{\mathbf{M}}}\ \underset{(n\times 1)}{\ddot{\underline{\mathbf{q}}}} + \underset{(n\times n)}{\underline{\mathbf{K}}}\ \underset{(n\times 1)}{\underline{\mathbf{q}}} = \underset{(n\times 1)}{\underline{\mathbf{0}}} \tag{5.15}$$

have one and only one solution (uniqueness theorem) which satisfies the initial conditions

$$\underline{\mathbf{q}}(t)\Big|_{t_o=0} = \underbrace{\underline{\mathbf{q}}_o}_{\substack{\text{initial}\\ \text{displacements}}} \,, \qquad \dot{\underline{\mathbf{q}}}(t)\Big|_{t_o=0} = \underbrace{\underline{\mathbf{v}}_o}_{\substack{\text{initial}\\ \text{velocities}}} \,. \tag{5.16}$$

Assume that

$$\underset{(n\times 1)}{\underline{\mathbf{q}}} = \underset{(n\times 1)}{\underline{\mathbf{A}}} \exp(i\omega t) \,, \tag{5.17a}$$

where i is the imaginary unit, ω is an unknown parameter (the natural circular frequencies sought), and $\underline{\mathbf{A}}$ is the unknown amplitude vector. It is obvious that

$$\dot{\underline{\mathbf{q}}} = i\omega\underline{\mathbf{A}}\exp(i\omega t) = i\omega\,\underline{\mathbf{q}}\,, \qquad \ddot{\underline{\mathbf{q}}} = -\omega^2\underline{\mathbf{A}}\exp(i\omega t) = -\omega^2\underline{\mathbf{q}}\,. \tag{5.17b}$$

Upon substitution of $\underline{\mathbf{q}}$ and $\ddot{\underline{\mathbf{q}}}$ into the equations of motion (5.15), we obtain

$$\Big(\underset{(n\times n)}{\underline{\mathbf{K}}} - \lambda\underset{(n\times n)}{\underline{\mathbf{M}}}\Big)\underset{(n\times 1)}{\underline{\mathbf{A}}} = \underset{(n\times 1)}{\underline{\mathbf{0}}}\,, \qquad \lambda = \omega^2 \tag{5.18}$$

which is a generalized algebraic eigenvalue problem with $\lambda = \omega^2$ as an eigenvalue and $\underline{\mathbf{A}}$ as an eigenvector. Solution for $\underline{\mathbf{A}}$ (different from the trivial one) exists if and only if

$$p_n(\lambda) = \left|\underline{\mathbf{K}} - \lambda\underline{\mathbf{M}}\right| = 0\,, \tag{5.19}$$

where $p_n(\lambda)$ is a polynomial of degree n. Let λ_ℓ be ($\lambda_1 \le \lambda_2 \le \lambda_3 \le \cdots \le \lambda_n$) the ℓth root (the ℓth eigenvalue) and $\underline{\mathbf{A}}_\ell$ be the corresponding solution (the ℓth eigenvector or eigenmatrix).

Remark 5.8 According to a fundamental algebraic theorem there exist n (not necessarily different) roots of the equation $p_n(\lambda) = 0$.

Let ν be an arbitrary scalar ($\nu \neq 0$). It is clear that

$$\left(\underline{\mathbf{K}} - \lambda_\ell\underline{\mathbf{M}}\right)\left(\nu\underline{\mathbf{A}}_\ell\right) = \left(\underline{\mathbf{K}} - \omega_\ell^2\underline{\mathbf{M}}\right)\left(\nu\underline{\mathbf{A}}_\ell\right) = \underline{\mathbf{0}} \tag{5.20}$$

which shows that $\nu\underline{\mathbf{A}}_\ell$ is also a solution of Eq. (5.18). If we have already determined $\omega_\ell = \pm\sqrt{\lambda_\ell}$ and $\underline{\mathbf{A}}_\ell$ then

$$\underline{\mathbf{q}}_\ell = \underline{\mathbf{A}}_\ell\exp(i\omega_\ell t) = \underline{\mathbf{A}}_\ell(\cos\omega_\ell t + i\sin\omega_\ell t) \tag{5.21a}$$

or

$$\underline{\mathbf{q}}_\ell = \underline{\mathbf{A}}_\ell\exp(-i\omega_\ell t) = \underline{\mathbf{A}}_\ell(\cos\omega_\ell t - i\sin\omega_\ell t) \tag{5.21b}$$

are solutions to the equations of motion (5.15). Since both the real part and the imaginary parts of the solutions (5.21) satisfy Eq. (5.15) it follows that the general and real solution of this equation (under the condition $\lambda_\ell \neq \lambda_k$; $\ell \neq k$, $\ell, k = 1, 2, 3, \ldots, n$) assumes the following form:

$$\underline{\mathbf{q}} = \sum_{\ell=1}^{n}\left(a_\ell\,\underline{\mathbf{A}}_\ell\cos\omega_\ell t + b_\ell\,\underline{\mathbf{A}}_\ell\sin\omega_\ell t\right), \tag{5.22}$$

where the undetermined constants of integration a_ℓ and b_ℓ can be determined from the initial conditions (5.16).

Since the mass matrix $\underline{\mathbf{M}}$ is positive definite it has an inverse $\underline{\mathbf{M}}^{-1}$. Multiplying Eq. (5.15) from the left by $\underline{\mathbf{M}}^{-1}$, we obtain

$$\underbrace{\underline{\mathbf{M}}^{-1}\underline{\mathbf{M}}}_{\underline{\mathbf{I}}}\,\underline{\ddot{\mathbf{q}}} + \underbrace{\underline{\mathbf{M}}^{-1}\underline{\mathbf{K}}}_{\underline{\mathbf{D}}}\,\underline{\mathbf{q}} = \underline{\mathbf{I}}\,\underline{\ddot{\mathbf{q}}} + \underline{\mathbf{D}}\,\underline{\mathbf{q}} = \underline{\mathbf{0}}, \tag{5.23}$$

where $\underset{(n\times n)}{\underline{\mathbf{I}}}$ is the unit matrix and $\underset{(n\times n)}{\underline{\mathbf{D}}}$ is referred to as dynamic matrix.

After substituting the solution (5.17a) for $\underline{\mathbf{q}}$ and its second time derivative $(5.17\text{b})_2$ into Eq. (5.23) we arrive at the result that $\underline{\mathbf{A}}$ satisfies the equation

$$\Big(\underset{(n\times n)}{\underline{\mathbf{D}}} - \lambda\underset{(n\times n)}{\underline{\mathbf{I}}}\Big)\underset{(n\times 1)}{\underline{\mathbf{A}}} = \underset{(n\times 1)}{\underline{\mathbf{0}}}\,, \qquad \lambda = \omega^2 \tag{5.24}$$

which is nothing but the well-known algebraic eigenvalue problem.

5.2.2 *Calculation of the Eigenvectors*

It is possible to find the eigenvectors by utilizing the adjugate matrix of the equation system (5.24). For brevity we shall introduce the following notation for the coefficient matrix:

$$\underline{\mathbf{B}} = \underline{\mathbf{D}} - \lambda\underline{\mathbf{I}}. \tag{5.25}$$

The inverse of $\underline{\mathbf{B}}$ in terms of the adjugate matrix is given by

$$\underline{\mathbf{B}}^{-1} = \frac{1}{|\underline{\mathbf{B}}|}\,\mathrm{adj}\,\underline{\mathbf{B}}\,, \qquad \det\underline{\mathbf{B}} = |\underline{\mathbf{B}}| = B\,. \tag{5.26}$$

Multiplying now throughout from the left side by $B\,\underline{\mathbf{B}}$, we have

$$B\,\underline{\mathbf{B}}\,\underline{\mathbf{B}}^{-1} = B\,\underline{\mathbf{I}} = \underline{\mathbf{B}}\,\mathrm{adj}\,\underline{\mathbf{B}} \tag{5.27}$$

which can be rewritten in terms of $\underline{\mathbf{D}}$ and $\lambda\underline{\mathbf{I}}$

$$\underline{\mathbf{I}}\underbrace{|\underline{\mathbf{D}} - \lambda\underline{\mathbf{I}}|}_{B} = (\underline{\mathbf{D}} - \lambda\underline{\mathbf{I}})\,\mathrm{adj}\,(\underline{\mathbf{D}} - \lambda\underline{\mathbf{I}})\,. \tag{5.28}$$

Since the determinant B is zero if $\lambda = \lambda_\ell$ it is worth substituting λ_ℓ for λ in the above equation. We obtain

$$\underset{(n\times n)}{(\underline{\mathbf{D}} - \lambda_\ell\underline{\mathbf{I}})}\,\mathrm{adj}\underset{(n\times n)}{(\underline{\mathbf{D}} - \lambda_\ell\underline{\mathbf{I}})} = \underset{(n\times n)}{\underline{\mathbf{0}}}\,, \quad \ell = 1, 2, \ldots, n. \tag{5.29a}$$

Recalling that the ℓth eigenvector (the ℓth mode shape) is a solution of the equation

$$\underset{(n\times n)}{\left(\underline{\mathbf{D}} - \lambda_\ell \underline{\mathbf{I}}\right)} \underset{(n\times 1)}{\underline{\mathbf{A}}_\ell} = \underset{(n\times 1)}{\underline{\mathbf{0}}} \tag{5.29b}$$

we should recognize each column in the adjugate matrix adj $\left(\underline{\mathbf{D}} - \lambda_\ell \underline{\mathbf{I}}\right)$ is directly proportional to the eigenvector $\underline{\mathbf{A}}_\ell$.

5.2.3 *Orthogonality of the Eigenvectors*

Now we shall show that two eigenvectors corresponding to two different eigenvalues are mutually orthogonal with respect to the mass and stiffness matrices. Let λ_ℓ and λ_k be two different eigenvalues. The corresponding eigenvectors $\underline{\mathbf{A}}_\ell$ and $\underline{\mathbf{A}}_k$ satisfy the equations

$$(\underline{\mathbf{K}} - \lambda_\ell \underline{\mathbf{M}})\underline{\mathbf{A}}_\ell = \underline{\mathbf{0}} \tag{5.30a}$$

and

$$(\underline{\mathbf{K}} - \lambda_k \underline{\mathbf{M}})\underline{\mathbf{A}}_k = \underline{\mathbf{0}}\,. \tag{5.30b}$$

Left multiplying Eq. [(5.30a)] ((5.30b)) by $[\underline{\mathbf{A}}_k^T]$ $(\underline{\mathbf{A}}_\ell^T)$, we obtain

$$\left(\underline{\mathbf{A}}_k, \underline{\mathbf{A}}_\ell\right)_K = \lambda_\ell \left(\underline{\mathbf{A}}_k, \underline{\mathbf{A}}_\ell\right)_M \tag{5.31a}$$

and

$$\left(\underline{\mathbf{A}}_k, \underline{\mathbf{A}}_\ell\right)_K = \lambda_k \left(\underline{\mathbf{A}}_k, \underline{\mathbf{A}}_\ell\right)_M, \tag{5.31b}$$

where we have taken into account that

$$\left(\underline{\mathbf{A}}_k, \underline{\mathbf{A}}_\ell\right)_K = \left(\underline{\mathbf{A}}_\ell, \underline{\mathbf{A}}_k\right)_K, \qquad \left(\underline{\mathbf{A}}_k, \underline{\mathbf{A}}_\ell\right)_M = \left(\underline{\mathbf{A}}_\ell, \underline{\mathbf{A}}_k\right)_M \tag{5.32}$$

which is a consequence of the fact that the matrices $\underline{\mathbf{M}}$ and $\underline{\mathbf{K}}$ are symmetric. Thus, subtracting Eq. (5.31b) from Eq. (5.31a) the left sides cancel out and we get

$$0 = \underbrace{(\lambda_\ell - \lambda_k)}_{\neq 0} \left(\underline{\mathbf{A}}_k, \underline{\mathbf{A}}_\ell\right)_M. \tag{5.33}$$

Since $\lambda_\ell \neq \lambda_k$ the above product vanishes if and only if the general orthogonality condition

$$\boxed{\left(\underline{\mathbf{A}}_k, \underline{\mathbf{A}}_\ell\right)_M = 0} \tag{5.34a}$$

is fulfilled.

It follows from Eqs. (5.32) with regard to the orthogonality condition (5.34) that the other orthogonality condition

$$\boxed{\left(\underline{\mathbf{A}}_k, \underline{\mathbf{A}}_\ell\right)_K = 0} \tag{5.34b}$$

also holds.

Remark 5.9 If the product

$$\left(\underline{\mathbf{A}}_k, \underline{\mathbf{A}}_\ell\right)_I = \underline{\mathbf{A}}_k^T \underline{\mathbf{A}}_\ell = 0 \tag{5.35}$$

taken on the unit matrix $\underline{\mathbf{I}}$ is zero then we say that the column matrices $\underline{\mathbf{A}}_k$ and $\underline{\mathbf{A}}_\ell$ are orthogonal in the classical sense.

Equations (5.34a) and (5.34b) are the general orthogonality conditions we wanted to prove.

For $k = \ell$, Eqs. (5.31) result in the following formula:

$$\left(\underline{\mathbf{A}}_\ell, \underline{\mathbf{A}}_\ell\right)_K = \lambda_\ell \left(\underline{\mathbf{A}}_\ell, \underline{\mathbf{A}}_\ell\right)_M . \tag{5.36}$$

The length (or norm) of $\underline{\mathbf{A}}_\ell$ with respect to the mass matrix $\underline{\mathbf{M}}$ is defined as

$$\left|\underline{\mathbf{A}}_\ell\right| = \sqrt{\left(\underline{\mathbf{A}}_\ell, \underline{\mathbf{A}}_\ell\right)_M}. \tag{5.37}$$

It is often useful to normalize the eigenvectors. This can be done by dividing the eigenvector $\underline{\mathbf{A}}_\ell$ with its length:

$$\underline{\psi}_\ell = \frac{\underline{\mathbf{A}}_\ell}{\left|\underline{\mathbf{A}}_\ell\right|} = \frac{\underline{\mathbf{A}}_\ell}{\sqrt{\left(\underline{\mathbf{A}}_\ell, \underline{\mathbf{A}}_\ell\right)_M}} . \tag{5.38}$$

For the normalized eigenvectors, it holds that

$$\left(\underline{\psi}_k, \underline{\psi}_\ell\right)_M = \delta_{k\ell}, \tag{5.39a}$$

where

$$\delta_{k\ell} = \begin{cases} 1 \text{ if } k = \ell \\ 0 \text{ if } k \neq \ell \end{cases} \qquad k, \ell = 1, 2, 3, \ldots, n \tag{5.39b}$$

is the Kronecker delta. Equation (5.39a) is a consequence of the orthogonality condition (5.34a). Let us now divide Eq. (5.31a) by the norm (5.38). If we take the relations (5.39a) and (5.39b) into account we get

$$\left(\underline{\psi}_k, \underline{\psi}_\ell\right)_K = \lambda_\ell \, \delta_{k\ell} . \tag{5.39c}$$

For the sake of our later considerations, we shall introduce the following notations:

$$\left(\underline{\mathbf{A}}_\ell, \underline{\mathbf{A}}_\ell\right)_M = m_{\ell\ell}\,, \qquad \left(\underline{\mathbf{A}}_\ell, \underline{\mathbf{A}}_\ell\right)_K = k_{\ell\ell}\,. \tag{5.40}$$

Exercise 5.2 Consider the two degree of freedom system shown in Fig. 5.3. Determine the corresponding eigenvalues and eigenvectors.

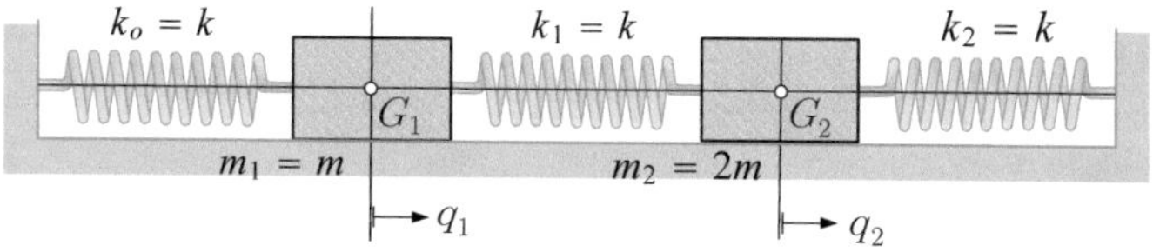

Fig. 5.3 Spring-mass system with two degrees of freedom

It follows from (4.58c) that the equations of motion in matrix notation are of the form

$$\underbrace{\begin{bmatrix} m & 0 \\ 0 & 2m \end{bmatrix}}_{\underset{(2\times2)}{\underline{\mathbf{M}}}} \underbrace{\begin{bmatrix} \ddot{q}_1 \\ \ddot{q}_2 \end{bmatrix}}_{\underset{(2\times1)}{\ddot{\underline{\mathbf{q}}}}} + \underbrace{\begin{bmatrix} 2k & -k \\ -k & 2k \end{bmatrix}}_{\underset{(2\times2)}{\underline{\mathbf{K}}}} \underbrace{\begin{bmatrix} q_1 \\ q_2 \end{bmatrix}}_{\underset{(2\times1)}{\underline{\mathbf{q}}}} = \begin{bmatrix} 0 \\ 0 \end{bmatrix}. \tag{5.41}$$

After substituting the solution and its second derivative from (4.58c) and left multiplying then by

$$\underline{\mathbf{M}}^{-1} = \begin{bmatrix} 1/m & 0 \\ 0 & 1/2m \end{bmatrix}, \tag{5.42}$$

we obtain

$$\left(\underline{\mathbf{D}} - \lambda\,\underline{\mathbf{I}}\right)\underline{\mathbf{A}} = \begin{bmatrix} 2\frac{k}{m} - \lambda & -\frac{k}{m} \\ -\frac{k}{2m} & \frac{k}{m} - \lambda \end{bmatrix} \begin{bmatrix} A_1 \\ A_2 \end{bmatrix} = \begin{bmatrix} 0 \\ 0 \end{bmatrix}, \qquad \lambda = \omega^2\,. \tag{5.43}$$

The characteristic equation is the determinant of the coefficient matrix

$$\left|\underline{\mathbf{D}} - \lambda\,\underline{\mathbf{I}}\right| = \lambda^2 - 3\frac{k}{m}\lambda + \frac{3}{2}\left(\frac{k}{m}\right)^2 = \left(\lambda - \frac{k}{m}\frac{\sqrt{3}+3}{2}\right)\left(\lambda + \frac{k}{m}\frac{\sqrt{3}-3}{2}\right) = 0$$

from where

$$\lambda_1 = -\frac{k}{m}\frac{\sqrt{3}-3}{2} = 0.633\,97\,\frac{k}{m}, \qquad \lambda_2 = \frac{k}{m}\frac{\sqrt{3}+3}{2} = 2.366\,03\frac{k}{m}$$

are the roots. There are two possibilities for calculating the eigenvectors. (a) We can find them from Eq. (5.43) if we back-substitute λ_1 and λ_2 (the solution steps are basically the same as those in Exercise 4.7). (b) We can also use the adjugate matrix $\operatorname{adj}(\underline{\mathbf{D}} - \lambda_\ell\,\underline{\mathbf{I}})$ for their calculation. Here we will select the second possibility so that we can show how to apply the adjugate matrix:

$$\operatorname{adj}\,(\underline{\mathbf{D}}-\lambda_\ell\,\underline{\mathbf{I}})=\operatorname{adj}\begin{bmatrix} 2\frac{k}{m}-\lambda_\ell & -\frac{k}{m} \\ -\frac{k}{2m} & \frac{k}{m}-\lambda_\ell \end{bmatrix}=\begin{bmatrix} \frac{k}{m}-\lambda_\ell & \frac{k}{m} \\ \frac{k}{2m} & 2\frac{k}{m}-\lambda_\ell \end{bmatrix}, \quad \ell=1,2\,.$$

If $\ell = 1$ we get

$$\operatorname{adj}\,(\underline{\mathbf{D}}-\lambda_\ell\,\underline{\mathbf{I}}) = \frac{k}{m}\begin{bmatrix} 1-\frac{3-\sqrt{3}}{2} & 1 \\ 0.5 & 2-\frac{3-\sqrt{3}}{2} \end{bmatrix} = \frac{k}{m}\begin{bmatrix} 0.366\,03 & 1.000\,00 \\ 0.500\,00 & 1.366\,02 \end{bmatrix},$$

where

$$\sqrt{\left[1-(3-\sqrt{3})/2\right]^2+0.5^2} = 0.619\,66 \text{ and } \sqrt{\left[2-(3-\sqrt{3})/2\right]^2+1} = 1.692\,93$$

are the norms of the two columns. Now we can normalize each column to unity. We obtain

$$\frac{k}{m}\begin{bmatrix} 0.590\,69 & 0.590\,69 \\ 0.806\,89 & 0.806\,89 \end{bmatrix} \text{ which means that } \underline{\mathbf{A}}_1 = \begin{bmatrix} \overset{1}{A}_1 \\ \overset{1}{A}_2 \end{bmatrix} = \frac{k}{m}\begin{bmatrix} 0.590\,69 \\ 0.806\,89 \end{bmatrix}.$$

For $\ell = 2$, we find in the same manner that

$$\underline{\mathbf{A}}_2 = \begin{bmatrix} \overset{2}{A}_1 \\ \overset{2}{A}_1 \end{bmatrix} = \frac{k}{m}\begin{bmatrix} -0.939\,071 \\ 0.343\,724 \end{bmatrix}.$$

With the two eigenvectors

$$(\underline{\mathbf{A}}_1, \underline{\mathbf{A}}_2)_M = \left(\frac{k}{m}\right)^2 \begin{bmatrix} 0.590\,69 & 0.806\,89 \end{bmatrix} \begin{bmatrix} m & 0 \\ 0 & 2m \end{bmatrix} \begin{bmatrix} -0.939\,07 \\ 0.343\,72 \end{bmatrix} =$$

$$= \frac{k^2}{m}\begin{bmatrix} 0.590\,69 & 0.806\,89 \end{bmatrix}\begin{bmatrix} -0.939\,07 \\ 0.687\,46 \end{bmatrix} = \frac{k^2}{m}(-0.590\,69 \times 0.939\,07 + $$
$$+ 0.806\,89 \times 0.687\,46) = 0$$

is their product on $\underline{\mathbf{M}}$. This result shows that the orthogonality condition (5.34a) is also fulfilled.

5.2.4 Repeated Roots

Assume that repeated roots are found when we solve the characteristic equation which can be given in terms of the distinct roots as the product

$$p_n(\lambda) = \left|\underline{\mathbf{K}} - \lambda\underline{\mathbf{M}}\right| = \\ = a\,(\lambda-\lambda_1)^{d_1}\,(\lambda-\lambda_2)^{d_2}\,(\lambda-\lambda_3)^{d_3}\ldots\left(\lambda-\lambda_\mu\right)^{d_\mu} = 0,\quad a = \text{constant},\qquad (5.44)$$

where the number d_k $(k = 1, 2, \ldots, \mu)$, $\mu \le n$ is the algebraic multiplicity of the root λ_k. If there are repeated roots at least one multiplicity is greater than one. It is also well known that the multiplicities satisfy the condition $d_1 + d_2 + d_3 + \cdots + d_\mu = n$. We assume that the number of linearly independent eigenvectors which belong to a repeated root is the same as the algebraic multiplicity of the root itself.[1] For a repeated root, we shall show that it is always possible to find such eigenvectors which are mutually orthogonal to each other on $\underline{\mathbf{M}}$.

Let λ_L be a repeated root of (5.44) with a multiplicity r $(2 \le r < n)$. The corresponding normalized eigenvectors are denoted by $\underline{\psi}_{1L}, \underline{\psi}_{2L}, \ldots, \underline{\psi}_{rL}$ where the second subscript L shows that these eigenvectors belong to the repeated eigenvalue λ_L. Since the eigenvectors $\underline{\psi}_{1L}, \underline{\psi}_{2L}, \ldots, \underline{\psi}_{rL}$ are normalized it holds that

$$\left(\underline{\psi}_{kL}, \underline{\psi}_{\ell L}\right)_M = \delta_{k\ell}\,. \qquad (5.45)$$

The linear combinations

$$\underline{\hat{\psi}}_{1L} = \sum_{k=1}^{r} c_{1k}\underline{\psi}_{kL},\quad \underline{\hat{\psi}}_{2L} = \sum_{k=1}^{r} c_{2k}\underline{\psi}_{kL},\quad \ldots,\quad \underline{\hat{\psi}}_{rL} = \sum_{k=1}^{r} c_{rk}\underline{\psi}_{kL}, \qquad (5.46)$$

in which the constants c_{1k}, c_{2k}, etc. are, in general, non-zero constants are clearly eigenvectors.

The normalized eigenvectors $\underline{\psi}_{kL}(x)\; k \in 1, 2, \ldots, r$ are not orthogonal to each other. In the sequel, we shall construct an orthogonal set by selecting the coefficients c_{1k}, c_{2k}, etc. appropriately. The procedure we detail below is called Gram–Schmidt orthogonalization [2]. Let

$$\underline{\hat{\psi}}_{1L} = \underline{\psi}_{1L} \qquad (5.47a)$$

by the use of which we define $\underline{\hat{\psi}}^*_{2L}$ in the following manner:

$$\underline{\hat{\psi}}^*_{2L} = \underline{\psi}_{2L} - \underline{\hat{\psi}}_{1L}(\underline{\psi}_{2L}, \underline{\hat{\psi}}_{1L})_M\,. \qquad (5.47b)$$

[1]If the matrix in question is symmetric then it is a normal matrix [1] for which the assumption is fulfilled.

In this equation, the term $\hat{\underline{\psi}}_{1L}(\underline{\psi}_{2L}, \hat{\underline{\psi}}_{1L})_M$ is the projection of $\underline{\psi}_{2L}$ on $\hat{\underline{\psi}}_{1L}$. Since

$$(\hat{\underline{\psi}}^*_{2L}, \hat{\underline{\psi}}_{1L})_M = (\underline{\psi}_{2L}, \hat{\underline{\psi}}_{1L})_M - \underbrace{(\underline{\psi}_{1L}, \hat{\underline{\psi}}_{1L})_M}_{=1}(\underline{\psi}_{2L}, \hat{\underline{\psi}}_{1L})_M = 0$$

it follows that $\hat{\underline{\psi}}^*_{2L}$ and

$$\hat{\underline{\psi}}_{2L} = \hat{\underline{\psi}}^*_{2L}/(\hat{\underline{\psi}}^*_{2L}, \hat{\underline{\psi}}^*_{2L})_M, \quad \left[(\hat{\underline{\psi}}_{2L}, \hat{\underline{\psi}}_{2L})_M = 1\right] \tag{5.47c}$$

are both orthogonal to $\hat{\underline{\psi}}_{1L}$.

If the multiplicity $r = 2$ then $\hat{\underline{\psi}}_{1L}$ and $\hat{\underline{\psi}}_{2L}$ are the two sought eigenfunctions.

If the multiplicity r is greater than two we proceed in a manner similar to the definition of $\hat{\underline{\psi}}_{2L}$. Let

$$\hat{\underline{\psi}}^*_{3L} = \underline{\psi}_{3L} - \hat{\underline{\psi}}_{1L}\big(\underline{\psi}_{3L}, \hat{\underline{\psi}}_{1L}\big)_M - \hat{\underline{\psi}}_{2L}\big(\underline{\psi}_{3L}, \hat{\underline{\psi}}_{2L}\big)_M \tag{5.48a}$$

in which the column matrices $\hat{\underline{\psi}}_{1L}\big(\underline{\psi}_{3L}, \hat{\underline{\psi}}_{1L}\big)_M$ and $\hat{\underline{\psi}}_{2L}\big(\underline{\psi}_{3L}, \hat{\underline{\psi}}_{2L}\big)_M$ are the projections of $\underline{\psi}_{3L}$ on $\hat{\underline{\psi}}_{1L}$ and $\hat{\underline{\psi}}_{2L}$. Since

$$\big(\hat{\underline{\psi}}^*_{3L}, \hat{\underline{\psi}}_{1L}\big)_M = \big(\underline{\psi}_{3L}, \hat{\underline{\psi}}_{1L}\big)_M - \underbrace{\big(\hat{\underline{\psi}}_{1L}, \hat{\underline{\psi}}_{1L}\big)_M}_{=1}\big(\underline{\psi}_{3L}, \hat{\underline{\psi}}_{1L}\big)_M - \underbrace{\big(\hat{\underline{\psi}}_{1L}, \hat{\underline{\psi}}_{2L}\big)_M}_{=0}\big(\underline{\psi}_{3L}, \hat{\underline{\psi}}_{2L}\big)_M = 0,$$

$$\big(\hat{\underline{\psi}}^*_{3L}, \hat{\underline{\psi}}_{2L}\big)_M = \big(\underline{\psi}_{3L}, \hat{\underline{\psi}}_{2L}\big)_M - \underbrace{\big(\hat{\underline{\psi}}_{1L}, \hat{\underline{\psi}}_{2L}\big)_M}_{=0}\big(\underline{\psi}_{3L}, \hat{\underline{\psi}}_{1L}\big)_M - \underbrace{\big(\hat{\underline{\psi}}_{2L}, \hat{\underline{\psi}}_{2L}\big)_M}_{=1}\big(\underline{\psi}_{3L}, \hat{\underline{\psi}}_{2L}\big)_M = 0$$

we can come to the conclusion that $\hat{\underline{\psi}}^*_{3L}$ and

$$\hat{\underline{\psi}}_{3L} = \hat{\underline{\psi}}^*_{3L}/\big(\hat{\underline{\psi}}^*_{3L}, \hat{\underline{\psi}}^*_{3L}\big)_M, \quad \left[\big(\hat{\underline{\psi}}_{3L}, \hat{\underline{\psi}}_{3L}\big)_M = 1\right] \tag{5.48b}$$

are orthogonal to $\hat{\underline{\psi}}_{1L}$ and $\hat{\underline{\psi}}_{2L}$.

The steps leading to (5.47a), (5.47c), and (5.48b) can easily be generalized:

$$\begin{array}{|c|}\hline \hat{\underline{\psi}}^*_{i+1,L} = \underline{\psi}_{i+1,L} - \sum_{k=1}^{i} \hat{\underline{\psi}}_{kL} \left(\underline{\psi}_{i+1,L}, \hat{\underline{\psi}}_{kL} \right)_M , \\ \hat{\underline{\psi}}_{i+1,L} = \hat{\underline{\psi}}^*_{i+1,L} / \left(\hat{\underline{\psi}}^*_{i+1,L}, \hat{\underline{\psi}}^*_{i+1,L} \right)_M , \qquad \left[\left(\hat{\underline{\psi}}_{i+1,L}, \hat{\underline{\psi}}_{i+1,L} \right)_M = 1 \right] . \\ \hline \end{array} \tag{5.49}$$

We have already mentioned that the above procedure is attributed to Gram [3–5] and Schmidt [6, 7]. It is also worth citing the review paper by Björck et al. [8] here.

5.2.5 The Natural Frequencies Are Real Numbers

We prove that the natural frequencies are real numbers. Suppose first that the eigenvalue λ_ℓ is not real, i.e., it is a complex number and we will show that this assumption leads to a contradiction. If λ_ℓ is complex then so is the corresponding eigenvector since the matrix $\underline{\mathbf{K}}$ is a real matrix in equation

$$\underline{\mathbf{K}}\,\underline{\mathbf{A}}_\ell = \lambda_\ell \,\underline{\mathbf{M}}\,\underline{\mathbf{A}}_\ell \tag{5.50}$$

to be satisfied by $\underline{\mathbf{A}}_\ell$—see (5.18) for a comparison. Let λ_ℓ be of the form

$$\lambda_\ell = a + ib, \tag{5.51a}$$

where a and b are non-zero real numbers. It is obvious that the complex conjugate of λ_ℓ, i.e., the number

$$\bar{\lambda}_\ell = a - ib \tag{5.51b}$$

is also an eigenvalue and the complex conjugate of $\underline{\mathbf{A}}_\ell$ is an eigenvector. Consequently, it holds that

$$\underline{\mathbf{K}}\,\bar{\underline{\mathbf{A}}}_\ell = \bar{\lambda}_\ell \,\underline{\mathbf{M}}\,\bar{\underline{\mathbf{A}}}_\ell \,. \tag{5.52}$$

Left multiplying (5.50) and (5.52) by $\bar{\underline{\mathbf{A}}}_\ell^T$ and $\underline{\mathbf{A}}_\ell^T$ and recalling the notational convention (5.13), we can write that

$$\left(\bar{\underline{\mathbf{A}}}_\ell, \underline{\mathbf{A}}_\ell\right)_K = \lambda_\ell \left(\bar{\underline{\mathbf{A}}}_\ell, \underline{\mathbf{A}}_\ell\right)_M \tag{5.53a}$$

and

$$\left(\underline{\mathbf{A}}_\ell, \bar{\underline{\mathbf{A}}}_\ell\right)_K = \bar{\lambda}_\ell \left(\underline{\mathbf{A}}_\ell, \bar{\underline{\mathbf{A}}}_\ell\right)_M , \tag{5.53b}$$

where for symmetry reasons it holds that

$$\left(\bar{\underline{\mathbf{A}}}_\ell, \underline{\mathbf{A}}_\ell\right)_K = \left(\underline{\mathbf{A}}_\ell, \bar{\underline{\mathbf{A}}}_\ell\right)_K \quad \text{and} \quad \left(\bar{\underline{\mathbf{A}}}_\ell, \underline{\mathbf{A}}_\ell\right)_M = \left(\underline{\mathbf{A}}_\ell, \bar{\underline{\mathbf{A}}}_\ell\right)_M . \tag{5.54}$$

Subtract (5.53b) from (5.53a). If we take (5.51) and (5.54) into account, we get

$$\left(\lambda_\ell - \bar{\lambda}_\ell\right)\left(\underline{\mathbf{A}}_\ell, \bar{\underline{\mathbf{A}}}_\ell\right)_M = 2bi\,\left(\underline{\mathbf{A}}_\ell, \bar{\underline{\mathbf{A}}}_\ell\right)_M = 0$$

in which $\left(\underline{\mathbf{A}}_\ell, \bar{\underline{\mathbf{A}}}_\ell\right)_M \neq 0$ since the mass matrix is positive definite. Hence,

$$b = 0 \tag{5.55}$$

and that was to be proved.

5.2.6 *Uncoupled Equations of Motion*

We have already seen in Exercise 4.6 that the equations of motion, depending on how we select the coordinate system, can be statically or dynamically coupled. We shall now investigate the issue how to transform the equations of motion to an uncoupled form. It is obvious on the base of (5.21a) and (5.39) that the matrix of the generalized displacements can be given in the form

$$\underline{\mathbf{q}} = x_1(t)\underline{\mathbf{A}}_1 + x_2(t)\underline{\mathbf{A}}_2 + \cdots + x_n(t)\underline{\mathbf{A}}_n = \sum_{\ell=1}^{n} x_\ell(t)\underline{\mathbf{A}}_\ell \tag{5.56}$$

which is a linear combination of the eigenvectors. By introducing matrix notations:

$$\underset{(1\times n)}{\underline{\mathbf{x}}^T} = [\,x_1\,|\,x_2\,|\,x_3\,|\cdots|\,x_n\,]\,, \tag{5.57}$$

$$\underset{(n\times n)}{\underline{\boldsymbol{\mathcal{A}}}} = \Bigg[\underbrace{\underline{\mathbf{A}}_1}_{(n\times 1)}\ \underbrace{\underline{\mathbf{A}}_2}_{(n\times 1)}\ \cdots\ \underbrace{\underline{\mathbf{A}}_n}_{(n\times 1)}\Bigg] = \begin{bmatrix} A_{11} & A_{12} & A_{13} & \cdots & A_{1n} \\ A_{21} & A_{22} & A_{23} & \cdots & A_{2n} \\ \vdots & \vdots & \vdots & \ddots & \vdots \\ A_{n1} & A_{n2} & A_{n3} & \cdots & A_{nn} \end{bmatrix}, \tag{5.58}$$

where $A_{k\ell}$ is the kth element of the eigenvector $\underline{\mathbf{A}}_\ell$ and we can rewrite Eq. (5.56) as a matrix equation

$$\underset{(n\times 1)}{\underline{\mathbf{q}}} = \underset{(n\times n)}{\underline{\boldsymbol{\mathcal{A}}}}\ \underset{(n\times 1)}{\underline{\mathbf{x}}}\ . \tag{5.59}$$

If we substitute this solution into the differential equation (5.5), we get

$$\underset{(n\times n)}{\underline{\mathbf{M}}}\;\underset{(n\times n)}{\underline{\boldsymbol{\mathcal{A}}}}\;\underset{(n\times 1)}{\underline{\ddot{\mathbf{x}}}} + \underset{(n\times n)}{\underline{\mathbf{K}}}\;\underset{(n\times n)}{\underline{\boldsymbol{\mathcal{A}}}}\;\underset{(n\times 1)}{\underline{\mathbf{x}}} = \underset{(n\times 1)}{\underline{\mathbf{0}}} . \tag{5.60a}$$

Left multiplying the above equation by $\underset{(n\times n)}{\underline{\boldsymbol{\mathcal{A}}}^T}$, we get

$$\underset{(n\times n)}{\underline{\boldsymbol{\mathcal{A}}}^T}\;\underset{(n\times n)}{\underline{\mathbf{M}}}\;\underset{(n\times n)}{\underline{\boldsymbol{\mathcal{A}}}}\;\underset{(n\times 1)}{\underline{\ddot{\mathbf{x}}}} + \underset{(n\times n)}{\underline{\boldsymbol{\mathcal{A}}}^T}\;\underset{(n\times n)}{\underline{\mathbf{K}}}\;\underset{(n\times n)}{\underline{\boldsymbol{\mathcal{A}}}}\;\underset{(n\times 1)}{\underline{\mathbf{x}}} = \underset{(n\times 1)}{\underline{\mathbf{0}}} \tag{5.60b}$$

in which

$$\underset{(n\times n)}{\underline{\boldsymbol{\mathcal{A}}}^T}\;\underset{(n\times n)}{\underline{\mathbf{M}}}\;\underset{(n\times n)}{\underline{\boldsymbol{\mathcal{A}}}} = \underbrace{\begin{bmatrix} \underline{\mathbf{A}}_1^T \\ \underline{\mathbf{A}}_2^T \\ \vdots \\ \underline{\mathbf{A}}_n^T \end{bmatrix}}_{(n\times n)} \underset{(n\times n)}{\underline{\mathbf{M}}} \underbrace{\left[\underline{\mathbf{A}}_1\, \underline{\mathbf{A}}_2 \cdots \underline{\mathbf{A}}_n\right]}_{(n\times n)} \underset{(5.13)}{= \uparrow =}$$

$$= \begin{bmatrix} (\underline{\mathbf{A}}_1, \underline{\mathbf{A}}_1)_M & (\underline{\mathbf{A}}_1, \underline{\mathbf{A}}_2)_M & \cdots & (\underline{\mathbf{A}}_1, \underline{\mathbf{A}}_n)_M \\ (\underline{\mathbf{A}}_2, \underline{\mathbf{A}}_1)_M & (\underline{\mathbf{A}}_2, \underline{\mathbf{A}}_2)_M & \cdots & (\underline{\mathbf{A}}_2, \underline{\mathbf{A}}_n)_M \\ \vdots & \vdots & \ddots & \vdots \\ (\underline{\mathbf{A}}_n, \underline{\mathbf{A}}_1)_M & (\underline{\mathbf{A}}_n, \underline{\mathbf{A}}_2)_M & \cdots & (\underline{\mathbf{A}}_n, \underline{\mathbf{A}}_n)_M \end{bmatrix} \underset{(5.39a)}{= \uparrow =}$$

$$= \begin{bmatrix} (\underline{\mathbf{A}}_1, \underline{\mathbf{A}}_1)_M & 0 & \cdots & 0 \\ 0 & (\underline{\mathbf{A}}_2, \underline{\mathbf{A}}_2)_M & \cdots & 0 \\ \vdots & \vdots & \ddots & \vdots \\ 0 & 0 & \cdots & (\underline{\mathbf{A}}_n, \underline{\mathbf{A}}_n)_M \end{bmatrix} \underset{(5.40)_1}{= \uparrow =}$$

$$= \begin{bmatrix} m_{11} & 0 & \cdots & 0 \\ 0 & m_{22} & \cdots & 0 \\ \vdots & \vdots & \ddots & \vdots \\ 0 & 0 & \cdots & m_{nn} \end{bmatrix} \tag{5.61a}$$

is the first term on the left side. After performing similar steps we obtain for the second term on the left side that

$$\underset{(n\times n)}{\boldsymbol{\mathcal{A}}^T}\ \underset{(n\times n)}{\underline{\mathbf{K}}}\ \underset{(n\times n)}{\boldsymbol{\mathcal{A}}} = \underbrace{\begin{bmatrix} \underline{\mathbf{A}}_1^T \\ \underline{\mathbf{A}}_2^T \\ \vdots \\ \underline{\mathbf{A}}_n^T \end{bmatrix}}_{(n\times n)} \underset{(n\times n)}{\underline{\mathbf{M}}} \underbrace{\left[\underline{\mathbf{A}}_1\ \underline{\mathbf{A}}_2 \cdots \underline{\mathbf{A}}_n\right]}_{(n\times n)} = \underset{(5.13)}{\uparrow} =$$

$$= \begin{bmatrix} (\underline{\mathbf{A}}_1, \underline{\mathbf{A}}_1)_K & 0 & \cdots & 0 \\ 0 & (\underline{\mathbf{A}}_2, \underline{\mathbf{A}}_2)_K & \cdots & 0 \\ \vdots & \vdots & \ddots & \vdots \\ 0 & 0 & \cdots & (\underline{\mathbf{A}}_n, \underline{\mathbf{A}}_n)_K \end{bmatrix} = \underset{(5.40)_2}{\uparrow} =$$

$$= \begin{bmatrix} k_{11} & 0 & \cdots & 0 \\ 0 & k_{22} & \cdots & 0 \\ \vdots & \vdots & \ddots & \vdots \\ 0 & 0 & \cdots & k_{nn} \end{bmatrix}. \tag{5.61b}$$

This means that the equation of motion (5.60b) is an uncoupled differential equation system:

$$\begin{bmatrix} m_{11} & 0 & \cdots & 0 \\ 0 & m_{22} & \cdots & 0 \\ \vdots & \vdots & \ddots & \vdots \\ 0 & 0 & \cdots & m_{nn} \end{bmatrix} \begin{bmatrix} \ddot{x}_1 \\ \ddot{x}_2 \\ \vdots \\ \ddot{x}_n \end{bmatrix} + \begin{bmatrix} k_{11} & 0 & \cdots & 0 \\ 0 & k_{22} & \cdots & 0 \\ \vdots & \vdots & \ddots & \vdots \\ 0 & 0 & \cdots & k_{nn} \end{bmatrix} \begin{bmatrix} x_1 \\ x_2 \\ \vdots \\ x_n \end{bmatrix} = \begin{bmatrix} 0 \\ 0 \\ \vdots \\ 0 \end{bmatrix} \tag{5.62}$$

constituted by the independent differential equations of order two

$$m_{\ell\ell}\, \ddot{x}_\ell + k_{\ell\ell}\, x_\ell = 0\,, \qquad \ell = 1, 2, \ldots, n, \tag{5.63}$$

where

$$m_{\ell\ell} = (\underline{\mathbf{A}}_\ell, \underline{\mathbf{A}}_\ell)_M\,, \qquad k_{\ell\ell} = (\underline{\mathbf{A}}_\ell, \underline{\mathbf{A}}_\ell)_K\,. \tag{5.64}$$

If we recall Eq. (5.36) which relates the products of eigenvectors on the mass and stiffness matrices to each other we obtain

$$\underbrace{(\underline{\mathbf{A}}_\ell, \underline{\mathbf{A}}_\ell)_K}_{k_{\ell\ell}} = \lambda_\ell\, \underbrace{(\underline{\mathbf{A}}_\ell, \underline{\mathbf{A}}_\ell)_M}_{m_{\ell\ell}} \tag{5.65}$$

from where

$$\lambda_\ell = \omega_{n\ell}^2 = \frac{k_{\ell\ell}}{m_{\ell\ell}}\,. \tag{5.66}$$

Consequently, (5.63) can also be given in the form

$$\ddot{x}_\ell + \omega_{n\ell}^2 x_\ell = 0\,, \qquad \ell = 1, 2, \ldots, n \tag{5.67}$$

which is the equation of motion of an undamped single degree of freedom system.

5.3 Rayleigh Quotient

5.3.1 Definition and Properties

Let $\underline{\mathbf{M}}$ and $\underline{\mathbf{K}}$ be the mass and stiffness matrices of an n degree of freedom system. Further let $\underline{\mathbf{A}}$ ($|\underline{\mathbf{A}}| \neq 0$) be an arbitrary column matrix of size $n \times 1$. We assume that $\underline{\mathbf{A}}$ is the amplitude of the generalized displacements. The Rayleigh quotient[2] is defined by the equation [11, 12]

$$\mathcal{R} = \frac{\left(\underline{\mathbf{A}}, \underline{\mathbf{A}}\right)_K}{\left(\underline{\mathbf{A}}, \underline{\mathbf{A}}\right)_M}\,. \tag{5.68a}$$

Remark 5.10 If the motion is harmonic with the natural circular frequency ω—this is, the case, in general, for undamped free vibrations—then the maximum kinetic and potential energy of the system are given by

$$\mathcal{E}_{\max} = \frac{1}{2}\omega^2 \left(\underline{\mathbf{A}}, \underline{\mathbf{A}}\right)_M\,, \qquad \mathcal{U}_{\max} = \frac{1}{2}\left(\underline{\mathbf{A}}, \underline{\mathbf{A}}\right)_K\,.$$

Since $\mathcal{E}_{\max} = \mathcal{U}_{\max}$ we can write

$$\omega^2 = \mathcal{R} = \frac{\left(\underline{\mathbf{A}}, \underline{\mathbf{A}}\right)_K}{\left(\underline{\mathbf{A}}, \underline{\mathbf{A}}\right)_M} \tag{5.68b}$$

which is nothing but the Rayleigh quotient. This explains the background of definition (5.68a).

[2] Also known as the Rayleigh–Ritz ratio, named after Lord Rayleigh (John William Strutt (1842–1919)) and Walther Ritz (1878–1909) [9, 10].

The Rayleigh quotient has the following properties:

1. Because $(\underline{\mathbf{A}}, \underline{\mathbf{A}})_K \geq 0$ and $(\underline{\mathbf{A}}, \underline{\mathbf{A}})_M > 0$ the Rayleigh quotient is a real nonnegative quantity.
2. The Rayleigh quotient is independent of $|\underline{\mathbf{A}}|$ (the length of $\underline{\mathbf{A}}$). Really if we give $\underline{\mathbf{A}}$ in the form

$$\underline{\mathbf{A}} = \nu\,\underline{\mathbf{e}}\,, \qquad |\underline{\mathbf{e}}| = (\underline{\mathbf{e}}, \underline{\mathbf{e}})_I = 1, \qquad \nu = |\underline{\mathbf{A}}| = \sqrt{(\underline{\mathbf{A}}, \underline{\mathbf{A}})_I},$$

we get

$$\mathcal{R} = \frac{(\underline{\mathbf{A}}, \underline{\mathbf{A}})_K}{(\underline{\mathbf{A}}, \underline{\mathbf{A}})_M} = \frac{(\underline{\mathbf{e}}, \underline{\mathbf{e}})_K\, \nu^2}{(\underline{\mathbf{e}}, \underline{\mathbf{e}})_M\, \nu^2} = \frac{(\underline{\mathbf{e}}, \underline{\mathbf{e}})_K}{(\underline{\mathbf{e}}, \underline{\mathbf{e}})_M}\,. \tag{5.69}$$

(Consider the hypersphere of unit radius and centered at the origin of the n-dimensional space. Then $\omega^2 = \lambda = \mathcal{R}$ is a function of $\underline{\mathbf{e}}$, i.e., it varies on the surface of the hypersphere.)

3. The Rayleigh quotient is bounded—it has both a lower limit and an upper one.
   - Since $\underline{\mathbf{M}}$ is positive definite and its elements are all finite it follows that $(\underline{\mathbf{e}}, \underline{\mathbf{e}})_M$ is bounded.
   - Since $\underline{\mathbf{K}}$ is positive semidefinite and its elements are all finite it follows that $(\underline{\mathbf{e}}, \underline{\mathbf{e}})_K$ is also bounded.
   - Consequently,

$$\omega^2 = \lambda = \mathcal{R} = \frac{(\underline{\mathbf{e}}, \underline{\mathbf{e}})_K}{(\underline{\mathbf{e}}, \underline{\mathbf{e}})_M}$$

   is bounded and has lower and upper limits.

4. The [lower] {upper} limit of the Rayleigh quotient is the [smallest] {greatest} eigenvalue $[\lambda_1]\{\lambda_n\}$.
   The system of eigenvectors $\underline{\psi}_\ell$ $(\ell = 1, 2, \ldots, n)$ is complete (constitutes a basis in the n-dimensional space). Hence, any $\underline{\mathbf{A}}$ can be given as a linear combination of the eigenvectors $\underline{\psi}_\ell$:

$$\underline{\mathbf{A}} = c_1\underline{\psi}_1 + c_2\underline{\psi}_2 + c_3\underline{\psi}_3 + \cdots + c_n\underline{\psi}_n, \tag{5.70}$$

   where the weight c_ℓ is the component of $\underline{\mathbf{A}}$ in the direction $\underline{\psi}_\ell$. Recalling that

$$\left(\underline{\psi}_\ell, \underline{\psi}_\ell\right)_M = 1, \quad \left(\underline{\psi}_k, \underline{\psi}_\ell\right)_M = 0 \;\; (k \neq \ell) \;\text{ and }\; \left(\underline{\psi}_\ell, \underline{\psi}_\ell\right)_K = \lambda_\ell$$

   for the Rayleigh quotient, we get

$$\omega^2 = \lambda = \mathcal{R} = \frac{\left(\underline{\mathbf{A}}, \underline{\mathbf{A}}\right)_K}{\left(\underline{\mathbf{A}}, \underline{\mathbf{A}}\right)_M} =$$

$$= \frac{\left(c_1\underline{\psi}_1 + c_2\underline{\psi}_2 + \cdots + c_n\underline{\psi}_n, c_1\underline{\psi}_1 + c_2\underline{\psi}_2 + \cdots + c_n\underline{\psi}_n\right)_K}{\left(c_1\underline{\psi}_1 + c_2\underline{\psi}_2 + \cdots + c_n\underline{\psi}_n, c_1\underline{\psi}_1 + c_2\underline{\psi}_2 + \cdots + c_n\underline{\psi}_n\right)_M} =$$

$$= \frac{c_1^2\left(\underline{\psi}_1, \underline{\psi}_1\right)_K + c_2^2\left(\underline{\psi}_2, \underline{\psi}_2\right)_K + \cdots + c_n^2\left(\underline{\psi}_n, \underline{\psi}_n\right)_K}{c_1^2\left(\underline{\psi}_1, \underline{\psi}_1\right)_M + c_2^2\left(\underline{\psi}_2, \underline{\psi}_2\right)_M + \cdots + c_n^2\left(\underline{\psi}_n, \underline{\psi}_n\right)_M} =$$

$$= \frac{\lambda_1 c_1^2 + \lambda_2 c_2^2 + \cdots + \lambda_n c_n^2}{c_1^2 + c_2^2 + \cdots + c_n^2}$$

from where it follows that

$$\omega^2 = \lambda = \mathcal{R} = \lambda_1 \frac{c_1^2 + \frac{\lambda_2}{\lambda_1}c_2^2 + \cdots + \frac{\lambda_n}{\lambda_1}c_n^2}{c_1^2 + c_2^2 + \cdots + c_n^2} \tag{5.71a}$$

or

$$\omega^2 = \lambda = \mathcal{R} = \lambda_n \frac{\frac{\lambda_1}{\lambda_n}c_1^2 + \frac{\lambda_2}{\lambda_n}c_2^2 + \cdots + c_n^2}{c_1^2 + c_2^2 + \cdots + c_n^2}. \tag{5.71b}$$

Equation (5.71a) proves that

$$\omega_{\min}^2 = \lambda_{\min} = \mathcal{R}_{\min} = \lambda_1 . \tag{5.72a}$$

We get in the same manner from Eq. (5.71b) that

$$\omega_{\max}^2 = \lambda_{\max} = \mathcal{R}_{\max} = \lambda_n . \tag{5.72b}$$

5. The Rayleigh quotient has a further and very useful property. Let $\underline{\psi}_\ell$ ($\ell = 1, 2, \ldots, n$) be an eigenvector. Further let

$$\underline{\mathbf{A}} = \underline{\psi}_\ell + \varepsilon\,\underline{\mathbf{x}} \tag{5.73}$$

be an approximation of $\underline{\psi}_\ell$ where (a) $|\varepsilon| \ll 1$ and (b) $\left(\underline{\psi}_\ell, \underline{\mathbf{x}}\right)_K = \left(\underline{\psi}_\ell, \underline{\mathbf{x}}\right)_M = 0$—this assumption does not violate generality. Then

$$\lambda\left(\underline{\mathbf{A}}\right) = \mathcal{R}\left(\underline{\mathbf{A}}\right) = \lambda_\ell + O\left(\varepsilon^2\right) \tag{5.74}$$

which means the error we have made is of order two. Really we can write

$$\lambda\left(\underline{\mathbf{A}}\right)=\mathcal{R}\left(\underline{\mathbf{A}}\right)=\frac{\left(\underline{\psi}_{\ell}+\varepsilon\,\underline{\mathbf{x}},\,\underline{\psi}_{\ell}+\varepsilon\,\underline{\mathbf{x}}\right)_K}{\left(\underline{\psi}_{\ell}+\varepsilon\,\underline{\mathbf{x}},\,\underline{\psi}_{\ell}+\varepsilon\,\underline{\mathbf{x}}\right)_M}=$$

$$=\frac{\overbrace{\left(\underline{\psi}_{\ell},\underline{\psi}_{\ell}\right)_K}^{\lambda_\ell}+2\varepsilon\overbrace{\left(\underline{\mathbf{x}},\underline{\psi}_{\ell}\right)_K}^{=0}+\varepsilon^2\left(\underline{\mathbf{x}},\,\underline{\mathbf{x}}\right)_K}{\underbrace{\left(\underline{\psi}_{\ell},\underline{\psi}_{\ell}\right)_M}_{=1}+2\varepsilon\underbrace{\left(\underline{\mathbf{x}},\underline{\psi}_{\ell}\right)_M}_{=0}+\varepsilon^2\left(\underline{\mathbf{x}},\,\underline{\mathbf{x}}\right)_M}=$$

$$=\frac{\lambda_\ell+\varepsilon^2\left(\underline{\mathbf{x}},\,\underline{\mathbf{x}}\right)_K}{1+\varepsilon^2\left(\underline{\mathbf{x}},\,\underline{\mathbf{x}}\right)_M}=\underset{\frac{1}{1+\varepsilon^2\left(\underline{\mathbf{x}},\underline{\mathbf{x}}\right)_M}\approx 1-\varepsilon^2\left(\underline{\mathbf{x}},\underline{\mathbf{x}}\right)_M}{\uparrow}\approx$$

$$\approx\left(\lambda_\ell+\varepsilon^2\left(\underline{\mathbf{x}},\,\underline{\mathbf{x}}\right)_K\right)\left(1-\varepsilon^2\left(\underline{\mathbf{x}},\,\underline{\mathbf{x}}\right)_M\right)=$$

$$=\lambda_\ell+\varepsilon^2\left[\left(\underline{\mathbf{x}},\,\underline{\mathbf{x}}\right)_K-\lambda_\ell\left(\underline{\mathbf{x}},\,\underline{\mathbf{x}}\right)_M-\varepsilon^2\left(\underline{\mathbf{x}},\,\underline{\mathbf{x}}\right)_K\left(\underline{\mathbf{x}},\,\underline{\mathbf{x}}\right)_M\right]=\lambda_\ell+O\left(\varepsilon^2\right).$$

5.3.2 *Matrix Iteration*

Our aim is to find the first eigenvalue and eigenvector by the use of an iteration procedure. Later on we shall generalize the procedure for the higher eigenvalues and eigenvectors. The iteration procedure starts by assuming an initial value (a first approximation) for the displacement $\underline{\mathbf{A}}$.

INITIALIZATION

Setting the initial value

$\underset{(0)}{\underline{\mathbf{A}}_1}=\text{input}$

Normalization

$\underset{(0)}{\breve{A}_1}=\sqrt{\left(\underset{(0)}{\underline{\mathbf{A}}_1},\underset{(0)}{\underline{\mathbf{A}}_1}\right)_M}$ $\qquad$ $\underset{(0)}{\underline{\psi}_1}=\underset{(0)}{\underline{\mathbf{A}}_1}/\underset{(0)}{\breve{A}_1}$

Estimation

$\underset{(0)}{\lambda_1}=\underset{(0)}{\omega_1^2}=\left(\underset{(0)}{\underline{\psi}_1},\underset{(0)}{\underline{\psi}_1}\right)_K$

ITERATION STEPS

Step 1.

Solution for $\underset{(1)}{\underline{\mathbf{A}}_1}$

$\underline{\mathbf{K}}\,\underset{(1)}{\underline{\mathbf{A}}_1}=\underline{\mathbf{M}}\,\underset{(0)}{\underline{\psi}_1}$

Normalization

$\underset{(1)}{\breve{A}_1}=\sqrt{\left(\underset{(1)}{\underline{\mathbf{A}}_1},\underset{(1)}{\underline{\mathbf{A}}_1}\right)_M}$ $\qquad$ $\underset{(1)}{\underline{\psi}_1}=\underset{(1)}{\underline{\mathbf{A}}_1}/\underset{(1)}{\breve{A}_1}$

Estimation

$\underset{(1)}{\lambda_1}=\underset{(1)}{\omega_1^2}=\left(\underset{(1)}{\underline{\psi}_1},\underset{(1)}{\underline{\psi}_1}\right)_K$

Step 2.

Solution for $\underset{(2)}{\underline{\mathbf{A}}_1}$

$$\underline{\mathbf{K}}\,\underset{(2)}{\underline{\mathbf{A}}_1} = \underline{\mathbf{M}}\,\underset{(1)}{\underline{\psi}_1}$$

Normalization

$$\underset{(2)}{\check{A}_1} = \sqrt{\left(\underset{(2)}{\underline{\mathbf{A}}_1}, \underset{(2)}{\underline{\mathbf{A}}_1}\right)_M} \qquad \underset{(2)}{\underline{\psi}_1} = \underset{(2)}{\underline{\mathbf{A}}_1} / \underset{(2)}{\check{A}_1}$$

Estimation

$$\underset{(2)}{\lambda_1} = \underset{(2)}{\omega_1^2} = \left(\underset{(2)}{\underline{\psi}_1}, \underset{(2)}{\underline{\psi}_1}\right)_K$$

Step $\underline{\ell}$

Solution for $\underset{(\ell)}{\underline{\mathbf{A}}_1}$

$$\underline{\mathbf{K}}\,\underset{(\ell)}{\underline{\mathbf{A}}_1} = \underline{\mathbf{M}}\,\underset{(\ell-1)}{\underline{\psi}_1}$$

Normalization

$$\underset{(\ell)}{\check{A}_1} = \sqrt{\left(\underset{(\ell)}{\underline{\mathbf{A}}_1}, \underset{(\ell)}{\underline{\mathbf{A}}_1}\right)_M} \qquad \underset{(\ell)}{\underline{\psi}_1} = \underset{(\ell)}{\underline{\mathbf{A}}_1} / \underset{(\ell)}{\check{A}_1}$$

Estimation

$$\underset{(\ell)}{\lambda_1} = \underset{(\ell)}{\omega_1^2} = \left(\underset{(\ell)}{\underline{\psi}_1}, \underset{(\ell)}{\underline{\psi}_1}\right)_K$$

Convergence is deemed to have been reached if the inequalities

$$\left| 1 - \frac{\underset{(\ell-1)}{\lambda_1}}{\underset{(\ell)}{\lambda_1}} \right| < \varepsilon_\lambda \quad \text{and} \quad \left| 1 - \frac{\underset{(\ell-1)}{\check{A}_1}}{\underset{(\ell)}{\check{A}_1}} \right| < \varepsilon_A$$

are satisfied where ε_λ and ε_A are the prescribed relative errors.

We shall now proceed and generalize the previous algorithm with the aim of determining the second eigenvalue and eigenvector. To this end, we introduce new mass and stiffness matrices in such a way that they will be orthogonal to the first eigenvector of the system. First we normalize the eigenvector $\underline{\mathbf{A}}_1$ to unity—after we have performed the previous iteration $\underline{\mathbf{A}}_1$ and λ_1 are known with a high accuracy:

$$\hat{\underline{\mathbf{A}}}_1 = \frac{\underline{\mathbf{A}}_1}{\sqrt{\left(\underline{\mathbf{A}}_1, \underline{\mathbf{A}}_1\right)_I}} = \frac{\underline{\mathbf{A}}_1}{|\underline{\mathbf{A}}_1|}\,. \tag{5.75}$$

Making use of $\hat{\underline{\mathbf{A}}}_1$ we define the new mass and stiffness matrices by the following relations:

$$\underset{(n\times n)}{\underline{\mathbf{M}}_2} = \underset{(n\times n)}{\underline{\mathbf{M}}_1} - \underset{(n\times n)}{\underline{\mathbf{M}}_1}\,\underbrace{\underset{(n\times 1)}{\hat{\underline{\mathbf{A}}}_1}\,\underset{(1\times n)}{\hat{\underline{\mathbf{A}}}_1^T}}_{(n\times n)} \qquad \underset{(n\times n)}{\underline{\mathbf{M}}_1} = \underset{(n\times n)}{\underline{\mathbf{M}}}\,; \tag{5.76a}$$

$$\underset{(n\times n)}{\underline{\mathbf{K}}_2} = \underset{(n\times n)}{\underline{\mathbf{K}}_1} - \underset{(n\times n)}{\underline{\mathbf{K}}_1}\,\underbrace{\underset{(n\times 1)}{\hat{\underline{\mathbf{A}}}_1}\,\underset{(1\times n)}{\hat{\underline{\mathbf{A}}}_1^T}}_{(n\times n)} \qquad \underset{(n\times n)}{\underline{\mathbf{K}}_1} = \underset{(n\times n)}{\underline{\mathbf{K}}}\,. \tag{5.76b}$$

Since the set of eigenvectors is complete any $\underline{\mathbf{A}}$ can be given in the form

$$\underline{\mathbf{A}} = \beta\hat{\underline{\mathbf{A}}}_1 + \gamma\left(\underline{\mathbf{A}}_2 + c_3\underline{\mathbf{A}}_3 + c_4\underline{\mathbf{A}}_4 + \cdots\right),$$

where β, γ, c_3, c_4, etc. are scalars and it holds that

$$\left(\beta\hat{\underline{\mathbf{A}}}_1, \gamma\left(\underline{\mathbf{A}}_2 + c_3\,\underline{\mathbf{A}}_3 + c_4\,\underline{\mathbf{A}}_4 + \cdots\right)\right)_M = \\ = \left(\beta\hat{\underline{\mathbf{A}}}_1, \gamma\left(\underline{\mathbf{A}}_2 + c_3\,\underline{\mathbf{A}}_3 + c_4\,\underline{\mathbf{A}}_3 + \cdots\right)\right)_K = 0\,.$$

Let us denote the sum $\underline{\mathbf{A}}_2 + c_3\,\underline{\mathbf{A}}_3 + c_4\,\underline{\mathbf{A}}_4 + \cdots$ by $\tilde{\underline{\mathbf{A}}}$. It is easy to check that

$$\underline{\mathbf{M}}_2\,\underline{\mathbf{A}} = \underline{\mathbf{M}}_2\left(\beta\hat{\underline{\mathbf{A}}}_1 + \gamma\tilde{\underline{\mathbf{A}}}\right) = \\ = \underbrace{\beta(\underline{\mathbf{M}}_1\hat{\underline{\mathbf{A}}}_1 - \underline{\mathbf{M}}_1\hat{\underline{\mathbf{A}}}_1\underbrace{\hat{\underline{\mathbf{A}}}_1^T\hat{\underline{\mathbf{A}}}_1}_{=1})}_{\beta\underline{\mathbf{M}}_2\hat{\underline{\mathbf{A}}}_1=\mathbf{0}} + \gamma\underline{\mathbf{M}}_2\tilde{\underline{\mathbf{A}}} = \gamma\underline{\mathbf{M}}_2\tilde{\underline{\mathbf{A}}}\,.$$

Hence, the mass matrix $\underline{\mathbf{M}}_2$ is orthogonal to $\hat{\underline{\mathbf{A}}}_1$ (or $\underline{\psi}_1$).

We remark that the same statement is valid for the stiffness matrix $\underline{\mathbf{K}}_2$, i.e., $\underline{\mathbf{K}}_2$ is also orthogonal to $\hat{\underline{\mathbf{A}}}_1$ and $\underline{\psi}_1$. In the light of this result, it is obvious that the procedure we have just established for the first eigenvalue and eigenvector is also applicable to determining the second eigenvalue and eigenvector. Here we repeat the ℓth step only:

Step ℓ

Solution for $\underset{(\ell)}{\underline{\mathbf{A}}_2}$

$$\underline{\mathbf{K}}_2\,\underset{(\ell)}{\underline{\mathbf{A}}_2} = \underline{\mathbf{M}}_2\,\underset{(\ell-1)}{\underline{\psi}_1}$$

Normalization

$$\underset{(\ell)}{\check{A}_2} = \sqrt{\left(\underset{(\ell)}{\underline{\mathbf{A}}_2}, \underset{(\ell)}{\underline{\mathbf{A}}_2}\right)_M} \qquad \underset{(\ell)}{\underline{\psi}_2} = \underset{(\ell)}{\underline{\mathbf{A}}_2}/\underset{(\ell)}{\check{A}_2}$$

Estimation

$$\underset{(\ell)}{\lambda_2} = \underset{(\ell)}{\omega_2}^2 = \left(\underset{(\ell)}{\underline{\psi}_2}, \underset{(\ell)}{\underline{\psi}_2}\right)_K$$

5.3.3 *Convergence of the Iteration Algorithm*

The convergence properties of the matrix iteration are discussed below. It follows from Eq. (5.30a) that the normalized eigenvector $\underline{\psi}_\ell$ satisfies the equation

$$\underline{\mathbf{K}}\,\underline{\psi}_\ell = \lambda_\ell\,\underline{\mathbf{M}}\,\underline{\psi}_\ell \qquad \lambda_\ell = \omega_\ell^2. \tag{5.77}$$

If we multiply this equation by $\underline{\mathbf{K}}^{-1}$ we get a relationship we shall utilize in the sequel:

$$\frac{\underline{\psi}_\ell}{\lambda_\ell} = \underline{\mathbf{K}}^{-1}\,\underline{\mathbf{M}}\,\underline{\psi}_\ell\,. \tag{5.78}$$

It is well known that the set of eigenvectors is complete, i.e., any vector in the n-dimensional space can be given as a linear combination of the eigenvectors $\underline{\psi}_1$, $\underline{\psi}_2, \ldots, \underline{\psi}_n$. Hence, for the initial value of the first eigenvector, we can write

$$\begin{gathered}\underset{(0)}{\underline{\mathbf{A}}_1} = c_{10}\underline{\psi}_1 + c_{20}\underline{\psi}_2 + \cdots + c_{2n}\underline{\psi}_n\,, \qquad \underset{(0)}{\check{A}_1} = \sqrt{\big(\underset{(0)}{\underline{\mathbf{A}}_1}, \underset{(0)}{\underline{\mathbf{A}}_1}\big)_M} \\ \underset{(0)}{\underline{\psi}_1} = \frac{1}{\underset{(0)}{\check{A}_1}}\underset{(0)}{\underline{\mathbf{A}}_1} = \frac{1}{\underset{(0)}{\check{A}_1}}\Big(c_{10}\underline{\psi}_1 + c_{20}\underline{\psi}_2 + \cdots + c_{2n}\underline{\psi}_n\Big)\,,\end{gathered} \tag{5.79}$$

where on the base of (5.39b)

$$c_{\ell 0} = \big(\underset{(0)}{\underline{\mathbf{A}}_1}, \underline{\psi}_\ell\big)_K \tag{5.80}$$

is the ℓth constant in the linear combination giving $\underset{(0)}{\underline{\mathbf{A}}_1}$.

For the first approximation, the following equation system should be solved for $\underset{(1)}{\underline{\mathbf{A}}_1}$:

$$\underline{\mathbf{K}}\,\underset{(1)}{\underline{\mathbf{A}}_1} = \underline{\mathbf{M}}\,\underset{(0)}{\underline{\psi}_1}.$$

The solution is of the form

$$\begin{aligned}\underset{(1)}{\underline{\mathbf{A}}_1} &= \underline{\mathbf{K}}^{-1}\,\underline{\mathbf{M}}\,\underset{(0)}{\underline{\psi}_1} = \underline{\mathbf{K}}^{-1}\,\underline{\mathbf{M}}\,\frac{1}{\underset{(0)}{\check{A}_1}}\Big(c_{10}\underline{\psi}_1 + c_{20}\underline{\psi}_2 + \cdots + c_{n0}\underline{\psi}_n\Big) = \\ &= \frac{1}{\underset{(0)}{\check{A}_1}}\Big(c_{10}\underbrace{\underline{\mathbf{K}}^{-1}\,\underline{\mathbf{M}}\,\underline{\psi}_1}_{\frac{\underline{\psi}_1}{\lambda_1}} + c_{20}\underbrace{\underline{\mathbf{K}}^{-1}\,\underline{\mathbf{M}}\,\underline{\psi}_2}_{\frac{\underline{\psi}_2}{\lambda_2}} + \cdots + c_{n0}\underbrace{\underline{\mathbf{K}}^{-1}\,\underline{\mathbf{M}}\,\underline{\psi}_n}_{\frac{\underline{\psi}_n}{\lambda_n}}\Big)\,.\end{aligned} \tag{5.81}$$

With

$$c_{\ell 1} = \frac{1}{\underset{(0)}{\check{A}_1}}c_{\ell 0} \tag{5.82}$$

for the final form of the first approximation we can write

$$\underset{(1)}{\underline{\mathbf{A}}}_1 = \frac{c_{11}}{\lambda_1}\underline{\psi}_1 + \frac{c_{21}}{\lambda_2}\underline{\psi}_2 + \cdots + \frac{c_{n1}}{\lambda_n}\underline{\psi}_n\,, \qquad \underset{(1)}{\breve{A}}_1 = \sqrt{\big(\underset{(1)}{\underline{\mathbf{A}}}_1, \underset{(1)}{\underline{\mathbf{A}}}_1\big)_M}\,,$$
$$\underset{(1)}{\underline{\psi}}_1 = \frac{1}{\underset{(1)}{\breve{A}}_1}\underset{(1)}{\underline{\mathbf{A}}}_1 = \frac{1}{\underset{(1)}{\breve{A}}_1}\left(\frac{c_{11}}{\lambda_1}\underline{\psi}_1 + \frac{c_{21}}{\lambda_2}\underline{\psi}_2 + \cdots + \frac{c_{n1}}{\lambda_n}\underline{\psi}_n\right). \tag{5.83}$$

As regards the second approximation we have to solve the equation system

$$\underline{\mathbf{K}}\,\underset{(2)}{\underline{\mathbf{A}}}_1 = \underline{\mathbf{M}}\,\underset{(1)}{\underline{\psi}}_1$$

for $\underset{(1)}{\underline{\mathbf{A}}}_2$. The solution is of the form

$$\begin{aligned}\underset{(2)}{\underline{\mathbf{A}}}_1 &= \underline{\mathbf{K}}^{-1}\,\underline{\mathbf{M}}\,\underset{(1)}{\underline{\psi}}_1 = \underline{\mathbf{K}}^{-1}\,\underline{\mathbf{M}}\,\frac{1}{\underset{(1)}{\breve{A}}_1}\left(\frac{c_{11}}{\lambda_1}\underline{\psi}_1 + \frac{c_{21}}{\lambda_2}\underline{\psi}_2 + \cdots + \frac{c_{11}}{\lambda_n}\underline{\psi}_n\right) = \\ &= \frac{1}{\underset{(1)}{\breve{A}}_1}\left(\frac{c_{11}}{\lambda_1}\underbrace{\underline{\mathbf{K}}^{-1}\,\underline{\mathbf{M}}\,\underline{\psi}_1}_{\frac{\underline{\psi}_1}{\lambda_1}} + \frac{c_{21}}{\lambda_2}\underbrace{\underline{\mathbf{K}}^{-1}\,\underline{\mathbf{M}}\,\underline{\psi}_2}_{\frac{\underline{\psi}_2}{\lambda_2}} + \cdots + \frac{c_{n1}}{\lambda_n}\underbrace{\underline{\mathbf{K}}^{-1}\,\underline{\mathbf{M}}\,\underline{\psi}_n}_{\frac{\underline{\psi}_n}{\lambda_n}}\right) = \\ &= \frac{c_{12}}{\lambda_1^2}\underline{\psi}_1 + \frac{c_{22}}{\lambda_2^2}\underline{\psi}_2 + \cdots + \frac{c_{n2}}{\lambda_n^2}\underline{\psi}_n\,,\end{aligned} \tag{5.84}$$

where

$$c_{\ell 2} = \frac{1}{\underset{(1)}{\breve{A}}_1}c_{\ell 1}\,. \tag{5.85}$$

It is now obvious that the i-t approximation assumes the following form:

$$\underset{(i)}{\underline{\mathbf{A}}}_1 = \frac{c_{1i}}{\lambda_1^i}\underline{\psi}_1 + \frac{c_{2i}}{\lambda_2^i}\underline{\psi}_2 + \cdots + \frac{c_{ni}}{\lambda_n^i}\underline{\psi}_n\,, \qquad c_{\ell i} = \frac{1}{\underset{(i)}{\breve{A}}_1}c_{\ell,i-1}\,. \tag{5.86}$$

Here the constants $c_{\ell i}$ are limited and $\lambda_1 < \lambda_2 < \cdots < \lambda_{n-1} < \lambda_n$. Hence, for a sufficiently large i it holds that

$$\frac{c_{1i}}{\lambda_1^i} \gg \frac{c_{2i}}{\lambda_2^i} \gg \frac{c_{3i}}{\lambda_3^i} \gg \cdots \gg \frac{c_{ni}}{\lambda_n^i} \tag{5.87}$$

which means that the very first term is the significant one in the right side of $(5.86)_1$, and consequently

$$\underset{(i)}{\underline{\mathbf{A}}}_1$$

tends to be the first eigenvector.

If $\underset{(i)}{\underline{\mathbf{A}}_1}$ tends to be the first eigenvector the Rayleigh quotient tends to be the first eigenvalue.

If we want to determine the second eigenvector and eigenvalue we have to solve the eigenvalue problem

$$\underline{\mathbf{K}}_2\,\underline{\psi}_\ell = \lambda_\ell\,\underline{\mathbf{M}}_2\,\underline{\psi}_\ell \qquad \lambda_\ell = \omega_\ell^2 \tag{5.88}$$

for which λ_2 is the smallest eigenvalue and $\underline{\psi}_2$ is the corresponding eigenvector. It is obvious the previous proof remains valid for this problem word by word, i.e., the matrix iteration yields the second eigenvector and the second eigenvalue.

By now there is no need for a further explanation to see that the previous results remain valid for eigenvalues and eigenvectors of order higher than two.

5.3.4 Properties of the Iteration Algorithm

The iteration algorithm we have presented is called matrix iteration (or inverse matrix iteration). Its main advantage is that it converges to an eigenvalue quickly, since both the eigenvalue and the eigenvector that belongs to it are improved in each iteration step. It is worth mentioning here that the convergence rate for the eigenvalue is cubic which means that the number of correct digits in the eigenvalues triples in each iteration step.

- The algorithm we have presented in Sect. 5.3.2 assumes that the eigenvalues (natural frequencies) are distinct, that is, there are no repeated roots of the characteristic Eq. (5.44).
- The algorithm requires less iteration steps if the eigenvalues $\lambda_1 < \lambda_2 < \lambda_3 < \dots < \lambda_n$ are well separated from each other.
- The number of iteration steps to be performed for finding λ_1 depends on how the trial eigenvector (the initial value of the first eigenvector) resembles the first eigenvector. It is worth, however, emphasizing again that the convergence for the eigenvalue is very fast. Solution to Problem 5.5 in Sect. C.5 clearly shows that.
- In case the trial vector (the initial value) is parallel to an eigenvector the solution algorithm does not converge since in each iteration step the approximation of the eigenvector remains parallel to the initial value.

5.3.5 Calculation of the Largest Eigenvalue

The matrix iteration in Sect. 5.3.2 can be applied in a slightly modified form to determine the largest eigenvalue and the corresponding eigenvector. Consider the eigenvalue problem

$$\underline{\mathbf{M}}\,\underline{\mathfrak{A}}_\ell = \nu_\ell\,\underline{\mathbf{K}}\,\underline{\mathfrak{A}}_\ell\,, \qquad \ell = 1, 2, \dots, n \tag{5.89}$$

for which ν_ℓ is the eigenvalue, $\underline{\mathfrak{A}}_\ell$ is the eigenvector, and $\underline{\varphi}_\ell$ is the normalized eigenvector:

$$\underline{\varphi}_\ell = \frac{\underline{\mathfrak{A}}_\ell}{\sqrt{\left(\underline{\mathfrak{A}}_\ell, \underline{\mathfrak{A}}_\ell\right)_K}} . \tag{5.90}$$

Comparison of the eigenvalue Problem (5.89) to the original eigenvalue Problem (5.50) shows that they are equivalent and

$$\begin{gathered}\underline{\mathfrak{A}}_1 = \underline{\mathbf{A}}_n, \quad \underline{\mathfrak{A}}_2 = \underline{\mathbf{A}}_{n-1}, \ldots, \underline{\mathfrak{A}}_{n-1} = \underline{\mathbf{A}}_2, \quad \underline{\mathfrak{A}}_n = \underline{\mathbf{A}}_1; \\ \nu_1 = \frac{1}{\lambda_1}, \quad \nu_2 = \frac{1}{\lambda_{n-1}}, \ldots, \nu_{n-1} = \frac{1}{\lambda_2}, \quad \nu_n = \frac{1}{\lambda_1} .\end{gathered} \tag{5.91}$$

This means that the first eigenvector $\underline{\mathfrak{A}}_1$ is equal to the last one for the eigenvalue problem (5.50)—this is $\underline{\mathbf{A}}_n$ and the reciprocal of the first eigenvalue ν_1 is the eigenvalue λ_n we want to determine. We can now rewrite the matrix iteration into the following form:

INITIALIZATION

Setting the initial value

$$\underset{(0)}{\underline{\mathfrak{A}}_1} = \text{input}$$

Normalization

$$\underset{(0)}{\breve{\mathfrak{A}}_1} = \sqrt{\left(\underset{(0)}{\underline{\mathfrak{A}}_1}, \underset{(0)}{\underline{\mathfrak{A}}_1}\right)_K} \qquad \underset{(0)}{\underline{\varphi}_1} = \underset{(0)}{\underline{\mathfrak{A}}_1} / \underset{(0)}{\breve{\mathfrak{A}}_1}$$

Estimation

$$\underset{(0)}{\nu_1} = \frac{1}{\underset{(0)}{\lambda_n}} = \frac{1}{\underset{(0)}{\omega_n^2}} = \left(\underset{(0)}{\underline{\varphi}_1}, \underset{(0)}{\underline{\varphi}_1}\right)_M$$

ITERATION STEPS

Step 1.
Solution for $\underset{(1)}{\underline{\mathfrak{A}}_1}$

$$\underline{\mathbf{M}}\,\underset{(1)}{\underline{\mathfrak{A}}_1} = \underline{\mathbf{K}}\,\underset{(0)}{\underline{\varphi}_1}$$

Normalization

$$\underset{(1)}{\breve{\mathfrak{A}}_1} = \sqrt{\left(\underset{(1)}{\underline{\mathfrak{A}}_1}, \underset{(1)}{\underline{\mathfrak{A}}_1}\right)_K} \qquad \underset{(1)}{\underline{\varphi}_1} = \underset{(1)}{\underline{\mathfrak{A}}_1} / \underset{(1)}{\breve{\mathfrak{A}}_1}$$

Estimation

$$\underset{(1)}{\nu_1} = \frac{1}{\underset{(1)}{\lambda_n}} = \frac{1}{\underset{(1)}{\omega_n^2}} = \left(\underset{(1)}{\underline{\varphi}_1}, \underset{(1)}{\underline{\varphi}_1}\right)_M$$

Step ℓ
Solution for $\underset{(\ell)}{\underline{\mathfrak{A}}_1}$

$$\underline{\mathbf{M}}\,\underset{(\ell)}{\underline{\mathfrak{A}}_1} = \underline{\mathbf{K}}\,\underset{(\ell-1)}{\underline{\varphi}_1}$$

Normalization

$$\underset{(\ell)}{\breve{\mathfrak{A}}_1} = \sqrt{\left(\underset{(\ell)}{\underline{\mathfrak{A}}_1}, \underset{(\ell)}{\underline{\mathfrak{A}}_1}\right)_K} \qquad \underset{(\ell)}{\underline{\varphi}_1} = \underset{(\ell)}{\underline{\mathfrak{A}}_1} / \underset{(\ell)}{\breve{\mathfrak{A}}_1}$$

Estimation

$$\underset{(\ell)}{\nu_1} = \frac{1}{\underset{(\ell)}{\lambda_n}} = \frac{1}{\underset{(\ell)}{\omega_n^2}} = \left(\underset{(\ell)}{\underline{\varphi}_1}, \underset{(\ell)}{\underline{\varphi}_1}\right)_M$$

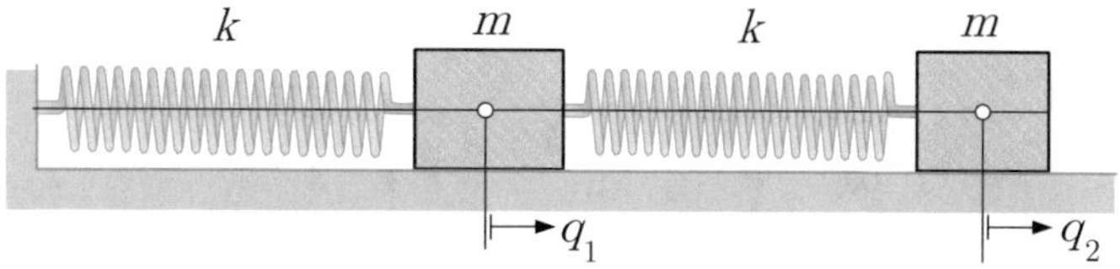

Fig. 5.4 Spring-mass system fixed at the left end

Exercise 5.3 Consider the system shown in Fig. 5.4. Estimate the first natural frequency and the first eigenvector.

This problem is the same as that we solved in Exercise 4.7. The mass matrix, the stiffness matrix, the first eigenvalue, and eigenvector are as follows:

$$\begin{gathered}\underline{\mathbf{M}} = m\begin{bmatrix}1 & 0\\ 0 & 1\end{bmatrix}, \qquad \underline{\mathbf{K}} = k\begin{bmatrix}2 & -1\\ -1 & 1\end{bmatrix},\\ \lambda_1 = \frac{k}{m}\frac{3-\sqrt{5}}{2} = \frac{k}{m}\mathbf{0.382\,01}, \quad \underline{\mathbf{A}}_1 = \begin{bmatrix}1\\ \frac{1+\sqrt{5}}{2}\end{bmatrix} = \begin{bmatrix}\mathbf{1.000\,00}\\ \mathbf{1.618\,03}\end{bmatrix}.\end{gathered} \tag{5.92}$$

Following the iteration algorithm given in Sect. 5.3.2 step by step yields:

INITIALIZATION

Setting the initial value:

$$\underset{(0)}{\underline{\mathbf{A}}_1} = \begin{bmatrix}1\\ 1\end{bmatrix}.$$

Normalization:

$$\underset{(0)}{\check{A}_1} = \sqrt{\left(\underset{(0)}{\underline{\mathbf{A}}_1}, \underset{(0)}{\underline{\mathbf{A}}_1}\right)_M} = \sqrt{2m}, \quad \underset{(0)}{\underline{\varphi}_1} = \underset{(0)}{\underline{\mathbf{A}}_1}/\underset{(0)}{\check{A}_1} = \frac{1}{\sqrt{2m}}\begin{bmatrix}1\\ 1\end{bmatrix}.$$

Estimation:

$$\underset{(0)}{\lambda_1} = \underset{(0)}{\omega_1^2} = \left(\underset{(0)}{\underline{\varphi}_1}, \underset{(0)}{\underline{\varphi}_1}\right)_K = \frac{k}{m}\begin{bmatrix}1 & 1\end{bmatrix}\begin{bmatrix}2 & -1\\ -1 & 1\end{bmatrix}\begin{bmatrix}1\\ 1\end{bmatrix} = \frac{k}{2m} = \frac{k}{m}0.5.$$

ITERATION STEPS

Step 1.

Solution for $\underset{(1)}{\underline{\mathbf{A}}_1}$:

$$\underline{\mathbf{K}}\,\underset{(1)}{\underline{\mathbf{A}}_1} = \underline{\mathbf{M}}\,\underset{(0)}{\underline{\varphi}_1} \quad\longrightarrow\quad k\begin{bmatrix}2 & -1\\ -1 & 1\end{bmatrix}\begin{bmatrix}\underset{(1)}{A_{11}}\\ \underset{(1)}{A_{21}}\end{bmatrix} = \sqrt{\frac{m}{2}}\begin{bmatrix}1 & 0\\ 0 & 1\end{bmatrix}\begin{bmatrix}1\\ 1\end{bmatrix},$$

$$\underset{(1)}{A_{11}} = \frac{2}{\sqrt{2}}\frac{\sqrt{m}}{k}, \qquad \underset{(1)}{A_{21}} = \frac{3}{\sqrt{2}}\frac{\sqrt{m}}{k}.$$

Normalization:

$$\check{A}_{\underset{(1)}{}1} = \sqrt{\left(\underset{(1)}{\mathbf{A}}_1, \underset{(1)}{\mathbf{A}}_1\right)_M} = \frac{m}{k}\sqrt{\frac{4+9}{2}},$$

$$\underset{(1)}{\underline{\varphi}}_1 = \underset{(1)}{\mathbf{A}}_1 / \underset{(1)}{\check{A}}_1 = \frac{1}{\sqrt{m}}\frac{1}{\sqrt{4+9}}\begin{bmatrix}2\\3\end{bmatrix}.$$

Estimation:

$$\underset{(1)}{\lambda}_1 = \underset{(1)}{\omega}_1^2 = \left(\underset{(1)}{\underline{\varphi}}_1, \underset{(1)}{\underline{\varphi}}_1\right)_K = \frac{k}{m}\frac{\begin{bmatrix}2\ 3\end{bmatrix}\begin{bmatrix}2 & -1\\-1 & 1\end{bmatrix}\begin{bmatrix}2\\3\end{bmatrix}}{4+9} = 0.384\,615\frac{k}{m},$$

$$\underset{(1)}{\mathbf{A}}_1 = \frac{2}{\sqrt{2}}\frac{\sqrt{m}}{k}\begin{bmatrix}1.0\\1.5\end{bmatrix}.$$

Step 2.
Solution for $\underset{(2)}{\mathbf{A}}_1$:

$$\underline{\mathbf{K}}\,\underset{(2)}{\mathbf{A}}_1 = \underline{\mathbf{M}}\underset{(1)}{\underline{\varphi}}_1 \quad\longrightarrow\quad k\begin{bmatrix}2 & -1\\-1 & 1\end{bmatrix}\begin{bmatrix}\underset{(2)}{A}_{11}\\ \underset{(2)}{A}_{21}\end{bmatrix} = \frac{\sqrt{m}}{\sqrt{4+9}}\begin{bmatrix}1 & 0\\0 & 1\end{bmatrix}\begin{bmatrix}2\\3\end{bmatrix},$$

$$\underset{(2)}{A}_{11} = \frac{5}{\sqrt{4+9}}\frac{\sqrt{m}}{k}, \qquad \underset{(2)}{A}_{21} = \frac{8}{\sqrt{4+9}}\frac{\sqrt{m}}{k}.$$

Normalization:

$$\underset{(2)}{\check{A}}_1 = \sqrt{\left(\underset{(2)}{\mathbf{A}}_1, \underset{(2)}{\mathbf{A}}_1\right)_M} = \frac{m}{k}\sqrt{\frac{25+64}{4+9}},$$

$$\underset{(2)}{\underline{\varphi}}_1 = \underset{(2)}{\mathbf{A}}_1 / \underset{(2)}{\check{A}}_1 = \frac{1}{\sqrt{m}}\frac{1}{\sqrt{25+64}}\begin{bmatrix}5\\8\end{bmatrix}.$$

Estimation:

$$\underset{(2)}{\lambda}_1 = \underset{(2)}{\omega}_1^2 = \left(\underset{(2)}{\underline{\varphi}}_1, \underset{(2)}{\underline{\varphi}}_1\right)_K =$$

$$= \frac{k}{m}\frac{1}{25+64}\times\begin{bmatrix}5\ 8\end{bmatrix}\begin{bmatrix}2 & -1\\-1 & 1\end{bmatrix}\begin{bmatrix}5\\8\end{bmatrix} = \mathbf{0.382\,02}\frac{k}{m}, \tag{5.93a}$$

$$\underset{(2)}{\mathbf{A}}_1 = \frac{5}{\sqrt{4+9}}\frac{\sqrt{m}}{k}\begin{bmatrix}\mathbf{1.0}\\\mathbf{1.6}\end{bmatrix}. \tag{5.93b}$$

It is worth comparing the results typeset in boldface letters in Eqs. (5.92) and (5.93).

Exercise 5.4 Estimate the second (the largest) eigenvector and eigenvalue for the system shown in Fig. 5.4. Since the spring–mass system is the same as that of Exercise 4.7 it holds that

$$\underline{\mathbf{M}} = m\begin{bmatrix} 2 & -1 \\ -1 & 1 \end{bmatrix}, \quad \underline{\mathbf{K}} = k\begin{bmatrix} 2 & -1 \\ -1 & 1 \end{bmatrix}, \quad \underline{\mathbf{M}}^{-1}\underline{\mathbf{K}} = \frac{k}{m}\begin{bmatrix} 2 & -1 \\ -1 & 1 \end{bmatrix}, \tag{5.94}$$

$$\lambda_2 = \frac{k}{m}\frac{3+\sqrt{5}}{2} = \frac{k}{m}\mathbf{2.618\,034}, \quad \underline{\mathfrak{A}}_1 = \underline{\mathbf{A}}_2 = \begin{bmatrix} 1 \\ \frac{1-\sqrt{5}}{2} \end{bmatrix} = \begin{bmatrix} \mathbf{1.000\,000} \\ \mathbf{-0.618\,034} \end{bmatrix} \tag{5.95}$$

$$\nu_1 = \frac{1}{\lambda_2} = \frac{m}{k}\frac{2}{3+\sqrt{5}} = \frac{m}{k}\mathbf{0.381\,966}. \tag{5.96}$$

For finding a good estimation, we shall apply the iteration algorithm given in Sect. 5.3.5:

INITIALIZATION

Setting the initial value:

$$\underset{(0)}{\underline{\mathfrak{A}}_1} = \begin{bmatrix} 1 \\ 1 \end{bmatrix}.$$

Normalization:

$$\underset{(0)}{\check{\mathfrak{A}}_1} = \sqrt{\left(\underset{(0)}{\underline{\mathfrak{A}}_1}, \underset{(0)}{\underline{\mathfrak{A}}_1}\right)_K} = \sqrt{k\,[1\;1]\begin{bmatrix} 2 & -1 \\ -1 & 1 \end{bmatrix}\begin{bmatrix} 1 \\ 1 \end{bmatrix}} = \sqrt{k},$$

$$\underset{(0)}{\underline{\varphi}_1} = \frac{\underset{(0)}{\underline{\mathfrak{A}}_1}}{\underset{(0)}{\check{\mathfrak{A}}_1}} = \frac{1}{\sqrt{k}}\begin{bmatrix} 1 \\ 1 \end{bmatrix}.$$

Estimation:

$$\underset{(0)}{\nu_1} = \frac{1}{\underset{(0)}{\lambda_2}} = \frac{1}{\underset{(0)}{\omega_2^2}} = \left(\underset{(0)}{\underline{\varphi}_1}, \underset{(0)}{\underline{\varphi}_1}\right)_M = \frac{m}{k}\,[1\;1]\begin{bmatrix} 1 & 0 \\ 0 & 1 \end{bmatrix}\begin{bmatrix} 1 \\ 1 \end{bmatrix} = \frac{2m}{k} = 2\frac{m}{k}.$$

ITERATION STEPS

Step 1.

Solution for $\underset{(1)}{\underline{\mathfrak{A}}_1}$:

$$\underline{\mathbf{M}}\,\underset{(1)}{\underline{\mathfrak{A}}_1} = \underline{\mathbf{K}}\,\underset{(0)}{\underline{\varphi}_1} \quad \rightarrow \quad \underset{(1)}{\underline{\mathfrak{A}}_1} = \underline{\mathbf{M}}^{-1}\underline{\mathbf{K}}\,\underset{(0)}{\underline{\varphi}_1} = \frac{\sqrt{k}}{m}\begin{bmatrix} 2 & -1 \\ -1 & 1 \end{bmatrix}\begin{bmatrix} 1 \\ 1 \end{bmatrix} = \frac{\sqrt{k}}{m}\begin{bmatrix} 1 \\ 0 \end{bmatrix},$$

$$\underset{(1)}{\check{\mathfrak{A}}_{11}} = \frac{\sqrt{k}}{m}, \qquad \underset{(1)}{\check{\mathfrak{A}}_{21}} = 0.$$

Normalization:

$$\underset{(1)}{\breve{\mathfrak{A}}_1}=\sqrt{\left(\underset{(1)}{\underline{\mathfrak{A}}_1},\underset{(1)}{\underline{\mathfrak{A}}_1}\right)_K}=\frac{k}{m}\sqrt{\begin{bmatrix}1&0\end{bmatrix}\begin{bmatrix}2&-1\\-1&1\end{bmatrix}\begin{bmatrix}1\\0\end{bmatrix}}=\frac{k}{m}\sqrt{2},$$

$$\underset{(1)}{\underline{\varphi}_1}=\frac{\underset{(1)}{\underline{\mathfrak{A}}_1}}{\underset{(1)}{\breve{\mathfrak{A}}_1}}=\frac{1}{\sqrt{2k}}\begin{bmatrix}1\\0\end{bmatrix}.$$

Estimation:

$$\underset{(1)}{\nu_1}=\frac{1}{\underset{(1)}{\lambda_2}}=\frac{1}{\underset{(1)}{\omega_2^2}}=\left(\underset{(1)}{\underline{\varphi}_1},\underset{(1)}{\underline{\varphi}_1}\right)_M=\frac{m}{k}\begin{bmatrix}1&0\end{bmatrix}\begin{bmatrix}1&0\\0&1\end{bmatrix}\begin{bmatrix}1\\0\end{bmatrix}=\frac{m}{2k}=0.5\frac{m}{k}.$$

Step 2.

Solution for $\underset{(2)}{\underline{\mathfrak{A}}_1}$:

$$\underline{\mathbf{M}}\,\underset{(2)}{\underline{\mathfrak{A}}_1}=\underline{\mathbf{K}}\,\underset{(1)}{\underline{\varphi}_1}\;\rightarrow\;\underset{(2)}{\underline{\mathfrak{A}}_1}=\underline{\mathbf{M}}^{-1}\underline{\mathbf{K}}\,\underset{(1)}{\underline{\varphi}_1}=$$

$$=\frac{1}{\sqrt{2}}\frac{\sqrt{k}}{m}\begin{bmatrix}2&-1\\-1&1\end{bmatrix}\begin{bmatrix}1\\0\end{bmatrix}=\frac{1}{\sqrt{2}}\frac{\sqrt{k}}{m}\begin{bmatrix}2\\-1\end{bmatrix},$$

$$\underset{(2)}{\breve{\mathfrak{A}}_{11}}=\sqrt{2}\frac{\sqrt{k}}{m},\qquad \underset{(2)}{\breve{\mathfrak{A}}_{21}}=-\frac{1}{\sqrt{2}}\frac{\sqrt{k}}{m}.$$

Normalization:

$$\underset{(2)}{\breve{\mathfrak{A}}_1}=\sqrt{\left(\underset{(2)}{\underline{\mathfrak{A}}_1},\underset{(2)}{\underline{\mathfrak{A}}_1}\right)_K}=\frac{k}{m}\sqrt{\begin{bmatrix}\sqrt{2}&-\frac{1}{\sqrt{2}}\end{bmatrix}\begin{bmatrix}2&-1\\-1&1\end{bmatrix}\begin{bmatrix}\sqrt{2}\\-\frac{1}{\sqrt{2}}\end{bmatrix}}=\frac{k}{m}\sqrt{6.5},$$

$$\underset{(2)}{\underline{\varphi}_1}=\frac{\underset{(2)}{\underline{\mathfrak{A}}_1}}{\underset{(2)}{\breve{\mathfrak{A}}_1}}=\frac{1}{\sqrt{13k}}\begin{bmatrix}2\\-1\end{bmatrix}.$$

Estimation:

$$\underset{(2)}{\nu_1}=\frac{1}{\underset{(2)}{\lambda_2}}=\frac{1}{\underset{(2)}{\omega_2^2}}=\left(\underset{(2)}{\underline{\varphi}_1},\underset{(2)}{\underline{\varphi}_1}\right)_M=$$

$$=\frac{m}{k}\frac{1}{13}\begin{bmatrix}2&-1\end{bmatrix}\begin{bmatrix}1&0\\0&1\end{bmatrix}\begin{bmatrix}2\\-1\end{bmatrix}=0.384\,615\,38\frac{m}{k}.$$

Step 3.
Solution for $\underset{(3)}{\underline{\mathfrak{A}}_1}$:

$$\mathbf{\underline{M}}\,\underset{(3)}{\underline{\mathfrak{A}}_1} = \mathbf{\underline{K}}\,\underset{(2)}{\underline{\varphi}_1} \quad \rightarrow \quad \underset{(3)}{\underline{\mathfrak{A}}_1} = \mathbf{\underline{M}}^{-1}\mathbf{\underline{K}}\,\underset{(2)}{\underline{\varphi}_1} =$$

$$= \frac{1}{\sqrt{13}}\frac{\sqrt{k}}{m}\begin{bmatrix} 2 & -1 \\ -1 & 1 \end{bmatrix}\begin{bmatrix} 2 \\ -1 \end{bmatrix} = \frac{1}{\sqrt{13}}\frac{\sqrt{k}}{m}\begin{bmatrix} 5 \\ -3 \end{bmatrix},$$

$$\underset{(3)}{\breve{\mathfrak{A}}_{11}} = \frac{5}{\sqrt{13}}\frac{\sqrt{k}}{m}, \qquad \underset{(3)}{\breve{\mathfrak{A}}_{21}} = -\frac{3}{\sqrt{13}}\frac{\sqrt{k}}{m}.$$

Normalization:

$$\underset{(3)}{\breve{\mathfrak{A}}_1} = \sqrt{\left(\underset{(3)}{\underline{\mathfrak{A}}_1}, \underset{(3)}{\underline{\mathfrak{A}}_1}\right)_K} = \frac{1}{\sqrt{13}}\frac{k}{m}\sqrt{\begin{bmatrix} 5 & -3 \end{bmatrix}\begin{bmatrix} 2 & -1 \\ -1 & 1 \end{bmatrix}\begin{bmatrix} 5 \\ -3 \end{bmatrix}} = \frac{k}{m}\frac{\sqrt{89}}{\sqrt{13}},$$

$$\underset{(2)}{\underline{\varphi}_1} = \frac{\underset{(2)}{\underline{\mathfrak{A}}_1}}{\underset{(2)}{\breve{\mathfrak{A}}_1}} = \frac{1}{\sqrt{89k}}\begin{bmatrix} 5 \\ -3 \end{bmatrix}.$$

Estimation:

$$\underset{(3)}{\nu_1} = \frac{1}{\underset{(2)}{\lambda_2}} = \frac{1}{\underset{(2)}{\omega_2^2}} = \left(\underset{(2)}{\underline{\varphi}_1}, \underset{(2)}{\underline{\varphi}_1}\right)_M =$$

$$= \frac{m}{k}\frac{1}{89}\begin{bmatrix} 5 & -3 \end{bmatrix}\begin{bmatrix} 1 & 0 \\ 0 & 1 \end{bmatrix}\begin{bmatrix} 5 \\ -3 \end{bmatrix} = 0.382\,022\frac{m}{k}.$$

For the sake of a comparison, we present the third estimation for the eigenvalue and the eigenvector (for the latter the first element is normalized to unity) in the following table.

Table 5.1 Estimated and exact values

	Eigenvalue $\nu_1 = 1/\lambda_2 = 1/\omega_2^2$	Eigenvector $\underline{\mathfrak{A}}_1^T = \underline{\mathbf{A}}_2^T$
Estimation	$0.382\,022\,\frac{m}{k}$	$\begin{bmatrix} 1.0 & -0.6000 \end{bmatrix}$
Exact value	$0.381\,966\,\frac{m}{k}$	$\begin{bmatrix} 1.0 & -0.6180 \end{bmatrix}$

Though the initial value for the eigenvector did not resemble the solution the results obtained in the third step are quite good.

5.4 Forced Vibrations

5.4.1 Forced Harmonic Vibrations

Assume that the vibrating system with n degrees of freedom is subjected to a harmonic force

$$\underset{(n\times 1)}{\underline{\mathbf{F}}_f} = \underset{(n\times 1)}{\underline{\mathbf{F}}_{fo}} \cos\omega_f t\,. \tag{5.97}$$

If the vibrations are undamped

$$\underset{(n\times n)}{\underline{\mathbf{M}}}\ \underset{(n\times 1)}{\ddot{\underline{\mathbf{q}}}} + \underset{(n\times n)}{\underline{\mathbf{K}}}\ \underset{(n\times 1)}{\underline{\mathbf{q}}} = \underset{(n\times 1)}{\underline{\mathbf{F}}_{fo}} \cos\omega_f t \tag{5.98}$$

is the equation of motion. The steady-state solution is sought in the form

$$\underset{(n\times 1)}{\underline{\mathbf{q}}} = \underset{(n\times 1)}{\underline{\mathbf{A}}_f} \cos\omega_f t, \tag{5.99}$$

where $\underline{\mathbf{A}}_f$ is the amplitude of the forced vibrations. Upon substitution of this solution into the equation of motion (5.97) we arrive at an inhomogeneous linear equation system for $\underline{\mathbf{A}}_f$:

$$(\underset{(n\times n)}{\underline{\mathbf{K}}} - \omega_f^2 \underset{(n\times n)}{\underline{\mathbf{M}}})\underset{(n\times 1)}{\underline{\mathbf{A}}_f} = \underset{(n\times 1)}{\underline{\mathbf{F}}_{fo}}\,. \tag{5.100}$$

Its solution is of the form

$$\underset{(n\times 1)}{\underline{\mathbf{A}}_f} = \frac{\mathrm{adj}[\underset{(n\times n)}{\underline{\mathbf{K}}} - \omega_f^2 \underset{(n\times n)}{\underline{\mathbf{M}}}]}{|\underset{(n\times n)}{\underline{\mathbf{K}}} - \omega_f^2 \underset{(n\times n)}{\underline{\mathbf{M}}}|}\underset{(n\times 1)}{\underline{\mathbf{F}}_{fo}} = \frac{\mathrm{adj}[\underset{(n\times n)}{\underline{\mathbf{K}}} - \omega_f^2 \underset{(n\times n)}{\underline{\mathbf{M}}}]}{\mathfrak{a}_n \prod_{\ell=1}^{n}\left(\omega_{n\ell}^2 - \omega_f^2\right)}\underset{(n\times 1)}{\underline{\mathbf{F}}_{fo}}, \tag{5.101}$$

where

$$|\underset{(n\times n)}{\underline{\mathbf{K}}} - \omega_f^2 \underset{(n\times n)}{\underline{\mathbf{M}}}| = \mathfrak{a}_n\left(\omega_f^2\right)^n + \mathfrak{a}_{n-1}\left(\omega_f^2\right)^{n-1} + \mathfrak{a}_{n-2}\left(\omega_f^2\right)^{n-2} + \cdots + \mathfrak{a}_1\omega_f^2 + \mathfrak{a}_0 = \mathfrak{a}_n \prod_{\ell=1}^{n}\left(\omega_{n\ell}^2 - \omega_f^2\right) \tag{5.102}$$

in which $\omega_{n\ell}^2$ is the ℓth root of the polynomial

$$\mathfrak{a}_n \left(\omega_f^2\right)^n + \mathfrak{a}_{n-1} \left(\omega_f^2\right)^{n-1} + \cdots + \mathfrak{a}_1 \omega_f^2 + \mathfrak{a}_0 = 0\,,$$

i.e., $\omega_{n\ell}$ is the ℓth natural circular frequency of the free vibrations. Resonance occurs if ω_f tends to $\omega_{n\ell}$ ($\ell = 1, 2, \ldots, n$).

5.4.2 *Nonharmonic Inhomogeneity*

Assume now that there are nonharmonic external and time-dependent forces in the system. If damping is neglected the equation of motion takes the form

$$\underset{(n\times n)}{\underline{\mathbf{M}}}\ \underset{(n\times 1)}{\underline{\ddot{\mathfrak{q}}}} + \underset{(n\times n)}{\underline{\mathbf{K}}}\ \underset{(n\times 1)}{\underline{\mathfrak{q}}} = \underset{(n\times 1)}{\underline{\mathbf{F}}_f}\,, \quad \underline{\mathbf{F}}_f^T = \left[\, f_1(t) \,|\, f_2(t) \,|\cdots|\, f_n(t) \,\right]. \tag{5.103}$$

Recalling that the system of the mass normalized eigenfunctions is complete in the n-dimensional space we can give the generalized displacement vector in terms of the eigenfunctions as

$$\underset{(n\times 1)}{\underline{\mathfrak{q}}} = \mathfrak{q}_1 \underline{\psi}_1 + \mathfrak{q}_2 \underline{\psi}_2 + \mathfrak{q}_3 \underline{\psi}_3 + \cdots + \mathfrak{q}_n \underline{\psi}_n = \sum_{\ell=1}^{n} \mathfrak{q}_\ell(t) \underline{\psi}_\ell \tag{5.104}$$

or in a more concise form

$$\underset{(n\times 1)}{\underline{\mathfrak{q}}} = \underset{(n\times n)}{\underline{\phi}}\ \underset{(n\times 1)}{\underline{\mathfrak{q}}} = \left[\ \underset{(n\times 1)}{\underline{\psi}_1} \middle| \underset{(n\times 1)}{\underline{\psi}_2} \middle| \cdots \middle| \underset{(n\times 1)}{\underline{\psi}_n}\ \right] \underset{(n\times 1)}{\underline{\mathfrak{q}}} =$$

$$= \begin{bmatrix} \psi_{11} & \psi_{12} & \psi_{13} & \cdots & \psi_{1n} \\ \psi_{21} & \psi_{22} & \psi_{23} & \cdots & \psi_{2n} \\ \vdots & \vdots & \vdots & \ddots & \vdots \\ \psi_{n1} & \psi_{n2} & \psi_{n3} & \cdots & \psi_{nn} \end{bmatrix} \begin{bmatrix} \mathfrak{q}_1 \\ \mathfrak{q}_2 \\ \vdots \\ \mathfrak{q}_n \end{bmatrix}, \tag{5.105}$$

where $\psi_{k\ell}$ is the kth element of the eigenvector $\underline{\psi}_\ell$. Equation (5.104) in terms of the new independent variable $\underline{\mathfrak{q}}$ assumes the form

$$\underset{(n\times n)}{\underline{\mathbf{M}}}\ \underset{(n\times n)}{\underline{\phi}}\ \underset{(n\times 1)}{\underline{\ddot{\mathfrak{q}}}} + \underset{(n\times n)}{\underline{\mathbf{K}}}\ \underset{(n\times n)}{\underline{\phi}}\ \underset{(n\times 1)}{\underline{\mathfrak{q}}} = \underset{(n\times 1)}{\underline{\mathbf{F}}_f}\,. \tag{5.106a}$$

If we left multiply the above equation by $\underset{(n\times n)}{\underline{\phi}^T}$ we get

$$\underset{(n\times n)}{\underline{\phi}^T}\ \underset{(n\times n)}{\underline{\mathbf{M}}}\ \underset{(n\times n)}{\underline{\phi}}\ \underset{(n\times 1)}{\underline{\ddot{\mathfrak{q}}}} + \underset{(n\times n)}{\underline{\phi}^T}\ \underset{(n\times n)}{\underline{\mathbf{K}}}\ \underset{(n\times n)}{\underline{\phi}}\ \underset{(n\times 1)}{\underline{\mathfrak{q}}} = \underset{(n\times n)}{\underline{\phi}^T}\ \underset{(n\times 1)}{\underline{\mathbf{F}}_f} \tag{5.106b}$$

in which

$$\underset{(n\times n)}{\underline{\phi}^T}\;\underset{(n\times n)}{\mathbf{M}}\;\underset{(n\times n)}{\underline{\phi}} = \underbrace{\begin{bmatrix} \underline{\psi}_1^T \\ \underline{\psi}_2^T \\ \vdots \\ \underline{\psi}_n^T \end{bmatrix}}_{(n\times n)} \underset{(n\times n)}{\mathbf{M}} \underbrace{\left[\underline{\psi}_1 | \underline{\psi}_2 | \cdots | \underline{\psi}_n\right]}_{(n\times n)} \underset{\text{notation (5.13)}}{=\uparrow=}$$

$$= \begin{bmatrix} \left(\underline{\psi}_1, \underline{\psi}_1\right)_M & \left(\underline{\psi}_1, \underline{\psi}_2\right)_M & \cdots & \left(\underline{\psi}_1, \underline{\psi}_n\right)_M \\ \left(\underline{\psi}_2, \underline{\psi}_1\right)_M & \left(\underline{\psi}_2, \underline{\psi}_2\right)_M & \cdots & \left(\underline{\psi}_2, \underline{\psi}_n\right)_M \\ \vdots & \vdots & \ddots & \vdots \\ \left(\underline{\psi}_n, \underline{\psi}_1\right)_M & \left(\underline{\psi}_n, \underline{\psi}_2\right)_M & \cdots & \left(\underline{\psi}_n, \underline{\psi}_n\right)_M \end{bmatrix} \underset{(5.39\text{a})}{=\uparrow=}$$

$$= \underbrace{\begin{bmatrix} 1 & 0 & \cdots & 0 \\ 0 & 1 & \cdots & 0 \\ \vdots & \vdots & \ddots & \vdots \\ 0 & 0 & \cdots & 1 \end{bmatrix}}_{\mathcal{I}} = \underset{(n\times n)}{\underline{\mathcal{I}}} \tag{5.107a}$$

is the unit matrix of size $(n \times n)$. As regards the product $\underset{(n\times n)}{\underline{\phi}^T}\;\underset{(n\times n)}{\mathbf{K}}\;\underset{(n\times n)}{\underline{\phi}}$, we get in a similar manner that

$$\underset{(n\times n)}{\underline{\phi}^T}\;\underset{(n\times n)}{\mathbf{K}}\;\underset{(n\times n)}{\underline{\phi}} = \underbrace{\begin{bmatrix} \underline{\psi}_1^T \\ \underline{\psi}_2^T \\ \vdots \\ \underline{\psi}_n^T \end{bmatrix}}_{(n\times n)} \underset{(n\times n)}{\mathbf{K}} \underbrace{\left[\underline{\psi}_1 | \underline{\psi}_2 | \cdots | \underline{\psi}_n\right]}_{(n\times n)} \underset{\text{notation (5.13)}}{=\uparrow=}$$

$$= \begin{bmatrix} \left(\underline{\psi}_1, \underline{\psi}_1\right)_K & 0 & \cdots & 0 \\ 0 & \left(\underline{\psi}_2, \underline{\psi}_2\right)_K & \cdots & 0 \\ \vdots & \vdots & \ddots & \vdots \\ 0 & 0 & \cdots & \left(\underline{\psi}_n, \underline{\psi}_n\right)_K \end{bmatrix} \underset{(5.39\text{c})}{=\uparrow=}$$

$$= \begin{bmatrix} \lambda_1 & 0 & \cdots & 0 \\ 0 & \lambda_2 & \cdots & 0 \\ \vdots & \vdots & \ddots & \vdots \\ 0 & 0 & \cdots & \lambda_n \end{bmatrix} = \underset{(n\times n)}{\underline{\mathcal{K}}} \,. \tag{5.107b}$$

After back-substituting relations (5.107) in the equation of motion (5.106b), we arrive at an uncoupled system of differential equations

$$\underset{(n\times n)}{\underline{\mathcal{I}}}\ \underset{(n\times 1)}{\underline{\ddot{\mathsf{q}}}} + \underset{(n\times n)}{\underline{\mathcal{K}}}\ \underset{(n\times 1)}{\underline{\mathsf{q}}} = \underset{(n\times n)}{\underline{\phi}^T}\ \underset{(n\times 1)}{\underline{\mathbf{F}}_f}\ . \tag{5.108}$$

Consequently,

$$\ddot{\mathsf{q}}_\ell + \lambda_\ell \mathsf{q}_\ell = Q_\ell = \underset{(1\times n)}{\underline{\psi}_\ell^T}\ \underset{(n\times 1)}{\underline{\mathbf{F}}_f}\ , \qquad \ell = 1, 2, \dots, n\ , \tag{5.109}$$

where Q_ℓ is defined by the matrix product on the right side. Since Q_ℓ is a function of time there arises the question how to find a solution for the inhomogeneous equation of motion (5.108). This issue will be considered in Remarks 5.11 and 5.12.

Remark 5.11 Let $\psi(t, \tau)$ be a function of two variables t and τ. We shall assume that $\psi(t, \tau)$ is differentiable continuously as many times as required. Further let $\tau_1(t)$ and $\tau_2(t)$ be two continuous time functions. Consider now the integral

$$\Psi(t) = \int_{\tau_1(t)}^{\tau_2(t)} \psi(t, \tau)\, \mathrm{d}\tau\ . \tag{5.110}$$

It is known that

$$\frac{\mathrm{d}\Psi(t)}{\mathrm{d}t} = \psi(t, \tau_2)\frac{\mathrm{d}\tau_2}{\mathrm{d}t} - \psi(t, \tau_1)\frac{\mathrm{d}\tau_1}{\mathrm{d}t} + \int_{\tau_1(t)}^{\tau_2(t)} \frac{\partial \psi(t, \tau)}{\partial t}\, \mathrm{d}\tau\ . \tag{5.111}$$

Remark 5.12 Assume that the particular solution of Eq. (5.108) is of the form

$$\mathsf{q}_{\ell\,\mathrm{part}} = \int_{\tau=0}^{t} Q_\ell(\tau) g(t-\tau)\, \mathrm{d}\tau, \tag{5.112}$$

where $g(t)$ is an unknown function. Substituting $\mathsf{q}_{\ell\,\mathrm{part}}$ and its second time derivative

$$\ddot{\mathsf{q}}_{\ell\,\mathrm{part}} \underset{(5.111)}{=\uparrow} \underset{\tau_1\equiv 0}{\uparrow=} \frac{\mathrm{d}}{\mathrm{d}t}\left\{ Q_\ell(\tau) g(t-\tau)|_{\tau=t} + \int_{\tau=0}^{t} Q_\ell(\tau)\frac{\mathrm{d}g(t-\tau)}{\mathrm{d}t}\,\mathrm{d}\tau \right\} =$$

$$= \frac{\mathrm{d}}{\mathrm{d}t}\left[Q_\ell(\tau) g(t-\tau)|_{\tau=t}\right] + \left[Q_\ell(\tau)\frac{\mathrm{d}g(t-\tau)}{\mathrm{d}t}\right]\Bigg|_{\tau=t} + \int_{\tau=0}^{t} Q_\ell(\tau)\frac{\mathrm{d}^2 g(t-\tau)}{\mathrm{d}t^2}\,\mathrm{d}\tau =$$

$$= g(t-\tau)|_{\tau=t}\frac{\mathrm{d}Q_\ell(t)}{\mathrm{d}t} + \left[\frac{\mathrm{d}g(t-\tau)}{\mathrm{d}t}\right]\Bigg|_{\tau=t} Q_\ell(t) + \int_{\tau=0}^{t} Q_\ell(\tau)\frac{\mathrm{d}^2 g(t-\tau)}{\mathrm{d}t^2}\,\mathrm{d}\tau$$

into Eq. (5.109) we get

$$\int_{\tau=0}^{t} Q_\ell(\tau)\left[\frac{\mathrm{d}^2 g(t-\tau)}{\mathrm{d}t^2}+\lambda_\ell\, g(t-\tau)\right]\mathrm{d}\tau+ \\ + g(t-\tau)|_{\tau=t}\frac{\mathrm{d}Q_\ell(t)}{\mathrm{d}t}+\left[\frac{\mathrm{d}g(t-\tau)}{\mathrm{d}t}\right]\Bigg|_{\tau=t} Q_\ell(t)=Q_\ell(t)\,.$$

This equation is satisfied if

$$\frac{\mathrm{d}^2 g}{\mathrm{d}t^2}+\lambda_\ell\, g=0 \quad\text{and}\quad g(0)=0\,, \qquad \left.\frac{\mathrm{d}g}{\mathrm{d}t}\right|_{t=0}=1\,,$$

i.e., if

$$g=\frac{1}{\sqrt{\lambda_\ell}}\sin\sqrt{\lambda_\ell}\,t\,. \tag{5.113}$$

Consequently,

$$\mathfrak{q}_{\ell\,\mathrm{part}}=\int_{\tau=0}^{t} Q_\ell(\tau)\frac{1}{\sqrt{\lambda_\ell}}\sin\sqrt{\lambda_\ell}\,(t-\tau)\,\mathrm{d}\tau \tag{5.114a}$$

is the particular solution and

$$\boxed{\mathfrak{q}_\ell=\underbrace{a_\ell\cos\sqrt{\lambda_\ell}\,t+b_\ell\sin\sqrt{\lambda_\ell}\,t}_{\text{transient function}}+\underbrace{\int_{\tau=0}^{t} Q_\ell(\tau)\frac{1}{\sqrt{\lambda_\ell}}\sin\sqrt{\lambda_\ell}\,(t-\tau)\,\mathrm{d}\tau}_{\text{particular solution}}} \tag{5.114b}$$

is the total solution in which a_ℓ and b_ℓ are undetermined integration constants. These can be calculated by using the initial conditions

$$\underline{\mathbf{q}}(t=0)=\underline{\mathbf{q}}(0)=\underline{\mathbf{q}}_o\,, \qquad \underline{\dot{\mathbf{q}}}(t=0)=\underline{\dot{\mathbf{q}}}(0)=\underline{\mathbf{v}}_o\,. \tag{5.115}$$

Recalling (5.105) we can write

$$\underset{(n\times 1)}{\underline{\mathbf{q}}}=\underset{(n\times n)}{\underline{\phi}}\ \underset{(n\times 1)}{\underline{\mathfrak{q}}}$$

from where left multiplying by $\underset{(n\times n)}{\underline{\phi}^T}\ \underset{(n\times n)}{\underline{\mathbf{M}}}$ and taking the relation

$$\underset{(n\times n)}{\underline{\phi}^T}\ \underset{(n\times n)}{\underline{\mathbf{M}}}\ \underset{(n\times n)}{\underline{\phi}}=\underset{(n\times n)}{\underline{\mathcal{I}}}$$

into account, we obtain

$$\underset{(n\times n)}{\underline{\phi}^T}\ \underset{(n\times n)}{\underline{\mathbf{M}}}\ \underset{(n\times 1)}{\underline{\mathbf{q}}}=\underset{(n\times n)}{\underline{\phi}^T}\ \underset{(n\times n)}{\underline{\mathbf{M}}}\ \underset{(n\times n)}{\underline{\phi}}\ \underset{(n\times 1)}{\underline{\mathfrak{q}}}=\underset{(n\times 1)}{\underline{\mathfrak{q}}} \tag{5.116}$$

Hence,

$$\underline{\mathbf{q}}(t=0) = \underline{\mathbf{q}}_o = \underline{\phi}^T \underline{\underline{\mathbf{M}}}\, \underline{\mathbf{q}}_o \quad \text{and} \quad \underline{\dot{\mathbf{q}}}(t=0) = \underline{\dot{\mathbf{q}}}_o = \underline{\phi}^T \underline{\underline{\mathbf{M}}}\, \underline{\mathbf{v}}_o\,. \tag{5.117}$$

If $t = 0$ then $\mathfrak{q}_{\ell\,\text{part}} = \dot{\mathfrak{q}}_{\ell\,\text{part}} = 0$. Consequently,

$$\begin{aligned} \mathfrak{q}_\ell(t=0) &= a_\ell = \underset{(1\times n)}{\underline{\psi}_\ell^T} \underset{(n\times n)}{\underline{\underline{\mathbf{M}}}} \underset{(n\times 1)}{\underline{\mathbf{q}}_o}\,, \\ \dot{\mathfrak{q}}_\ell(t=0) &= \sqrt{\lambda_\ell}\, b_\ell = \underset{(1\times n)}{\underline{\psi}_\ell^T} \underset{(n\times n)}{\underline{\underline{\mathbf{M}}}} \underset{(n\times 1)}{\underline{\mathbf{v}}_o}\,. \end{aligned} \tag{5.118}$$

In many cases, it is sufficient to determine and then to utilize the first k eigenvectors and natural frequencies.

5.5 Problems

Problem 5.1 Given the flexibility matrix of the simply supported uniform beam shown in Fig. 5.5:

$$\underset{(3\times 3)}{\mathbf{f}} = \begin{bmatrix} f_{11} & f_{12} & f_{13} \\ f_{21} & f_{22} & f_{23} \\ f_{31} & f_{32} & f_{33} \end{bmatrix} = \frac{a^3}{12IE} \begin{bmatrix} 9 & 13 & 7 \\ 13 & 16 & 11 \\ 7 & 11 & 9 \end{bmatrix}.$$

Here I is the moment of inertia and E is the modulus of elasticity. Determine the stiffness matrix and then the equations of motion under the following conditions: (a) the mass of the beam is much less than the masses m_1, m_2, and m_3; (b) there is no excitation and system is undamped.

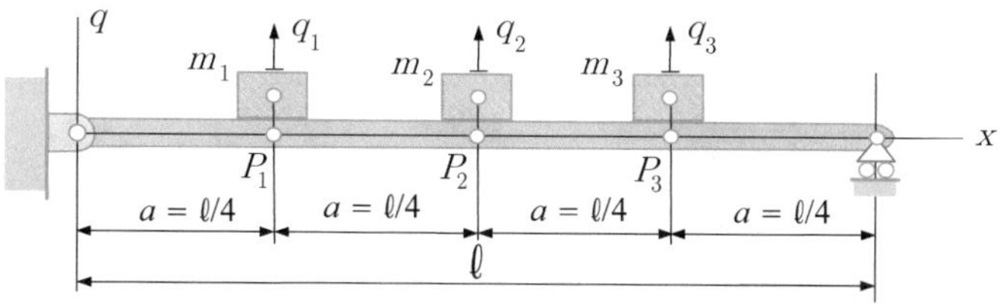

Fig. 5.5 Simply supported beam with three concentrated masses

Problem 5.2 Determine the eigenvalues (natural frequencies) of the system shown in Fig. 5.5. Assume that $m_1 = m_2 = m_3 = m$. (Hint: The equation of motion is given in a matrix form by Eq. (C.5.13).)

Problem 5.3 Given the flexibility matrix of the cantilever beam shown in Fig. 5.6:

$$\underset{(3\times3)}{\mathbf{f}} = \frac{a^3}{6IE}\begin{bmatrix} 2 & 5 & 8 \\ 5 & 16 & 28 \\ 8 & 28 & 54 \end{bmatrix}.$$

Here I is the moment of inertia and E is the modulus of elasticity. Determine the stiffness matrix and then the equations of motion under the following conditions: (a) the mass of the beam is much less than the masses m_1, m_2, and m_3; (b) there is no excitation and the system is undamped.

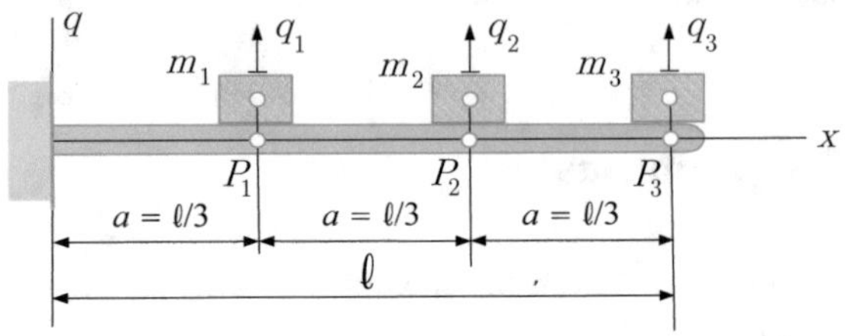

Fig. 5.6 Cantilever beam with three concentrated masses

Problem 5.4 Determine the eigenvalues (natural frequencies) of the cantilever beam shown in Fig. 5.5. Assume that $m_1 = m_2 = m_3 = m$. (Hint: The equation of motion is given in matrix form by Eq. (C.5.15).)

Problem 5.5 Find the equations of motion for the system shown in Fig. 5.7. Use the matrix iteration detailed in Sect. 5.3.2 for determining the three eigenvalues (natural frequencies) and eigenvectors.

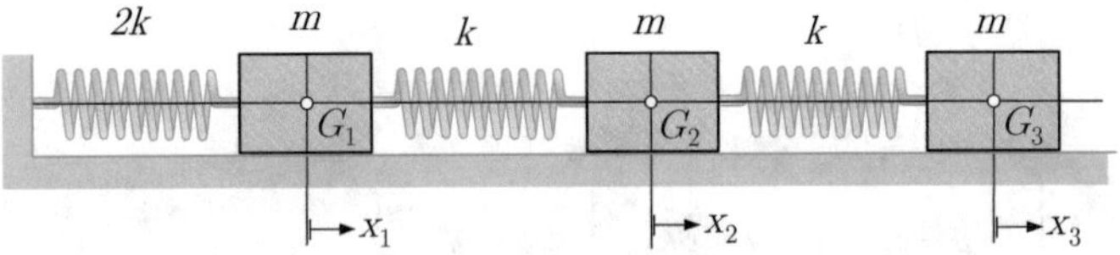

Fig. 5.7 Spring-mass system with three degrees of freedom

Problem 5.6 Determine the highest eigenvalue and eigenvector for Problem 5.5 (for the previous problem) by using the matrix iteration detailed in Sect. 5.3.5.

References

1. R.A. Horn, C.R. Johnson, *Matrix Analysis* (Cambridge University Press, Cambridge, 1985)
2. G.B. Arfken, H.J. Weber, *Mathematical Methods for Physicists*, 7th edn. (Elsevier Academic Press, Cambridge, 2005)
3. J.P. Gram, *Om Rakkeudviklinger, bestemte ved mindste Kvadraters Methode (On series expansions, determined by the method of least squares)* (Host, Copenhagen, 1879)
4. J.P. Gram, Om Beregning af en Bevoxnings ved Hjælp af Provetrær. Tiddskrift Skorburg **6**, 137–198 (1883)
5. J.P. Gram, Über die Entwickelung reeller Functionen in Reihen mittelst der Methode der kleinsten Quadrate. J. Reine Angew. Math. **94**, 41–73 (1883)
6. E. Schmidt, Zur Theorie der linearen und nichtlinearen Integralgleichungen I, Entwicklung willkülicher Funktionen nach Systemen vorgeschriebener. Math. Ann. **63**, 433–476 (1907)
7. E. Schmidt, Zur Theorie der linearen und nichtlinearen Integralgleichungen II, Aufliisung der allgemeinen linearen Integralgleichung. Math. Ann. **64**, 161–174 (1907)
8. S.J. Leon, A. Björck, W. Gander, Gram–Schmidt orthogonalization: 100 years and more. Numer. Linear Algebra Appl. **1**, 1–40 (2010)
9. J.W. Strutt, L. Rayleigh, On the calculation of Chladni's figures for a square plate. Philos. Mag. Ser. 6 **22**, 225–229 (1911)
10. R. Walther, Über eine neue Methode zur Lösung gewisser Variationsprobleme der mathematischen Physik (On a new method for the solution of certain variational problems of mathematical physics). J. Reine Angew. Math. **135**, 1–61 (1909)
11. J.W. Strutt, L. Rayleigh, *The Theory of Sound*, vol. 1 (Macmillen and Co., New York, 1877)
12. J.W. Strutt, L. Rayleigh, *The Theory of Sound*, vol. 2 (Macmillen and Co., New York, 1878)

Chapter 6
Some Special Problems of Rotational Motion

6.1 Flywheels—Rotational Speed Fluctuation

A flywheel is the wheel on the end of a crankshaft or a rotating shaft—see Fig. 6.1 which shows a typical flywheel. A flywheel has, in general, three functions:

1. Storing and providing kinetic energy, for instance, in reciprocating engines since the energy source is intermittent in them.
2. Moderating speed fluctuations of a shaft (or shafts) in an engine through its inertia. Any sudden increase due to a change in the loading of the system can be evened out if we apply a flywheel.

Fig. 6.1 Flywheel

G. Szeidl and L. P. Kiss, *Mechanical Vibrations*, Foundations of Engineering Mechanics,
https://doi.org/10.1007/978-3-030-45074-8_6

3. Controlling the orientation of a mechanical system. In such applications, the angular momentum of a flywheel is purposely transferred as a torque to the attaching mechanical system when energy is transferred to or from the flywheel, thereby causing the attaching system to rotate into a desired position.[1]

In the present book, we shall deal with the first two functions of the flywheel only.

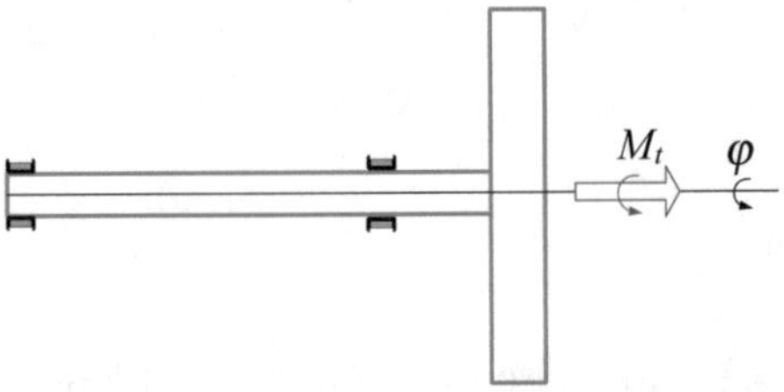

Fig. 6.2 Rotating shaft with a flywheel

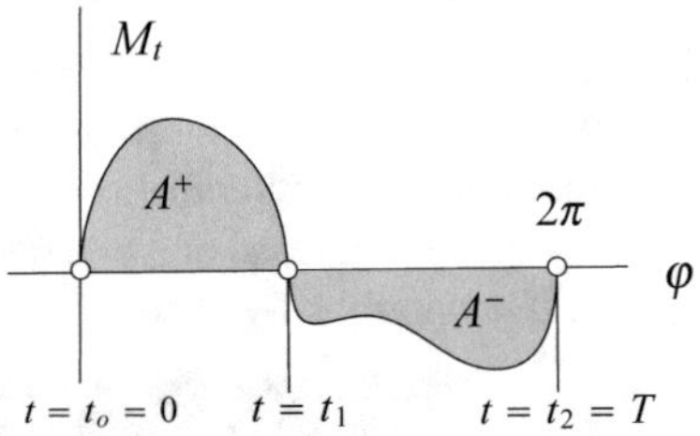

Fig. 6.3 Torque against the angle of rotation

Figure 6.2 shows a simplified model of a rotating shaft with the flywheel attached to the right end of the shaft. Assume that the torque exerted on the flywheel is a periodic function of the angle of rotation, that is, it satisfies the relation

$$M_t(\varphi) = M_t(\varphi + 2\pi)\,. \tag{6.1}$$

Further we shall also assume that the total area $A^+ + A^- = 0$—see Fig. 6.3. In other words,

$$\int_0^{2\pi} M_t(\varphi)\mathrm{d}\varphi = 0\,. \tag{6.2}$$

Making use of the principle of work and energy (1.58a) we can write

$$\mathcal{E}_1 - \mathcal{E}_o = \frac{1}{2} J_w \left(\omega_1^2 - \omega_o^2\right) = \frac{1}{2}\left(\omega_1 + \omega_o\right)\left(\omega_1 - \omega_o\right) J_w = \int_0^{\varphi(t_1)} M_t(\varphi)\,\mathrm{d}\varphi = A^+\,, \tag{6.3a}$$

[1]Item 3 and Fig. 6.1 are taken from WIKIPEDIA.

$$\mathcal{E}_2 - \mathcal{E}_1 = \frac{1}{2} J_w \left(\omega_2^2 - \omega_1^2\right) = \frac{1}{2} (\omega_2 + \omega_1)(\omega_2 - \omega_1) J_w = \int_{\varphi(t_1)}^{\varphi(t_2)} M_t(\varphi)\,\mathrm{d}\varphi = A^-, \tag{6.3b}$$

where J_w is the moment of inertia of the flywheel and shaft with respect to the axis of rotation, ω is the angular velocity of the flywheel, $\omega_o = \omega(t = t_o = 0)$, $\omega_1 = \omega(t = t_1)$, $\omega_2 = \omega(t = t_2)$. If we add (6.3a) to (6.3b), we get that

$$\frac{1}{2} J_w \left(\omega_2^2 - \omega_0^2\right) = 0$$

thus

$$\omega_2 = \omega_0 \,. \tag{6.4}$$

Since $A^+ > 0$ and $A^- < 0$ it follows that

$$\omega_1 = \omega_{\max} \quad \text{and} \quad \omega_2 = \omega_{\min} \tag{6.5}$$

are the maximum and minimum of the angular velocity. Let ω_{mean} be the mean value of ω:

$$\omega_{\text{mean}} = \frac{1}{2} (\omega_{\max} + \omega_{\min}) \,. \tag{6.6}$$

The coefficient of fluctuation of rotational speed δ is defined by the relation

$$\delta = \frac{\omega_{\max} - \omega_{\min}}{\omega_{\text{mean}}} . \tag{6.7}$$

By utilizing (6.5), (6.6), and (6.7) we obtain from (6.3a) that

$$\omega_{\text{mean}}\, \omega_{\text{mean}}\, \delta\, J_w = A^+$$

or

$$\boxed{\delta = \frac{A^+}{\omega_{\text{mean}}^2 J_w}} \,. \tag{6.8}$$

In other words, the greater the moment of inertia J_w for a given ω_{mean} and A^+ the smaller the coefficient of fluctuation of rotational speed.

Table 6.1 Values for the fluctuation of rotational speed

Device name	Fluctuation of rotational speed
Spinning machinery	0.02–0.10
Pump (rotary or motor)	0.03–0.05
Machine tools	0.025–0.02
Car engine	0.0003–0.005
Direct current motor	0.0005–0.001
Direct drive electric machines	0.0001–0.0002

Table 6.1 shows the possible values for the fluctuation of rotational speed.

6.2 The Effect of a Change in the Load Torque on the Rotational Speed of Shafts

The rotational motion is said to be stable if a small change in the load torque results in only a small change in the angular velocity of the steady-state rotational motion. Figure 6.4 shows a model of the rotating shaft, where M_d is the driving moment and M_ℓ is the load torque. The angular velocity and acceleration are denoted by ω and α. It is not difficult to check that the equation of motion is of the form

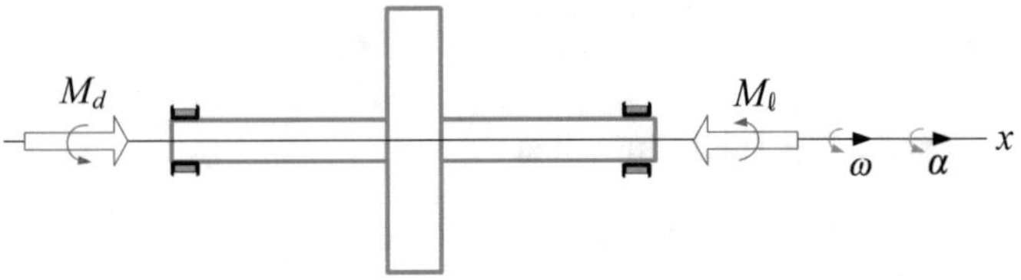

Fig. 6.4 Rotating shaft model

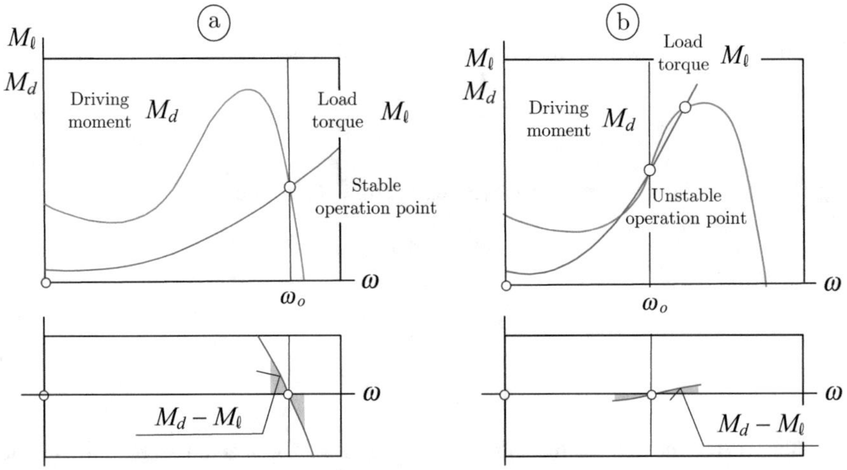

Fig. 6.5 M_ℓ and M_d against the angular velocity

$$J_s\alpha = J_s\dot{\omega} = M_d - M_\ell \tag{6.9}$$

where J_s is the moment of inertia of the shaft and flywheel with respect to the axis x. For steady-state motion

$$M_d = M_\ell\,.$$

Hence,

$$\alpha = 0 \quad \text{and} \quad \omega = \omega_o\,.$$

Here ω_o is the angular velocity of the steady-state rotation. Figures 6.5.a and 6.5.b represent the functions $M_d(\omega)$ and $M_\ell(\omega)$. The intersection point of the two curves is the operating point (or working point) since the angular acceleration is zero while the angular velocity is constant at this point: $\omega = \omega_o = \text{constant}$.

Assume that the angular velocity changes for some reasons in the neighborhood of the operating point. Then, it holds that

$$J_s\alpha = M_d - M_\ell = \Delta M\,. \tag{6.10}$$

If ω suddenly changes in the case (a) then

$$J_s\alpha = \begin{cases} \Delta M \;< 0 \;\; \alpha < 0 \;\; [\text{if } \omega \text{ is increased}] \\ \Delta M \;> 0 \;\; \alpha > 0 \;\; (\text{if } \omega \text{ is decreased}) \end{cases}$$

consequently, ω returns to the steady-state value: the operating point is then said to be stable.

If ω suddenly changes in the case (b) then

$$J_s\alpha = \begin{cases} \Delta M \;> 0 \;\; \alpha > 0 \;\; [\text{if } \omega \text{ is increased}] \\ \Delta M \;< 0 \;\; \alpha < 0 \;\; (\text{if } \omega \text{ is decreased}) \end{cases}$$

consequently ω does not return to the steady-state value: the operating point is then said to be unstable.

6.3 Rotating Shafts—Balancing

6.3.1 Determination of the Reactions

Figure 6.6 depicts a rotating shaft of mass m. We shall assume that

(a) The shaft is a rigid body.
(b) The coordinate axis x coincides with the centerline of the shaft, the axis y is vertical.
(c) The Greek coordinate system is attached rigidly to the shaft (it rotates together with the shaft). Its origin is the mass center G and the coordinate axis ξ is parallel to the axis x.
(d) The mass center G is not located on the axis x.
(e) The position vector $\boldsymbol{\rho}_G$ lies in the coordinate plane $(\eta\zeta)$.
(f) The two bearings are smooth, their middle planes intersect the axis of rotation x at the points A and B; the unknown reaction forces $\mathbf{F}_A$ and $\mathbf{F}_B$ pass through the points A and B.
(g) The angular velocity $\omega = \omega_\xi = \omega_x$ is constant.

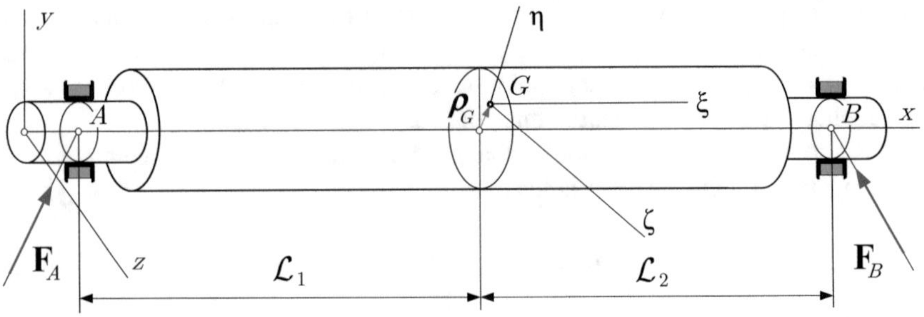

Fig. 6.6 Rotating shaft model for calculating reaction forces

Since the effective forces are equivalent to the external forces it follows that the resultants should be equal:

$$\mathcal{K} = \mathbf{F}_A + \mathbf{F}_B + \mathbf{W}\,, \qquad \mathbf{W} = -mg\,\mathbf{i}_y\,, \tag{6.11a}$$

where m is the mass of the shaft. The moments of the two force systems about the points A and B should also be equal:

$$\mathcal{D}_A = \mathbf{M}_A\,, \qquad \mathcal{D}_B = \mathbf{M}_B\,. \tag{6.11b}$$

In (6.11b)

$$\mathcal{D}_A = \mathcal{D}_G + \mathcal{K} \times \mathbf{r}_{GA} \tag{6.12}$$

is the moment of the effective forces about the point A. It also holds:

(i)

$$\mathcal{K} = m\mathbf{a}_G = -m\omega^2 \boldsymbol{\rho}_G\,, \tag{6.13a}$$

(ii)

$$\mathbf{r}_{GA} = -\boldsymbol{\rho}_G - \mathcal{L}_1 \mathbf{i}_\xi, \tag{6.13b}$$

(iii) and on the basis of Eq. (1.37)

$$\underset{(3\times1)}{\underline{\mathbf{D}}_G} = \underbrace{\underset{(3\times3)}{\underline{\mathbf{J}}_G}\;\underset{(3\times1)}{\underline{\boldsymbol{\alpha}}}}_{=0} + \underset{(3\times3)}{\underline{\boldsymbol{\omega}}_\times}\;\underbrace{\underset{(3\times3)}{\underline{\mathbf{J}}_G}\;\underset{(3\times1)}{\underline{\boldsymbol{\omega}}}}_{\underset{(3\times1)}{\underline{\mathbf{H}}_G}} = \underset{(3\times3)}{\underline{\boldsymbol{\omega}}_\times}\;\underset{(3\times1)}{\underline{\mathbf{H}}_G} =$$

$$= \underbrace{\begin{bmatrix} 0 & 0 & 0 \\ 0 & 0 & -\omega_\xi \\ 0 & \omega_\xi & 0 \end{bmatrix}}_{\underset{(3\times3)}{\underline{\boldsymbol{\omega}}_\times}} \underbrace{\begin{bmatrix} J_\xi & -J_{\xi\eta} & -J_{\xi\zeta} \\ -J_{\eta\xi} & J_\eta & -J_{\eta\zeta} \\ -J_{\zeta\xi} & -J_{\zeta\eta} & J_\zeta \end{bmatrix}}_{\underset{(3\times3)}{\underline{\mathbf{J}}_G}} \underbrace{\begin{bmatrix} \omega_\xi \\ 0 \\ 0 \end{bmatrix}}_{\underset{(3\times1)}{\underline{\boldsymbol{\omega}}}} =$$

$$= \underbrace{\begin{bmatrix} 0 & 0 & 0 \\ 0 & 0 & -\omega_\xi \\ 0 & \omega_\xi & 0 \end{bmatrix}}_{\underset{(3\times3)}{\underline{\boldsymbol{\omega}}_\times}} \underbrace{\begin{bmatrix} J_\xi \omega_\xi \\ -J_{\eta\xi}\omega_\xi \\ -J_{\zeta\xi}\omega_\xi \end{bmatrix}}_{\underset{(3\times1)}{\underline{\mathbf{H}}_G}}$$

or in vectorial notation

$$\mathcal{D}_G = \boldsymbol{\omega} \times \mathbf{H}_G = \omega_\xi^2\, \mathbf{i}_\xi \times \left(J_\xi \mathbf{i}_\xi - J_{\eta\xi}\mathbf{i}_\eta - J_{\zeta\xi}\mathbf{i}_\zeta\right) = \\ = \omega_\xi^2\, \mathbf{i}_\xi \times \left(-J_{\eta\xi}\mathbf{i}_\eta - J_{\zeta\xi}\mathbf{i}_\zeta\right), \quad (6.13c)$$

(iv) and finally

$$\mathcal{K} \times \mathbf{r}_{GA} = -m\omega_\xi^2 \boldsymbol{\rho}_G \times \left(-\boldsymbol{\rho}_G - \mathcal{L}_1 \mathbf{i}_\xi\right) = m\omega_\xi^2 \mathcal{L}_1 \boldsymbol{\rho}_G \times \mathbf{i}_\xi\,. \quad (6.13d)$$

Making use of relations (6.13) Eq. (6.12) yields

$$\mathcal{D}_A = \omega_\xi^2\, \mathbf{i}_\xi \times \left(-m\mathcal{L}_1 \boldsymbol{\rho}_G - J_{\eta\xi}\mathbf{i}_\eta - J_{\zeta\xi}\mathbf{i}_\zeta\right)\,. \quad (6.14)$$

We get in a similar manner that

$$\mathcal{D}_B = \mathcal{D}_G + \mathcal{K} \times \mathbf{r}_{GB}, \quad (6.15)$$

where

$$\mathcal{K} \times \mathbf{r}_{GB} = -m\omega_\xi^2 \boldsymbol{\rho}_G \times \left(-\boldsymbol{\rho}_G + \mathcal{L}_2 \mathbf{i}_\xi\right) = -m\omega^2 \mathcal{L}_1 \boldsymbol{\rho}_G \times \mathbf{i}_\xi\,. \quad (6.16)$$

Comparison of Eqs. (6.15), (6.13c) and (6.16) leads to the following result:

$$\mathcal{D}_B = \omega_\xi^2\, \mathbf{i}_\xi \times \left(m\mathcal{L}_2 \boldsymbol{\rho}_G - J_{\eta\xi}\mathbf{i}_\eta - J_{\zeta\xi}\mathbf{i}_\zeta\right)\,. \quad (6.17)$$

On the other hand,

$$\mathbf{M}_A = \mathbf{r}_{AB} \times \mathbf{F}_B + \mathbf{r}_{AG} \times \mathbf{W} = \\ = \left(\mathcal{L}_1 + \mathcal{L}_2\right) \mathbf{i}_\xi \times \mathbf{F}_B + \left(\mathcal{L}_1 \mathbf{i}_\xi + \boldsymbol{\rho}_G\right) \times \left(-mg\mathbf{i}_y\right) \quad (6.18)$$

and

$$\mathbf{M}_B = \mathbf{r}_{BA} \times \mathbf{F}_A + \mathbf{r}_{BG} \times \mathbf{W} = \\ = -\left(\mathcal{L}_1 + \mathcal{L}_2\right) \mathbf{i}_\xi \times \mathbf{F}_A + \left(-\mathcal{L}_2 \mathbf{i}_\xi + \boldsymbol{\rho}_G\right) \times \left(-mg\mathbf{i}_y\right)\,. \quad (6.19)$$

Upon substitution of (6.14) and (6.18) into equation $\mathcal{D}_A = \mathbf{M}_A$, we get

$$\omega_\xi^2\, \mathbf{i}_\xi \times \left(-m\mathcal{L}_1 \boldsymbol{\rho}_G - J_{\eta\xi}\mathbf{i}_\eta - J_{\zeta\xi}\mathbf{i}_\zeta\right) = \\ = \left(\mathcal{L}_1 + \mathcal{L}_2\right)\mathbf{i}_\xi \times \mathbf{F}_B + \left(\mathcal{L}_1 \mathbf{i}_\xi + \boldsymbol{\rho}_G\right) \times \left(-mg\mathbf{i}_y\right)\,.$$

Cross-multiplying from the right by $\mathbf{i}_\xi$ and expanding then the triple cross products by taking the relations

$$\left(-m\mathcal{L}_1 \boldsymbol{\rho}_G - J_{\eta\xi}\mathbf{i}_\eta - J_{\zeta\xi}\mathbf{i}_\zeta\right) \cdot \mathbf{i}_\xi = 0 \quad \text{and} \quad \mathbf{F}_B \cdot \mathbf{i}_\xi = 0$$

into account, we find

$$\omega_\xi^2\left(-m\mathcal{L}_1\boldsymbol{\rho}_G - J_{\eta\xi}\mathbf{i}_\eta - J_{\zeta\xi}\mathbf{i}_\zeta\right) = (\mathcal{L}_1 + \mathcal{L}_2)\,\mathbf{F}_B + \mathcal{L}_1\left(-mg\mathbf{i}_y\right)$$

from where

$$\boxed{\mathbf{F}_B = \underbrace{\frac{\mathcal{L}_1}{\mathcal{L}_1+\mathcal{L}_2}mg\mathbf{i}_y}_{\text{independent of}\,\omega_\xi} + \underbrace{\frac{\omega_\xi^2}{\mathcal{L}_1+\mathcal{L}_2}\left(-m\mathcal{L}_1\boldsymbol{\rho}_G - J_{\eta\xi}\mathbf{i}_\eta - J_{\zeta\xi}\mathbf{i}_\zeta\right)}_{\text{rotates with the angular velocity}\,\omega_\xi}} \tag{6.20}$$

is the reaction force at B.

Let us proceed with equation $\boldsymbol{\mathcal{D}}_B = \mathbf{M}_B$. Substituting (6.14) and (6.19), we obtain

$$\omega_\xi^2\,\mathbf{i}_\xi \times \left(m\mathcal{L}_2\boldsymbol{\rho}_G - J_{\eta\xi}\mathbf{i}_\eta - J_{\zeta\xi}\mathbf{i}_\zeta\right) = \\ = -\left(\mathcal{L}_1+\mathcal{L}_2\right)\mathbf{i}_\xi \times \mathbf{F}_A + \left(-\mathcal{L}_2\mathbf{i}_\xi + \boldsymbol{\rho}_G\right) \times \left(-mg\mathbf{i}_y\right).$$

Cross-multiplying from the right by $\mathbf{i}_\xi$ we shall find an equation for $\mathbf{F}_A$ if we utilize that now

$$\left(m\mathcal{L}_2\boldsymbol{\rho}_G - J_{\eta\xi}\mathbf{i}_\eta - J_{\zeta\xi}\mathbf{i}_\zeta\right)\cdot\mathbf{i}_\xi = 0 \qquad \text{and} \qquad \mathbf{F}_A\cdot\mathbf{i}_\xi = 0\,.$$

This way we have

$$\omega_\xi^2\left(m\mathcal{L}_2\boldsymbol{\rho}_G - J_{\eta\xi}\mathbf{i}_\eta - J_{\zeta\xi}\mathbf{i}_\zeta\right) = -\left(\mathcal{L}_1+\mathcal{L}_2\right)\mathbf{F}_A + \mathcal{L}_2\left(mg\mathbf{i}_y\right)$$

from where

$$\boxed{\mathbf{F}_A = \underbrace{\frac{\mathcal{L}_2}{\mathcal{L}_1+\mathcal{L}_2}mg\mathbf{i}_y}_{\text{independent of}\,\omega_\xi} - \underbrace{\frac{\omega_\xi^2}{\mathcal{L}_1+\mathcal{L}_2}\left(m\mathcal{L}_2\boldsymbol{\rho}_G - J_{\eta\xi}\mathbf{i}_\eta - J_{\zeta\xi}\mathbf{i}_\zeta\right)}_{\text{rotates with the angular velocity}\,\omega_\xi}} \tag{6.21}$$

is the reaction force at A.

Let $e = \left|\boldsymbol{\rho}_G\right|$ be the length of the position vector $\boldsymbol{\rho}_G$. If $e = 0$ then $\mathbf{a}_G = \mathbf{0}$. Hence, it follows from Eq. (6.11a)—$m\mathbf{a}_G = \mathbf{0}$—that

$$\mathbf{F}_A + \mathbf{F}_B - mg\,\mathbf{i}_y = \mathbf{0}\,. \tag{6.22}$$

This means that the resultant of the external forces acting on the shaft vanishes: we say that the shaft is statically balanced.

Remark 6.1 If the mass center of a body rotating about a fixed axis is on the axis of rotation then the body is said to be statically balanced: in this case, the resultant of the external forces acting on the body vanishes.

For statically balanced shafts $e = 0$. Hence

$$\mathbf{F}_B = \frac{\mathcal{L}_1}{\mathcal{L}_1 + \mathcal{L}_2} mg\mathbf{i}_y - \frac{\omega_\xi^2}{\mathcal{L}_1 + \mathcal{L}_2}\left(J_{\eta\xi}\mathbf{i}_\eta + J_{\zeta\xi}\mathbf{i}_\zeta\right) \tag{6.23}$$

and

$$\mathbf{F}_A = \frac{\mathcal{L}_2}{\mathcal{L}_1 + \mathcal{L}_2} mg\mathbf{i}_y + \frac{\omega_\xi^2}{\mathcal{L}_1 + \mathcal{L}_2}\left(J_{\eta\xi}\mathbf{i}_\eta + J_{\zeta\xi}\mathbf{i}_\zeta\right) \tag{6.24}$$

are the two reactions.

Remark 6.2 If $e = 0$ and $J_{\eta\xi} = J_{\zeta\xi} = 0$ the axis of rotation is a principal axis of inertia. Then the shaft is said to be dynamically balanced.

For dynamically balanced shafts, the reaction forces

$$\mathbf{F}_B = \frac{\mathcal{L}_1}{\mathcal{L}_1 + \mathcal{L}_2} mg\mathbf{i}_y \quad \text{and} \quad \mathbf{F}_A = \frac{\mathcal{L}_2}{\mathcal{L}_1 + \mathcal{L}_2} mg\mathbf{i}_y \tag{6.25}$$

are independent of time (they do not rotate together with the shaft).

6.3.2 Balancing in Two Planes

By [adding](removing) masses [to the shaft](from the shaft) in planes S_1 and S_2—see Fig. 6.7—we can eliminate dynamic unbalance: we can set $J_{\eta\xi}$ and $J_{\zeta\xi}$ to zero. Now it is our aim to clarify how to balance the shaft dynamically, i.e., to establish the equations that show the effects of the various parameters on the solution. We shall

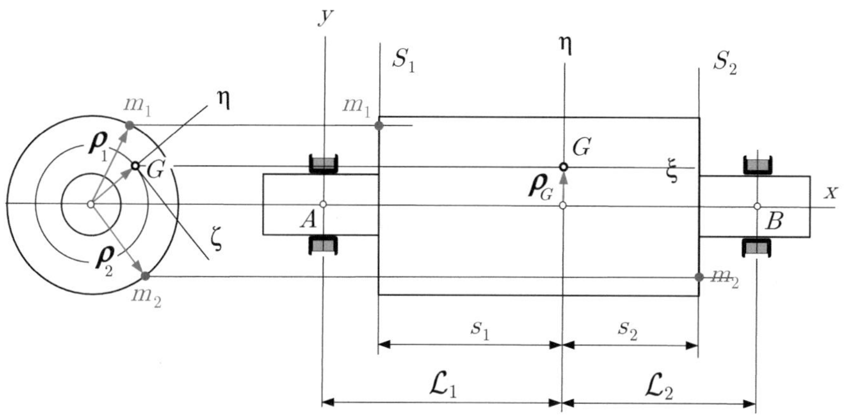

Fig. 6.7 Rotating shaft—balancing in two planes

use the notations of Fig. 6.7, which are basically the same as those in Fig. 6.6. The new notations are as follows: m_1 and m_2 are the masses (to be added to the shaft or to be removed from the shaft) in the planes S_1 and S_2 which are perpendicular to the axis of rotation. The horizontal distances of the two planes from the coordinate plane $(\eta\zeta)$ in the absolute, i.e., the Latin and the Greek (the rotating), coordinate systems are naturally the same and are denoted by s_1 and s_2. The position vectors $\boldsymbol{\rho}_1$ and $\boldsymbol{\rho}_2$ are those of the masses m_1 and m_2.

The products $m_1\boldsymbol{\rho}_1$ and $m_2\boldsymbol{\rho}_2$ are called unbalances.

We shall derive equations for the unbalances from the conditions the dynamically balanced shaft should satisfy. The line of thought is based on the equations of dynamic equilibrium.

The system of effective forces can be replaced at the mass center G by the resultant

$$\hat{\mathcal{K}} = \underbrace{-m\omega_\xi^2\boldsymbol{\rho}_G}_{\mathcal{K}\text{ – see (6.13a)}} + \underbrace{-m_1\omega_\xi^2\boldsymbol{\rho}_1}_{\mathcal{K}_1} + \underbrace{-m_2\omega_\xi^2\boldsymbol{\rho}_2}_{\mathcal{K}_2} = \mathcal{K} + \mathcal{K}_1 + \mathcal{K}_2 \tag{6.26}$$

and the moment resultant

$$\hat{\mathcal{D}}_G = \underbrace{\mathcal{D}_G}_{\text{see (6.13c)}} + \mathbf{r}_1 \times \mathcal{K}_1 + \mathbf{r}_2 \times \mathcal{K}_2,$$

where

$$\mathbf{r}_1 = -\boldsymbol{\rho}_G - s_1\mathbf{i}_\xi + \boldsymbol{\rho}_1 \quad \text{and} \quad \mathbf{r}_2 = -\boldsymbol{\rho}_G + s_2\mathbf{i}_\xi + \boldsymbol{\rho}_2 \,.$$

Consequently,

$$\begin{aligned}\hat{\mathcal{D}}_G = \omega_\xi^2\,\mathbf{i}_\xi \times \left(-J_{\eta\xi}\mathbf{i}_\eta - J_{\zeta\xi}\mathbf{i}_\zeta\right) + \left(-\boldsymbol{\rho}_G - s_1\mathbf{i}_\xi + \boldsymbol{\rho}_1\right) \times \left(-m_1\omega_\xi^2\boldsymbol{\rho}_1\right) + \\ + \left(-\boldsymbol{\rho}_G + s_2\mathbf{i}_\xi + \boldsymbol{\rho}_2\right) \times \left(-m_2\omega_\xi^2\boldsymbol{\rho}_2\right) .\end{aligned} \tag{6.27}$$

If the shaft is dynamically balanced then $\hat{\mathcal{K}} = \mathbf{0}$, i.e.,

$$m_1\boldsymbol{\rho}_1 + m_2\boldsymbol{\rho}_2 + m\boldsymbol{\rho}_G = \mathbf{0} \tag{6.28a}$$

and $\hat{\mathcal{D}}_G = \mathbf{0}$. Thus

$$\mathbf{i}_\xi \times \left(-J_{\eta\xi}\mathbf{i}_\eta - J_{\zeta\xi}\mathbf{i}_\zeta\right) + m_1\boldsymbol{\rho}_G \times \boldsymbol{\rho}_1 + m_1 s_1\mathbf{i}_\xi \times \boldsymbol{\rho}_1 + m_2\boldsymbol{\rho}_G \times \boldsymbol{\rho}_2 - m_2 s_2\mathbf{i}_\xi \times \boldsymbol{\rho}_2 = \mathbf{0} \,.$$

Cross-multiplying from the right by $\mathbf{i}_\xi$ and after expanding the triple cross products taking into account that $\mathbf{i}_\eta$, $\mathbf{i}_\zeta$, $\boldsymbol{\rho}_1$, $\boldsymbol{\rho}_2$, and $\boldsymbol{\rho}_G$ are perpendicular to $\mathbf{i}_\xi$, we obtain

$$-J_{\eta\xi}\mathbf{i}_\eta - J_{\zeta\xi}\mathbf{i}_\zeta + m_1 s_1\boldsymbol{\rho}_1 - m_2 s_2\boldsymbol{\rho}_2 = \mathbf{0} \,. \tag{6.28b}$$

Equations (6.28a) and (6.28b) can be solved for the unknown unbalances.

After multiplying Eq. (6.28a) by s_2 and then adding it to Eq. (6.28b), we get

$$m s_2 \boldsymbol{\rho}_G + m_1 \boldsymbol{\rho}_1 (s_1 + s_2) - J_{\eta\xi} \mathbf{i}_\eta - J_{\zeta\xi} \mathbf{i}_\zeta = \mathbf{0}$$

from where

$$\boxed{m_1 \boldsymbol{\rho}_1 = \frac{1}{s_1 + s_2} \left(J_{\eta\xi} \mathbf{i}_\eta + J_{\zeta\xi} \mathbf{i}_\zeta - m s_2 \boldsymbol{\rho}_G \right) .} \tag{6.29a}$$

If we multiply Eq. (6.28a) by s_1 and subtract Eq. (6.28b) from the result we arrive at the relation

$$m s_1 \boldsymbol{\rho}_G + m_2 \boldsymbol{\rho}_2 (s_1 + s_2) + J_{\eta\xi} \mathbf{i}_\eta + J_{\zeta\xi} \mathbf{i}_\zeta = \mathbf{0} .$$

Hence,

$$\boxed{m_2 \boldsymbol{\rho}_2 = -\frac{1}{s_1 + s_2} \left(J_{\eta\xi} \mathbf{i}_\eta + J_{\zeta\xi} \mathbf{i}_\zeta + m s_1 \boldsymbol{\rho}_G \right) .} \tag{6.29b}$$

Remark 6.3 In principle, we know m, $\boldsymbol{\rho}_G$, $J_{\eta\xi}$, and $J_{\zeta\xi}$. With the knowledge of these quantities we can select the locations of the planes S_1 and S_2 freely. As soon as we have selected s_1 and s_2 the unbalances $m_1 \boldsymbol{\rho}_1$ and $m_2 \boldsymbol{\rho}_2$ are uniquely determined.

Remark 6.4 It may occur that m_1 (or m_2) is negative (mass should be removed). To avoid a negative mass the locations of the planes S_1 and S_2 should be selected carefully when we design the rotating shaft.

6.3.3 Elastic Shafts, Stability of Rotation. Laval's Theorem

Since a rotating shaft is, in general, unbalanced the forces of inertia (the centrifugal force) will cause it to bend out. When the shaft rotates at an angular velocity equal to the natural circular frequency of transverse oscillations this vibration becomes large and shows up as a whirling of the shaft. Our aim is to examine this phenomenon. First, we shall consider a simple (one degree of freedom) model.

Assumptions and notations:

(i) The mass of the shaft can be neglected.
(ii) $a = b$—the midplane of the disk (or the gear) is a plane of symmetry of the structure.
(iii) The mass and the moment of inertia of the disk (with respect to the vertical axis of symmetry) are m and J_x.
(iv) The mass center G does not coincide with the geometric center C. The radius from C to the mass center G is denoted by e. This distance is, however, much smaller than the radius of the disk.

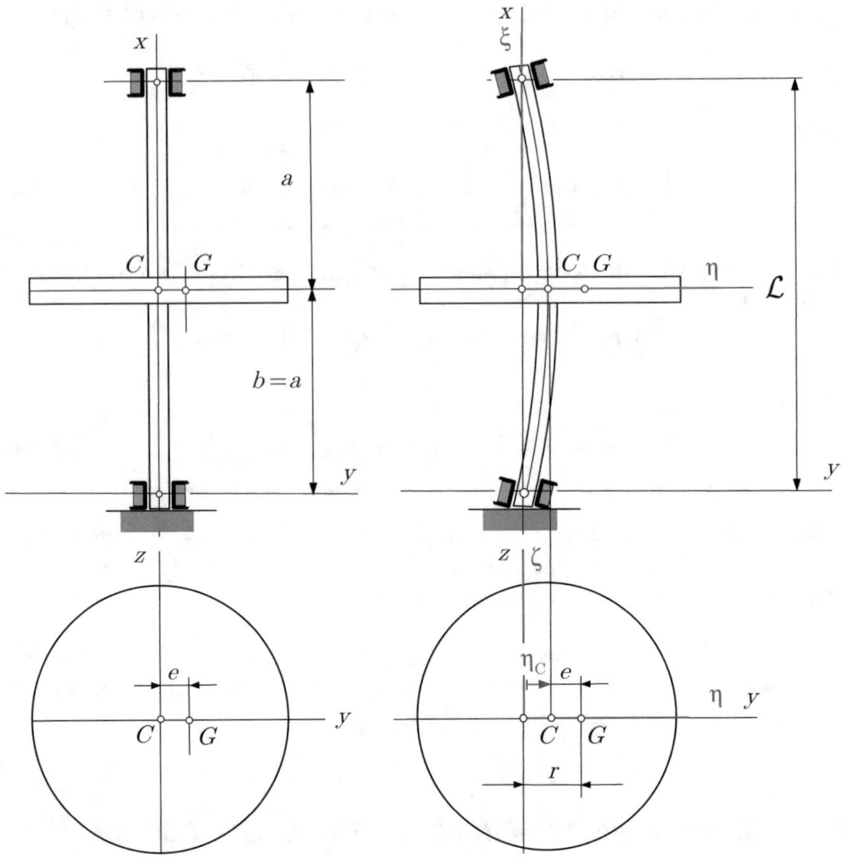

Fig. 6.8 Laval's model for investigating stability of rotation

(v) The Latin coordinate system (xyz) is the absolute coordinate system. The Greek coordinate system $(\xi\eta\zeta)$ is rigidly attached to the rotating shaft—it rotates together with the shaft. The coordinate axis ξ is perpendicular to the middle plane of the disk—for symmetry reasons, it is now parallel to the coordinate axis x.

(vi) The angular velocity $\omega = \omega_x = \omega_\xi$ is constant.

(vii) The radial displacement of the geometric center C in the Greek coordinate system is η_C.

(viii) The shaft is a simply supported beam with a circular cross section of diameter d, and E is the Young modulus.

For a simply supported shaft of circular cross section, the radial displacement of the geometrical center C due to a radial force F of the same direction is

$$\eta_C = \frac{F}{k} = Ff\,, \qquad f = \frac{1}{k} = \frac{\mathcal{L}^3}{48IE}\,, \qquad I = \frac{d^4\pi}{64}\,. \tag{6.30}$$

In the present case, F is the centrifugal force:

$$F = (\eta_C + e)\, m\omega^2 \,. \tag{6.31}$$

Consequently,

$$\eta_C = \frac{F}{k} = \frac{(\eta_C + e)\, m\omega^2}{k}$$

or

$$\eta_C \left(1 - \frac{\omega^2}{\frac{k}{m}}\right) = e\, \frac{m\omega^2}{k}$$

from where

$$|\eta_C| = \frac{\frac{\omega^2}{\frac{k}{m}}}{\left|1 - \frac{\omega^2}{\frac{k}{m}}\right|}\, e = |\eta_C| = \frac{\frac{\omega^2}{\omega_{cr}^2}}{\left|1 - \frac{\omega^2}{\omega_{cr}^2}\right|}\, e = \frac{\chi^2}{|1 - \chi^2|}\, e \tag{6.32a}$$

in which

$$\frac{k}{m} = \omega_{cr}^2 \quad \text{and} \quad \chi = \sqrt{\frac{\omega^2}{\omega_{cr}^2}} \tag{6.32b}$$

are the critical angular velocity and a dimensionless parameter. It follows from (6.32a) that

$$\frac{|\eta_C + e|}{e} = \frac{r}{e} = \frac{1}{|1 - \chi^2|}, \tag{6.33}$$

where r is the total radial displacement of the mass center G—see Fig. 6.8. The quotient r/e is a function of the dimensionless parameter χ, which is shown in Fig. 6.9. Laval (1845–1913) observed that the rotational motion is smoother if $\chi > \sqrt{2}$, i.e., if $\omega > \sqrt{2}\,\omega_{cr}$.

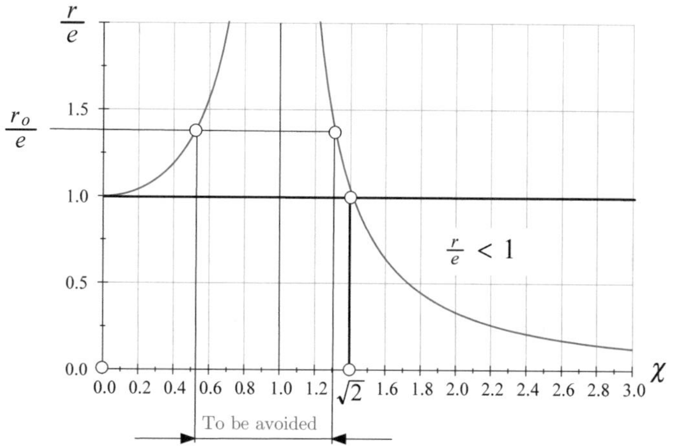

Fig. 6.9 Dimensionless radial displacement against χ

Accordingly, Laval's theorem: *(a) the critical angular velocity is given by*

$$\boxed{\omega_{cr} = \sqrt{\frac{k}{m}}\,;} \tag{6.34}$$

(b) the total radial displacement r is less than e if $\omega > \sqrt{2}\omega_{cr}$.
If $r = r_o$ is a prescribed value and

(i) $\omega < \omega_{cr}$ then $\chi < 1$ and

$$\frac{e}{r_o} = 1 - \frac{\omega^2}{\omega_{cr}^2}\,, \qquad \omega = \omega_{cr}\sqrt{1 - \frac{e}{r_o}}\,; \tag{6.35a}$$

(ii) $\omega > \omega_{cr}$ then $\chi > 1$ and

$$\frac{e}{r_o} = \frac{\omega^2}{\omega_{cr}^2} - 1\,, \qquad \omega = \omega_{cr}\sqrt{1 + \frac{e}{r_o}}\,. \tag{6.35b}$$

For more details concerning the behavior of rotating shafts, the reader is referred to the following works: [1, 2].

6.3.4 Stability of Rotation. Gyroscopic Effects

If the arrangement is not symmetric, i.e., $a \neq b$ the middle plane of the disk will not be perpendicular to the axis ξ. In order to take this fact into account, a more accurate model is needed. Assume that the shaft is simply supported, its cross section is circular with a diameter d, $I = d^4\pi/64$ and E is Young's modulus of the shaft. Let f_{11} and φ_{11} be the deflection and the angle of rotation at G due to a unit force and a unit couple both acting at G—see Fig. 6.10.

The deflection f_{11m} and angle of rotation φ_{11m} at G caused by a unit couple and a unit force at G satisfy the symmetry condition $f_{11m} = \varphi_{11m}$. It can be shown that

$$f_{11} = \frac{a^2 b^2}{3IE\mathcal{L}}\,, \quad f_{11m} = \varphi_{11m} = \frac{ab\,(b-a)}{3IE\mathcal{L}}\,, \quad \varphi_{11m} = \frac{a^2 - ab - b^2}{3IE\mathcal{L}}\,. \tag{6.36a}$$

For $a = b$, it follows that

$$f_{11} = \frac{\mathcal{L}^3}{48IE}\,, \qquad f_{11m} = \varphi_{11m} = 0\,, \qquad \varphi_{11m} = -\frac{\mathcal{L}}{12IE}\,. \tag{6.36b}$$

If the beam is subjected to the force F and couple M then

$$y_G = f_{11}F + f_{11m}M \quad \text{and} \quad \varphi_G = \varphi_{11}F + \varphi_{11m}M \tag{6.37}$$

are the deflection and angle of rotation at G.

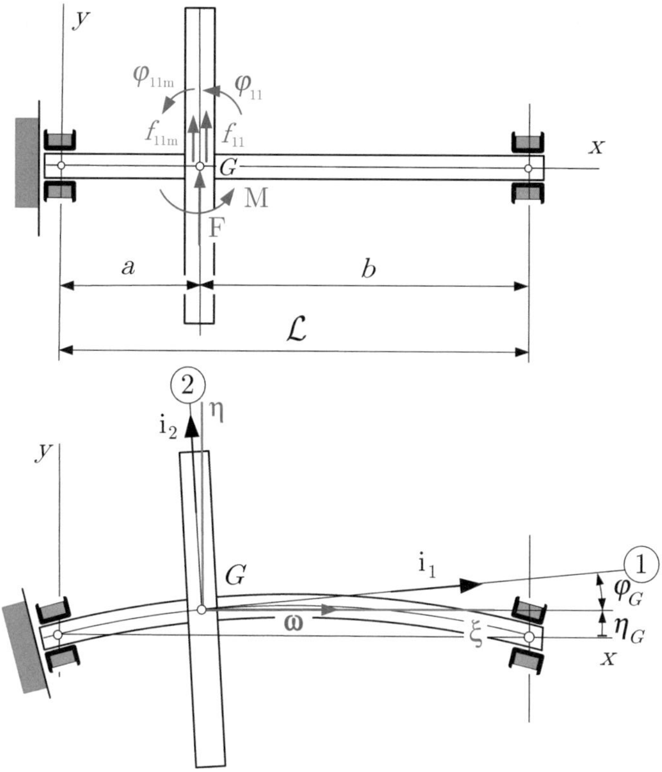

Fig. 6.10 Deformed rotating shaft

Figure 6.10 shows the deformed rotating shaft assuming that (a) $e = 0$, (b) the angular velocity $\boldsymbol{\omega} = \omega_x\,\mathbf{i}_x = \omega\,\mathbf{i}_x$ is constant. The coordinate system (xyz) is the absolute coordinate system. The Greek coordinate system $(\xi\eta\zeta)$ with origin at G is attached to the disk (gear) in such a manner that the coordinate axis ξ remains parallel to the coordinate axis ξ when the shaft rotates. The principal axes (123) of the disk (gear) at G are given by the unit vectors $\mathbf{i}_1$, $\mathbf{i}_2$ and $\mathbf{i}_3 = \mathbf{i}_1 \times \mathbf{i}_2$—the latter is not seen in Fig. 6.10 since that shows the shaft at such a point of time when $\mathbf{i}_3$ is perpendicular to the plane of Fig. 6.10. The moments of inertia with respect to these axes are denoted in the usual way by $J_1 > J_2 = J_3$. It is clear that this coordinate system rotates also with the shaft.

The following transformations are performed in the coordinate system of the principal axes. Since

$$\boldsymbol{\omega} = \omega\,\mathbf{i}_x = \omega\,(\cos\varphi_G\,\mathbf{i}_1 - \sin\varphi_G\,\mathbf{i}_2) \tag{6.38}$$

it follows that

$$\underset{(3\times1)}{\underline{\mathbf{H}}_G} = \underset{(3\times3)}{\underline{\mathbf{J}}_G}\;\underset{(3\times1)}{\underline{\boldsymbol{\omega}}} = \omega\begin{bmatrix} J_1 & 0 & 0\\ 0 & J_2 & 0\\ 0 & 0 & J_3\end{bmatrix}\begin{bmatrix}\cos\varphi_G\\ -\sin\varphi_G\\ 0\end{bmatrix} = \omega\begin{bmatrix}J_1\cos\varphi_G\\ -J_2\sin\varphi_G\\ 0\end{bmatrix}. \tag{6.39}$$

If we take into account that $\boldsymbol{\alpha} = \dot{\boldsymbol{\omega}} = \mathbf{0}$ and $|\varphi| \ll 1$ we obtain from (1.37) (or (1.38)) that

$$\begin{aligned}\mathcal{D}_G = \boldsymbol{\omega} \times \boldsymbol{H}_G =& \omega^2 (\cos\varphi_G\,\mathbf{i}_1 - \sin\varphi_G\,\mathbf{i}_2) \times (J_1 \cos\varphi_G\,\mathbf{i}_1 - J_2 \sin\varphi_G\,\mathbf{i}_2) = \\ &= \omega^2 (-J_2 \sin\varphi_G \cos\varphi_G + J_1 \sin\varphi_G \cos\varphi_G)\,\mathbf{i}_3 = \\ &= \omega^2 \underbrace{(J_1 - J_2)}_{J_d} \underbrace{\sin\varphi_G}_{\approx\varphi_G} \underbrace{\cos\varphi_G}_{\approx 1}\,\mathbf{i}_3 = \omega^2 J_d\,\varphi_G\,\mathbf{i}_3\,.\end{aligned} \tag{6.40}$$

The shaft is elastic. The system of effective forces at G is equivalent to the following force couple system:

$$\mathbf{F} = -F\,\mathbf{i}_\eta = -\underbrace{m\eta_G\omega^2}_{F}\,\mathbf{i}_\eta\,, \qquad \mathbf{M}_G = \underbrace{\omega^2 J_d\,\varphi_G}_{-M}\,\mathbf{i}_3, \tag{6.41}$$

where F is the resultant of the forces of inertia and M is the moment due to the forces of inertia. Upon substitution of F and M into (6.37), we obtain

$$\begin{aligned}\eta_G &= f_{11}F + f_{11\,m}\,M = f_{11}m\omega^2\eta_G - f_{11\,m}\omega^2 J_d\,\varphi_G\,,\\ \varphi_G &= \varphi_{11}F + \varphi_{11\,m}\,M = \varphi_{11}m\omega^2\eta_G - \varphi_{11\,m}\,\omega^2 J_d\,\varphi_G\end{aligned}$$

or

$$\begin{aligned}\left(1 - f_{11}m\omega^2\right)\eta_G + f_{11\,m}\,\omega^2 J_d\,\varphi_G &= 0\,,\\ -\varphi_{11}m\omega^2\eta_G + \left(1 + \varphi_{11\,m}\,\omega^2 J_d\right)\varphi_G &= 0.\end{aligned} \tag{6.42}$$

Let us introduce the following notations: $f_{11}m = 1/\omega^2_{cr\,o}$, $\omega^2/\omega^2_{cr\,o} = \chi^2$ (if $a = b$ we get $k = 1/f_{11} = 1/f$ and $\omega^2_{cr\,o} = \omega^2_{cr}$ see (6.32b)). Making use of these notations, we obtain the following equation system from (6.42):

$$\begin{aligned}(1 - \chi^2)\eta_G + \frac{f_{11\,m}}{f_{11}}\frac{J_d}{m}\chi^2\varphi_G &= 0\,,\\ -\frac{\varphi_{11}}{f_{11}}\chi^2\eta_G + \left(1 + \frac{\varphi_{11\,m}}{f_{11}}\frac{J_d}{m}\chi^2\right)\varphi_G &= 0\,.\end{aligned} \tag{6.43}$$

Solutions different from the trivial ones for η_G and φ_G exist if and only if

$$\begin{aligned}d = \begin{vmatrix} 1 - \chi^2 & \dfrac{f_{11\,m}}{f_{11}}\dfrac{J_d}{m}\chi^2 \\ -\dfrac{\varphi_{11}}{f_{11}}\chi^2 & 1 + \dfrac{\varphi_{11\,m}}{f_{11}}\dfrac{J_d}{m}\chi^2 \end{vmatrix} &= \\ = (1 - \chi^2)\left(1 + \frac{\varphi_{11\,m}}{f_{11}}\frac{J_d}{m}\chi^2\right) + \frac{\varphi_{11}}{f_{11}}\frac{\varphi_{11\,m}}{f_{11}}\frac{J_d}{m}\chi^4 &= 0\end{aligned} \tag{6.44a}$$

from where we get

$$(\varphi_{11}\varphi_{11m} - f_{11}\varphi_{11m})\, J_d \chi^4 + \left(f_{11}\varphi_{11m} J_d - f_{11}^2 m\right)\chi^2 + f_{11}^2 m = 0\,. \qquad (6.44b)$$

The above equation can easily be solved for the unknown χ.

6.4 Problems

Problem 6.1 Find the critical angular velocity of the rotating shaft shown in Fig. 6.11. Assume that (a) the eccentricity of the mass center of the rotating disk $e = 0$, (b) the angular velocity $\boldsymbol{\omega} = \omega_x\, \mathbf{i}_x = \omega\, \mathbf{i}_x$ is constant.

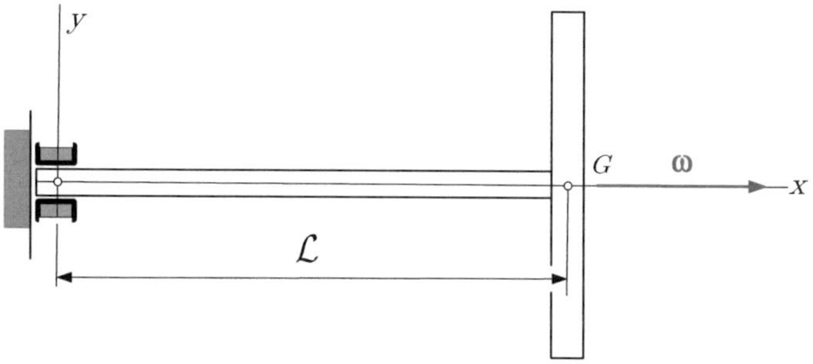

Fig. 6.11 Disk at the left end of the rotaing shaft

Problem 6.2 Find the critical angular velocity of the rotating shaft shown in Fig. 6.10 if $b = 4a$.

References

1. A. Föppl, *Das Problem der Lavalschen Turbinenwelle* (Der Civilingenieur, 1895), pp. 333–342
2. G.L. Robert, J.P. Vincent, *Dynamics of Rotating Shafts*. The Shock and Vibration Monograph Series (1969)

Chapter 7
Systems with Infinite Degrees of Freedom

7.1 Equilibrium Equations for Spatial Beams

First, we shall consider such problems for which the physical quantities depend only on one spatial coordinate and time. These investigations are based on the equilibrium conditions a spatial beam should satisfy. Figure 7.1 shows the centerline AB of a beam portion before deformation and the beam portion together with its centerline after deformation. The arc coordinate $(\xi)[s]$ is measured on the centerline of the beam (before)[after] deformation. Let us denote the arc coordinate on the left cross section (on cross section A) by s_o, on the right cross section (on cross section B) by s. We remark that the arc coordinate s will be regarded as if it were a parameter for the further investigations. The unit tangent to the deformed centerline is denoted by $\mathbf{t}(s)$. We shall assume that the loads exerted on the beam are equivalent to a distributed

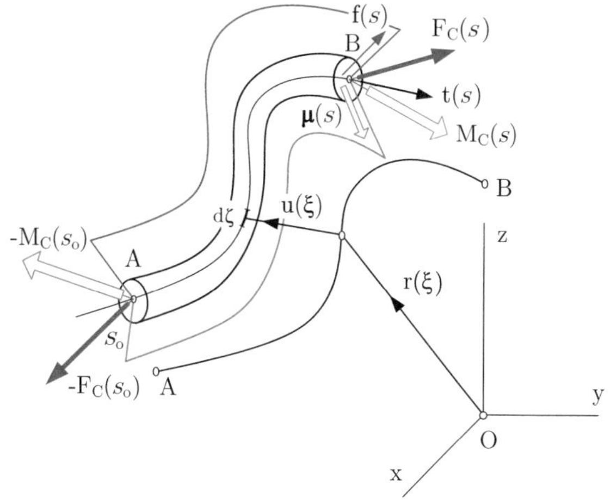

Fig. 7.1 A beam portion with the external and internal forces acting on it

G. Szeidl and L. P. Kiss, *Mechanical Vibrations*, Foundations of Engineering Mechanics,
https://doi.org/10.1007/978-3-030-45074-8_7

force system $\mathbf{f}(s)$ and a distributed moment load $\boldsymbol{\mu}(s)$ both acting on the deformed centerline. The resultant and moment resultant of the inner forces (stresses) at the center of cross sections A and B are denoted by $-\mathbf{F}_C(s_o)$, $-\mathbf{M}_C(s_o)$ and $\mathbf{F}_C(s)$, $\mathbf{M}_C(s_o)$. Since the beam is in equilibrium, the total system of forces acting on the beam should have a zero resultant and a zero moment resultant (say) about the origin O.

The resultant is zero if

$$-\mathbf{F}_C(s_o) + \int_{s_o}^{s} \mathbf{f}(\xi)\mathrm{d}\xi + \mathbf{F}_C(s) = \mathbf{0}\,. \tag{7.1a}$$

Let $\mathbf{u}(s)$ be the displacement on the centerline. A vanishing moment resultant (taken about the origin) follows from the following equation:

$$-(\mathbf{r}(s_o) + \mathbf{u}(s_o)) \times \mathbf{F}_C(s_o) - \mathbf{M}_C(s_o) + \int_{s_o}^{s} [(\mathbf{r}(\xi) + \mathbf{u}(\xi)) \times \mathbf{f}(\xi) + \boldsymbol{\mu}(\xi)]\,\mathrm{d}\xi +$$
$$+ (\mathbf{r}(s) + \mathbf{u}(s)) \times \mathbf{F}_S(s) + \mathbf{M}_S(s) = \mathbf{0}\,. \tag{7.1b}$$

Let us derive the above equations with respect to s. Since the quantities regarded at s_o are constants (s_o is fixed), we get by taking into account that the derivative of an integral with respect to the upper limit is the integrand taken at the upper limit that

$$\boxed{\frac{\mathrm{d}\mathbf{F}_C(s)}{\mathrm{d}s} + \mathbf{f}(s) = \mathbf{0}} \tag{7.2a}$$

and

$$\frac{\mathrm{d}\mathbf{M}_C(s)}{\mathrm{d}s} + \frac{\mathrm{d}\,(\mathbf{r}(s)+\mathbf{u}(s))}{\mathrm{d}s} \times \mathbf{F}_C(s) + \boldsymbol{\mu}(s) + (\mathbf{r}(s)+\mathbf{u}(s)) \times \underbrace{\left[\frac{\mathrm{d}\mathbf{F}_C(s)}{\mathrm{d}s} + \mathbf{f}(s)\right]}_{=\mathbf{0}} = \mathbf{0},$$

or

$$\boxed{\frac{\mathrm{d}\mathbf{M}_C(s)}{\mathrm{d}s} + \frac{\mathrm{d}\,(\mathbf{r}(s) + \mathbf{u}(s))}{\mathrm{d}s} \times \mathbf{F}_C(s) + \boldsymbol{\mu}(s) = \mathbf{0}\,.} \tag{7.2b}$$

Equations (7.2a) and (7.2b) are the equilibrium conditions sought. It is obvious that they are valid for the deformed beam. For small displacements and deformations, however, we do not make a difference between the initial and deformed states of the beam. Then

$$\frac{\mathrm{d}\,(\mathbf{r}(s) + \mathbf{u}(s))}{\mathrm{d}s} \approx \frac{\mathrm{d}\mathbf{r}(s)}{\mathrm{d}s} = \mathbf{t}(s) \tag{7.3}$$

and Eq. (7.2b) simplifies to

$$\boxed{\frac{\mathrm{d}\mathbf{M}_C(s)}{\mathrm{d}s} + \mathbf{t}(s) \times \mathbf{F}_C(s) + \boldsymbol{\mu}(s) = \mathbf{0}\,.} \tag{7.4}$$

7.2 Longitudinal Vibrations of Rods

7.2.1 Equations of Motion

Figure 7.2 shows the portion of a uniform rod with length x. The rod (or bar) is made of homogeneous material. If the vibrations are longitudinal, the resultant of the inner forces is an axial force

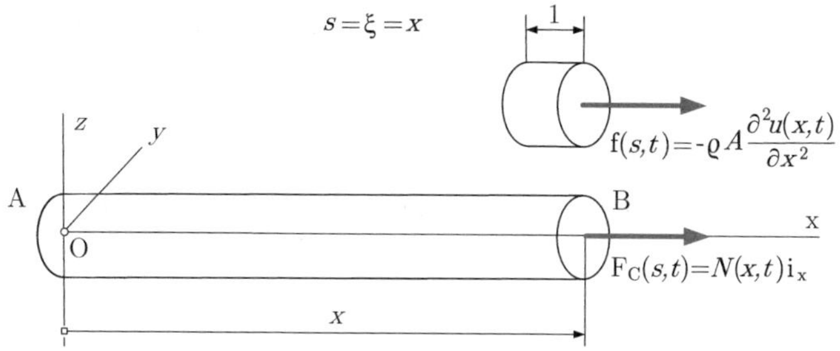

Fig. 7.2 Uniform rod and its element

$$\mathbf{F}_C(s,t) = \mathbf{F}_C(x,t) = N(x,t)\mathbf{i}_x \tag{7.5}$$

and

$$\begin{aligned} \mathbf{M}_C(s,t) = \mathbf{M}_C(x,t) = \boldsymbol{\mu}(s,t) = \boldsymbol{\mu}(x,t) = \mathbf{0}\,, \\ \mathbf{t} = \mathbf{i}_x\,, \quad \mathbf{t}(s) \times \mathbf{F}_C(s,t) = \mathbf{i}_x \times N(x,t)\mathbf{i}_x = \mathbf{0}\,. \end{aligned} \tag{7.6}$$

Thus Eq. (7.4) is an identity.

As regards Eq. (7.2a), the distributed load for a unit length $\mathbf{f}(s) = \mathbf{f}(x)$ is the force of inertia. For longitudinal vibrations, each cross section moves horizontally as if it were a rigid body. Therefore $\mathbf{u} = u(x)\mathbf{i}_x$ is the displacement field in the rod and

$$\mathbf{f}(s) = -\underbrace{\rho A}_{\mathrm{dm}} \frac{\partial^2 u(x,t)}{\partial t^2} \tag{7.7}$$

is the force of inertia for a unit length where ρ is the density and A is the area of the cross section.

Upon substitution of Eqs. (7.5) and (7.7) into Eq. (7.2a), we get

$$\frac{\partial N}{\partial x} - \rho A \frac{\partial^2 u}{\partial t^2} = 0\,. \tag{7.8}$$

It is known from mechanics of materials that the axial force N can be given in terms of the normal stress σ_{xx}:

$$N = A\sigma_{xx}, \tag{7.9}$$

where according to Hooke's law

$$\sigma_{xx} = E\varepsilon_{xx} = E\frac{\partial u}{\partial x}. \tag{7.10}$$

Here E is Young's modulus and $\varepsilon_{xx} = \partial u/\partial x$ is the axial strain. Consequently

$$N = AE\frac{\partial u}{\partial x}. \tag{7.11}$$

Upon substitution of N into (7.8), we obtain the equation of motion for the axial displacement u:

$$\boxed{\frac{\partial}{\partial x}\left(AE\frac{\partial u}{\partial x}\right) = \rho A\frac{\partial^2 u}{\partial t^2}.} \tag{7.12}$$

This equation is a partial differential equation of order two.

Boundary conditions should be imposed on u or N.

In addition, the solution for u should also satisfy the initial conditions prescribed on u and its time derivative $\dot{u}$.

7.2.2 Solution by Separation of Variables

After rearranging Eq. (7.12) we have

$$\boxed{c^2\frac{\partial^2 u}{\partial x^2} = \frac{\partial^2 u}{\partial t^2}, \qquad c^2 = \frac{E}{\rho} > 0.} \tag{7.13}$$

We seek the solution in the form (separation of variables)

$$u(x,t) = U(x)\gamma(t). \tag{7.14}$$

Substituting this solution into Eq. (7.13) yields

$$\frac{c^2}{U(x)}\frac{\mathrm{d}^2 U(x)}{\mathrm{d}x^2} = \frac{1}{\gamma(t)}\frac{\mathrm{d}^2\gamma(t)}{\mathrm{d}t^2} = -\omega^2. \tag{7.15}$$

Since the left side of this equation is independent of t, whereas the right side is independent of x we can come to the conclusion that each side must be the same constant. Letting this constant be $-\omega^2$ (ω is an unknown parameter here), we arrive at two ordinary differential equations: one equation for $U(x)$, the other equation for $\gamma(t)$:

$$-\frac{\mathrm{d}^2 U(x)}{\mathrm{d}x^2} = \lambda U(x), \quad \lambda = \left(\frac{\omega}{c}\right)^2 = \frac{\rho\omega^2}{E}; \qquad \frac{\mathrm{d}^2\gamma(t)}{\mathrm{d}t^2} + \omega^2\gamma(t) = 0\,. \tag{7.16}$$

The general solutions are given by the equations

$$U(x) = \mathcal{A}\sin\sqrt{\lambda}\,x + \mathcal{B}\cos\sqrt{\lambda}\,x\,, \qquad \gamma(t) = \mathcal{C}\sin\omega t + \mathcal{D}\cos\omega t\,, \tag{7.17}$$

where the unknown integration constants $(\mathcal{A}, \mathcal{B})$ and $[\mathcal{C}, \mathcal{D}]$ depend on the (boundary) and [initial] conditions.

It follows from (7.14) and (7.11) that $U(x)$ and $AE\,U^{(1)}(x)$ are, in fact, the amplitudes of the longitudinal motion and the axial force along the rod.

The boundary conditions that can be imposed are categorized as essential and natural boundary conditions. As we have already mentioned

(i) either the amplitude of the motion, i.e., $U(x)$ can be prescribed at an end of the rod (this type of boundary conditions is the essential boundary condition) or
(ii) the amplitude of the axial force, i.e., $N(x) = AE\,U^{(1)}(x)$ can be prescribed at the same end of the rod (this type of boundary conditions is called natural boundary condition).

Exercise 7.1 Consider a uniform rod fixed at its left end and free at its right end—see Fig. 7.3. Determine the solution if $u(x, 0) = \mathfrak{u}(x)$ and $\dot{u}(x, 0) = \mathfrak{g}(x)$ are the initial conditions.

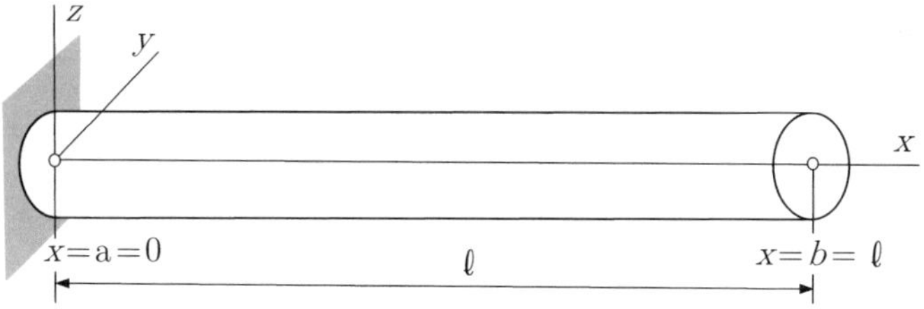

Fig. 7.3 Uniform rod fixed at its left end

It is evident that $u(0, t) = U(0)\gamma(t) = 0$ and $N(\ell, t) = AE\,U^{(1)}(\ell)\gamma(t) = 0$. Consequently, differential equation $(7.16)_1$ is associated with the boundary conditions

$$\begin{aligned} U(0) &= \left(\mathcal{A}\sin\sqrt{\lambda}\,x + \mathcal{B}\cos\sqrt{\lambda}\,x\right)\Big|_{x=0} = \mathcal{B} = 0\,, \\ U^{(1)}(\ell) &= \sqrt{\lambda}\left(\mathcal{A}\cos\sqrt{\lambda}\,x - \mathcal{B}\sin\sqrt{\lambda}\,x\right)\Big|_{x=\ell} = \sqrt{\lambda}\,\mathcal{A}\cos\sqrt{\lambda}\,\ell = 0\,. \end{aligned} \tag{7.18}$$

Since $\mathcal{A}$ and λ cannot be zero for a nontrivial solution concerning $U(x)$, we have

$$\cos \sqrt{\lambda}\,\ell = 0\,. \tag{7.19}$$

This equation is called frequency equation. It is satisfied if

$$\sqrt{\lambda_k}\,\ell = \pm(2k-1)\frac{\pi}{2}\,, \qquad k = 1, 2, 3, \ldots \tag{7.20a}$$

are the roots from where

$$\lambda_k = \frac{\pi^2}{4\ell^2}(2k-1)^2\,, \qquad \omega_k \underset{(7.16)_2}{=} c\sqrt{\lambda_k} = \sqrt{\frac{E}{\rho}\lambda_k} = \sqrt{\frac{E}{\rho}}\frac{\pi}{2\ell}(2k-1)\,. \tag{7.20b}$$

Observe that the number of roots λ_k is infinite.

We remark that $\mathcal{A}$ can be arbitrary since the differential equation $(7.16)_1$, is homogeneous. In what follows we shall set it to 1.

The roots λ_k are called eigenvalues, and the values of ω are the natural frequencies or eigenfrequencies. The reason for calling ω_k the kth natural frequency is that solution $(7.17)_3$ describes a harmonic motion with this natural frequency.

The function

$$U_k(x) = \sin\sqrt{\lambda_k}x = \sin\frac{(2k-1)\pi}{2\ell}x \tag{7.21}$$

is the kth eigenfunction, or the kth normal mode. For $k = 1$ we speak about fundamental frequency and fundamental mode.

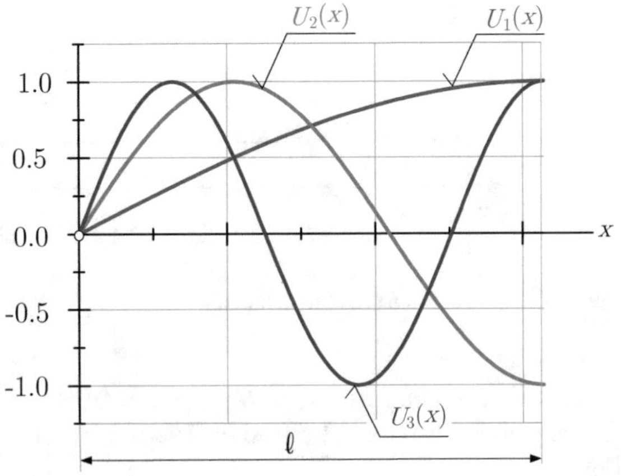

Fig. 7.4 Three eigenfunctions of the longitudinally vibrating rod

Figure 7.4 shows the first three eigenfunctions.

The eigenfunctions $U_k(x)$ have the following properties:

$$\int_0^\ell U_k(x)\, U_n(x)\mathrm{d}x = \int_0^\ell \sin\lambda_k x\ \sin\lambda_n x\ \mathrm{d}x =$$
$$= \int_0^\ell \sin\frac{(2k-1)\pi}{2\ell}x\ \sin\frac{(2n-1)\pi}{2\ell}x\ \mathrm{d}x = \frac{\ell}{2}\begin{cases}1 \text{ if } k=n,\\ 0 \text{ if } k\neq n.\end{cases} \quad k,n = 1,2,3\ldots \tag{7.22}$$

If the functions $U_k(x)$ and $U_n(x)$ satisfy this equation—known as orthogonality condition—then we say that they are orthogonal to each other. This result means that the eigenfunctions with different subscripts are orthogonal functions.

The functions

$$\psi_k(x) = \sqrt{\frac{2}{\ell}}\sin\frac{(2k-1)\pi}{2\ell}x \tag{7.23}$$

are called normalized eigenfunctions.

Recalling (7.14), $(7.17)_2$, and (7.21), we can give the solution that belongs to ω_k in the form

$$u_k(x,t) = U_k(x)\gamma_k(t) = \underbrace{\sin\lambda_k x}_{U_k(x)}\ \underbrace{(\mathcal{C}_k\sin\omega_k t + \mathcal{D}_k\cos\omega_k t)}_{\gamma_k(t)}. \tag{7.24}$$

The function $u_k(x,t)$ is the kth mode of vibration.

The points at which $U_k(x)$ (or $u_k(x,t)$) is equal to zero are the nodal points or simply nodes.

It is obvious that

$$\dot{\gamma}_k(t) = \omega_k\,(\mathcal{C}_k\cos\omega_k t - \mathcal{D}_k\sin\omega_k t)\ . \tag{7.25}$$

The total solution can be obtained by the use of the principle of superposition:

$$\boxed{\begin{aligned}u(x,t) &= \sum_{k=1}^{\infty} u_k(x,t) = \sum_{k=1}^{\infty} U_k(x)\,\gamma_k(t) =\\ &= \sum_{k=1}^{\infty}\sin\lambda_k x\ (\mathcal{C}_k\sin\omega_k t + \mathcal{D}_k\cos\omega_k t)\,.\end{aligned}} \tag{7.26}$$

Its time derivative is given by

$$\dot{u}(x,t) = \sum_{k=1}^{\infty}\dot{u}_k(x,t) = \sum_{k=1}^{\infty} U_k(x)\,\dot{\gamma}_k(t) =$$
$$= \sum_{k=1}^{\infty}\omega_k\ \sin\lambda_k x\ (\mathcal{C}_k\cos\omega_k t - \mathcal{D}_k\sin\omega_k t)\,. \tag{7.27}$$

The unknown integration constants can be determined by using the initial conditions:

$$\mathfrak{u}(x) = u(x,0) = \sum_{k=1}^{\infty} U_k(x)\,\gamma_k(0) = \sum_{k=1}^{\infty} \mathcal{D}_k U_k(x)\,, \tag{7.28a}$$

$$\mathfrak{g}(x) = \dot{u}(x,0) = \sum_{k=1}^{\infty} U_k(x)\,\dot{\gamma}_k(0) = -\sum_{k=1}^{\infty} \omega_k \mathcal{C}_k U_k(x)\,. \tag{7.28b}$$

Multiply throughout by $U_n(x)$ ($n = 1, 2, 3, \ldots$) and integrate the result obtained with respect to x from zero to ℓ. We have

$$\int_0^{\ell} \mathfrak{u}(x) U_n(x)\,\mathrm{d}x = \sum_{k=1}^{\infty} \mathcal{D}_k \int_0^{\ell} U_k(x)\,U_n(x)\mathrm{d}x = \underset{(7.22)}{\uparrow} = \frac{\ell \mathcal{D}_k}{2} \begin{cases} 1 \text{ if } k = n \\ 0 \text{ if } k \neq n \end{cases},$$

$$\int_0^{\ell} \mathfrak{g}(x) U_n(x)\,\mathrm{d}x = -\sum_{k=1}^{\infty} \omega_k \mathcal{C}_k \int_0^{\ell} U_k(x)\,U_n(x)\mathrm{d}x = \underset{(7.22)}{\uparrow} =$$

$$= -\frac{\ell \omega_k \mathcal{C}_k}{2} \begin{cases} 1 \text{ if} k = n \\ 0 \text{ if} k \neq n \end{cases}.$$

Hence

$$\boxed{\mathcal{D}_n = \frac{2}{\ell} \int_0^{\ell} \mathfrak{u}(x) U_n(x)\,\mathrm{d}x\,, \qquad \mathcal{C}_n = \frac{2}{\ell \omega_n} \int_0^{\ell} \mathfrak{g}(x) U_n(x)\,\mathrm{d}x\,.} \tag{7.29}$$

Exercise 7.2 Assume that the rod shown in Fig. 7.3 is subjected to an axial force N_o at its free end (at the right end). Determine the solution of the equation of motion if the force is suddenly removed.

Due to the force N_o the axial strain is constant

$$\varepsilon_{xx} = \frac{\mathrm{d}u(x)}{\mathrm{d}x} = \frac{\sigma_{xx}}{E} = \frac{N_o}{AE}$$

from where

$$\mathfrak{u}(x) = u(x,0) = \frac{N_o x}{AE} \tag{7.30}$$

is the initial displacement field in the rod. When the force is suddenly removed, the rod is at rest; therefore, the initial velocity $\mathfrak{g}(x)$ is zero. We can now utilize Eq. (7.29) to determine the integration constants $\mathcal{D}_n$ and $\mathcal{C}_n$. We get

$$\mathcal{D}_n = \frac{2}{\ell}\int_0^\ell \mathfrak{u}(x)U_n(x)\,\mathrm{d}x = \underset{(7.30),\,(7.21)}{\uparrow} = \frac{2N_o}{AE\ell}\int_0^\ell x\sin\frac{(2n-1)\pi}{2\ell}x\,\mathrm{d}x = \\ = \frac{8N_o\ell}{AE}\frac{(-1)^n}{\pi^2(2n-1)^2} \tag{7.31a}$$

and

$$\mathcal{C}_n = \frac{2}{\ell\omega_n}\int_0^\ell \mathfrak{g}(x)U_n(x)\,\mathrm{d}x = 0\,. \tag{7.31b}$$

With the integration constants equation (7.26) yields

$$u(x,t) = \underset{(7.20\mathrm{b})_2}{\uparrow} = \frac{8N_o\ell}{AE\pi^2}\sum_{n=1}^{\infty}\frac{(-1)^n}{(2n-1)^2}\cos\sqrt{\frac{E}{\rho}}\frac{\pi}{2\ell}(2n-1)t. \tag{7.32}$$

7.3 Transverse Vibration of a String

7.3.1 *Equations of Motion*

Figure 7.5 shows a string (or cable). We shall assume that

(i) the string is uniform (the cross section is constant),
(ii) it is made of homogeneous and linearly elastic material for which E is Young's modulus, and ρ is the density,
(iii) the string is flexible, i.e., it has no resistance to bending ($\mathbf{M}_C = \mathbf{0}$ in Eq. (7.2b)),
(iv) its statical load is vertical ($\mathbf{f}(s) = f(x)\mathbf{i}_z$—for free vibrations $f(x) = 0$),
(v) each cross section moves vertically ($\mathbf{u} = w(x,t)\mathbf{i}_z$ is the displacement field),
(vi) the displacements and deformations are small,
(vii) the string is uniformly stretched by a constant axial force (tensile force) N, and
(viii) its weight can be neglected.

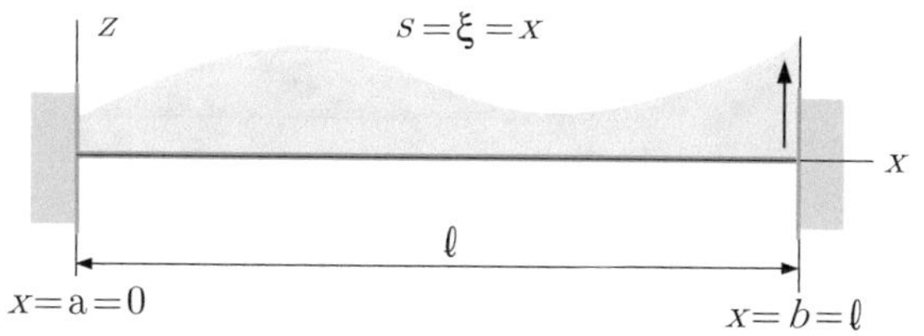

Fig. 7.5 String fixed at both ends

Under these conditions

$$\mathbf{r}(s) + \mathbf{u}(s) = x\mathbf{i}_x + w(x,t)\mathbf{i}_z, \quad \boldsymbol{\mu} = \mathbf{0}, \quad \mathbf{f}(x,t) = -\rho A \frac{\partial^2 w(x,t)}{\partial t^2}\mathbf{i}_z \tag{7.33}$$

are the position vector of the deformed string, the distributed moment load, and the forces of inertia per unit length. Hence, Eq. (7.2b) yields

$$\frac{\mathrm{d}\mathbf{M}_C(s)}{\mathrm{d}s} + \frac{\mathrm{d}\,(\mathbf{r}(s)+\mathbf{u}(s))}{\mathrm{d}s} \times \mathbf{F}_C(s) + \boldsymbol{\mu}(s) = \frac{\partial\,(x\mathbf{i}_x + w(x,t)\mathbf{i}_z)}{\partial x} \times \mathbf{F}_C(x,t) =$$
$$= \left(\mathbf{i}_x + \frac{\partial w(x,t)}{\partial x}\mathbf{i}_z\right) \times \mathbf{F}_C(x,t) = \mathbf{0} \tag{7.34a}$$

in which the inner force $\mathbf{F}_C(x,t)$ is resolved into a horizontal (the constant tensile force) and a vertical component:

$$\mathbf{F}_C(x,t) = N\mathbf{i}_x + F_{Cz}(x,t)\mathbf{i}_z\,. \tag{7.34b}$$

Let us derive Eq. (7.34a) with respect to x. We get

$$\frac{\partial^2 w(x,t)}{\partial x^2}\mathbf{i}_z \times (N\mathbf{i}_x + F_{Cz}(x,t)\mathbf{i}_z) + \left(\mathbf{i}_x + \frac{\partial w(x,t)}{\partial x}\mathbf{i}_z\right) \times \frac{\partial \mathbf{F}_C(x,t)}{\partial x} = \mathbf{0}. \tag{7.35}$$

If we recall Eq. (7.2a) and take into account that for free vibrations $\mathbf{f}(s) = \mathbf{f}(x,t)$ is given by $(7.33)_3$, we get

$$\frac{\partial \mathbf{F}_C(x,t)}{\partial x} = \rho A \frac{\partial^2 w(x,t)}{\partial t^2}\mathbf{i}_z\,. \tag{7.36}$$

Upon substitution of this relation into Eq. (7.35)

$$N\frac{\partial^2 w}{\partial x^2}\mathbf{i}_y - \rho A \frac{\partial^2 w(x,t)}{\partial t^2}\mathbf{i}_y$$

is the result. Hence,

$$\boxed{c^2 \frac{\partial^2 w}{\partial x^2} = \frac{\partial^2 w}{\partial t^2}, \qquad c^2 = \frac{N}{\rho A} > 0} \tag{7.37}$$

is the equation of motion.

We remark that boundary conditions should be imposed on w.

In addition, the solution for w should also satisfy the initial conditions prescribed on w and its time derivative $\dot{w}$.

We seek the solution again in the form of a product (separation of variables)

$$W(x,t) = W(x)\gamma(t)\,. \tag{7.38}$$

Substituting this product into (7.37), we obtain

$$\frac{c^2}{W(x)}\frac{\mathrm{d}^2 W(x)}{\mathrm{d}x^2} = \frac{1}{\gamma(t)}\frac{\mathrm{d}^2\gamma(t)}{\mathrm{d}t^2} = -\omega^2 . \tag{7.39}$$

Here, the left side is independent of t, whereas the right side is independent of x; hence, each side must be the same constant. Denoting this constant by $-\omega^2$ (ω is an unknown parameter here), we arrive at two ordinary differential equations:

$$-\frac{\mathrm{d}^2 W(x)}{\mathrm{d}x^2} = \lambda W(x), \quad \lambda = \left(\frac{\omega}{c}\right)^2 = \frac{\rho A\omega^2}{N}; \quad \frac{\mathrm{d}^2\gamma(t)}{\mathrm{d}t^2} + \omega^2\gamma(t) = 0 . \tag{7.40}$$

The general solutions are given by the equations

$$W(x) = \mathcal{A}\sin\sqrt{\lambda}\,x + \mathcal{B}\cos\sqrt{\lambda}\,x , \qquad G(t) = \mathcal{C}\sin\omega t + \mathcal{D}\cos\omega t . \tag{7.41}$$

The unknown integration constants ($\mathcal{A}$, $\mathcal{B}$) and [$\mathcal{C}$, $\mathcal{D}$] can be determined by using the boundary and initial conditions.

Assume that the string is fixed at both ends. Then, differential equation (7.37) is associated with the following boundary conditions:

$$w(0,t) = W(0) = 0, \qquad w(\ell,t) = W(\ell) = 0. \tag{7.42}$$

Let $\mathfrak{w}(x)$ and $\mathfrak{g}(x)$ $x \in [0,\ell]$ be given functions. The initial conditions

$$w(x,0) = \mathfrak{w}(x), \qquad \dot{w}(x,0) = \mathfrak{g}(x) \tag{7.43}$$

should also be satisfied by the solution $w(x,t)$.

Remark 7.1 Compare now the equations that describe the longitudinal vibrations of a rod and the transverse vibrations of a string. As regards the equations of motion, we can get (7.37) from (7.12) if we write w for u and $c^2 = N/\rho E$ for $c^2 = E/\rho$.

The differential equation for the amplitude function $(7.40)_1$ and the solution $(7.41)_1$ can be obtained from $(7.16)_1$ and $(7.17)_1$ if we write W for U.

The differential equations $(7.40)_2$, $(7.16)_2$ for the time function $\gamma(t)$ and the solutions $(7.41)_2$, $(7.17)_2$ coincide formally with each other.

Note that we have denoted the unknown integration constants in the same way for both problems, i.e., by $\mathcal{A}$, $\mathcal{B}$, $\mathcal{C}$, and $\mathcal{D}$.

Exercise 7.3 Assume that the string is fixed at its ends. Determine the solution if the initial conditions are given by Eq. (7.41).

After substituting solution (7.41) into boundary conditions (7.42), we get

$$\begin{aligned} W(0) &= \left(\mathcal{A}\sin\sqrt{\lambda}\,x + \mathcal{B}\cos\sqrt{\lambda}\,x\right)\Big|_{x=0} = \mathcal{B} = 0 , \\ W(\ell) &= \sqrt{\lambda}\left(\mathcal{A}\sin\sqrt{\lambda}\,x + \mathcal{B}\cos\sqrt{\lambda}\,x\right)\Big|_{x=\ell} = \sqrt{\lambda}\,\mathcal{A}\sin\sqrt{\lambda}\,\ell = 0 . \end{aligned} \tag{7.44}$$

Since $\mathcal{A} \neq 0$ in $(7.44)_2$

$$\sin \sqrt{\lambda}\,\ell = 0 \tag{7.45}$$

is the frequency equation. Hence

$$\sqrt{\lambda_k}\,\ell = \pm k\pi\,, \qquad k = 1, 2, 3, \ldots \tag{7.46a}$$

are the roots from where

$$\lambda_k = \frac{k^2\,\pi^2}{\ell^2}\,, \qquad \omega_k \underset{(7.40)_2}{\overset{\uparrow}{=}} c\sqrt{\lambda_k} = \frac{kc\pi}{\ell} = \frac{k\pi}{\ell}\sqrt{\frac{N}{\rho A}}\,. \tag{7.46b}$$

The number of roots λ_k is naturally infinite. The function

$$W_k(x) = \sin \sqrt{\lambda_k} x = \sin \frac{k\pi}{\ell} x \tag{7.47}$$

is the kth eigenfunction (or the kth normal mode).

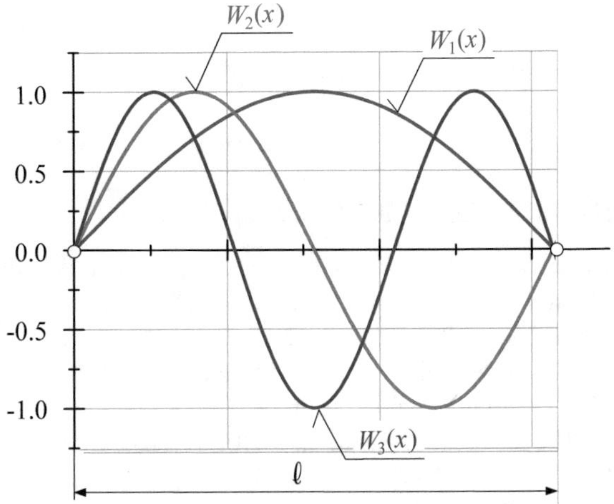

Fig. 7.6 Three eigenfunctions of the vibrating string

Figure 7.6 shows the first three eigenfunctions. It can be checked with ease that the eigenfunctions satisfy the following orthogonality condition:

$$\int_0^\ell W_k(x)\,W_n(x)\,\mathrm{d}x = \int_0^\ell \sin \lambda_k x\, \sin \lambda_n x\,\mathrm{d}x =$$

$$= \int_0^\ell \sin \frac{k\pi}{\ell} x\, \sin \frac{n\pi}{\ell} x\,\mathrm{d}x = \begin{cases} \ell/2 \text{ if } k = n, \\ 0 \ \text{ if } k \neq n. \end{cases} \qquad k, n = 1, 2, 3 \ldots \tag{7.48}$$

The normalized eigenfunctions are given by the equation

$$\psi_k(x) = \sqrt{\frac{2}{\ell}}\sin\frac{k\pi}{\ell}x\,. \tag{7.49}$$

Since $\mathcal{B} = 0$, it follows from (7.38) and (7.41) that the solution corresponding to ω_k is of the form

$$w_k(x,t) = W_k(x)\gamma_k(t) = \underbrace{\sin\lambda_k x}_{W_k(x)}\,\underbrace{(\mathcal{C}_k\sin\omega_k t + \mathcal{D}_k\cos\omega_k t)}_{\gamma_k(t)}\,. \tag{7.50}$$

The function $w_k(x,t)$ is the kth mode of vibration, and the points at which $W_k(x) = w_k(x,t) = 0$ are the nodal points.

By applying the principle of superposition, we get the total solution

$$\boxed{w(x,t) = \sum_{k=1}^{\infty} w_k(x,t) = \sum_{k=1}^{\infty}\sin\lambda_k x\ (\mathcal{C}_k\sin\omega_k t + \mathcal{D}_k\cos\omega_k t)} \tag{7.51}$$

and its time derivative as well:

$$\dot{w}(x,t) = \sum_{k=1}^{\infty}\dot{w}_k(x,t) = \sum_{k=1}^{\infty}\omega_k\ \sin\lambda_k x\ (\mathcal{C}_k\cos\omega_k t - \mathcal{D}_k\cos\omega_k t)\,. \tag{7.52}$$

The initial conditions

$$\mathfrak{w}(x) = w(x,0) = \sum_{k=1}^{\infty} W_k(x)\,\gamma_k(0) = \sum_{k=1}^{\infty}\mathcal{D}_k W_k(x)\,, \tag{7.53a}$$

$$\mathfrak{g}(x) = \dot{w}(x,0) = \sum_{k=1}^{\infty} W_k(x)\,\dot{\gamma}_k(0) = -\sum_{k=1}^{\infty}\omega_k\mathcal{C}_k W_k(x) \tag{7.53b}$$

make it possible to determine the unknown integration constants if we take the orthogonality conditions (7.48) into account in the following integrals:

$$\int_0^\ell \mathfrak{w}(x)W_n(x)\,\mathrm{d}x = \sum_{k=1}^{\infty}\mathcal{D}_k\int_0^\ell W_k(x)\,W_n(x)\mathrm{d}x = \underset{(7.48)}{\uparrow} = \frac{\ell\mathcal{D}_k}{2}\begin{cases}1 \text{ if } k = n\\ 0 \text{ if } k \neq n\end{cases},$$

$$\int_0^\ell \mathfrak{g}(x)W_n(x)\,\mathrm{d}x = -\sum_{k=1}^{\infty}\omega_k\mathcal{C}_k\int_0^\ell W_k(x)\,W_n(x)\mathrm{d}x = \underset{(7.48)}{\uparrow} =$$

$$= -\frac{\ell\omega_k\mathcal{C}_k}{2}\begin{cases}1 \text{ if } k = n\\ 0 \text{ if } k \neq n\end{cases}.$$

Hence

$$\mathcal{D}_n = \frac{2}{\ell}\int_0^\ell \mathfrak{w}(x)W_n(x)\,\mathrm{d}x\,, \qquad \mathcal{C}_n = \frac{2}{\ell\omega_n}\int_0^\ell \mathfrak{g}(x)W_n(x)\,\mathrm{d}x\,. \tag{7.54}$$

7.4 Torsional Vibrations

Figure 7.7 represents a shaft. We shall assume that

(i) the shaft is uniform (the cross section is constant),
(ii) it is made of homogeneous and linearly elastic material for which G is the shear modulus of elasticity and ρ is the density,
(iii) its statical load is a distributed moment load $\boldsymbol{\mu}(s) = \mu_x(x)\mathbf{i}_x$ acting on its longitudinal axis (for free vibrations $\mu_x(x) = 0$),
(iv) each cross section rotates in its own plane as if it were a rigid body—the rotation vector is denoted by $\varphi(x)\mathbf{i}_x$, and
(v) the rotations and deformations are small.

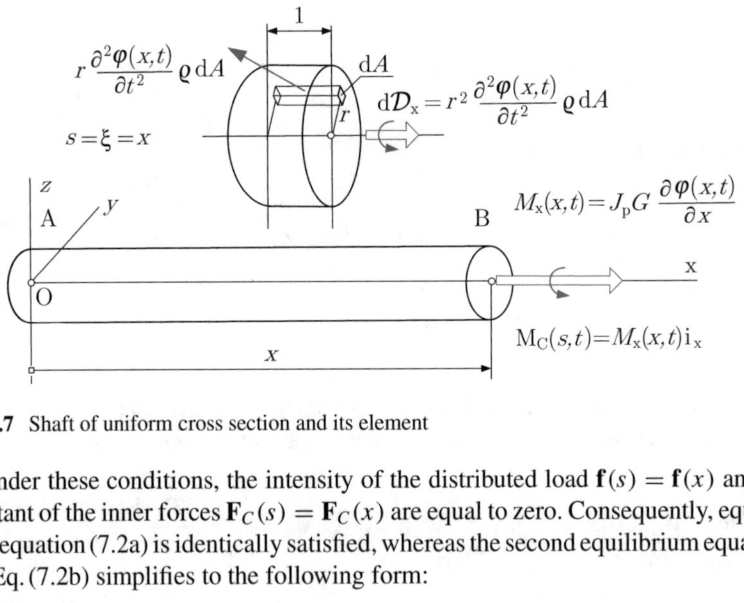

Fig. 7.7 Shaft of uniform cross section and its element

Under these conditions, the intensity of the distributed load $\mathbf{f}(s) = \mathbf{f}(x)$ and the resultant of the inner forces $\mathbf{F}_C(s) = \mathbf{F}_C(x)$ are equal to zero. Consequently, equilibrium equation (7.2a) is identically satisfied, whereas the second equilibrium equation, i.e., Eq. (7.2b) simplifies to the following form:

$$\frac{\mathrm{d}\mathbf{M}_C(x)}{\mathrm{d}x} + \boldsymbol{\mu}(x) = \mathbf{0}\,. \tag{7.55}$$

For the vibration problems considered, $\mathbf{M}_C = M_x(x,t)\mathbf{i}_x$ is the torsional moment which is related to the angle of rotation $\varphi(x)$ via the equation

$$M_x(x,t) = GJ_p \frac{\mathrm{d}\varphi(x,t)}{\mathrm{d}x}, \tag{7.56}$$

where J_p is the areal polar moment of inertia. It is clear from Fig. 7.7 that the moment of the effective forces for a unit length of the axis of rotation can be given as

$$\mathcal{D}_x = \int_A \mathcal{D}_x = \underbrace{\int_A r^2 \rho \, \mathrm{d}A}_{I_p} \frac{\partial^2 \varphi}{\partial t^2}, \tag{7.57}$$

where I_p is the mass polar moment of inertia. If the diameter of the shaft is constant $I_p = \rho J_p$. The distributed moment load is due to the effective forces and in accordance with the d'Alembert principle can be given in the following form:

$$\mu_x(x,t) = -I_p \frac{\partial^2 \varphi}{\partial t^2} = -J_p \rho \frac{\partial^2 \varphi}{\partial t^2}. \tag{7.58}$$

Since both $\mathbf{M}_C$ and $\boldsymbol{\mu}$ are parallel to the axis x from Eq. (7.55), we get

$$\frac{\mathrm{d}M_x(x,t)}{\mathrm{d}x} + \mu_x(x,t) = 0. \tag{7.59}$$

Substitution of (7.56) and (7.58) into this equation results in the equation of motion:

$$\boxed{\frac{\partial}{\partial x}\left(GJ_p \frac{\partial \varphi}{\partial x}\right) = I_p \frac{\partial^2 \varphi}{\partial t^2}.} \tag{7.60}$$

If the diameter of the shaft is constant this equations becomes a bit simpler

$$\boxed{c^2 \frac{\partial^2 \varphi}{\partial x^2} = \frac{\partial^2 \varphi}{\partial t^2}, \qquad c^2 = \frac{G}{\rho} > 0.} \tag{7.61}$$

It should be mentioned that boundary conditions have to be imposed on φ or torsional moment M_x.

In addition, the solution for φ should satisfy the initial conditions prescribed on φ and its time derivative $\dot{\varphi}$.

Assume that

$$\varphi(x,t) = \Phi(x)\gamma(t). \tag{7.62}$$

Then we get in the same manner as earlier—see, for instance, the steps leading to Eq. (7.40)—that

$$-\frac{\mathrm{d}^2 \Phi(x)}{\mathrm{d}x^2} = \lambda \Phi(x), \quad \lambda = \left(\frac{\omega}{c}\right)^2 = \frac{\rho \omega^2}{G}; \qquad \frac{\mathrm{d}^2 \gamma(t)}{\mathrm{d}t^2} + \omega^2 \gamma(t) = 0. \tag{7.63}$$

The general solutions are given by the equations

$$\Phi(x) = \mathcal{A}\sin\sqrt{\lambda}x + \mathcal{B}\cos\sqrt{\lambda}x\,, \qquad \gamma(t) = \mathcal{C}\sin\omega t + \mathcal{D}\cos\omega t\,. \tag{7.64}$$

Remark 7.2 Let us introduce the notations $y(x) = U(x) = W(x) = \Phi(x)$ and operators $K(y) = -y^{(2)}(x)$ and $M(y) = y(x)$. Further let $x = a$ and $x = b$ be those x coordinates that belong to the left and right extremities of the rod (or shaft): $b > a$ and $b - a = \ell$ is the length of the rod (or shaft). Making use of the notations introduced differential equations $(7.16)_1$, $(7.40)_1$ and $(7.63)_1$ can be rewritten in a common form:

$$K[y(x)] = \lambda M[y(x)]\,. \tag{7.65}$$

This differential equation is, in general, associated with the following boundary conditions:

$$\left.\begin{array}{lll} y(a) = 0\,, & y(b) = 0 & \text{(the ends of the rod or shaft are fixed); or} \\ y(a) = 0\,, & y^{(1)}(b) = 0 & \text{(the left end is fixed, the right end is free); or} \\ y^{(1)}(a) = 0\,, & y^{(1)}(b) = 0 & \text{(the ends of the rod or shaft are free).} \end{array}\right\} \tag{7.66}$$

Differential equation (7.65) and boundary conditions $(7.66)_1$ (or $(7.66)_2$, or $(7.66)_3$) constitute a boundary value problem for which both the differential equation and the boundary conditions are homogeneous, and λ is an unknown parameter. This problem is also referred to as an eigenvalue problem with lambda as the eigenvalue.

Table 7.1 shows some typical values for the three eigenvalue problems. The second and third rows in the table are based on the solution of Exercises 7.1 and 7.3. Data in the fourth row are taken from the solution of Problem 7.2.

Table 7.1 Characteristic values for the eigenvalue problems considered

Boundary conditions	Frequency equation	Normalized eigenfunctions (mode shapes)	Eigen-values λ_k	Natural frequencies ω_k
$y(0)=0$ $y(\ell)=0$	$\sin\sqrt{\lambda}\ell=0$	$\psi_k(x) = \sqrt{\frac{2}{\ell}}\sin\frac{k\pi}{\ell}x$	$\frac{k^2\pi^2}{\ell^2}$	$\frac{ck\pi}{\ell}$
$y(0)=0$ $y^{(1)}(\ell)=0$	$\cos\sqrt{\lambda}\ell=0$	$\psi_k(x)=\sqrt{\frac{2}{\ell}}\sin\frac{(2k-1)\pi}{2\ell}x$	$\frac{\pi^2(2k-1)^2}{4\ell^2}$	$\frac{c\pi(2k-1)}{2\ell}$
$y^{(1)}(0)=0$ $y^{(1)}(\ell)=0$	$\sin\sqrt{\lambda}\ell=0$	$\psi_k(x) = \sqrt{\frac{2}{\ell}}\sin\frac{k\pi}{\ell}x$	$\frac{k^2\pi^2}{\ell^2}$	$\frac{ck\pi}{\ell}$

7.5 Flexural Vibrations of Beams

7.5.1 Equilibrium Equations

7.5.1.1 Assumptions

Figure 7.8 depicts a beam. We shall assume that the followings are true:

(i) the beam is uniform (the cross section is constant),
(ii) the coordinate plane xz is a plane of symmetry,
(iii) the beam is made of homogeneous and linearly elastic material for which E is Young's modulus, ν is the Poisson number, and G is the shear modulus of elasticity,
(iv) the displacements and deformations are small,
(v) for the arc coordinates it holds that $s = \xi = x$, $\mathrm{d}s = \mathrm{d}\xi = \mathrm{d}x$,
(vi) the vertical static load $\mathbf{f}(s) = f_z(x)\mathbf{i}_z$ is exerted on the centerline x (under this load the deformed centerline remains in the coordinate plane xz)—for free vibrations $f_z(x) = 0$,
(vii) the distributed moment load $\boldsymbol{\mu} = \mu_y(x)\mathbf{i}_y$ is zero,
(viii) the beam is preloaded by a constant axial force N which is (positive) [negative] if it is a (tensile) [compressive] force, and
(ix) it will be assumed that the displacement field of the beam is of the form

$$\mathbf{u} = (u(x) + \varphi_y(x)z)\mathbf{i}_x + w(x)\mathbf{i}_z\,, \qquad \varphi_y(x) = -\mathrm{d}w/\mathrm{d}x, \tag{7.67}$$

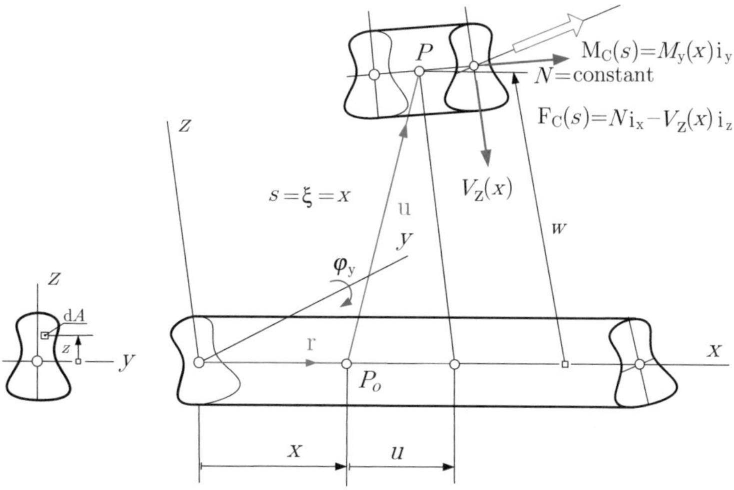

Fig. 7.8 Beam and its element with the inner forces

which means that the cross section of the beam rotates as if it were a rigid body and remains perpendicular to the deformed centerline[1]—φ_y is the rotation of the centerline about the direction y,

(x) since the deformations are small the derivative $\mathrm{d}u/\mathrm{d}x$ satisfies the inequality

$$|\mathrm{d}u/\mathrm{d}x| \ll 1$$

which means that it can be neglected when it is compared to 1.

7.5.1.2 Equilibrium Conditions

Under these conditions

$$\mathbf{F}_C = N\mathbf{i}_x - V_z(x)\mathbf{i}_z\,, \qquad \mathbf{M}_C = M_y(x)\mathbf{i}_y \tag{7.68}$$

are the resultant and moment resultant of the inner forces acting on a cross section. The bending moment M_y is positive if its vector points in the positive y direction, the shear force V_z is positive if it points in the negative z direction—see Fig. 7.8 for details. With $\mathbf{F}_C$ Eq. (7.2a) yields

$$-\frac{\mathrm{d}V_z}{\mathrm{d}x} + f_z = 0\,. \tag{7.69}$$

As regards Eq. (7.2a), we may write

$$\begin{aligned}\frac{\mathrm{d}\mathbf{M}_C(x)}{\mathrm{d}x} &+ \frac{\mathrm{d}\,(\mathbf{r}(x) + \mathbf{u}(x))}{\mathrm{d}x} \times \mathbf{F}_C(x) + \boldsymbol{\mu}(x) = \\ &= \frac{\mathrm{d}M_y}{\mathrm{d}x}\mathbf{i}_y + \frac{\mathrm{d}\left[(x+u)\,\mathbf{i}_x + w\mathbf{i}_z\right]}{\mathrm{d}x} \times (N\mathbf{i}_x - V_z(x)\mathbf{i}_z) + \boldsymbol{\mu}(x) = \mathbf{0}\end{aligned} \tag{7.70}$$

where

$$\frac{\mathrm{d}}{\mathrm{d}x}(x+u) = 1 + \frac{\mathrm{d}u}{\mathrm{d}x} \approxeq 1\,, \quad N = \text{constant}\,, \qquad \text{and} \quad \boldsymbol{\mu} = \mathbf{0}\,. \tag{7.71}$$

Calculate now the cross products in Eq. (7.70) and dot multiply the result by $\mathbf{i}_y$. We have

$$\frac{\mathrm{d}M_y}{\mathrm{d}x} + N\frac{\mathrm{d}w}{\mathrm{d}x} + V_z = 0\,. \tag{7.72}$$

After deriving this equation with respect to x we can substitute (7.69) for $\mathrm{d}V_z/\mathrm{d}x$. The result is

[1] This assumption is known as Euler–Bernoulli beam hypothesis.

$$\frac{\mathrm{d}^2 M_y}{\mathrm{d}x^2} + \frac{\mathrm{d}}{\mathrm{d}x}\left(N\frac{\mathrm{d}w}{\mathrm{d}x}\right) + f_z = 0\,. \tag{7.73}$$

This equation is the equilibrium equation the moment resultant of the internal forces (the bending moment M_y) should satisfy if the beam is preloaded by an axial force N [1, 2].

We shall denote the aerial moment of inertia of the cross section A with respect to the centroidal axis y by I. It is obvious on the basis of Fig. 7.8 that

$$I = \int_A z^2\,\mathrm{d}A\,.$$

In accordance with assumption (ix), the positive φ_y rotates the axis z (the centerline) in clockwise sense as shown in Fig. 7.8. The bending moment M_y and the shear force V_z—see Eq. (7.72)—can all be given in terms of the vertical displacement w [3]. The corresponding equations are as follows:

$$M_y = IE\frac{\mathrm{d}\varphi_y}{\mathrm{d}x} = -IE\frac{\mathrm{d}^2 w}{\mathrm{d}x^2}\,, \quad V_z = IE\frac{\mathrm{d}^3 w}{\mathrm{d}x^3} - N\frac{\mathrm{d}w}{\mathrm{d}x}\,. \tag{7.74}$$

Upon substitution of $(7.74)_2$ for M_y in Eq. (7.73), we have

$$\boxed{\frac{\mathrm{d}^2}{\mathrm{d}x^2}\left(IE\frac{\mathrm{d}^2 w}{\mathrm{d}x^2}\right) - \frac{\mathrm{d}}{\mathrm{d}x}\left(N\frac{\mathrm{d}w}{\mathrm{d}x}\right) = f_z\,.} \tag{7.75a}$$

If the beam is a uniform one and the axial force is independent of x equilibrium equation (7.75a) is simplified:

$$\boxed{IE\frac{\mathrm{d}^4 w}{\mathrm{d}x^4} - N\frac{\mathrm{d}^2 w}{\mathrm{d}x^2} = f_z\,.} \tag{7.75b}$$

Equations (7.75) make it possible to determine the deflection $w(x)$ if the beam is preloaded by the axial force N.

7.5.1.3 Boundary Conditions

It is obvious that Eqs. (7.75) should be associated with appropriate boundary conditions which depend on the supports at the endpoints of the beam. If an endpoint of the beam

(a) cannot move vertically then the deflection w is zero.
(b) cannot rotate then the rotation φ_y is zero.
(c) can freely rotate then the bending moment M_y is zero.

(d) is free then the shear force V_z is zero.
(e) is (vertically) [rotationally] restrained by a (spring) [torsional spring] the sum of the (forces) [moments] shown in Fig. 7.9 should be zero—the spring constants are denoted by $k_{z\ell}$, k_{zr} and $k_{\gamma\ell}$, $k_{\gamma r}$, respectively.

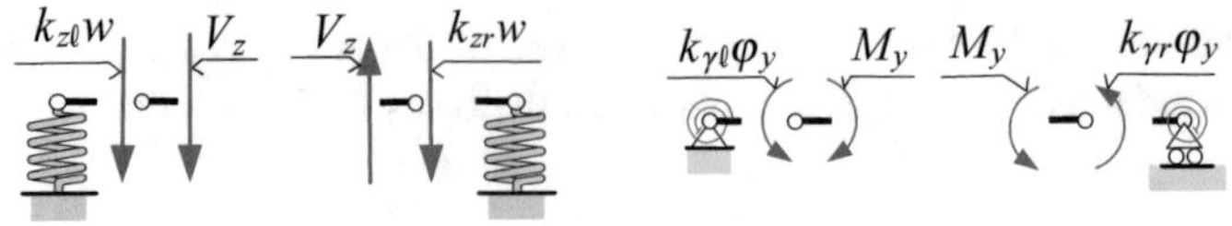

Fig. 7.9 Spring supported beam ends with inner forces

In accordance with these rules, Table 7.2 shows the boundary conditions for the beams that are (i) simply supported, (ii) fixed at the left end and hinged at the right end, (iii) fixed at both ends, and (iv) fixed at the left end and free at the right end.

The axial force N is rigid which means that it is always parallel to the axis x.

If the left or right endpoint of the beam is restrained (vertically) (rotationally) by a (spring) [torsional spring], then the boundary conditions are of the form

$$V_z + k_{z\ell} w = 0\,, \quad V_z - k_{zr} w = 0\,; \quad M_y - k_{\gamma\ell}\varphi_y = 0\,, \quad M_y + k_{\gamma r}\varphi_y = 0\,. \tag{7.76}$$

Table 7.2 Boundary conditions for some beams

Support arrangements	Boundary conditions	
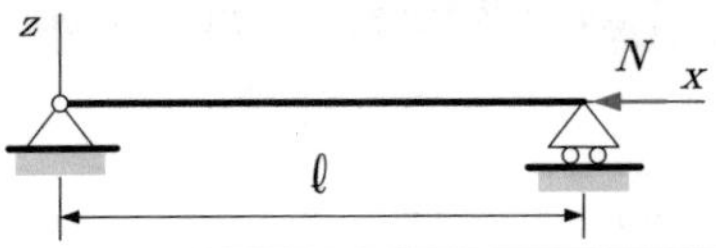	$w(0) = 0$ $M_y(0) = 0$	$w(\ell) = 0$ $M_y(\ell) = 0$
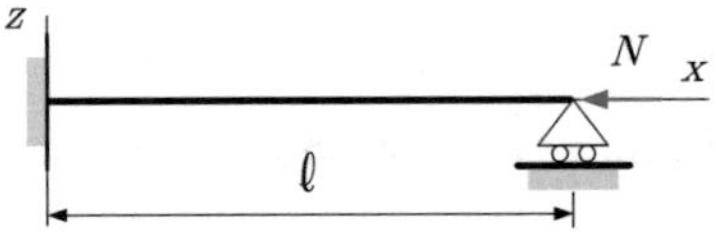	$w(0) = 0$ $\varphi_y(0) = 0$	$w(\ell) = 0$ $M_y(\ell) = 0$
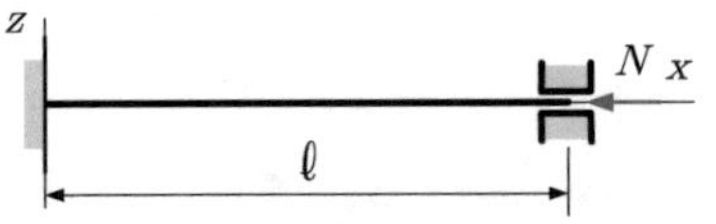	$w(0) = 0$ $\varphi_y(0) = 0$	$w(\ell) = 0$ $\varphi_y(\ell) = 0$
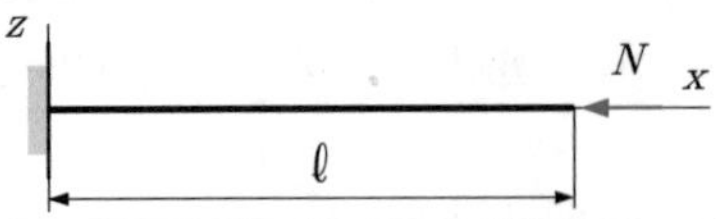	$w(0) = 0$ $\varphi_y(0) = 0$	$M_y(\ell) = 0$ $V_z(\ell) = 0$

7.5.2 *Equation of Motion and Solutions for the Axially Unloaded Beam*

7.5.2.1 Equation of Motion

If the beam is unloaded, the axial force N is zero and the distributed load f_z is the force of inertia:

$$f_z = -\rho A \frac{\partial^2 w}{\partial t^2}\,. \tag{7.77}$$

Upon substitution of Eq. (7.77) into equilibrium equation (7.75a) we get the equation of motion

$$\frac{\partial^2}{\partial x^2}\left(IE\frac{\partial^2 w}{\partial x^2}\right) + \rho A \frac{\partial^2 w}{\partial t^2} = 0, \tag{7.78}$$

which can be rewritten in the following form:

$$c^2\frac{\partial^4 w}{\partial x^4} + \frac{\partial^2 w}{\partial t^2} = 0\,, \qquad c^2 = \frac{IE}{\rho A} > 0 \tag{7.79}$$

if the beam is uniform.

7.5.2.2 Solutions

By applying the method of separation of variables, we may write

$$w(x,t) = W(x)\gamma(t)\,. \tag{7.80}$$

Thus

$$\frac{c^2}{W(x)}\frac{\mathrm{d}^4 W(x)}{\mathrm{d}x^4} = -\frac{1}{\gamma(t)}\frac{\mathrm{d}^2\gamma(t)}{\mathrm{d}t^2} = \omega^2 \tag{7.81}$$

in which ω^2 is a constant (It will turn out later that ω is the circular frequency of the vibrations). Equation (7.81) can be rewritten in the form

$$\frac{\mathrm{d}^4 W(x)}{\mathrm{d}x^4} = \lambda W(x), \quad \lambda = \left(\frac{\omega}{c}\right)^2 = \frac{\rho A\omega^2}{IE} = \beta^4; \quad \frac{\mathrm{d}^2\gamma(t)}{\mathrm{d}t^2} + \omega^2\gamma(t) = 0. \tag{7.82}$$

The general solution of Eq. $(7.82)_3$ is given by

$$\gamma(t) = \mathcal{C}\sin\omega t + \mathcal{D}\cos\omega t\,. \tag{7.83}$$

As regards Eq. $(7.82)_1$, we shall assume that

$$W(x) = \mathcal{A}\,e^{\chi x} \tag{7.84}$$

is a particular solution where $\mathcal{A}$ and χ are constants. If we substitute it into $(7.82)_1$, we get the characteristic equation

$$\chi^4 - \lambda = \chi^4 - \beta^4 = 0. \tag{7.85}$$

Hence

$$\chi_1 = \beta, \quad \chi_2 = -\beta, \quad \chi_3 = i\beta, \quad \chi_4 = -i\beta; \quad \beta = \sqrt[4]{\lambda} = \sqrt{\frac{\omega}{c}} = \sqrt[4]{\frac{\rho A \omega^2}{IE}} \tag{7.86a}$$

in terms of which

$$\begin{aligned} W(x) &= \mathcal{A}_1\, e^{\beta x} + \mathcal{A}_2\, e^{-\beta x} + \mathcal{A}_3\, e^{i\beta x} + \mathcal{A}_4\, e^{-i\beta x} = \\ &= \mathcal{A}_1\, e^{\beta x} + \mathcal{A}_2\, e^{-\beta x} + \mathcal{A}_3\, (\cos\beta x + i \sin\beta x) + \mathcal{A}_4\, (\cos\beta x - i \sin\beta x) \end{aligned} \tag{7.86b}$$

is the complex general solution to the differential equation $(7.82)_1$ in which $\mathcal{A}_1$, $\mathcal{A}_2$, $\mathcal{A}_3$, and $\mathcal{A}_4$ are undetermined integration constants. Since the real and imaginary parts in Eq. (7.86b) are real and linearly independent particular solutions of the differential equation $(7.82)_1$ it follows that the real general solution can be given as a linear combination of these particular solutions:

$$W(x) = \mathcal{A}_1 \cos\beta x + \mathcal{A}_2 \sin\beta x + \mathcal{A}_3 \cosh\beta x + \mathcal{A}_4 \sinh\beta x\,, \tag{7.87}$$

or

$$\begin{aligned} W(x) = \mathcal{A}_1 \frac{1}{2}(\cosh\beta x + \cos\beta x) + \mathcal{A}_2 \frac{1}{2}(\sinh\beta x + \sin\beta x) + \\ + \mathcal{A}_3 \frac{1}{2}(\cosh\beta x - \cos\beta x) + \mathcal{A}_4 \frac{1}{2}(\sinh\beta x - \sin\beta x) = \\ = \mathcal{A}_1\, S(\beta x) + \mathcal{A}_2\, T(\beta x) + \mathcal{A}_3\, U(\beta x) + \mathcal{A}_4\, V(\beta x) \end{aligned} \tag{7.88a}$$

where the functions

$$\begin{aligned} S(\beta x) &= \frac{1}{2}(\cosh\beta x + \cos\beta x)\,, \\ T(\beta x) &= \frac{1}{2}(\sinh\beta x + \sin\beta x)\,, \\ U(\beta x) &= \frac{1}{2}(\cosh\beta x - \cos\beta x)\,, \\ V(\beta x) &= \frac{1}{2}(\sinh\beta x - \sin\beta x) \end{aligned} \tag{7.88b}$$

are called Krylov–Duncan functions [4–6].

We remark that $\mathcal{A}_1$, $\mathcal{A}_2$, $\mathcal{A}_3$, and $\mathcal{A}_4$ are, in each case, different constants. They can be determined by utilizing the boundary conditions.

7.5.2.3 Properties of the Krylov–Duncan Functions

It can be checked by making paper-and-pencil calculations that the Krylov–Duncan functions and their derivatives at $x = 0$ form a unit matrix shown in Table 7.3:

Table 7.3 The unit matrix that belongs to the Krylov–Duncan functions

Krylov–Duncan functions and their derivatives at $x = 0$			
$S(0) = 1$	$\left.\frac{\mathrm{d}}{\beta \mathrm{d}x} S(\beta x)\right\|_{x=0} = 0$	$\left.\frac{\mathrm{d}^2}{\beta^2 \mathrm{d}x^2} S(\beta x)\right\|_{x=0} = 0$	$\left.\frac{\mathrm{d}^3}{\beta^3 \mathrm{d}x^3} S(\beta x)\right\|_{x=0} = 0$
$T(0) = 0$	$\left.\frac{\mathrm{d}}{\beta \mathrm{d}x} T(\beta x)\right\|_{x=0} = 1$	$\left.\frac{\mathrm{d}^2}{\beta^2 \mathrm{d}x^2} T(\beta x)\right\|_{x=0} = 0$	$\left.\frac{\mathrm{d}^3}{\beta^3 \mathrm{d}x^3} T(\beta x)\right\|_{x=0} = 0$
$U(0) = 0$	$\left.\frac{\mathrm{d}}{\beta \mathrm{d}x} U(\beta x)\right\|_{x=0} = 0$	$\left.\frac{\mathrm{d}^2}{\beta^2 \mathrm{d}x^2} U(\beta x)\right\|_{x=0} = 1$	$\left.\frac{\mathrm{d}^3}{\beta^3 \mathrm{d}x^3} U(\beta x)\right\|_{x=0} = 0$
$V(0) = 0$	$\left.\frac{\mathrm{d}}{\beta \mathrm{d}x} V(\beta x)\right\|_{x=0} = 0$	$\left.\frac{\mathrm{d}^2}{\beta^2 \mathrm{d}x^2} V(\beta x)\right\|_{x=0} = 0$	$\left.\frac{\mathrm{d}^3}{\beta^3 \mathrm{d}x^3} V(\beta x)\right\|_{x=0} = 1$

For the sake of completeness, the first four derivatives of the Krylov–Duncan functions are presented in Table 7.4.

Table 7.4 Derivative rules

Derivatives of the Krylov–Duncan functions				
$S(\beta x)$	$S^{(1)}(\beta x) = \beta V(\beta x)$	$S^{(2)}(\beta x) = \beta^2 U(\beta x)$	$S^{(3)}(\beta x) = \beta^3 T(\beta x)$	$S^{(4)}(\beta x) = \beta^4 S(\beta x)$
$T(\beta x)$	$T^{(1)}(\beta x) = \beta S(\beta x)$	$T^{(2)}(\beta x) = \beta^2 V(\beta x)$	$T^{(3)}(\beta x) = \beta^3 U(\beta x)$	$T^{(4)}(\beta x) = \beta^4 T(\beta x)$
$U(\beta x)$	$U^{(1)}(\beta x) = \beta T(\beta x)$	$U^{(2)}(\beta x) = \beta^2 S(\beta x)$	$U^{(3)}(\beta x) = \beta^3 V(\beta x)$	$U^{(4)}(\beta x) = \beta^4 U(\beta x)$
$V(\beta x)$	$V^{(1)}(\beta x) = \beta U(\beta x)$	$V^{(2)}(\beta x) = \beta^2 T(\beta x)$	$V^{(3)}(\beta x) = \beta^3 S(\beta x)$	$V^{(4)}(\beta x) = \beta^4 V(\beta x)$

Making use of the formulae detailed in Table 7.3, the integration constants $\mathcal{A}_1$, $\mathcal{A}_2$, $\mathcal{A}_3$, and $\mathcal{A}_4$ in (7.88a) can all be given in terms of the initial parameters:

$$\begin{aligned} \mathcal{A}_1 &= W(\beta x)\big|_{x=0}, & \mathcal{A}_2 &= \frac{1}{\beta}\left.\frac{\mathrm{d}W(\beta x)}{\mathrm{d}x}\right|_{x=0}, \\ \mathcal{A}_3 &= \frac{1}{\beta^2}\left.\frac{\mathrm{d}^2 W(\beta x)}{\mathrm{d}x^2}\right|_{x=0}, & \mathcal{A}_4 &= \frac{1}{\beta^3}\left.\frac{\mathrm{d}^3 W(\beta x)}{\mathrm{d}x^3}\right|_{x=0}. \end{aligned} \tag{7.89}$$

On the basis of Table 7.4, we can give the solution and its derivatives in an ordered form:

$$\begin{aligned} W(x) &= \mathcal{A}_1\, S(\beta x) + \mathcal{A}_2\, T(\beta x) + \mathcal{A}_3\, U(\beta x) + \mathcal{A}_4\, V(\beta x)\,, \\ \frac{\mathrm{d}W(x)}{\mathrm{d}x} &= \beta\,(\mathcal{A}_1\, V(\beta x) + \mathcal{A}_2\, S(\beta x) + \mathcal{A}_3\, T(\beta x) + \mathcal{A}_4\, U(\beta x))\,, \\ \frac{\mathrm{d}^2W(x)}{\mathrm{d}x^2} &= \beta^2\,(\mathcal{A}_1\, U(\beta x) + \mathcal{A}_2\, V(\beta x) + \mathcal{A}_3\, S(\beta x) + \mathcal{A}_4\, T(\beta x))\,, \\ \frac{\mathrm{d}^3W(x)}{\mathrm{d}x^3} &= \beta^3\,(\mathcal{A}_1\, T(\beta x) + \mathcal{A}_2\, U(\beta x) + \mathcal{A}_3\, V(\beta x) + \mathcal{A}_4\, S(\beta x))\,. \end{aligned} \tag{7.90}$$

For the sake of our later considerations, we give some useful relations here:

$$\begin{aligned} T^2(\beta x) - V^2(\beta x) &= \sinh\beta x \sin\beta x\,, \\ U(\beta x)T(\beta x) - V(\beta x)S(\beta x) &= \frac{1}{2}\,(\cosh\beta x \sin\beta x - \cos\beta x \sinh\beta x)\,, \\ U^2(\beta x) - V(\beta x)T(\beta x) &= \frac{1}{2}\,(1 - \cosh\beta x \cos\beta x)\,, \\ S^2(\beta x) - T(\beta x)V(\beta x) &= \frac{1}{2}\,(\cosh\beta x \cos\beta x + 1)\,. \end{aligned} \tag{7.91}$$

7.5.2.4 Boundary Conditions

By utilizing the data in Table 7.2 and Eqs. $(7.74)_{1,2}$, we can give the common boundary conditions in terms of the deflection W:

1. For a fixed (clamped) end, the deflection and the rotation are equal to zero:

$$W = 0\,, \qquad \varphi_y = -\frac{\mathrm{d}W}{\mathrm{d}x} = 0\,. \tag{7.92a}$$

2. For a pinned endpoint, the deflection and the bending moment are equal to zero:

$$W = 0\,, \qquad M_y = -IE\frac{\mathrm{d}^2W}{\mathrm{d}x^2} = 0\,. \tag{7.92b}$$

3. For a free end, the bending moment and the shear force are equal to zero:

$$M_y = -IE\frac{\mathrm{d}^2W}{\mathrm{d}x^2} = 0\,, \qquad V_z = IE\frac{\mathrm{d}^3W}{\mathrm{d}x^3} = 0\,. \tag{7.92c}$$

We shall assume in accordance with Table 7.2 that $x = 0$ at the left end of the beam with length ℓ. Comparison of the boundary conditions considered at the left end of the beam, the first column in Table 7.3, and Eqs. (7.90) shows that two from the integration constants $\mathcal{A}_1$, $\mathcal{A}_2$, $\mathcal{A}_3$, and $\mathcal{A}_4$ are always zero.

7.5.2.5 Eigenfrequencies of Simply Supported (Pinned-Pinned) Beams

By utilizing the boundary conditions $(7.92a)_1$, $(7.92b)_2$ taken at the left end of the beam, Eqs. $(7.90)_{1,3}$, and the first column in Table 7.3, we get

$$W(\beta x)|_{x=0} = \mathcal{A}_1 = 0\,, \qquad \left.\frac{\mathrm{d}^2 W(\beta x)}{\mathrm{d}x^2}\right|_{x=0} = \beta^2 \mathcal{A}_3 = 0, \tag{7.93}$$

which shows that the integration constants $\mathcal{A}_1$ and $\mathcal{A}_3$ are equal to zero. Hence

$$W(\beta x) = \mathcal{A}_2\, T(\beta x) + \mathcal{A}_4\, V(\beta x) \tag{7.94}$$

is the solution. Boundary conditions $(7.92a)_1$, $(7.92b)_2$ taken now at the left end of the beam yield two homogeneous linear equations for the integration constants $\mathcal{A}_2$ and $\mathcal{A}_4$:

$$\begin{aligned} W(\beta x)|_{x=\ell} &= \mathcal{A}_2\, T(\beta\ell) + \mathcal{A}_4\, V(\beta\ell) = 0\,, \\ \left.\frac{\mathrm{d}^2 W(\beta x)}{\mathrm{d}x^2}\right|_{x=\ell} &= \beta^2 \left(\mathcal{A}_2\, V(\beta\ell) + \mathcal{A}_4\, T(\beta\ell)\right) = 0 \end{aligned}$$

or

$$\begin{aligned} \mathcal{A}_2\, T(\beta\ell) + \mathcal{A}_4\, V(\beta\ell) &= 0\,, \\ \mathcal{A}_2\, V(\beta\ell) + \mathcal{A}_4\, T(\beta\ell) &= 0\,. \end{aligned} \tag{7.95}$$

Solutions for $\mathcal{A}_2$ and $\mathcal{A}_4$ different from the trivial ones exist if and only if the determinant of this equation system vanishes:

$$\begin{vmatrix} T(\beta\ell) & V(\beta\ell) \\ V(\beta\ell) & T(\beta\ell) \end{vmatrix} = T^2(\beta\ell) - V^2(\beta\ell) = \underset{(7.91)_1}{\uparrow} = \sinh\beta\ell\,\sin\beta\ell = 0\,. \tag{7.96}$$

This equation is the frequency equation (or characteristic determinant). Its roots are given by

$$\beta_k\ell = k\pi\,, \qquad k = 1, 2, 3, \ldots \tag{7.97}$$

With β_n Eq. $(7.82)_2$ results in two formulae, one for the eigenvalue λ the other for the eigenfrequency ω:

$$\lambda_k = (\beta_k)^4 = \left(\frac{k\pi}{\ell}\right)^4, \qquad \omega_k = (\beta_k\ell)^2\sqrt{\frac{IE}{\rho A\ell^4}} = \left(\frac{k\pi}{\ell}\right)^2\sqrt{\frac{IE}{\rho A}}\,. \tag{7.98}$$

From now on the subscript $_k$ shows that the quantity considered belongs to the root β_k.

Since determinant (7.96) is zero, Eqs. (7.95) are not independent. Consequently, the integration constants are also not independent. From Eq. (7.95)$_1$, we get

$$\mathcal{A}_{4k} = -\mathcal{A}_{2k}\,\frac{T(\beta_k\ell)}{V(\beta_k\ell)} = -\mathcal{A}_{2k}\,\frac{\sinh\beta_k x + \sin\beta_k x}{\sinh\beta_k x - \sin\beta_k x} = -\mathcal{A}_{2k}\,.$$

Hence

$$\begin{aligned} W(\beta_k x) &= \mathcal{A}_{2k}\,\left(T_k(\beta_k x) - V_k(\beta_k x)\right) = \\ &= \mathcal{A}_{2k}\left(\frac{1}{2}\,(\sinh\beta_k x + \sin\beta_k x) - \frac{1}{2}\,(\sinh\beta_k x - \sin\beta_k x)\right) = \\ &= \mathcal{A}_{2k}\sin\beta_k x = \mathcal{A}_{2k}\sin\frac{k\pi}{\ell}x = \mathcal{A}_{2k}W_k(x), \end{aligned} \tag{7.99}$$

where the function

$$W_k(x) = \sin\frac{k\pi}{\ell}x \tag{7.100}$$

is the kth eigenfunction (or the kth normal mode) that belongs to the eigenvalue $\lambda_k = (\beta_k)^4$.

Remark 7.3 Note that eigenfunction (7.100) coincides with eigenfunction (7.47) which describes the kth normal mode of a vibrating string.

Let us assume that the initial conditions are given by the following equations:

$$w(x,0) = \mathfrak{w}(x)\,, \qquad \dot{w}(x,0) = \mathfrak{g}(x), \tag{7.101}$$

where $\mathfrak{w}(x)$ and $\mathfrak{g}(x)$ are given functions. By repeating the line of thought presented in Exercise 7.3, it is easy to prove that

$$w(x,t) = \sum_{k=1}^{\infty}\sin\lambda_k x\;(\mathcal{C}_k\sin\omega_k t + \mathcal{D}_k\cos\omega_k t) \tag{7.102}$$

is the total solution where

$$\mathcal{D}_k = \frac{2}{\ell}\int_0^\ell \mathfrak{w}(x)W_k(x)\,\mathrm{d}x \quad \text{and} \quad \mathcal{C}_k = \frac{2}{\ell\omega_k}\int_0^\ell \mathfrak{g}(x)W_k(x)\,\mathrm{d}x\,. \tag{7.103}$$

Table 7.5 contains the frequency equations, the eigenfunctions, and the values (formulae) of the product $\beta_k\ell$ for six support arrangements. Proofs have been given only for simply supported and fixed-pinned beams in Sect. 7.5.2.5 and Exercise 7.4. Proofs for the other four cases are left to Problem 7.4.

Table 7.5 Frequency equations, eigenfunctions and the values of $\beta_k \ell$

Beam with its supports	Frequency equation	Eigenfunction $W_k(x)$	Value of $\beta_k \ell$ $k = 1, 2, \ldots$
Simply supported 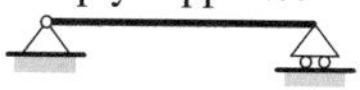	$\sin \beta_k \ell = 0$	$\sin \beta_k x$	$\beta_k \ell = k\pi$
Fixed-pinned 	$\tan \beta_k \ell = \tanh \beta_k \ell$	$\cosh \beta_k x - \cos \beta_k x -$ $-b_k (\sinh \beta_k x - \sin \beta_k x)$ $b_k = \dfrac{\cosh \beta_k \ell - \cos \beta_k \ell}{\sinh \beta_k \ell - \sin \beta_k \ell}$	$\beta_1 \ell = 3.92660$ $\beta_2 \ell = 7.06858$ $\beta_k \ell \approx \frac{\pi}{4} + k\pi$
Fixed-fixed 	$\cosh \beta_k \ell \cos \beta_k \ell = 1$	$\cosh \beta_k x - \cos \beta_k x -$ $-b_k (\sinh \beta_k x - \sin \beta_k x)$ $b_k = \dfrac{\cosh \beta_k \ell - \cos \beta_k \ell}{\sinh \beta_k \ell - \sin \beta_k \ell}$	$\beta_1 \ell = 4.73004$ $\beta_2 \ell = 7.85321$ $\beta_k \ell \approx \frac{\pi}{2} + k\pi$
Fixed-free 	$\cosh \beta_k \ell \cos \beta_k \ell = -1$	$\cosh \beta_k x - \cos \beta_k x -$ $-b_k (\sinh \beta_k x - \sin \beta_k x)$ $b_k = \dfrac{\cosh \beta_k \ell + \cos \beta_k \ell}{\sinh \beta_k \ell + \sin \beta_k \ell}$	$\beta_1 \ell = 1.87510$ $\beta_2 \ell = 4.69409$ $\beta_k \ell \approx (k - 0.5)\pi$
Pinned-free 	$\tan \beta_k \ell = \tanh \beta_k \ell$	$\sinh \beta_k x + b_k \sin \beta_k x$ $b_k = \dfrac{\sinh \beta_k \ell}{\sin \beta_k \ell}$	$\beta_1 \ell = 3.92660$ $\beta_2 \ell = 7.06858$ $\beta_k \ell \approx \frac{\pi}{4} + k\pi$ 0 for rigid body rotation
Free-free 	$\cosh \beta_k \ell \cos \beta_k \ell = 1$	$\cosh \beta_k x + \cos \beta_k x -$ $-b_k (\sinh \beta_k x + \sin \beta_k x)$ $b_k = \dfrac{\cosh \beta_k \ell - \cos \beta_k \ell}{\sinh \beta_k \ell - \sin \beta_k \ell}$	$\beta_1 \ell = 4.73014$ $\beta_2 \ell = 7.85321$ $\beta_k \ell \approx \frac{\pi}{2} + k\pi$ 0 for rigid body motion

It is also worth to mention that the relative error of the approximate relations presented in the last column of Table 7.5 reaches its maximum for the fixed-free beam if $k = 3$:

$$\delta_{max} = \left[1 - \frac{(\beta_3 \ell)_{\text{exact}}}{(\beta_3 \ell)_{\text{approximate}}} \right] \times 100 =$$

$$= \left(1 - \frac{7.854557}{2.5 \times \pi} \right) \times 100 = \left(1 - \frac{7.854557}{7.853982} \right) \times 100 = 0.0098\% . \quad (7.104)$$

Exercise 7.4 Find the eigenfrequencies of a beam with length ℓ if the beam is fixed at the left end ($x = 0$) and pinned at the right end ($x = \ell$)—see Table 7.2 which shows, among others, this beam as well.

The solution is based on Sect. 7.5.2.5. Making use of the boundary conditions

$$W(\beta x)|_{x=0} = \mathcal{A}_1 = 0\,, \quad \left.\frac{\mathrm{d}W(\beta x)}{\mathrm{d}x}\right|_{x=0} = \beta \mathcal{A}_2 = 0\,, \tag{7.105a}$$

$$W(\beta x)|_{x=\ell} = \mathcal{A}_3\, U(\beta\ell) + \mathcal{A}_4\, V(\beta\ell) = 0\,, \tag{7.105b}$$

$$\left.\frac{\mathrm{d}^2 W(\beta x)}{\mathrm{d}x^2}\right|_{x=\ell} = \beta^2\,(\mathcal{A}_3\, S(\beta\ell) + \mathcal{A}_4\, T(\beta\ell)) = 0, \tag{7.105c}$$

we obtain the following equation system for the non-zero integration constants $\mathcal{A}_3$ and $\mathcal{A}_4$:

$$\begin{aligned} \mathcal{A}_3\, U(\beta\ell) + \mathcal{A}_4\, V(\beta\ell) &= 0\,, \\ \mathcal{A}_3\, S(\beta\ell) + \mathcal{A}_4\, T(\beta\ell) &= 0\,. \end{aligned} \tag{7.106}$$

We can get non-zero solutions for the integration constants $\mathcal{A}_3$ and $\mathcal{A}_4$ if the frequency equation (characteristic determinant) is zero:

$$\begin{vmatrix} U(\beta\ell) & V(\beta\ell) \\ S(\beta\ell) & T(\beta\ell) \end{vmatrix} = U(\beta\ell)T(\beta\ell) - V(\beta\ell)S(\beta\ell) = \underset{(7.91)_2}{\uparrow} =$$

$$= \frac{1}{2}\,(\cosh\beta\ell \sin\beta\ell - \cos\beta\ell \sinh\,\beta\ell) = 0\,. \tag{7.107}$$

It follows from here that

$$\tan\beta\ell = \tanh\beta\ell\,. \tag{7.108}$$

The roots of this equation are given by

$$\beta_k \ell \approx \frac{\pi}{4} + k\pi\,. \tag{7.109a}$$

For completeness, we list the first four roots here:

$$\beta_1\ell = 3.926602\,, \quad \beta_2\ell = 7.068583\,, \quad \beta_3\ell = 10.210176\,, \quad \beta_4\ell = 13.351768\,. \tag{7.109b}$$

According to Eq. (7.98), the kth eigenfrequency of the beam is

$$\omega_k = (\beta_k)^2 \sqrt{\frac{IE}{\rho A}}\,. \tag{7.110}$$

The integration constants $\mathcal{A}_3$ and $\mathcal{A}_4$ are not independent of each other. Equation (7.106)$_1$ yields

$$\mathcal{A}_{4k} = -\mathcal{A}_{3k}\frac{U(\beta_k\ell)}{V(\beta_k\ell)} = -\mathcal{A}_{3k}\frac{\cosh\beta_k\ell - \cos\beta_k\ell}{\sinh\beta_k\ell - \sin\beta_k\ell}\,. \tag{7.111}$$

Consequently,

$$\begin{aligned} W(\beta_n x) &= \mathcal{A}_{3k}\left(U(\beta_k x) - \frac{U(\beta_k\ell)}{V(\beta_k\ell)}V(\beta_k x)\right) = \\ &= \mathcal{A}_{3k}\frac{1}{2}\left(\cosh\beta_k x - \cos\beta_k x - \frac{\cosh\beta_k\ell - \cos\beta_k\ell}{\sinh\beta_k\ell - \sin\beta_k\ell}(\sinh\beta_k x - \sin\beta_k x)\right) = \\ &= \mathcal{A}_{3k}W_k(x), \end{aligned} \tag{7.112}$$

where

$$\begin{aligned} W_k(x) &= U(\beta_k x) - \frac{U(\beta_k\ell)}{V(\beta_k\ell)}V(\beta_k x) = \\ &= \cosh\beta_k x - \cos\beta_k x - \frac{\cosh\beta_k\ell - \cos\beta_k\ell}{\sinh\beta_k\ell - \sin\beta_k\ell}(\sinh\beta_k x - \sin\beta_k x) \end{aligned} \tag{7.113}$$

is the kth eigenfunction.

7.5.3 Orthogonality of the Eigenfunctions

Let us introduce the notations $y(x) = W(x)$, $y_k(x) = W_k(x)$ and the operators $K(y) = y^{(4)}(x)$ and $M(y) = y$. Further, let $x = a = 0$ and $x = b = \ell$ be those x coordinates that belong to the left and right extremities of the beam. Making use of the notations introduced differential equation (7.82) can be rewritten in the following form:

$$K[y(x)] = \lambda M[y(x)]\,. \tag{7.114}$$

This differential equation is, in general, associated with the following boundary conditions:

$$\left.\begin{array}{l} y(a) = 0\,,\ y^{(2)}(a) = 0 \\ y(b) = 0\,,\ y^{(2)}(b) = 0 \end{array}\right\} \quad \text{(simply supported beam)}, \tag{7.115a}$$

$$\left.\begin{array}{l} y(a) = 0\,,\ y^{(1)}(a) = 0 \\ y(b) = 0\,,\ y^{(2)}(b) = 0 \end{array}\right\} \quad \text{(fixed-pinned beam)}, \tag{7.115b}$$

$$\left.\begin{array}{l} y(a) = 0\,,\ y^{(1)}(a) = 0 \\ y(b) = 0\,,\ y^{(1)}(b) = 0 \end{array}\right\} \quad \text{(fixed-fixed beam)}, \tag{7.115c}$$

$$\left.\begin{aligned} y(a)=0\,,\ y^{(1)}(a)=0 \\ y^{(2)}(b)=0\,,\ y^{(3)}(b)=0 \end{aligned}\right\} \qquad \text{(fixed-free beam)}, \tag{7.115d}$$

$$\left.\begin{aligned} y(a)=0\,,\ y^{(2)}(a)=0 \\ y^{(2)}(b)=0\,,\ y^{(3)}(b)=0 \end{aligned}\right\} \qquad \text{(pinned-free beam)}, \tag{7.115e}$$

$$\left.\begin{aligned} y(a)^{(2)}=0\,,\ y^{(3)}(a)=0 \\ y^{(2)}(b)=0\,,\ y^{(3)}(b)=0 \end{aligned}\right\} \qquad \text{(free-free beam)}. \tag{7.115f}$$

Differential equation (7.114) and boundary conditions (7.115) constitute six boundary value problems for which both the differential equation and the boundary conditions are homogeneous, and λ is an unknown parameter. These problems are referred to, in accordance with all that has been said so far, as eigenvalue problems with lambda as the eigenvalue.

The integrals

$$\left(\int_a^b y_n(x)K[y_k(x)]\,\mathrm{d}x=(y_n,\,y_k)_K\right) \tag{7.116a}$$

$$\left[\int_a^b y_n(x)M[y_k(x)]\,\mathrm{d}x=(y_n,\,y_k)_M\right] \tag{7.116b}$$

define the products of the eigenfunctions $y_n(x)$ and $y_k(x)$ taken on the operator $(K(y))$ $[M(y)]$.

As regards the product $(y_n,\,y_k)_K$, we may write

$$\begin{aligned} (y_n,\,y_k)_K &= \int_a^b y_n(x)\frac{\mathrm{d}^4 y_k(x)}{\mathrm{d}x^4}\mathrm{d}x = y_n(x)\frac{\mathrm{d}^3 y_k(x)}{\mathrm{d}x^3}\bigg|_a^b - \int_a^b \frac{\mathrm{d}y_n(x)}{\mathrm{d}x}\frac{\mathrm{d}^3 y_k(x)}{\mathrm{d}x^3}\mathrm{d}x = \\ &= \underbrace{y_n(x)\frac{\mathrm{d}^3 y_k(x)}{\mathrm{d}x^3}\bigg|_a^b}_{=0} - \underbrace{\frac{\mathrm{d}y_n(x)}{\mathrm{d}x}\frac{\mathrm{d}^2 y_k(x)}{\mathrm{d}x^2}\bigg|_a^b}_{=0} + \int_a^b \frac{\mathrm{d}^2 y_n(x)}{\mathrm{d}x^2}\frac{\mathrm{d}^2 y_k(x)}{\mathrm{d}x^2}\mathrm{d}x = \\ &= \int_a^b \frac{\mathrm{d}^2 y_n(x)}{\mathrm{d}x^2}\frac{\mathrm{d}^2 y_k(x)}{\mathrm{d}x^2}\mathrm{d}x\,, \qquad k,n=1,2,3,\ldots \end{aligned} \tag{7.117}$$

in which due to boundary conditions (7.115) the underbraced terms are equal to zero. Consequently, it holds that

$$(y_n,\,y_k)_K = \int_a^b \frac{\mathrm{d}^2 y_n(x)}{\mathrm{d}x^2}\frac{\mathrm{d}^2 y_k(x)}{\mathrm{d}x^2}\mathrm{d}x = \int_a^b \frac{\mathrm{d}^2 y_k(x)}{\mathrm{d}x^2}\frac{\mathrm{d}^2 y_n(x)}{\mathrm{d}x^2}\mathrm{d}x = (y_k,\,y_n)_K\,. \tag{7.118}$$

Hence, the product $(y_n,\,y_k)_K$ is a commutative operation:

$$(y_n,\,y_k)_K = (y_k,\,y_n)_K\,.$$

According to its definition

$$(y_n, y_k)_M = \int_a^b y_n(x) y_k(x) \mathrm{d}x = (y_k, y_n)_M \,, \tag{7.119}$$

which shows that the product $(y_n, y_k)_M$ is also a commutative operation: $(y_n, y_k)_M = (y_k, y_n)_M$.

Since the eigenfunctions $y_k(x)$ and $y_n(x)$ satisfy differential equation (7.114), we may write

$$\frac{\mathrm{d}^4 y_k(x)}{\mathrm{d}x^4} = \lambda_k y_k(x) \qquad \text{and} \qquad \frac{\mathrm{d}^4 y_n(x)}{\mathrm{d}x^4} = \lambda_n y_n(x) \,. \tag{7.120}$$

Multiply the first equation by $y_n(x)$, the second by $y_k(x)$ then integrate the result over the interval $[a, b]$. With regard to definitions (7.116), we have

$$(y_n, y_k)_K = \lambda_k \, (y_n, y_k)_M \,, \tag{7.121a}$$

$$(y_k, y_n)_K = \lambda_n \, (y_k, y_n)_M \,. \tag{7.121b}$$

Subtract now Eq. (7.121b) from Eq. (7.121a). In a view of the commutativity of the products $(y_n, y_k)_K$ and $(y_n, y_k)_M$, we obtain

$$(\lambda_k - \lambda_n) \, (y_k, y_n)_M = (\lambda_k - \lambda_n) \int_a^b y_n(x) y_k(x) \mathrm{d}x = 0, \tag{7.122}$$

where we have assumed that $\lambda_k - \lambda_n \neq 0$ if $k \neq n$. This means that

$$\int_a^b y_n(x) y_k(x) \mathrm{d}x = 0 \,, \qquad k \neq n \,. \tag{7.123}$$

If this equation is satisfied then we say—in accordance with what we said concerning Eq. (7.22)—that the functions $y_n(x)$ and $y_k(x)$ are orthogonal to each other.

We define the norm of the eigenfunction $y_k(x)$ by the following equation:

$$|y_k(x)| = \sqrt{\int_a^b y_k(x) y_k(x) \, \mathrm{d}x} \,. \tag{7.124}$$

Comparison of Eqs. (7.122) and (7.124) yields

$$\boxed{\int_a^b y_n(x) y_k(x) \mathrm{d}x = \begin{cases} |y_k(x)|^2 & \text{if } k = n, \\ 0 & \text{if } k \neq n. \end{cases} \qquad k, n = 1, 2, 3 \ldots} \tag{7.125}$$

Equation

$$\psi_k(x) = \frac{y_k(x)}{|y_k(x)|} \tag{7.126}$$

defines the normalized eigenfunction $\psi_k(x)$. Since $\psi_k(x)$ is normalized, it holds that

$$\int_a^b \psi_k(x)\psi_k(x)\,\mathrm{d}x = 1\,. \tag{7.127}$$

Hence

$$\boxed{\int_a^b \psi_n(x)\psi_k(x)\mathrm{d}x = \begin{cases} 1 \text{ if } k = n, \\ 0 \text{ if } k \neq n. \end{cases} \quad k, n = 1, 2, 3 \ldots} \tag{7.128}$$

It follows from Eqs. (7.117) and (7.119) that

$$(y_n, y_n)_K > 0\,, \qquad (y_n, y_n)_M > 0\,. \tag{7.129}$$

If we make k and n equal in (7.121), we get

$$(y_n, y_n)_K = \lambda_n\,(y_n, y_n)_M\,. \tag{7.130}$$

Consequently

$$\lambda_n = \frac{(y_n, y_n)_K}{(y_n, y_n)_M} > 0, \tag{7.131}$$

which shows that the non-zero eigenvalues are all positive quantities.

7.5.4 *Forced Vibration of Beams*

If the beam is subjected to a time-dependent transverse load and vibrates—we assume that the axial force in the beam is zero—then the total load consists of two parts, the first is the force of inertia the second is the time-dependent transverse load:

$$f_y = -\rho A \frac{\partial^2 w}{\partial t^2} + f_y(x, t)\,. \tag{7.132}$$

If we substitute f_y into equilibrium equation (7.75a), we get the equation of motion for the case of forced vibrations:

$$\frac{\partial^2}{\partial x^2}\left(IE\frac{\partial^2 w}{\partial x^2}\right) + \rho A \frac{\partial^2 w}{\partial t^2} = f_y(x, t)\,. \tag{7.133a}$$

For a uniform beam, this equation is a bit simpler:

$$IE\frac{\partial^4 w}{\partial x^4} + \rho A \frac{\partial^2 w}{\partial t^2} = f_y(x, t)\,. \tag{7.133b}$$

When seeking solution to this problem, we shall utilize some results presented in Sect. 5.4.2 which is devoted to a similar problem, i.e., to the forced vibrations of systems with finite degrees of freedom.

Since the set of normalized eigenfunctions $\psi_k(x)$ is complete in the Infinite-dimensional function space, we shall assume that the solution is of the form

$$w(x,t) = \mathrm{q}_1(t)\psi_1(x) + \mathrm{q}_2(t)\psi_2(x) + \mathrm{q}_3(t)\psi_3(x) + \cdots = \\ = \sum_{k=1}^{\infty} \mathrm{q}_k(t)\psi_k(x), \tag{7.134}$$

where the functions $\mathrm{q}_k(t)$ are the unknowns. Substituting solution (7.134) into Eq. (7.133b), we get

$$IE \sum_{k=1}^{\infty} \frac{\mathrm{d}^4\psi_k(x)}{\mathrm{d}x^4}\mathrm{q}_k(t) + \rho A \sum_{k=1}^{\infty} \psi_k(x)\frac{\mathrm{d}^2\mathrm{q}_k(t)}{\mathrm{d}t^2} = f_y(x,t)\,. \tag{7.135}$$

Since the eigenfunctions $\psi_k(x)$ satisfy Eq. (7.82), it also holds that

$$\frac{\mathrm{d}^4\psi_k(x)}{\mathrm{d}x^4} - \lambda_k\psi_k(x) = 0\,, \qquad \lambda_k = \frac{\rho A\omega_k^2}{IE}\,. \tag{7.136}$$

We can rewrite Eq. (7.135) if we take these relations into account:

$$\sum_{k=1}^{\infty} \omega_k^2\psi_k(x)\mathrm{q}_k(t) + \sum_{k=1}^{\infty} \psi_k(x)\frac{\mathrm{d}^2\mathrm{q}_k(t)}{\mathrm{d}t^2} = \frac{f_y(x,t)}{\rho A}\,. \tag{7.137}$$

Multiply this equation by $\psi_n(x)$ and integrate the result over the interval $[a = 0, b = \ell]$. We get

$$\sum_{k=1}^{\infty} \omega_k^2 \int_a^b \psi_n(x)\psi_k(x)\mathrm{d}x\, \mathrm{q}_k(t) + \sum_{k=1}^{\infty} \int_a^b \psi_n(x)\psi_k(x)\mathrm{d}x\, \frac{\mathrm{d}^2\mathrm{q}_k(t)}{\mathrm{d}t^2} = \\ = \int_a^b \frac{f_y(x,t)}{\rho A}\psi_n(x)\,\mathrm{d}x\,. \tag{7.138}$$

Let us introduce the notations

$$\omega_n^2 = \hat{\lambda}_n\,, \qquad Q_n(t) = \int_a^b \frac{f_y(x,t)}{\rho A}\psi_n(x)\,\mathrm{d}x \tag{7.139}$$

and take relations (7.128) into account. Then, we obtain a differential equation for the function $\mathrm{q}_n(t)$:

$$\ddot{\mathrm{q}}_n(t) + \hat{\lambda}_n\mathrm{q}_n(t) = Q_n(t), \tag{7.140}$$

where $Q_n(t)$ is the generalized force that belongs to the function $\mathfrak{q}_n$. This equation coincides formally with Eq. (5.109) we derived for vibrating systems with finite degrees of freedom. Hence, its solution is given by Eq. (5.114b) which can be rewritten in the following form:

$$\begin{aligned}\mathfrak{q}_n = & \underbrace{a_n \cos\sqrt{\hat{\lambda}_n}\, t + b_n \sin\sqrt{\hat{\lambda}_n}\, t}_{\text{transient function}} + \underbrace{\int_{\tau=0}^{t} Q_n(\tau) \frac{1}{\sqrt{\hat{\lambda}_n}} \sin\sqrt{\hat{\lambda}_n}\,(t-\tau)\,\mathrm{d}\tau}_{\text{particular solution}} = \\ & = a_n \cos\omega_n t + b_n \sin\omega_n t + \frac{1}{\omega_n}\int_{\tau=0}^{t} Q_n(\tau)\sin\omega_n (t-\tau)\,\mathrm{d}\tau\,. \end{aligned} \tag{7.141}$$

Here, a_n and b_n are unknown integration constants. They can be determined by using the initial conditions.

Exercise 7.5 The beam shown in Fig. 7.10 is subjected to a harmonic excitation force

$$f_z(x,t) = F_1\,\delta(x-\ell_1)\sin\omega_f t, \tag{7.142}$$

where $\delta(\ell_1 - x)$ is the Dirac function. Find the particular solution, i.e., the steady-state vibrations of the beam.

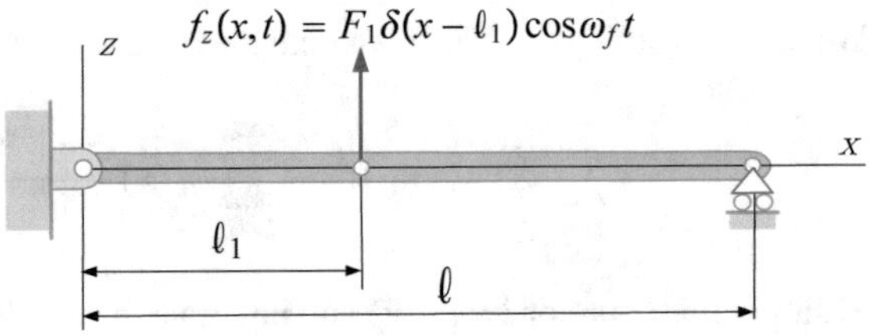

Fig. 7.10 Beam subjected to harmonic excitation

In view of (7.100), it is not too difficult to check that the normalized eigenfunctions for a simply supported beam are given by the following equation:

$$\psi_n(x) = \sqrt{\frac{2}{\ell}}\sin\frac{n\pi}{\ell}x\,. \tag{7.143}$$

Making use of Eq. (7.139), we can determine the generalized force $Q_n(t)$ for the function $\mathfrak{q}_n$:

$$\begin{aligned}Q_n(t) = \int_{a=0}^{b=\ell}\frac{f_y(x,t)}{\rho A}\psi_n(x)\,\mathrm{d}x = \frac{F_1}{\rho A}\sqrt{\frac{2}{\ell}}\int_0^{\ell}\delta(x-\ell_1)\cos\omega_f t\,\sin\frac{n\pi}{\ell}x,\,\mathrm{d}x = \\ = \frac{F_1}{\rho A}\sqrt{\frac{2}{\ell}}\sin\frac{n\pi\ell_1}{\ell}\cos\omega_f t\,. \end{aligned} \tag{7.144}$$

It follows from (7.141) that

$$\mathfrak{q}_{n\,\text{part}} = \frac{1}{\omega_n}\int_{\tau=0}^{t} Q_n(\tau)\sin\omega_n\,(t-\tau)\,\mathrm{d}\tau =$$
$$= \frac{F_1}{\rho A}\sqrt{\frac{2}{\ell}}\sin\frac{n\pi\ell_1}{\ell}\frac{1}{\omega_n}\int_0^t \cos\omega_f\,\tau\sin\omega_n\,(t-\tau)\,\mathrm{d}\tau,$$

where

$$\frac{1}{\omega_n}\int_0^t \cos\omega_f\,\tau\sin\omega_n\,(t-\tau)\,\mathrm{d}\tau = \frac{\cos\omega_f\,t}{\omega_n^2-\omega_f^2} + \frac{\cos\omega_n\,t}{\omega_n^2-\omega_f^2}\frac{\omega_f}{\omega_n}\,.$$

Note that the second term can be dropped since that is a solution to the homogeneous part of Eq. (7.140). Hence

$$\mathfrak{q}_{n\,\text{part}} = \frac{F_1}{\rho A}\sqrt{\frac{2}{\ell}}\sin\frac{n\pi\ell_1}{\ell}\frac{\cos\omega_f\,t}{\omega_n^2-\omega_f^2}\,.$$

It is worth mentioning that the same result can be obtained for $\mathfrak{q}_{n\,\text{part}}$ if we consider an undamped single degree of freedom system subjected to the harmonic excitation (7.144).

With $\mathfrak{q}_{n\,\text{part}}$, we may give the total solution:

$$w(x,t) = \frac{2F_1}{\rho A\ell}\sum_{n=1}^{\infty}\frac{1}{\omega_n^2-\omega_f^2}\sin\frac{n\pi\ell_1}{\ell}\sin\frac{n\pi x}{\ell}\cos\omega_f\,t\,.$$

7.5.5 Vibration of Axially Loaded Beams

Assume that the axial force N is constant (independent of x). Then, with regard to (7.77), the equation of motion follows from Eq. (7.75b):

$$IE\frac{\partial^4 w}{\partial x^4} \mp N\frac{\partial^2 w}{\partial x^2} + \rho A\frac{\partial^2 w}{\partial t^2} = 0, \tag{7.145}$$

where N is now the magnitude of the axial force. Hence, the sign of N in the above equation is (negative) [positive] if the axial force is (tensile) [compressive] force. It is obvious that the solution has the same mathematical form as that of Eq. (7.78) which is given by Eq. (7.80). Thus

$$\frac{c^2}{W(x)}\left(\frac{\mathrm{d}^4W(x)}{\mathrm{d}x^4} \mp \frac{N}{IE}\frac{\mathrm{d}^2W(x)}{\mathrm{d}x^2}\right) = -\frac{1}{\gamma(t)}\frac{\mathrm{d}^2\gamma(t)}{\mathrm{d}t^2} = \tilde{\omega}^2\,,\quad c^2 = \frac{IE}{\rho A}, \tag{7.146}$$

from where we have

$$\frac{\mathrm{d}^4 W(x)}{\mathrm{d}x^4} \mp \frac{N}{IE}\frac{\mathrm{d}^2 W(x)}{\mathrm{d}x^2} = \left(\frac{\tilde{\omega}}{c}\right)^2 W(x) = \lambda W(x)\,,$$
$$\lambda = \frac{\rho A\tilde{\omega}^2}{IE} = \beta^4\,; \qquad \frac{\mathrm{d}^2\gamma(t)}{\mathrm{d}t^2} + \tilde{\omega}^2\gamma(t) = 0\,. \tag{7.147}$$

We seek a particular solution to Eq. (7.147)$_1$ in the form

$$W(x) = \mathcal{A}\, e^{\chi x}\,. \tag{7.148}$$

After substituting it into Eq. (7.147)$_1$, we get

$$\chi^4 \mp \frac{N}{IE}\chi^2 - \frac{\rho A\tilde{\omega}^2}{IE} = 0\,. \tag{7.149}$$

If the axial force is compressive, the roots for χ are

$$\chi_1 = \beta_{1c} = \sqrt{-\frac{N}{2IE} + \sqrt{\left(\frac{N}{2IE}\right)^2 + \frac{\rho A\tilde{\omega}^2}{IE}}}\,, \qquad \chi_2 = -\beta_{1c} \tag{7.150a}$$

and

$$\chi_3 = i\beta_{2c} = i\sqrt{\frac{N}{2IE} + \sqrt{\left(\frac{N}{2IE}\right)^2 + \frac{\rho A\tilde{\omega}^2}{IE}}}\,, \qquad \chi_3 = -i\beta_{2c}\,. \tag{7.150b}$$

If the axial force is tensile, the roots for χ are different:

$$\chi_1 = \beta_{1t} = \beta_{2c} = \sqrt{\frac{N}{2IE} + \sqrt{\left(\frac{N}{2IE}\right)^2 + \frac{\rho A\tilde{\omega}^2}{IE}}}\,, \qquad \chi_2 = -\beta_{2c} \tag{7.151a}$$

and

$$\chi_3 = i\beta_{2t} = i\beta_{1c} = i\sqrt{-\frac{N}{2IE} + \sqrt{\left(\frac{N}{2IE}\right)^2 + \frac{\rho A\tilde{\omega}^2}{IE}}}\,, \quad \chi_4 = -i\beta_{2t} = -i\beta_{1c}\,. \tag{7.151b}$$

We can now express the solution for W both for the case of

(a) compressive axial force:

$$W(x) = \mathcal{A}_1 \cosh\beta_{1c}x + \mathcal{A}_2 \sinh\beta_{1c}x + \mathcal{A}_3 \cos\beta_{2c}x + \mathcal{A}_4 \sin\beta_{2c}x\,, \tag{7.152a}$$

(a) tensile axial force:

$$\begin{aligned}W(x) =&\mathcal{A}_1 \cosh\beta_{1t}x + \mathcal{A}_2 \sinh\beta_{1t}x + \mathcal{A}_3 \cos\beta_{2t}x + \mathcal{A}_4 \sin\beta_{2t}x = \\ =&\mathcal{A}_1 \cosh\beta_{2c}x + \mathcal{A}_2 \sinh\beta_{2c}x + \mathcal{A}_3 \cos\beta_{1c}x + \mathcal{A}_4 \sin\beta_{1c}x\,.\end{aligned} \tag{7.152b}$$

The simply supported uniform beam shown in Fig. 7.11 is subjected to an axial force N. In the sequel, it is our aim to find a relation between the eigenfrequencies $\tilde{\omega}_k$ and the axial force N [7, 8]. The boundary conditions the solution for W should satisfy are given in Table 7.2 where relation (7.74) between M_y and W should also be taken into account:

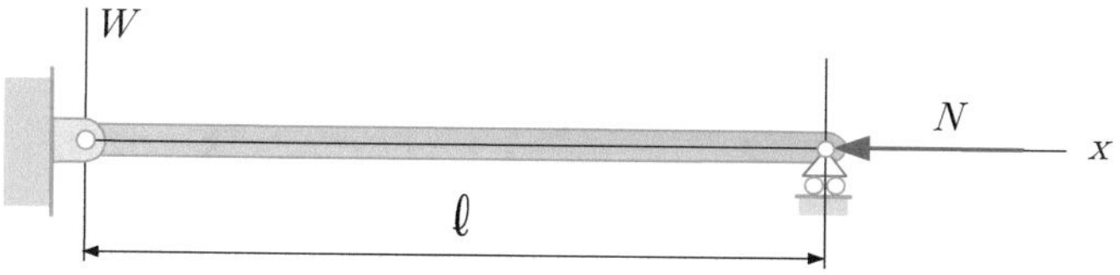

Fig. 7.11 Simply supported beam and the axial force acting on it

$$W(0) = 0\,, \qquad M_y(0) = -IE\frac{\mathrm{d}^2W}{\mathrm{d}x^2}\bigg|_{x=0} = 0\,, \tag{7.153a}$$

$$W(\ell) = 0\,, \qquad M_y(\ell) = -IE\frac{\mathrm{d}^2W}{\mathrm{d}x^2}\bigg|_{x=\ell} = 0\,. \tag{7.153b}$$

First, we shall consider the case of the compressive axial force. After substituting solution (7.152a) into boundary conditions (7.153a), we have

$$\begin{aligned}W(0) &= \mathcal{A}_1 + \mathcal{A}_3 = 0\,,\\ \frac{\mathrm{d}^2W}{\mathrm{d}x^2}\bigg|_{x=0} &= \beta_{1c}^2\mathcal{A}_1 - \beta_{2c}^2\mathcal{A}_4 = 0\,.\end{aligned}$$

Since the determinant of this homogeneous linear equation system is not zero, it follows that

$$\mathcal{A}_1 = \mathcal{A}_3 = 0\,. \tag{7.154}$$

Boundary conditions (7.153b) result in the following equations:

$$\begin{aligned}W(\ell) &= \mathcal{A}_2 \sinh\beta_{1c}\ell + \mathcal{A}_4 \sin\beta_{2c}\ell = 0\,,\\ \frac{\mathrm{d}^2W}{\mathrm{d}x^2}\bigg|_{x=\ell} &= \beta_{1c}^2\mathcal{A}_2 \sinh\beta_{1c}\ell - \beta_{2c}^2\mathcal{A}_3 \sin\beta_{2c}\ell = 0\,.\end{aligned}$$

Solutions for $\mathcal{A}_2$ and $\mathcal{A}_4$ different from the trivial ones can be obtained from this equation system if

$$\sinh\beta_{1c}\ell\,\sin\beta_{2c}\ell = 0\,. \tag{7.155a}$$

If the axial force is a tensile one, a similar line of thought leads to the following characteristic determinant:

$$\sinh \beta_{2c}\ell \, \sin \beta_{1c}\ell = 0 \,. \tag{7.155b}$$

Since ($\sinh \beta_{2c}\ell$) [$\sinh \beta_{1c}$] is not zero if ($\beta_{2c}\ell \neq 0$) [$\beta_{1c} \neq 0$] it follows from Eqs. (7.155) that

$$\sin \beta_{1c}\ell = 0 \,, \qquad \sinh \beta_{2c}\ell = 0 \,. \tag{7.156}$$

Hence,

$$\beta_{1c}\ell = k\pi \,, \qquad \beta_{2c}\ell = k\pi \qquad k = 1, 2, 3, \dots \tag{7.157}$$

After substituting β_{1c} and β_{2c} and squaring the result, we get

$$-\frac{N}{2IE} + \sqrt{\left(\frac{N}{2IE}\right)^2 + \frac{\rho A \tilde{\omega}_k^2}{IE}} = \left(\frac{k\pi}{\ell}\right)^2, \quad \frac{N}{2IE} + \sqrt{\left(\frac{N}{2IE}\right)^2 + \frac{\rho A \tilde{\omega}_k^2}{IE}} = \left(\frac{k\pi}{\ell}\right)^2,$$

or

$$\sqrt{\left(\frac{N}{2IE}\right)^2 + \frac{\rho A \tilde{\omega}_k^2}{IE}} = \left(\frac{k\pi}{\ell}\right)^2 \pm \frac{N}{2IE},$$

where the last sign on the right side is (positive) [negative] if N is (tensile) [compressive]. Square this equation and divide it by $(k\pi/\ell)^4$. We obtain

$$\boxed{\frac{\tilde{\omega}_k^2}{\dfrac{IE}{\rho A}\left(\dfrac{k\pi}{\ell}\right)^4} = 1 \pm \frac{N}{EI\left(\dfrac{k\pi}{\ell}\right)^2} \,.} \tag{7.158}$$

It is worth rewriting this equation by taking the following facts into account:

(a) According to Eq. (7.98), the square of the kth eigenfrequency of the unloaded and simply supported beam is

$$\omega_k^2 = \frac{IE}{\rho A}\left(\frac{k\pi}{\ell}\right)^4 . \tag{7.159a}$$

(b) As regards the stability problem of the simply supported beam, the kth critical compressive force (buckling load) is given by the following equation:

$$N_{k\,\mathrm{crit}} = EI\left(\frac{k\pi}{\ell}\right)^2 . \tag{7.159b}$$

(c) The strains due to the axial force and its critical values on the longitudinal axis of the beam can be calculated as

$$\varepsilon_x = \frac{N}{AE}\,; \qquad \varepsilon_{x\,k\,\mathrm{crit}} = \frac{N_{k\,\mathrm{crit}}}{AE}\,. \tag{7.159c}$$

Making use of Eqs. (7.159), we may rewrite Eq. (7.158) in the following form:

$$\boxed{\frac{\tilde{\omega}_k^2}{\omega_k^2} = 1 \pm \frac{N}{N_{k\,\mathrm{crit}}} = 1 \pm \frac{\varepsilon_x}{\varepsilon_{x\,k\,\mathrm{crit}}}\,.} \tag{7.160}$$

Remark 7.4 If IE tends to zero and the last sign is positive in Eq. (7.158)—then the axial force is tensile—we have

$$\tilde{\omega}_k = \frac{k\pi}{\ell}\sqrt{\frac{N}{\rho A}}\,. \tag{7.161}$$

This result coincides with the eigenfrequencies of a taut string—compare Eqs. (7.161) and $(7.46\mathrm{b})_2$.

Remark 7.5 Equation (7.160) shows that the square of the eigenfrequencies of the simply supported beam subjected to an axial force is a linear function of the force. Figure 7.12 depicts the quotient $\tilde{\omega}_1^2/\omega_1^2$ against the quotient $N/N_{1\,\mathrm{crit}}$. It is clear from the figure that $\tilde{\omega}_1^2 = \omega_1^2$ if $N = 0$. If the force is compressive, then $\tilde{\omega}_1^2 = 0$ for $N = N_{1\,\mathrm{crit}}$.

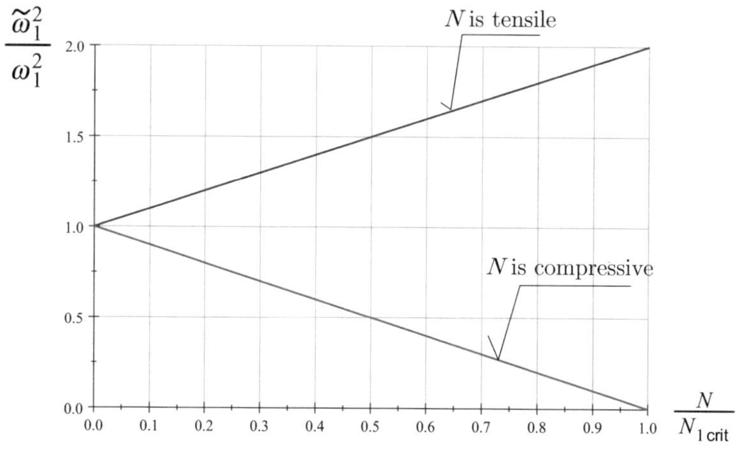

Fig. 7.12 The effect of the axial force on the first circular frequencies

It is also an important question what happens to the simply supported beam if the compressive force is greater than $N_{1\,\mathrm{crit}}$. In this respect, we refer the reader to book [1].

7.6 Problems

Problem 7.1 The uniform rod shown in Fig. 7.13 is fixed at its ends. The rod vibrates longitudinally. Assume that $u(x, 0) = \mathrm{u}(x)$ and $\dot{u}(x, 0) = \mathrm{g}(x)$ are the initial conditions. Determine the eigenfunctions, the natural frequencies, and the function that describes the motion.

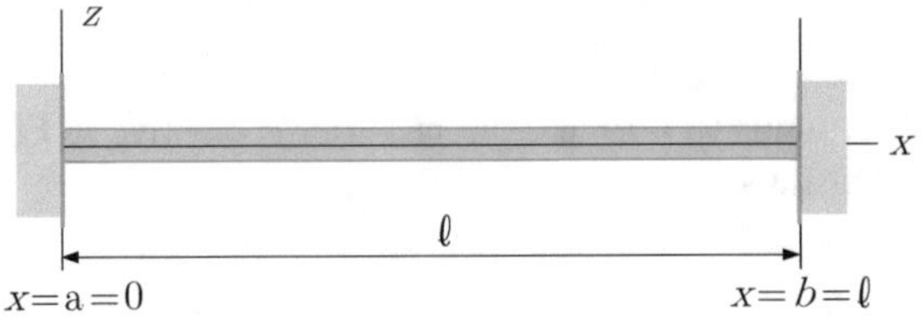

Fig. 7.13 Uniform rod fixed at its ends

Problem 7.2 The uniform rod shown in Fig. 7.14 is free at its ends. We assume that the rod vibrates longitudinally. Determine the eigenfunctions and the natural frequencies. What is the solution if the rod is subjected to the same tensile forces N_o at its ends (the rod is in equilibrium) and the forces are suddenly removed.

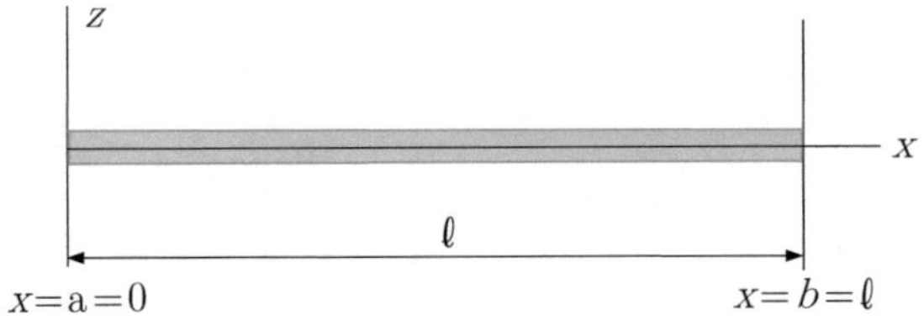

Fig. 7.14 Uniform rod free at its ends

Problem 7.3 Prove that ω^2 (or λ) is a positive quantity in Eq. (7.82)$_2$.

Problem 7.4 Derive the frequency equation, the eigenfunctions, and the eigenvalues (or the products $\beta_k \ell$) for fixed-fixed, fixed-free, pinned-free, and free-free beams.

Problem 7.5 Given the flexibility matrix of the simply supported uniform beam shown in Fig. 7.15,

$$\underset{(3\times3)}{\mathbf{f}} = \begin{bmatrix} f_{11} & f_{12} & f_{13} \\ f_{21} & f_{22} & f_{23} \\ f_{31} & f_{32} & f_{33} \end{bmatrix} = \frac{\ell^3}{1944IE} \begin{bmatrix} 12.5 & 19.5 & 8.5 \\ 19.5 & 40.5 & 19.5 \\ 8.5 & 19.5 & 12.5 \end{bmatrix}.$$

Let ℓ, I, E, A, and ρ be the length of the beam, the area moment of inertia, the modulus of elasticity, the cross-sectional area, and the density. Determine the stiffness matrix and then the natural frequencies provided that $m_1 = m_2 = m_3 = m/3 = \rho A\ell/3$ where m is the total mass of the beam and compare the results obtained with those valid for a continuous beam.

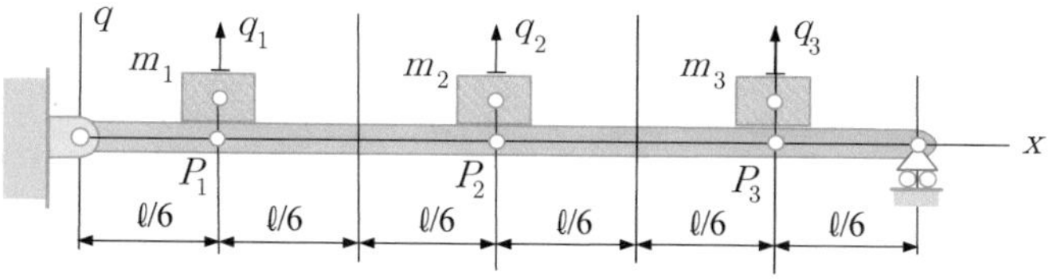

Fig. 7.15 Beam with three concentrated messes

References

1. L.N. Virgin, *Vibration of Axially Loaded Structures* (Cambridge University Press, Cambridge, 2007)
2. A. Bokaian, Natural frequencies of beams under compressive axial loads. J. Sound Vib. **126**, 49–65 (1988). https://doi.org/10.1016/0022-460X(88)90397-5
3. S.P. Timoshenko, D.H. Young, *Elements of Strength of Materials* (Van Nostrand Reinhold, New York, 1977)
4. A.N. Krylov, *Vibration of Ships* (in Russian) (GI Red Sudistroit Lit., Moscow, 1936)
5. W.J. Duncan, Free and forced oscillations of continuous beams treatment by the admittance method. Lond. Edinb. Dublin Philos. Mag. J. Sci. Ser. 7 **34**(228), 49–63 (1943). https://doi.org/10.1080/14786444308521329
6. W.J. Duncan, *Mechanical Admittances and their Applications to Oscillation Problems* (H.M.S.O, London, 1947)
7. A.E. Galef, Bending frequencies of compressed beams. J. Accustical Soc. Am. **44**, 643 (1968). https://doi.org/10.1121/1.1911144.339
8. V. Birman, I. Elishakoff, J. Singer, On the effect of axial compression on the bounds of simple harmonic motion. Isr. J. Technol. **20**, 254–258 (1982)

Chapter 8
Eigenvalue Problems of Ordinary Differential Equations

8.1 Differential Equations and Boundary Conditions

Consider the ordinary differential equation[1]

$$K\,[y] = \lambda\, M\,[y]\,, \tag{8.1a}$$

where $y(x)$ is the unknown function and λ is a parameter (the eigenvalue sought). The differential operators $K\,[y]$ and $M\,[y]$ are defined by the following relationships:

$$\begin{aligned} K\,[y] &= \sum_{n=0}^{\kappa}(-1)^n\left[f_n(x)y^{(n)}(x)\right]^{(n)}\,, \qquad \frac{\mathrm{d}^n(\ldots)}{\mathrm{d}x^n} = (\ldots)^{(n)}\,;\\ M\,[y] &= \sum_{n=0}^{\mu}(-1)^n\left[g_n(x)y^{(n)}(x)\right]^{(n)}\,, \qquad \kappa > \mu \geq 1 \end{aligned} \tag{8.1b}$$

in which the real function $(f_\nu(x))$ $[g_\nu(x)]$ is differentiable continuously (κ) $[\mu]$ times and

$$f_\kappa(x) \neq 0\,, \qquad g_\mu(x) \neq 0 \qquad \text{if } x \in [a, b]\,. \tag{8.1c}$$

Note that 2κ, which is the order of the differential operator on the left side of (8.1a), is greater than 2μ, which is the order of the differential operator on the right side.

Let $x \in [a, b]$, $a > b$, $b - a = \ell$ be the interval in which we seek the solution of differential equation (8.1).

We shall assume that differential equation (8.1) is associated with 2κ homogeneous boundary conditions of the form

[1] The line of thought in section Eigenvalue problems of ordinary differential equations is based mainly on the book [1] by Lothar Collatz (1910–1990).

G. Szeidl and L. P. Kiss, *Mechanical Vibrations*, Foundations of Engineering Mechanics,
https://doi.org/10.1007/978-3-030-45074-8_8

$$U_r[y] = \sum_{n=1}^{2\kappa} \left[\alpha_{nr}\, y^{n-1}(a) + \beta_{nr}\, y^{n-1}(b) \right] = 0\,, \qquad r = 1, 2, \ldots, 2\kappa, \tag{8.2}$$

where α_{nr} and β_{nr} are real constants.

Remark 8.1 Differential equation (8.1) and boundary conditions (8.2) constitute an eigenvalue problem in which λ is the eigenvalue to be determined and the solution that belongs to λ is called eigenfunction. If $\mu = 0$ the right side of differential equation (8.1) is

$$M\,[y] = \lambda g_0(x) y(x) \tag{8.3}$$

and then the eigenvalue problem is called simple. The eigenvalue problems that provide the eigenfrequencies for the longitudinal and torsional vibrations of rods as well as for the transverse vibrations of strings and beams—we remind the reader of Remark 7.2 and Sect. 7.5.3 here—are all simple ones.

If $\mu > 0$ we speak about generalized eigenvalue problem.

Remark 8.2 The boundary conditions should be linearly independent of each other. It is obvious that a linear combination of the boundary conditions is also a boundary condition. By selecting suitable linear combinations, derivatives with an order higher than $\kappa - 1$ can be removed from some boundary conditions. If they are removed in every possible way we may have, altogether, say, e boundary conditions which do not involve derivatives higher than $\kappa - 1$. These boundary conditions are called essential boundary conditions. The further $2k - e$ boundary conditions are the natural boundary conditions [2].

The functions $u(x)$ and $v(x)$ ($u(x), v(x)$ are not identically equal to zero if $x \in [a, b]$) are called (a) admissible functions if they satisfy the essential boundary conditions, (b) comparison functions if they satisfy both the essential and the natural boundary conditions, and (c) eigenfunctions if they satisfy differential equation (8.1) and boundary conditions (8.2).

8.2 Self-adjointness

The integrals

$$(u, v)_K = \int_a^b u(x)\, K[v(x)]\, \mathrm{d}x\,, \qquad (u, v)_M = \int_a^b u(x)\, M[v(x)]\, \mathrm{d}x \tag{8.4}$$

taken on the set of the comparison functions $u(x)$, $v(x)$ are products defined on the operators K and M.

Remark 8.3 For the sake of comparison, it is worth recalling a similar definition set up for vibrating systems with finite degrees of freedom by Eq. (5.13). It will turn out that products (8.4) play such a role in the sequel which resembles that of the products (5.13) in the algebraic eigenvalue problems of oscillating systems with finite degree of freedom.

Let us now detail the product $(u, v)_K$. Making use of (8.1b), we may write

$$(u, v)_K = \int_a^b u(x) \sum_{n=0}^{\kappa} (-1)^n \left[f_n(x) v^{(n)}(x) \right]^{(n)} \, \mathrm{d}x \tag{8.5}$$

in which the integral

$$I_{uv} = (-1)^n \int_a^b u(x) \left[f_n(x) v^{(n)}(x) \right]^{(n)} \, \mathrm{d}x$$

can be manipulated into a more suitable form if we perform partial integrations. We get

$$\begin{aligned} I_{uv} &= (-1)^n \Big[u(x) \left[f_n(x)\, v^{(n)}(x) \right]^{(n-1)} \Big]_a^b + \\ &+ (-1)^{n-1} \int_a^b u(x) \left[f_n(x) v^{(n)}(x) \right]^{(n-1)} \mathrm{d}x = (-1)^n \Big[u(x) \left[f_n(x)\, v^{(n)}(x) \right]^{(n-1)} - \\ &- u(x)^{(1)} \left[f_n(x)\, v^{(n)}(x) \right]^{(n-2)} + u(x)^{(2)} \left[f_n(x)\, v^{(n)}(x) \right]^{(n-3)} - \cdots \Big]_a^b + \\ &+ \int_a^b u^{(n)}(x) f_n(x) v^{(n)}(x) \, \mathrm{d}x = \left[\sum_{r=0}^{n-1} (-1)^{(n+r)} u(x)^{(r)} \left[f_n(x)\, v^{(n)}(x) \right]^{(n-1-r)} \right]_a^b + \\ &+ \int_a^b u^{(n)}(x) f_n(x) v^{(n)}(x) \, \mathrm{d}x. \end{aligned} \tag{8.6}$$

Hence

$$\begin{aligned} (u, v)_K = &\left[\sum_{n=0}^{\kappa} \sum_{r=0}^{n-1} (-1)^{(n+r)} u(x)^{(r)} \left[f_n(x)\, v^{(n)}(x) \right]^{(n-1-r)} \right]_a^b + \\ &+ \sum_{n=0}^{\kappa} \int_a^b u^{(n)}(x) f_n(x) v^{(n)}(x) \, \mathrm{d}x \,. \end{aligned} \tag{8.7a}$$

It follows from Eq. (8.7a) that

$$\begin{aligned} (u, v)_M = &\left[\sum_{n=0}^{\mu} \sum_{r=0}^{n-1} (-1)^{(n+r)} u(x)^{(r)} \left[g_n(x)\, v^{(n)}(x) \right]^{(n-1-r)} \right]_a^b + \\ &+ \sum_{n=0}^{\mu} \int_a^b u^{(n)}(x) g_n(x) v^{(n)}(x) \, \mathrm{d}x \,. \end{aligned} \tag{8.7b}$$

These results are naturally valid for the products $(v, u)_K$ and $(v, u)_M$ if we write u for v and conversely v for u.

Eigenvalue problem (8.1), (8.2) is said to be self-adjoint if the products (8.4) are commutative, i.e., it holds that

$$(u, v)_K = (v, u)_K \qquad \text{and} \qquad (u, v)_M = (v, u)_M \,. \tag{8.8}$$

Conditions (8.8) are called conditions of self-adjointness.

It follows from Eqs. (8.7a) and (8.7b) that

$$(u, v)_K - (v, u)_K = \left[\sum_{n=0}^{\kappa} \sum_{r=0}^{n-1} (-1)^{(n+r)} u(x)^{(r)} \left[f_n(x)\, v^{(n)}(x) \right]^{(n-1-r)} - \right. \\ \left. - v(x)^{(r)} \left[f_n(x)\, u^{(n)}(x) \right]^{(n-1-r)} \right]_a^b = 0 \tag{8.9a}$$

and

$$(u, v)_M - (v, u)_M = \left[\sum_{n=0}^{\mu} \sum_{r=0}^{n-1} (-1)^{(n+r)} u(x)^{(r)} \left[g_n(x)\, v^{(n)}(x) \right]^{(n-1-r)} - \right. \\ \left. - v(x)^{(r)} \left[g_n(x)\, u^{(n)}(x) \right]^{(n-1-r)} \right]_a^b = 0 \tag{8.9b}$$

if the eigenvalue problem (8.1), (8.2) is self-adjoint.

Remark 8.4 According to Eqs. (8.9), self-adjointness of eigenvalue problem (8.1), (8.2) depends on what value the solutions have on the boundary of the interval $[a, b]$, i.e., on the boundary conditions.

Remark 8.5 The three eigenvalue problems defined by differential equation (7.65) and boundary conditions (7.66) are all self-adjoint. The same is valid for the six eigenvalue problems defined by differential equation (7.114) and boundary conditions (7.115).

8.3 Orthogonality of the Eigenfunctions in General Sense

Let us assume that the eigenvalue problem (8.1), (8.2) is self-adjoint. Further, let y_k and y_ℓ be two different solutions to the eigenvalue problem (two linearly independent eigenfunctions) which, therefore, satisfy the differential equations

$$K[y_k] = \lambda_k\, M[y_k]\,, \qquad K[y_\ell] = \lambda_\ell\, M[y_\ell] \tag{8.10}$$

and boundary conditions

$$U_r[y_k] = \sum_{n=1}^{2\kappa} \left[\alpha_{nr} \, y_k^{n-1}(a) + \beta_{nr} \, y_k^{n-1}(b) \right] = 0 \,,$$
$$U_r[y_\ell] = \sum_{n=1}^{2\kappa} \left[\alpha_{nr} \, y_\ell^{n-1}(a) + \beta_{nr} \, y_\ell^{n-1}(b) \right] = 0, \qquad r = 1, 2, \ldots, 2\kappa \tag{8.11}$$

where λ_k and λ_ℓ denote the eigenvalues that belong to the eigenfunctions y_k and y_ℓ. We shall assume that $\lambda_k \neq \lambda_\ell$.

Multiply Eq. (8.10)$_1$ by $y_\ell(x)$, Eq. (8.10)$_b$ by $y_k(x)$ then integrate the result over the interval $[a, b]$. With regard to definitions (8.4), we get

$$(y_\ell, y_k)_K = \lambda_k \, (y_\ell, y_k)_M \,, \qquad (y_k, y_\ell)_K = \lambda_\ell \, (y_k, y_\ell)_M \,. \tag{8.12}$$

Consider now the difference (8.12)$_1$–(8.12)$_2$. Since we have assumed that the eigenvalue problem (8.1), (8.2) is self-adjoint the products $(y_\ell, y_k)_K$ and $(y_\ell, y_k)_M$ are commutative. Hence, the difference is

$$(\lambda_k - \lambda_\ell) \, (y_k, y_\ell)_M = 0, \tag{8.13}$$

where $\lambda_k - \lambda_\ell \neq 0$. This means that

$$(y_k, y_\ell)_M = 0 \qquad k \neq \ell \,. \tag{8.14a}$$

In view of this equation, it follows from (8.12) that

$$(y_k, y_\ell)_K = 0 \qquad k \neq \ell \,. \tag{8.14b}$$

If Eqs. (8.14) are satisfied, then the functions $y_k(x)$ and $y_\ell(x)$ are orthogonal to each other in general sense.

For simple eigenvalue problems

$$M[y(x)] = g_0(x) \, y(x) \,. \tag{8.15}$$

Consequently, the orthogonality condition assumes the form

$$(y_k, y_\ell)_M = \int_a^b y_k(x) \, g_0(x) \, y_\ell(x) \, \mathrm{d}x = 0 \,. \tag{8.16}$$

For $g_0(x) > 0$, $x \in [a, b]$, the functions $\sqrt{g_0(x)} y_k(x)$ and $\sqrt{g_0(x)} y_\ell(x)$ are orthogonal in the traditional sense.

In view of Eq. (8.12), we may write on the basis of (8.14) that

$$(y_k, y_\ell)_K = \begin{cases} \lambda_n \, (y_k, y_\ell)_M & \text{if } k = \ell, \\ 0 & \text{if } k \neq \ell. \end{cases} \qquad k, \ell = 1, 2, 3, \ldots \tag{8.17}$$

Remark 8.6 Note that the line of thought in this subsection is basically the same as in Sect. 7.5.3.

8.4 On the Reality of the Eigenvalues

We shall prove that the eigenvalues of the self-adjoint eigenvalue problems are all real numbers. We assume that the eigenvalue λ_ℓ is complex, i.e., it can be given in the form $\lambda_\ell = \mathfrak{a} + i\mathfrak{b}$—$\mathfrak{a}$ and $\mathfrak{b}$ are real numbers—and we shall show that this assumption leads to a contradiction. Let $y_\ell(x)$ be the eigenfunction that belongs to the eigenvalue λ_ℓ. It is clear that λ_ℓ and $y_\ell(x)$ satisfy the differential equation

$$K[y_\ell(x)] = \lambda_\ell M[y_\ell(x)]\,. \tag{8.18}$$

If $y_\ell(x)$ is a real function, the left side of this equation is also real, whereas the right side is complex since λ is complex. This means that the eigenfunction which belongs to a complex eigenvalue should also be complex. Hence, $y_\ell(x)$ can be given in the form

$$y_\ell(x) = u_\ell(x) + i v_\ell(x), \tag{8.19}$$

where $u_\ell(x)$, $v_\ell(x)$ are real functions and i is the imaginary unit. Let us multiply Eq. (8.18) by the complex conjugate of the eigenfunction $y_\ell(x)$; then, integrate the result in the interval $[a, b]$. We get

$$(\bar{y}_k, y_\ell)_K = \lambda_\ell\,(\bar{y}_k, y_\ell)_M\,. \tag{8.20}$$

If we express the left side of this equation in terms of $u_\ell(x)$ and $v_\ell(x)$, we obtain

$$\begin{aligned}(\bar{y}_k, y_\ell)_K &= (u_\ell(x) - i v_\ell(x), u_\ell(x) + i v_\ell(x))_K = \\ (u_\ell(x), u_\ell(x))_K &+ \big[(u_\ell(x), v_\ell(x))_K - (v_\ell(x), u_\ell(x))_K\big] i + (v_\ell(x), v_\ell(x))_K = \\ &= (u_\ell(x), u_\ell(x))_K + (v_\ell(x), v_\ell(x))_K\,,\end{aligned} \tag{8.21}$$

where, because of the commutativity of the product $(u_\ell(x), v_\ell(x))_K$, the coefficient of the imaginary unit is zero. This result shows that the left side of Eq. (8.20) is not complex but a real number.

As regards the product $(\bar{y}_k, y_\ell)_M$, we may write by substituting M for K in the above equation that

$$(\bar{y}_k, y_\ell)_M = (u_\ell(x), u_\ell(x))_M + (v_\ell(x), v_\ell(x))_M\,, \tag{8.22}$$

which is also a real number. Consequently, we can rewrite (8.20) in the following form:

$$(u_\ell(x), u_\ell(x))_K + (v_\ell(x), v_\ell(x))_K = \mathfrak{a}\left[(u_\ell(x), u_\ell(x))_M + (v_\ell(x), v_\ell(x))_M\right] + \\ + i\mathfrak{b}\left[(u_\ell(x), u_\ell(x))_M + (v_\ell(x), v_\ell(x))_M\right], \quad (8.23)$$

where the left side is real, whereas the right side is complex which is, therefore, a contradiction. For

$$\left[(u_\ell(x), u_\ell(x))_M + (v_\ell(x), v_\ell(x))_M\right] \neq 0,$$

the equality in (8.23) is possible only if $\mathfrak{b} = 0$, i.e., if the eigenvalue λ_ℓ is a real number.

We can prove the reality of the eigenvalues more formally if we take into account that $\bar{y}_\ell$ and $\bar{\lambda}_\ell$ satisfies Eq. (8.18):

$$K[\bar{y}_\ell(x)] = \bar{\lambda}_\ell M[\bar{y}_\ell(x)]. \quad (8.24)$$

If we multiply (8.24) by $y_\ell(x)$, integrate the result in the interval $[a, b]$ and follow the steps leading to (8.23), we get

$$(u_\ell(x), u_\ell(x))_K + (v_\ell(x), v_\ell(x))_K = \mathfrak{a}\left[(u_\ell(x), u_\ell(x))_M + (v_\ell(x), v_\ell(x))_M\right] - \\ - i\mathfrak{b}\left[(u_\ell(x), u_\ell(x))_M + (v_\ell(x), v_\ell(x))_M\right]. \quad (8.25)$$

Let us now consider the difference of Eqs. (8.23) and (8.25). We have

$$2i\mathfrak{b}\left[(u_\ell(x), u_\ell(x))_M + (v_\ell(x), v_\ell(x))_M\right] = 0 \quad (8.26)$$

from where it follows that $\mathfrak{b} = 0$ if $\left[(u_\ell(x), u_\ell(x))_M + (v_\ell(x), v_\ell(x))_M\right] \neq 0$.

8.5 Boundary Expressions

For the comparison function $u(x) = v(x)$, we may rewrite (8.7a) and (8.7b) in the following forms:

$$(u, u)_K = \sum_{n=0}^{\kappa} \int_a^b u^{(n)}(x) f_n(x) u^{(n)}(x)\, \mathrm{d}x + K_0[u(x)] \quad (8.27a)$$

and

$$(u, u)_M = \sum_{n=0}^{\mu} \int_a^b u^{(n)}(x) g_n(x) u^{(n)}(x)\, \mathrm{d}x + M_0[u(x)], \quad (8.27b)$$

where

$$K_0[u(x)] = \left[\sum_{n=0}^{\kappa} \sum_{r=0}^{n-1} (-1)^{(n+r)} u(x)^{(r)} \left[f_n(x)\, u^{(n)}(x) \right]^{(n-1-r)} \right]_a^b \tag{8.28a}$$

and

$$M_0[u(x)] = \left[\sum_{n=0}^{\mu} \sum_{r=0}^{n-1} (-1)^{(n+r)} u(x)^{(r)} \left[g_n(x)\, u^{(n)}(x) \right]^{(n-1-r)} \right]_a^b . \tag{8.28b}$$

Here $K_0[u(x)]$ and $M_0[u(x)]$ are the Dirichlet boundary expressions [3].

8.6 Sign of Eigenvalues

The eigenvalue problem (8.1), (8.2) is said to be positive definite if the eigenvalues are positive, positive semidefinite if one eigenvalue is zero and the other eigenvalues are all positive, negative semidefinite if one eigenvalue is zero and the other eigenvalues are all negative, and finally negative definite if the eigenvalues are negative.

If we equalize ℓ and k in (8.12), we get

$$(y_\ell, y_\ell)_K = \lambda_k\, (y_\ell, y_\ell)_M . \tag{8.29}$$

Hence

$$\lambda_k = \frac{(y_\ell, y_\ell)_K}{(y_\ell, y_\ell)_M} = \frac{\displaystyle\int_a^b y_\ell(x)\, K[y_\ell(x)]\, \mathrm{d}x}{\displaystyle\int_a^b y_\ell(x)\, M[y_\ell(x)]\, \mathrm{d}x} . \tag{8.30}$$

This equation shows that the sign of λ_ℓ is a function of the products $(y_\ell, y_\ell)_K$ and $(y_\ell, y_\ell)_M$.

Assume that

$$(u, u)_K > 0 , \quad \text{and} \quad (u, u)_M > 0 \tag{8.31}$$

for any comparison function $u(x)$. Then the eigenvalue problem (8.1), (8.2) is positive definite (or full definite).

The Rayleigh quotient is defined by the equation

$$\mathcal{R}[u(x)] = \frac{(u, u)_K}{(u, u)_M} = \frac{\displaystyle\int_a^b u(x)\, K[u(x)]\, \mathrm{d}x}{\displaystyle\int_a^b u(x)\, M[u(x)]\, \mathrm{d}x} \tag{8.32}$$

in which $u(x)$ is a comparison function. Upon substitution of (8.27a) and (8.27b) for the numerator and denominator in (8.30), we get

$$\mathcal{R}[u(x)] = \frac{(u,u)_K}{(u,u)_M} = \frac{\displaystyle\sum_{n=0}^{\kappa}\int_a^b u^{(n)}(x)f_n(x)u^{(n)}(x)\,\mathrm{d}x + K_0[u(x)]}{\displaystyle\sum_{n=0}^{\mu}\int_a^b u^{(n)}(x)g_n(x)u^{(n)}(x)\,\mathrm{d}x + M_0[u(x)]}\,. \tag{8.33}$$

If the eigenvalue problem (8.1), (8.2) is self-adjoint, then

$$K_0[u(x)] = M_0[u(x)] = 0\,. \tag{8.34}$$

Assume that the eigenvalue problem (8.1), (8.2) is self-adjoint and

$$M[y] = (-1)^{\mu}[g_{\mu}(x)y^{(\mu)}]^{(\mu)}\,. \tag{8.35}$$

Then the eigenvalue problem belongs to the single term class of eigenvalue problems [3]. If $\mu = 0$ in (8.35), the considered self-adjoint eigenvalue problem is simple.

8.7 Determination of Eigenvalues

Let us denote the linearly independent particular solutions of the differential equation $K[y] = \lambda M[y]$ by $z_\ell(x,\lambda)$ $(\ell = 1, 2, \ldots, 2\kappa)$. With $z_\ell(x,\lambda)$, the general solution is of the form

$$y(x) = \sum_{\ell=1}^{2\kappa} \mathcal{A}_\ell z_\ell(x,\lambda), \tag{8.36}$$

where the undetermined integration constants $\mathcal{A}_\ell$ can be obtained from the boundary conditions:

$$\sum_{\ell=1}^{2\kappa} U_r[z_\ell(x,\lambda)]\mathcal{A}_\ell = 0\,, \qquad r = 1, 2, \ldots, 2\kappa\,. \tag{8.37}$$

Since these equations constitute a homogeneous linear equation system for the unknowns $\mathcal{A}_\ell$ solutions different from the trivial one (nontrivial solutions) exist if and only if the determinant of the system is zero:

$$\Delta(\lambda) = \det\big[U_r[z_\ell(x,\lambda)]\big] = 0\,. \tag{8.38}$$

This is the equation which should be solved for finding the eigenvalues λ. We remark that $\Delta(\lambda)$ is often referred to as characteristic determinant.

In the sequel, we shortly outline, without entering deeply into details, some properties the eigenvalues have.

Assume that the coefficients of the differential equation considered and the initial conditions—if there are initial conditions at all—are continuous and differentiable functions of the parameter λ. Then the solution is an analytical function of λ—see Sect. 5.7 in [1] or [4] for further details.

Differential equation (8.10) obviously meets the above conditions: (a) it is a linear function of λ and (b) there are no initial conditions. It then follows that $z_\ell(x, \lambda)$ is an analytical and single-valued (regular) function of λ, i.e., an entire function of λ for all fixed $x \in [a, b]$.

Hence, it holds that: If $\Delta(\lambda)$ is identically equal to zero then each λ is an eigenvalue. Otherwise, function $\Delta(\lambda)$ has an infinite sequence of isolated zero points which can be ordered according to their magnitudes:

$$0 \le |\lambda_1| \le |\lambda_2| \le |\lambda_3| \le \cdots \tag{8.39}$$

Exercise 8.1 Consider the eigenvalue problem

$$-y^{(2)}(x) = \lambda y(x)\,,$$
$$y(0) - y(1) = 0\,, \qquad y^{(1)}(0) + y^{(1)}(1) = 0\,.$$

Find the eigenvalues λ [1].

Making use of the notation $k^2 = \lambda$, we can give the general solution of this differential equation in the form

$$y(x) = \mathcal{A}_1 \cos kx + \mathcal{A}_2 \sin kx$$

in which $\mathcal{A}_1$ and $\mathcal{A}_2$ are undetermined integration constants. Consequently

$$\Delta(\lambda) = \begin{vmatrix} 1 - \cos k & -\sin k \\ -k \sin k & k\,(1 + \cos k) \end{vmatrix} = k\left(1 - \sin^2 k + \cos^2 k\right) = 0$$

is the characteristic determinant. This equation shows that any $\lambda = k^2$ is an eigenvalue.

Exercise 8.2 The transverse vibrations of a free-free beam are described, according to (7.82), by the differential equation

$$\frac{\mathrm{d}^4 W(x)}{\mathrm{d}x^4} = \lambda W(x)\,, \quad \lambda = \frac{\rho A \omega^2}{IE} = \beta^4, \tag{8.40}$$

which is associated with the following boundary conditions:

$$\begin{aligned} W^{(2)}(x)\big|_{x=0} &= 0\,, \qquad & W^{(3)}(x)\big|_{x=0} &= 0\,, \\ W^{(2)}(x)\big|_{x=\ell} &= 0\,, \qquad & W^{(3)}(x)\big|_{x=\ell} &= 0\,. \end{aligned} \tag{8.41}$$

Find the eigenvalues λ.

Since $W(x) =$ constant is such a solution for which $\Delta(\lambda)$ is identically equal to zero, it follows that $\lambda = 0$ is an eigenvalue. The non-constant solutions are given by (7.87), and it is shown in Sect. C.7—see the solution to Problem 7.4—that

$$\Delta(\lambda) = \cosh \beta_k \ell \cos \beta_k \ell - 1 = \cosh \sqrt[4]{\lambda_k}\, \ell \cos \sqrt[4]{\lambda_k}\, \ell - 1 = 0\,, \qquad k = 1, 2, 3, \ldots$$

is the characteristic determinant from where we get

$$\sqrt[4]{\lambda_1}\ell = 4.730141, \quad \sqrt[4]{\lambda_2}\ell = 7.853205, \qquad \sqrt[4]{\lambda_k}\ell \approx \frac{\pi}{2} + k\pi, \quad k = 3, 4, 5, \ldots$$

8.8 The Green Functions

Consider the inhomogeneous ordinary differential equation

$$L[y(x)] = r(x), \tag{8.42a}$$

where the differential operator of order κ is defined by the following equation:

$$L[y(x)] = \sum_{n=0}^{\kappa} p_n(x)\, y^{(n)}(x)\,. \tag{8.42b}$$

Here, $\kappa \geq 1$ is a natural number, and the functions $p_n(x)$ and $r(x)$ are continuous if $x \in [a, b]$ $(b > a,\ b - a = \ell)$ and $p_\kappa(x) \neq 0$. We shall assume that the inhomogeneous differential equation (8.42) is associated with the homogeneous boundary conditions

$$U_r[y] = \sum_{n=1}^{\kappa} \left[\alpha_{nr}\, y^{n-1}(a) + \beta_{nr}\, y^{n-1}(b)\right] = 0\,, \qquad r = 1, 2, 3, \ldots, \kappa \tag{8.43}$$

which are formally the same as boundary conditions (8.2).

Solution of the boundary value problem (8.42), (8.43) is sought in the form

$$y(x) = \int_a^b G(x, \xi) r(\xi)\, \mathrm{d}\xi, \tag{8.44}$$

where $G(x, \xi)$ is the Green[2] function [5–7] defined by the following four properties:

1. The Green function is a continuous function of x and ξ in each of the triangles $a \leq x \leq \xi \leq b$ and $a \leq \xi \leq x \leq b$. In addition, it is κ times differentiable with

---

[2]George Green (1793–1841).

respect to x and the derivatives

$$\frac{\partial^n G(x,\xi)}{\partial x^n} = G^{(n)}(x,\xi)\,, \qquad (n = 1, 2, \dots, \kappa)$$

are also continuous functions of x and ξ in the triangles $a \le x \le \xi \le b$ and $a \le \xi \le x \le b$.

2. Let ξ be fixed in $[a, b]$. The Green function and its derivatives

$$G^{(n)}(x,\xi) = \frac{\partial^n G(x,\xi)}{\partial x^n}\,, \qquad (n = 1, 2, \dots, \kappa - 2) \tag{8.45}$$

should be continuous for $x = \xi$:

$$\lim_{\varepsilon \to 0} \left[G^{(n)}(\xi + \varepsilon, \xi) - G^{(n)}(\xi - \varepsilon, \xi)\right] =$$
$$= \left[G^{(n)}(\xi + 0, \xi) - G^{(n)}(\xi - 0, \xi)\right] = 0 \quad (n = 0, 1, 2, \dots, \kappa - 2)\,. \tag{8.46a}$$

The derivative $G^{(\kappa-1)}(x,\xi)$ should, however, have a jump

$$\lim_{\varepsilon \to 0} \left[G^{(\kappa-1)}(\xi + \varepsilon, \xi) - G^{(\kappa-1)}(\xi - \varepsilon, \xi)\right] =$$
$$= \left[G^{(\kappa-1)}(\xi + 0, \xi) - G^{(\kappa-1)}(\xi - 0, \xi)\right] = \frac{1}{p_\kappa(\xi)} \tag{8.46b}$$

if $x = \xi$.

3. Let α be an arbitrary but finite non-zero constant. For a fixed $\xi \in [a, b]$, the product $G(x,\xi)\alpha$ as a function of x $(x \ne \xi)$ should satisfy the homogeneous differential equation

$$L\,[G(x,\xi)\alpha] = 0\,.$$

4. The product $G(x,\xi)\alpha$ as a function of x should satisfy the boundary conditions

$$U_r\,[G(x,\xi)\alpha] = 0\,, \qquad r = 1, 2, 3, \dots, \kappa\,.$$

Remark 8.7 The above definitions are basically the same as that of Bocher [5, 6] who generalized an earlier result concerning ordinary differential equations of order two.

If there exists the Green function defined above for the boundary value problem (8.42), (8.43) then the function $y(x)$ given by (8.44) satisfies differential equation (8.42) and boundary conditions (8.43).

Using (8.44), we can determine the first $\kappa - 1$ derivatives of $y(x)$:

$$y^{(1)}(x) = \int_a^b G^{(1)}(x,\xi) r(\xi)\, \mathrm{d}\xi, \;\; y^{(2)}(x) = \int_a^b G^{(2)}(x,\xi) r(\xi)\, \mathrm{d}\xi, \;\; \dots \tag{8.47a}$$

$$\ldots,\quad y^{(\kappa-1)}(x) = \int_a^b G^{(\kappa-1)}(x,\xi)r(\xi)\,\mathrm{d}\xi =$$
$$= \int_a^x G^{(\kappa-1)}(x,\xi)r(\xi)\,\mathrm{d}\xi + \int_x^b G^{(\kappa-1)}(x,\xi)r(\xi)\,\mathrm{d}\xi\,. \tag{8.47b}$$

When calculating the κ-th derivative, we have to take, however, into account the additive resolution of the $(\kappa-1)$-th derivative as well as discontinuity (8.46b). We have

$$y^{(\kappa)}(x) = \int_a^b G^{(\kappa)}(x,\xi)r(\xi)\,\mathrm{d}\xi +$$
$$+ G^{(\kappa-1)}(x,\xi)r(\xi)\big|_{\xi=x-0} - G^{(\kappa-1)}(x,\xi)r(\xi)\big|_{\xi=x+0} =$$
$$= \int_a^b G^{(\kappa)}(x,\xi)r(\xi)\,\mathrm{d}\xi + \left[G^{(\kappa-1)}(x,x-0) - G^{(\kappa-1)}(x,x+0)\right]r(x),$$

where $x > x-0$ and $x < x+0$. Hence

$$y^{(\kappa)}(x) = \int_a^b G^{(\kappa)}(x,\xi)r(\xi)\,\mathrm{d}\xi + \frac{r(x)}{p_\kappa(x)}\,. \tag{8.48}$$

If we substitute solution (8.44) and derivatives (8.47), (8.48) back into the differential equation (8.42), we get

$$L[y(x)] = \sum_{n=0}^{\kappa} p_n(x)\,y^{(n)}(x) = \int_a^b \left\{\sum_{n=0}^{\kappa} p_n(x)\,G^{(\kappa)}(x,\xi)\right\} r(\xi)\,\mathrm{d}\xi +$$
$$+ \frac{r(x)}{p_\kappa(x)}\,p_\kappa(x) = r(x)\,. \tag{8.49}$$

Here, the expression within the braces is zero due to Property 3 of the definition—Properties 1 and 2 were taken into account when we determined the derivatives of the Green function. Equation (8.49) shows that integral (8.44) is really a solution of differential equation (8.42). According to Property 4, the Green function satisfies the boundary conditions. Hence, integral (8.44) is really the solution of the boundary value problem (8.42), (8.43) as well.

8.9 Calculation of the Green Function

Let us denote the linearly independent particular solutions of the homogeneous ordinary differential equation

$$L[y(x)] = 0 \tag{8.50}$$

by

$$z_1(x)\,, z_2(x)\,, z_3(x)\,, \ldots, z_\kappa(x)\,. \tag{8.51}$$

Since the general solution is a linear combination of the particular solutions, it can be given in the following form (Fig. 8.1):

$$y(x) = \sum_{\ell=1}^{\kappa} \mathcal{A}_\ell\, z_\ell(x), \tag{8.52}$$

where the coefficients $\mathcal{A}_\ell$ are arbitrary integration constants. Since the Green function satisfies the homogeneous differential equation (8.50) in each of the triangular domains $a \le x \le \xi \le b$ and $a \le \xi \le x \le b$, it follows that it can be given as a

Fig. 8.1 Triangular domains

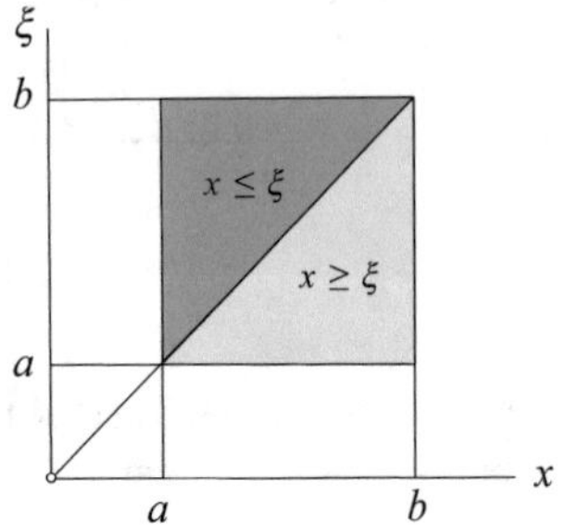

linear combination of the particular solutions $z_\ell(x)$, i.e., by Eq. (8.51). The integration constants $\mathcal{A}_\ell$ should, however, be different in the two triangular domains. For this reason, we shall assume that

$$\begin{aligned} G(x,\xi) &= \sum_{\ell=1}^{\kappa} \left(a_\ell(\xi) + b_\ell(\xi)\right) z_\ell(x)\,, \qquad x \le \xi\,; \\ G(x,\xi) &= \sum_{\ell=1}^{\kappa} \left(a_\ell(\xi) - b_\ell(\xi)\right) z_\ell(x)\,, \qquad x \ge \xi \end{aligned} \tag{8.53}$$

where the coefficients $a_\ell(\xi)$ and $b_\ell(\xi)$ are unknown functions. Continuity conditions (8.46a) yield the following equations:

$$\sum_{\ell=1}^{\kappa} b_\ell(\xi)\, z_\ell^{(n)}(\xi) = 0\,, \qquad n = 0, 1, 2, \ldots, \kappa - 2 \tag{8.54a}$$

As regards discontinuity condition (8.46b), we get

$$\frac{1}{p_\kappa(\xi)} = \left[G^{(\kappa-1)}(\xi+0,\xi) - G^{(\kappa-1)}(\xi-0,\xi)\right] = -2\sum_{\ell=1}^{\kappa} b_\ell(\xi)\, z_\ell^{(\kappa-1)}(\xi)\,.$$

Hence

$$\sum_{\ell=1}^{\kappa} b_\ell(\xi)\, z_\ell^{(\kappa-1)}(\xi) = -\frac{1}{2p_\kappa(\xi)}\,. \tag{8.54b}$$

Continuity and discontinuity conditions (8.46) have resulted in the inhomogeneous linear system of Eqs. (8.54) for the unknowns $b_\ell(\xi)$ ($\ell = 1, 2, \ldots, \kappa$). Its determinant assumes the following form:

$$\begin{vmatrix} z_1(\xi) & z_2(\xi) & \ldots & z_\kappa(\xi) \\ z_1^{(1)}(\xi) & z_2^{(1)}(\xi) & \ldots & z_\kappa^{(1)}(\xi) \\ \ldots & \ldots & \ldots & \ldots \\ z_1^{(\kappa-1)}(\xi) & z_2^{(\kappa-1)}(\xi) & \ldots & z_\kappa^{(\kappa-1)}(\xi) \end{vmatrix}\,. \tag{8.55}$$

This determinant is the Wronskian[3] [8] of differential equation (8.42) which is not zero since the particular solutions $z_\ell(x)$ are linearly independent. This means that we can always find a unique solution for the coefficients $b_\ell(\xi)$ in representation (8.53) of the Green function.

Making use of Property 4 of the Green function, we can find equations for the coefficients $a_\ell(\xi)$:

$$U_r[G] = U_r\left[\sum_{\ell=1}^{\kappa} (a_\ell(\xi) \pm b_\ell(\xi))\, z_\ell(x)\right] = 0\,, \qquad r = 1, 2, 3, \ldots, \kappa \tag{8.56}$$

where the sign is (positive) [negative] if the Green function in the boundary condition U_r is taken at $(x = a)$ $[x = b]$. Since the boundary conditions are linear functions of the solution $y(x)$—in the present case of the Green function—it follows that

$$\sum_{\ell=1}^{\kappa} a_\ell\, U_r[z_\ell] = \mp\sum_{\ell=1}^{\kappa} b_\ell\, U_r[z_\ell]\,, \qquad r = 1, 2, 3, \ldots, \kappa\,. \tag{8.57}$$

This inhomogeneous linear equation system has a unique solution for the coefficients $a_\ell(\xi)$ if and only if its determinant is not zero:

$$|U_r[z_\ell]| = \begin{vmatrix} U_1[z_1] & U_1[z_2] & \ldots & U_1[z_\kappa] \\ U_2[z_1] & U_2[z_2] & \ldots & U_2[z_\kappa] \\ \ldots & \ldots & \ldots & \ldots \\ U_\kappa[z_1] & U_\kappa[z_2] & \ldots & U_\kappa[z_\kappa] \end{vmatrix} \neq 0\,. \tag{8.58}$$

[3] Józef Höené-Wroński (1776–1853).

8.10 Symmetry of the Green Function

Consider the following two inhomogeneous boundary value problems:

$$L[u(x)] = r(x)\,, \qquad U_r[u(x)] = 0\,; \tag{8.59a}$$

$$L[v(x)] = s(x)\,, \qquad U_r[v(x)] = 0, \tag{8.59b}$$

where the differential operator L is given by Eq. (8.42b), $u(x)$ and $v(x)$ are the unknown functions, while $r(x)$ and $s(x)$ are continuous inhomogeneities in the interval $x \in [a, b]$. We shall assume that boundary value problems (8.59a) and (8.59b) are self-adjoint. Then

$$(u, v)_L - (v, u)_L = 0,$$

in which according to Eq. (8.44)

$$u(x) = \int_a^b G(x, \xi) r(\xi)\, \mathrm{d}\xi \quad \text{and} \quad v(x) = \int_a^b G(x, \xi) s(\xi)\, \mathrm{d}\xi\,.$$

Thus

$$(u, v)_L - (v, u)_L = \int_a^b (u\, L[v] - v\, L[u])\, \mathrm{d}x =$$

$$= \int_a^b \int_a^b s(x)\, [G(x, \xi) - G(\xi, x)]\, r(\xi)\, \mathrm{d}\xi \mathrm{d}x = 0\,. \tag{8.60}$$

Since both $r(x)$ and $s(x)$ are arbitrary continuous and non-zero functions in the interval $[a, b]$, it follows that the last integral in Eq. (8.60) can be zero if and only if

$$\boxed{G(x, \xi) = G(\xi, x)\,.} \tag{8.61}$$

In words: the Green function of self-adjoint boundary value problems is a symmetric function.

8.11 Calculation of the Green Functions for Some Practical Problems

8.11.1 Simple Beam Problems

Consider a uniform beam and assume that $N = 0$, which means that there is no axial force in the beam. Then according to Eq. (7.75b) the equilibrium problems of uniform beams are governed by the differential equation:

$$L(w) = \frac{d^4 w}{dx^4} = \frac{f_z(x)}{IE} = \tilde{f}_z(x), \tag{8.62}$$

where $f_z(x)$ is the intensity of the distributed load, I is the areal moment of inertia, and E is the Young's modulus. For a simply supported beam of length ℓ ($a = 0$, $b = \ell$), this equation is associated with the following boundary conditions:

$$U_1 = w(0) = 0\,, \quad U_2 = w^{(2)}(0) = 0\,, \quad U_3 = w(\ell) = 0\,, \quad U_4 = w^{(2)}(\ell) = 0. \tag{8.63}$$

Boundary value problem (8.62), (8.63) is self-adjoint. Our aim is to determine the Green function.

The linearly independent particular solutions of equation $L(w) = 0$ are given by

$$w_1 = 1, \quad w_2 = x, \quad w_3 = x^2, \quad w_4 = x^3\,. \tag{8.64}$$

We seek the Green function in the form given by Eq. (8.53) where the functions $b_1(\xi), \ldots, b_4(\xi)$ are solutions of the linear equation system (8.54):

$$\begin{bmatrix} 1 & \xi & \xi^2 & \xi^3 \\ 0 & 1 & 2\xi & 3\xi^2 \\ 0 & 0 & 2 & 6\xi \\ 0 & 0 & 0 & 6 \end{bmatrix} \begin{bmatrix} b_1 \\ b_2 \\ b_3 \\ b_4 \end{bmatrix} = \begin{bmatrix} 0 \\ 0 \\ 0 \\ -\frac{1}{2} \end{bmatrix}$$

from where we get

$$\begin{bmatrix} b_1 \\ b_2 \\ b_3 \\ b_4 \end{bmatrix} = \frac{1}{12} \begin{bmatrix} \xi^3 \\ -3\xi^2 \\ 3\xi \\ -1 \end{bmatrix}. \tag{8.65}$$

Boundary conditions (8.63) result in the following equation system:

$$a_1 w_1(0) + a_2 w_2(0) + a_3 w_3(0) + a_4 w_4(0) =$$
$$= -b_1 w_1(0) - b_2 w_2(0) - b_3 w_3(0) - b_4 w_4(0)\,,$$

$$a_1 w_1^{(2)}(0) + a_2 w_2^{(2)}(0) + a_3 w_3^{(2)}(0) + a_4 w_4^{(2)}(0) =$$
$$= -b_1 w_1^{(2)}(0) - b_2 w_2^{(2)}(0) - b_3 w_3^{(2)}(0) - b_4 w_4^{(2)}(0)\,,$$

$$a_1 w_1(\ell) + a_2 w_2(\ell) + a_3 w_3(\ell) + a_4 w_4(\ell) =$$
$$= b_1 w_1(\ell) + b_2 w_2(\ell) + b_3 w_3(\ell) + b_4 w_4(\ell)\,,$$

$$a_1 w_1^{(2)}(\ell) + a_2 w_2^{(2)}(\ell) + a_3 w_3^{(2)}(\ell) + a_4 w_4^{(2)}(\ell) =$$
$$= b_1 w_1^{(2)}(\ell) + b_2 w_2^{(2)}(\ell) + b_3 w_3^{(2)}(\ell) + b_4 w_4^{(2)}(\ell)$$

from where after substituting the particular solutions (8.64) and the results for the coefficients $b_1(\xi), \ldots, b_4(\xi)$ we obtain

$$\begin{bmatrix} 1 & 0 & 0 & 0 \\ 0 & 0 & 2 & 0 \\ 1 & \ell & \ell^2 & \ell^3 \\ 0 & 0 & 2 & 6\ell \end{bmatrix} \begin{bmatrix} a_1 \\ a_2 \\ a_3 \\ a_4 \end{bmatrix} = \frac{1}{12} \begin{bmatrix} -\xi^3 \\ -6\xi \\ \xi^3 - 3\xi^2\ell + 3\xi\ell^2 - \ell^3 \\ 6\xi - 6\ell \end{bmatrix}.$$

Thus

$$\begin{bmatrix} a_1 \\ a_2 \\ a_3 \\ a_4 \end{bmatrix} = \frac{1}{12\ell} \begin{bmatrix} -\xi^3\ell \\ \xi\left(2\xi^2 - 3\ell\xi + 4\ell^2\right) \\ -3\xi\ell \\ 2\xi - \ell \end{bmatrix}$$

and

$$G(x,\xi) = (a_1 + b_1) + (a_2 + b_2)\,x + (a_3 + b_3)\,x^2 + (a_4 + b_4)\,x^3 =$$
$$= \frac{x}{6\ell}\left(\xi^3 - 3\xi^2\ell + 2\xi\ell^2 + x^2\xi - x^2\ell\right) = \frac{x}{6\ell}\left(\ell - \xi\right)\left(2\ell\xi - \xi^2 - x^2\right) \qquad x \le \xi$$

Since boundary value problem (8.62), (8.63) is self-adjoint, the Green function is symmetric. Consequently, we can use symmetry Eq. (8.61) to give the Green function for $x \ge \xi$:

$$G(x,\xi) = \frac{\xi}{6\ell}\left(\ell - x\right)\left(2\ell x - \xi^2 - x^2\right) \qquad x \ge \xi\,.$$

Remark 8.8 Assume that $f_z = \delta\,(x - \xi)$ in boundary value problem (8.62), (8.63) for which we have determined the Green functions above. This means that the simply supported beam shown in Fig. 8.2 is subjected to a unit force at the point with coordinates ξ. It is also obvious that the right side of Eq. (8.62) is then

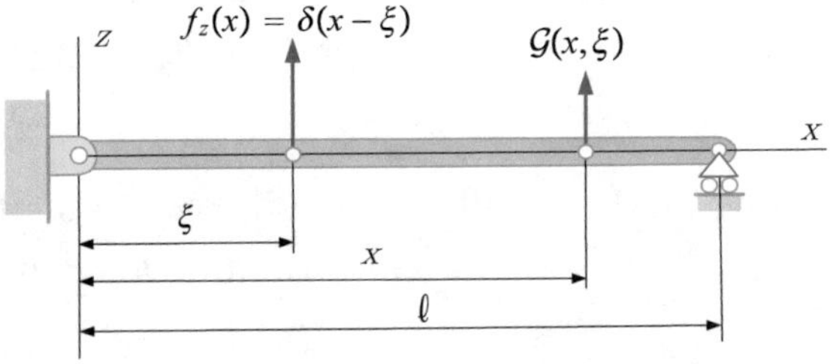

Fig. 8.2 Simply supported beam loaded by a vertical unit force

$$\tilde{f}_z(x) = \frac{\delta\,(x - \xi)}{IE}\,.$$

Making use of Eq. (8.44), we can now give the solution for the deflection $w(x)$ in the following form:

$$w(x) = \int_0^\ell G(x,\eta)\tilde{f}_z(\eta)\,\mathrm{d}\eta = \int_0^\ell \frac{1}{IE}G(x,\eta)\delta(\eta-\xi)\,\mathrm{d}\eta = \frac{1}{IE}G(x,\xi) = \mathcal{G}(x,\xi), \tag{8.66}$$

where we have taken integral (B.1.2) into account. It follows from this equation that (a) the Green function for the differential equation $IEw^{(4)}(x) = f_z(x)$, which is associated with boundary conditions (8.63) (simply supported beam), is $\mathcal{G}(x,\xi) = G(x,\xi)/IE$; (b) the Green function $\mathcal{G}(x,\xi)$ is the deflection $\mathcal{G}(x,\xi) = G(x,\xi)/IE$ of the beam at the point x due to a vertical unit force exerted on the beam at the point ξ. Thus, we can also calculate the Green function for beam problems by determining the deflection at the point x caused by the unit force exerted on the beam at the point ξ.

Table 8.1 Green's functions for some beams

Beam with its supports	Boundary conditions	Green's functions for DE $IEw^{(4)} = f_z$ ($\mathcal{G}(x,\xi)\ \forall x \le \xi$)
Simply supported 	$w(0) = w^{(2)}(0) = 0$ $w(\ell) = w^{(2)}(\ell) = 0$	$\frac{x}{6IE\ell}(\ell-\xi)(2\ell\xi-\xi^2-x^2)$
Fixed-pinned 	$w(0) = w^{(1)}(0) = 0$ $w(\ell) = w^{(2)}(\ell) = 0$	$\frac{x^2}{12IE\ell^3}(\ell-\xi)(6\xi\ell^2-2\ell^2x-3\xi^2\ell-2\ell\xi x+x\xi^2)$
Fixed-fixed 	$w(0) = w^{(1)}(0) = 0$ $w(\ell) = w^{(1)}(\ell) = 0$	$\frac{x^2}{6IE\ell^3}(\ell-\xi)^2(3\xi\ell-\ell x-2\xi x)$
Fixed-free 	$w(0) = w^{(1)}(0) = 0$ $w^{(2)}(\ell) = w^{(3)}(\ell) = 0$	$\frac{1}{6IE}x^2(3\xi-x)$

Table 8.1 contains the Green function $\mathcal{G}(x, \xi)$ for four typical support arrangements including the case of the simply supported beam. Calculation of the Green functions for fixed-pinned, fixed-fixed, and fixed-free beams is left for Problem 8.1, and the details are presented in Sect. C.8.

Given the Green functions $\mathcal{G}(x, \xi)$ for simply supported (pinned-pinned), fixed-pinned, fixed-fixed, and fixed-free beams—see Table 8.1 for details—solution for the deflection $w(x)$ due to the load f_z is given by the relation

$$w(x) = \int_0^\ell \mathcal{G}(x, \eta) f_z(\eta)\, \mathrm{d}\eta\,. \tag{8.67}$$

8.11.2 Preloaded Beams

Equilibrium problems of beams preloaded by an axial force are governed by the differential equation

$$L(w) = \frac{\mathrm{d}^4 w}{\mathrm{d}x^4} \mp \frac{N}{IE}\frac{\mathrm{d}^2 w}{\mathrm{d}x^2} = \frac{1}{IE} f_z, \tag{8.68}$$

which is associated with boundary conditions

$$w(0) = w^{(2)}(0) = w(\ell) = w^{(2)}(\ell) = 0 \tag{8.69}$$

for the simply supported beam shown in Fig. 8.3. It can easily be seen that boundary value problem (8.68), (8.69) is self-adjoint.

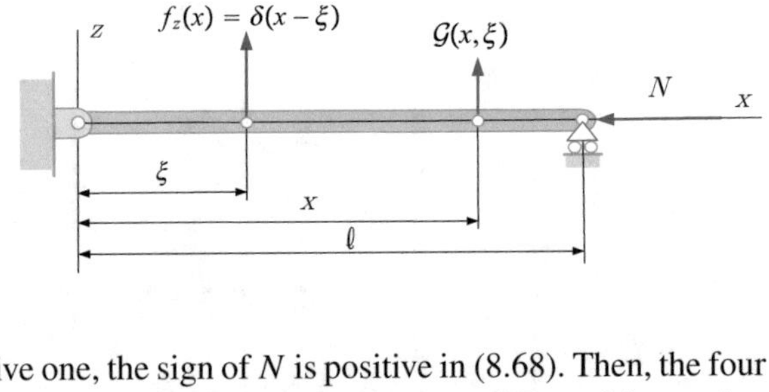

Fig. 8.3 Simply supported beam loaded by a vertical unit force—the preload is an axial force

If the force is a compressive one, the sign of N is positive in (8.68). Then, the four linearly independent particular solutions of the homogeneous differential equation $L(w) = 0$ are

$$w_1 = 1, \quad w_2 = x, \quad w_3 = \cos px, \quad w_4 = \sin px, \tag{8.70}$$

$$p = \sqrt{N/IE}\,, \quad N > 0\,. \tag{8.71}$$

Our aim is to find the Green function both for compressive and tensile axial forces. First, we shall assume that the axial force is a compressive one.

It follows on the basis of Remark 8.8 that the Green function is the deflection at the point $x \in [0, \ell]$ due to a unit force exerted on the beam at the point $\xi \in [0, \ell]$. Since the point ξ divides the interval $[0, \ell]$ into two parts, we shall assume that

$$\mathcal{G}(x, \xi) = \mathcal{A}_1(\xi) + \mathcal{A}_2(\xi)\, x + \mathcal{A}_3(\xi) \cos px + \mathcal{A}_4(\xi) \sin px, \qquad x \leq \xi; \tag{8.72a}$$

$$\mathcal{G}(x, \xi) = \mathcal{B}_1(\xi) + \mathcal{B}_2(\xi)\, x + \mathcal{B}_3(\xi) \cos px + \mathcal{B}_4(\xi) \sin px, \qquad x \geq \xi. \tag{8.72b}$$

The Green function should satisfy

(a) the boundary conditions at the left end of the beam:

$$\mathcal{G}(0, \xi) = 0 \qquad \text{(the deflection is zero)}, \tag{8.73a}$$

$$\mathcal{G}^{(2)}(0, \xi) = 0 \qquad \text{(the bending moment is zero)}; \tag{8.73b}$$

(b) the continuity and discontinuity conditions if $\xi = x$:

$$\mathcal{G}(\xi - 0, \xi) = \mathcal{G}(\xi + 0, \xi) \qquad \text{(continuity of the deflection)}, \tag{8.74a}$$

$$\mathcal{G}^{(1)}(\xi - 0, \xi) = \mathcal{G}^{(1)}(\xi + 0, \xi) \qquad \text{(continuity of the rotation—see } (7.74)_1), \tag{8.74b}$$

$$\mathcal{G}^{(2)}(\xi - 0, \xi) = \mathcal{G}^{(2)}(\xi + 0, \xi) \quad \text{(continuity of the bending moment—see } (7.74)_1), \tag{8.74c}$$

$$\mathcal{G}^{(3)}(\xi - 0, \xi) - \mathcal{G}^{(3)}(\xi + 0, \xi) = -1/IE \tag{8.74d}$$

(discontinuity condition by taking into account the fact that the right side of equation is $\delta(x - \xi)/IE$ for the Green function);

(c) and the boundary conditions at the right end of the beam:

$$\mathcal{G}(\ell, \xi) = 0 \qquad \text{(the deflection is zero)}, \tag{8.75a}$$

$$\mathcal{G}^{(2)}(\ell, \xi) = 0 \qquad \text{(the bending moment is zero)}. \tag{8.75b}$$

Substitution of (8.72) into Eqs. (8.73)–(8.75) results in a system of linear equations for the unknown functions $\mathcal{A}_1(\xi), \ldots, \mathcal{A}_4(\xi)$ and $\mathcal{B}_1(\xi), \ldots, \mathcal{B}_4(\xi)$. The coefficient matrix is of the form

$$\underline{\boldsymbol{\mathcal{A}}} = \begin{bmatrix} 1 & 0 & 1 & 0 & 0 & 0 & 0 & 0 \\ 0 & 0 & -p^2 & 0 & 0 & 0 & 0 & 0 \\ 1 & \xi & \cos p\xi & \sin p\xi & -1 & -\xi & -\cos p\xi & -\sin p\xi \\ 0 & 1 & -p\sin p\xi & p\cos p\xi & 0 & -1 & p\sin p\xi & -p\cos p\xi \\ 0 & 0 & -p^2\cos p\xi & -p^2\sin p\xi & 0 & 0 & p^2\cos p\xi & p^2\sin p\xi \\ 0 & 0 & p^3\sin p\xi & -p^3\cos p\xi & 0 & 0 & -p^3\sin p\xi & p^3\cos p\xi \\ 0 & 0 & 0 & 0 & 1 & \ell & \cos p\ell & \sin p\ell \\ 0 & 0 & 0 & 0 & 0 & 0 & -p^2\cos p\ell & -p^2\sin p\ell \end{bmatrix}$$

Utilizing the notations

$$\underline{\boldsymbol{\mathcal{X}}}^T = \left[\ \mathcal{A}_1 \middle| \mathcal{A}_2 \middle| \mathcal{A}_3 \middle| \mathcal{A}_4 \middle|\ \mathcal{B}_1 \middle| \mathcal{B}_2 \middle| \mathcal{B}_3 \middle| \mathcal{B}_4 \right],$$

and

$$\underline{\boldsymbol{\mathcal{F}}}^T = \left[\ 0|0|0|0|0|-1/IE|0|0\right],$$

the equation system to be solved can be given in the following form:

$$\underline{\boldsymbol{\mathcal{A}}}\,\underline{\boldsymbol{\mathcal{X}}} = \underline{\boldsymbol{\mathcal{F}}}. \tag{8.76}$$

Making use of the solutions

$$\mathcal{A}_1 = 0\,, \qquad \mathcal{A}_2 = \frac{\xi - \ell}{p^2 IE\ell}\,, \qquad \mathcal{A}_3 = 0\,, \qquad \mathcal{A}_4 = -\frac{\sin\left[p(\xi - \ell)\right]}{p^3 IE\sin p\ell} \tag{8.77}$$

we can determine the sought Green functions from Eq. (8.72a):

$$\mathcal{G}(x,\xi) = \frac{1}{p^3 IE\ell\sin p\ell}\left[xp\,(\xi - \ell)\sin p\ell - \ell\sin\left[p(\xi - \ell)\right]\sin px\right]\,, \qquad x \le \xi \tag{8.78}$$

The Green function for the case of $x \ge \xi$ can be calculated either by substituting the solutions

$$\mathcal{B}_1 = -\frac{\xi}{p^2 IE},\quad \mathcal{B}_2 = \frac{\xi}{p^2 IE\ell},\quad \mathcal{B}_3 = \frac{\sin p\xi}{p^3 IE},\quad \mathcal{B}_4 = -\frac{1}{p^3 IE\sin p\ell}\sin px\cos p\ell \tag{8.79}$$

into (8.72b), or by writing ξ for x and x for ξ in (8.78). Hence

$$\mathcal{G}(x,\xi) = \mathcal{D}_c\left[xp\,(\xi - \ell)\sin p\ell - \ell\sin\left[p(\xi - \ell)\right]\sin px\right]\,, \qquad x \le \xi; \tag{8.80a}$$

$$\mathcal{G}(x,\xi) = \mathcal{D}_c\left[\xi p\,(x - \ell)\sin p\ell - \ell\sin\left[p(\xi - \ell)\right]\sin p\xi\right]\,, \qquad x \ge \xi \tag{8.80b}$$

where

$$\mathcal{D}_c = 1/p^3 IE\ell\sin p\ell\,. \tag{8.80c}$$

If the force is a tensile one

$$w_1 = 1, \quad w_2 = x, \quad w_3 = \cosh px, \quad w_4 = \sinh px, \tag{8.81}$$

$$p = \sqrt{N/IE}, \quad N > 0 \tag{8.82}$$

are the four linearly independent particular solutions of the differential equation $L(w) = 0$. Then the Green function assumes the form

$$\mathcal{G}(x, \xi) = \mathcal{A}_1(\xi) + \mathcal{A}_2(\xi)\, x + \mathcal{A}_3(\xi) \cosh px + \mathcal{A}_4(\xi) \sinh px, \qquad x \le \xi; \tag{8.83a}$$

$$\mathcal{G}(x, \xi) = \mathcal{B}_1(\xi) + \mathcal{B}_2(\xi)\, x + \mathcal{B}_3(\xi) \cosh px + \mathcal{B}_4(\xi) \sinh px, \qquad x \ge \xi. \tag{8.83b}$$

Continuity and discontinuity conditions (8.73)–(8.75) yield again a system of linear equations for the unknowns

$$\mathcal{A}_1(\xi), \dots, \mathcal{A}_4(\xi) \quad \text{and} \quad \mathcal{B}_1(\xi), \dots, \mathcal{B}_4(\xi).$$

Making use of the solutions

$$\mathcal{A}_1 = 0, \quad \mathcal{A}_2 = -\frac{\xi - \ell}{p^2 IE\ell}, \quad \mathcal{A}_3 = 0, \quad \mathcal{A}_4 = \frac{\sinh[p(\xi - \ell)]}{p^3 IE \sinh p\ell}, \tag{8.84a}$$

$$\begin{aligned} \mathcal{B}_1 &= \frac{\xi}{p^2 IE}, \quad \mathcal{B}_2 = -\frac{\xi}{p^2 IE\ell}, \\ \mathcal{B}_3 &= -\frac{\sinh p\xi}{p^3 IE}, \quad \mathcal{B}_4 = \frac{1}{p^3 IE \sinh p\ell} \sinh p\xi \cosh p\ell \end{aligned} \tag{8.84b}$$

from (8.83), one obtains the Green function:

$$\mathcal{G}(x, \xi) = \mathcal{D}_t \left[xp(\ell - \xi) \sinh p\ell - \ell \sinh[p(\xi - \ell)] \sinh px \right], \quad x \le \xi. \tag{8.85a}$$

$$\mathcal{G}(x, \xi) = \mathcal{D}_t \left[\xi p(\ell - x) \sinh p\ell - \ell \sinh[p(\xi - \ell)] \sinh p\xi \right], \quad x \ge \xi, \tag{8.85b}$$

where

$$\mathcal{D}_t = 1/p^3 IE\ell \sinh p\ell. \tag{8.85c}$$

It can be checked with ease that Green functions (8.78) and (8.83) satisfy symmetry condition (8.61).

Tables 8.2 and 8.3 contain the Green functions of pinned-pinned, fixed-pinned, and fixed-fixed beams both for compressive and for tensile preloads.

Given the Green functions $\mathcal{G}(x, \xi)$ for the axially preloaded simply supported (pinned-pinned), fixed-pinned, and fixed-fixed beams in Tables 8.2 and 8.3 solution for the deflection $w(x)$ due to the transverse load f_z is given by Eq. (8.67).

Table 8.2 Green's functions for compressive preload

Beams preloaded by a compressive force	Boundary conditions	Green's functions for DE $w^4 + pw^{(2)} = f_z/IE$ ($\mathcal{G}(x, \xi)\ \forall x \leq \xi$)
Simply supported 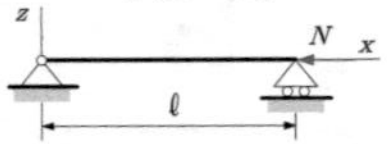	$w(0)=w^{(2)}(0)=0$ $w(\ell)=w^{(2)}(\ell)=0$	$\mathcal{D}_c\,[\xi p\,(x-\ell)\sin p\ell - \ell\sin[p(\xi-\ell)]\sin p\xi]$ $\mathcal{D}_c = 1/p^3 IE\ell\sin p\ell$
Fixed-pinned 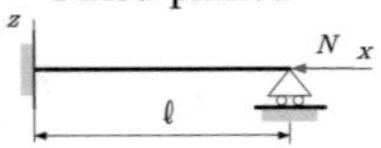	$w(0)=w^{(1)}(0)=0$ $w(\ell)=w^{(2)}(\ell)=0$	$D_c\{p[(\xi-\ell)\sin p\ell-\ell\sin[p(\xi-\ell)]](1+\cos px)-$ $-[p\,(\xi-\ell)\cos p\ell-\sin[p(\xi-\ell)]](\sin px-px)\}$ $D_c = 1/p^3 IE\,(p\ell\cos p\ell - \sin p\ell)$
Fixed-fixed 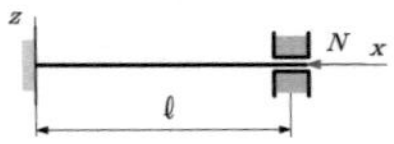	$w(0)=w^{(1)}(0)=0$ $w(\ell)=w^{(1)}(\ell)=0$	$\mathcal{D}_c\Big\{\big[-\xi p+\sin[p(\xi-\ell)]+\ell p\cos[p(\xi-\ell)]-\sin p\xi+$ $+\sin p\ell+(\xi-\ell)\,p\cos p\ell\big](1-\cos px)-$ $-\big[1-\cos[p(\xi-\ell)]+(\xi-\ell)\,p\sin p\ell-$ $-\cos p\ell+\cos p\xi\big](\sin px-px)\Big\}$ $\mathcal{D}_c = 1/p^3 IE\,(2\cos p\ell + \ell p\sin p\ell - 2)$

Table 8.3 Green's functions for tensile preload

Beams preloaded by a tensile force	Boundary conditions	Green's functions for DE $w^4 - pw^{(2)} = f_z/IE$ $(\mathcal{G}(x,\xi)\ \forall x \leq \xi)$
Simply supported 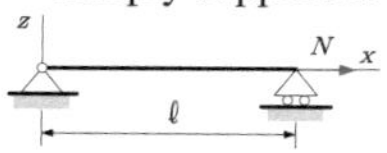	$w(0) = w^{(2)}(0) = 0$ $w(\ell) = w^{(2)}(\ell) = 0$	$\mathcal{D}_t\big[xp\,(\ell-\xi)\sinh p\ell - \ell\sinh[p(\xi-\ell)]\sinh p\xi\big]$ $\mathcal{D}_t = 1/p^3 IE\ell\sinh p\ell$
Fixed-pinned 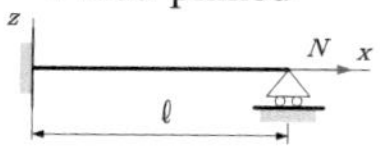	$w(0) = w^{(1)}(0) = 0$ $w(\ell) = w^{(2)}(\ell) = 0$	$D_t\Big\{p\big[(\xi-\ell)\sinh p\ell - \ell\sinh[p(\xi-\ell)]\big](1-\cosh px) -$ $-\big[\sinh[p(\xi-\ell)] - p\,(\xi-\ell)\cosh p\ell\big](\sinh px - px)\Big\}$ $D_t = 1/p^3 IE\,(p\ell\cosh p\ell - \sinh p\ell)$
Fixed-fixed 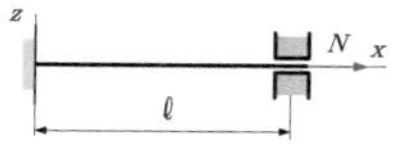	$w(0) = w^{(1)}(0) = 0$ $w(\ell) = w^{(1)}(\ell) = 0$	$\mathcal{D}_t\Big\{\big[-\xi p + \sinh[p(\xi-\ell)] + \ell p\cosh[p(\xi-\ell)] - \sinh p\xi +$ $+\sinh p\ell + (\xi-\ell)\,p\cosh p\ell\big](1-\cosh px) -$ $-\big[1-\cosh[p(\xi-\ell)] - (\xi-\ell)\,p\sinh p\ell -$ $-\cosh p\ell + \cosh p\xi\big](\sinh px - px)\Big\}$ $\mathcal{D}_t = 1/p^3 IE\,(2 - 2\cosh p\ell + \ell p\sinh p\ell)$

8.11.3 Preloaded Circular Plates

8.11.3.1 The Governing Equations of the Problem

Figure 8.4 shows (1) a clamped circular plate, (2) a simply supported circular plate, and (3) a simply supported circular plate which is elastically restrained on its boundary by a linear torsional spring with a spring constant k_γ. The thickness and the outer radius of the plates are denoted by $2b$ and $R_\circ \gg 2b$. The material of the plates is homogeneous and isotropic with Young's modulus E and Poisson's number ν. We shall assume that the plates are preloaded by a constant radial load f_R in their middle plane. The preload can be either compressive or tensile—Figure 8.4 depicts the case of compressive radial load. Moreover, the plates are subjected to a vertical distributed load p_z. It is a further assumption that the vertical load p_z, the deformations, and the stress state are all axisymmetric, i.e., they are functions of the radial coordinate R only.

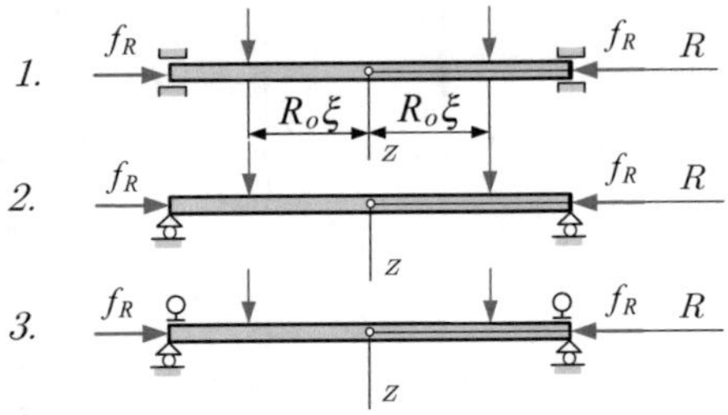

Fig. 8.4 Circular plates preloaded by a compressive radial load

For this reason, we shall use the cylindrical coordinate system (R, θ, z) when calculating the Green functions that belong to the boundary value problems of the three support arrangements shown in Fig. 8.4.

For our later considerations, we shall introduce some notations: $I_1 = 8b^3/12$ is the moment of inertia, $E_1 = E/(1-\nu^2)$ is the modified Young's modulus, $\mathcal{F} = R_o^2 f_R/I_1 E_1$ and $\mathcal{F}_\nu = \mathcal{F}/(1-\nu)$ are the dimensionless in-plane load and the modified dimensionless in-plane load, $\mathcal{K} = R_o k_\gamma / I_1 E_1$ and $\mathcal{K}_\nu = 1 - \mathcal{K}/(1-\nu)$ are the dimensionless spring constant and the modified dimensionless spring constant, and $r = R/R_o$ and ξ are dimensionless independent variables. The bending moment and shear force per unit length are denoted by M_R and Q_R.

The deflection w due to the load $p_z(r)$ acting perpendicularly to the middle plane of the plate should meet the differential equation [9, 10]

$$\tilde{\Delta}\tilde{\Delta}w \pm \mathcal{F}\tilde{\Delta}w = \frac{R_o^4}{I_1 E_1} p_z, \qquad \tilde{\Delta} = \frac{\mathrm{d}^2}{\mathrm{d}r^2} + \frac{1}{r}\frac{\mathrm{d}}{\mathrm{d}r} \tag{8.86}$$

where the sign preceding $\mathcal{F}$ is positive for compression and negative for tension. It is known [11] that

$$\varphi_\theta = -\frac{1}{R_o}\frac{\mathrm{d}w}{\mathrm{d}r}, \quad M_R = -\frac{I_1 E_1}{R_o^2}\left(\frac{\mathrm{d}^2 w}{\mathrm{d}r^2} + \frac{\nu}{r}\frac{\mathrm{d}w}{\mathrm{d}r}\right) \tag{8.87}$$

and

$$Q_R \frac{R_o^3}{I_1 E_1} = \frac{\mathrm{d}}{\mathrm{d}r}\left[\frac{\mathrm{d}^2}{\mathrm{d}r^2} + \frac{1}{r}\frac{\mathrm{d}}{\mathrm{d}r} \pm \mathcal{F}\right] w(r) \tag{8.88}$$

are the rotation about the axis θ, the bending moment, and the shear force in terms of the deflection.

Depending on what the supports are, Eq. (8.86) should be associated with appropriate boundary conditions. As regards the outer boundary, it is clear from Fig. 8.4 that for a clamped plate

$$w|_{r=1} = 0, \qquad \left.\frac{\mathrm{d}w}{\mathrm{d}r}\right|_{r=1} = 0 \tag{8.89}$$

are the boundary conditions. If the plate is simply supported then the boundary conditions are of the form

$$w|_{r=1} = 0, \qquad \left.\left(\frac{\mathrm{d}^2 w}{\mathrm{d}r^2} + \frac{\nu}{r}\frac{\mathrm{d}w}{\mathrm{d}r}\right)\right|_{r=1} = 0\,. \tag{8.90}$$

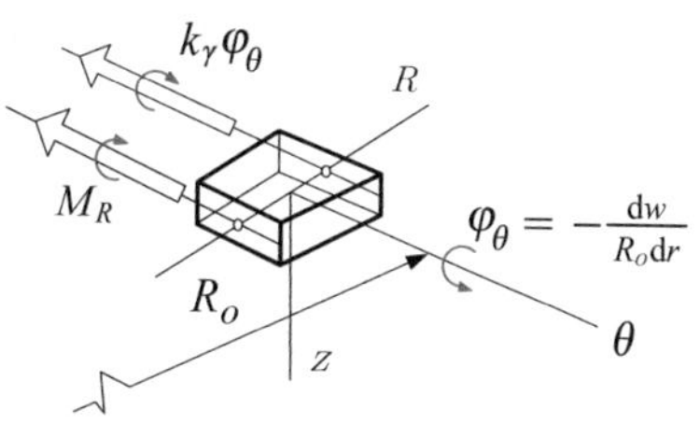

Fig. 8.5 Circular plate element

For the plate supported by a torsional spring, it follows from Figs. 8.4 and 8.5 that

$$\begin{gathered} w|_{r=1} = 0\,, \\ M_R|_{r=1} + k_\gamma \varphi_\theta|_{r=1} = \left.\left(\frac{\mathrm{d}^2 w}{\mathrm{d}r^2} + \frac{\nu}{r}\frac{\mathrm{d}w}{\mathrm{d}r}\right)\right|_{r=1} + \frac{R_o}{I_1 E_1} k_\gamma \left.\frac{\mathrm{d}w}{\mathrm{d}r}\right|_{r=1} = 0 \end{gathered} \tag{8.91}$$

are the boundary conditions. It is also clear that the deflection at the center of the plate should meet the conditions

$$w|_{r=0} = \text{finite}\,, \qquad \left.\frac{\mathrm{d}w}{\mathrm{d}r}\right|_{r=0} = 0\,. \tag{8.92}$$

8.11.3.2 Definition of the Green Function

Assume that the plate is subjected to a uniform load distributed on the circle with radius $R_o\xi$ (ξ is also a dimensionless coordinate)—see Fig. 8.4 . The resultant of the total load is assumed to be 1. The deflection due to the load at r is denoted by $G(r, \xi)$ and is referred to as the Green function. It is obvious that the Green function should satisfy the homogeneous equation in (8.86) if $0 \leq r < \xi$ and $\xi < r \leq 1$.

8.11.3.3 Green Functions for Compressive f_R

As is well known, the general solution of the homogeneous equation in (8.86) assumes the form

$$w(r) = c_1 + c_2 \ln r + c_3 J_o(\sqrt{\mathcal{F}}r) + c_4 Y_o(\sqrt{\mathcal{F}}r), \tag{8.93}$$

where c_1, c_2, c_3, and c_4 are undetermined constants of integration, while $J_o(\sqrt{\mathcal{F}}r)$ and $Y_o(\sqrt{\mathcal{F}}r)$ are Bessel functions. Since the Green function should meet conditions (8.92), it can be given as

$$\mathcal{G}(r,\xi) = \mathcal{A}_1 + \mathcal{A}_3 J_0(\sqrt{\mathcal{F}}r)\,, \qquad 0 \leq r \leq \xi\,, \tag{8.94a}$$

$$\mathcal{G}(r,\xi) = \mathcal{B}_1 + \mathcal{B}_2 \ln r + \mathcal{B}_3 J_0(\sqrt{\mathcal{F}}r) + \mathcal{B}_4 Y_0(\sqrt{\mathcal{F}}r)\,, \qquad \xi \leq r \leq 1, \tag{8.94b}$$

where the constants of integration $\mathcal{A}_1, \mathcal{A}_3$ and $\mathcal{B}_1, \mathcal{B}_2, \mathcal{B}_3, \mathcal{B}_4$ are to be determined from the continuity and discontinuity conditions prescribed at $r = \xi$ and from the boundary conditions which are imposed on the boundary $r = 1$.

Note that the continuity conditions

$$\mathcal{G}(\xi - 0, \xi) = \mathcal{G}(\xi + 0, \xi)\,, \tag{8.95a}$$

$$\mathcal{G}^{(1)}(\xi - 0, \xi) = \mathcal{G}^{(1)}(\xi + 0, \xi)\,, \tag{8.95b}$$

$$\mathcal{G}^{(2)}(\xi - 0, \xi) = \mathcal{G}^{(2)}(\xi + 0, \xi) \tag{8.95c}$$

and the discontinuity condition

$$2\pi R_o\xi\,[\mathcal{Q}_R(\xi + 0) - \mathcal{Q}_R(\xi - 0)] = 2\pi R_o\xi\mathcal{Q}_R(\xi + 0) = 1\,, \tag{8.96}$$

in which $\mathcal{Q}_R(\xi - 0) = 0$ from the vertical equilibrium are all independent of the supports. Here (a) the derivatives with respect to r are denoted in the same manner as the derivatives with respect to x in (8.45) and (b) according to (8.88) it holds that

$$\mathcal{Q}_R \frac{R_o^3}{I_1 E_1} = \frac{\mathrm{d}}{\mathrm{d}r}\left[\frac{\mathrm{d}^2}{\mathrm{d}r^2} + \frac{1}{r}\frac{\mathrm{d}}{\mathrm{d}r} + \mathcal{F}\right]\mathcal{G}(\xi, r)\,. \tag{8.97}$$

In the sequel, we shall determine the Green function for the elastically restrained and simply supported plate.

When calculating the derivatives of the Green functions, we have to utilize the derivatives of the Bessel functions $J_o(\sqrt{\mathcal{F}r})$ and $Y_o(\sqrt{\mathcal{F}r})$. The latter derivatives are given by Eqs. (B.2.7). Thus

$$\mathcal{G}^{(1)}(r,\xi) = -\mathcal{A}_3\sqrt{\mathcal{F}}J_1(\sqrt{\mathcal{F}r}) \qquad 0 \le r \le \xi\,, \tag{8.98a}$$

$$\mathcal{G}^{(1)}(r,\xi) = \mathcal{B}_2\frac{1}{r} - \mathcal{B}_3\sqrt{\mathcal{F}}J_1(\sqrt{\mathcal{F}r}) - \mathcal{B}_4\sqrt{\mathcal{F}}Y_1(\sqrt{\mathcal{F}r}) \qquad \xi \le r \le 1\,, \tag{8.98b}$$

$$\mathcal{G}^{(2)}(r,\xi) = \mathcal{A}_3\mathcal{F}\left[\frac{J_1\left(\sqrt{\mathcal{F}r}\right)}{\sqrt{\mathcal{F}r}} - J_0\left(\sqrt{\mathcal{F}r}\right)\right], \qquad 0 \le r \le \xi\,, \tag{8.99a}$$

$$\begin{aligned}\mathcal{G}^{(2)}(r,\xi) = -\mathcal{B}_2\frac{1}{r^2} + \mathcal{B}_3\mathcal{F}\left[\frac{J_1\left(\sqrt{\mathcal{F}r}\right)}{\sqrt{\mathcal{F}r}} - J_0\left(\sqrt{\mathcal{F}r}\right)\right] + \\ + \mathcal{B}_4\mathcal{F}\left(\frac{Y_1\left(\sqrt{\mathcal{F}r}\right)}{\sqrt{\mathcal{F}r}} - Y_0\left(\sqrt{\mathcal{F}r}\right)\right), \qquad \xi \le r \le 1\,, \end{aligned}\tag{8.99b}$$

$$\begin{aligned}\mathcal{G}^{(3)}(r,\xi) &= -\mathcal{A}_3\frac{\mathcal{F}^{3/2}}{4}\left[J_3(\sqrt{\mathcal{F}r}) - 3J_1(\sqrt{\mathcal{F}r})\right] = \\ &= \mathcal{A}_3\frac{\mathcal{F}^{3/2}}{4}\left[\frac{1}{\sqrt{\mathcal{F}r}}J_2(\sqrt{\mathcal{F}r}) - J_1\left(\sqrt{\mathcal{F}r}\right)\right], \qquad 0 \le r \le \xi \end{aligned}\tag{8.100a}$$

and

$$\begin{aligned}&\mathcal{G}^{(3)}(r,\xi) = \\ &= \frac{2}{r^3}\mathcal{B}_2 - \mathcal{B}_3\frac{\mathcal{F}^{3/2}}{4}\left[J_3(\sqrt{\mathcal{F}r}) - 3J_1(\sqrt{\mathcal{F}r})\right] - \mathcal{B}_4\frac{\mathcal{F}^{3/2}}{4}\left[Y_3(\sqrt{\mathcal{F}r}) - 3Y_1(\sqrt{\mathcal{F}r})\right] = \\ &\quad = \frac{2}{r^3}\mathcal{B}_2 + \mathcal{B}_3\frac{\mathcal{F}^{3/2}}{4}\left[\frac{1}{\sqrt{\mathcal{F}r}}J_2(\sqrt{\mathcal{F}r}) - J_1\left(\sqrt{\mathcal{F}r}\right)\right] + \\ &\quad + \mathcal{B}_4\frac{\mathcal{F}^{3/2}}{4}\left[\frac{1}{\sqrt{\mathcal{F}r}}Y_2(\sqrt{\mathcal{F}r}) - Y_1\left(\sqrt{\mathcal{F}r}\right)\right], \qquad \xi \le r \le 1 \end{aligned}\tag{8.100b}$$

are the derivatives of the Green function.

After substituting the Green function and its derivatives into the continuity condition (8.95a) and combining then the continuity conditions (8.95b) and (8.95c), we have

$$\mathcal{A}_1 + \mathcal{A}_3 J_0(\sqrt{\mathcal{F}}\xi) = \mathcal{B}_1 + \mathcal{B}_2 \ln \xi + \mathcal{B}_3 J_0(\sqrt{\mathcal{F}}\xi) + \mathcal{B}_4 Y_0(\sqrt{\mathcal{F}}\xi)\,, \tag{8.101a}$$

$$-\mathcal{A}_3\sqrt{\mathcal{F}} J_1\left(\sqrt{\mathcal{F}}\xi\right) = \mathcal{B}_2\frac{1}{\xi} - \mathcal{B}_3\sqrt{\mathcal{F}} J_1\left(\sqrt{\mathcal{F}}\xi\right) - \mathcal{B}_4\sqrt{\mathcal{F}} Y_1\left(\sqrt{\mathcal{F}}\xi\right)\,, \tag{8.101b}$$

$$\mathcal{A}_3 J_0(\sqrt{\mathcal{F}}\xi) = \mathcal{B}_3 J_0(\sqrt{\mathcal{F}}\xi) + \mathcal{B}_4 Y_0(\sqrt{\mathcal{F}}\xi)\,. \tag{8.101c}$$

It follows from (8.97) that

$$\begin{aligned}
\frac{R_K^3}{I_1 E_1} Q_R(\xi+0) &= \mathcal{B}_2 \frac{d}{dr}\left\{-\frac{1}{r^2} + \frac{1}{r^2} + \mathcal{F}\ln r\right\}\bigg|_{r=\xi} + \\
&+ \mathcal{B}_3 \frac{d}{dr}\left\{\left[\frac{\mathcal{F}}{2}\left[J_2(\sqrt{\mathcal{F}}r) - J_0(\sqrt{\mathcal{F}}r)\right] - \mathcal{F}\frac{J_1(\sqrt{\mathcal{F}}r)}{\sqrt{\mathcal{F}}r}\right] + \mathcal{F} J_0(\sqrt{\mathcal{F}}r)\right\}\bigg|_{r=\xi} \\
&+ \mathcal{B}_4 \frac{d}{dr}\left\{\left[\frac{\mathcal{F}}{2}\left[Y_2(\sqrt{\mathcal{F}}r) - Y_0(\sqrt{\mathcal{F}}r)\right] - \mathcal{F}\frac{Y_1(\sqrt{\mathcal{F}}r)}{\sqrt{\mathcal{F}}r}\right] + \mathcal{F} Y_0(\sqrt{\mathcal{F}}r)\right\}\bigg|_{r=\xi} = \\
&= \mathcal{B}_2 \frac{d}{dr}\left\{-\frac{1}{r^2} + \frac{1}{r^2} + \mathcal{F}\ln r\right\}\bigg|_{r=\xi} + \\
&+ \mathcal{B}_3 \underbrace{\frac{d}{dr}\left\{\left[\frac{\mathcal{F}}{2}\left[J_2(\sqrt{\mathcal{F}}r) + J_0(\sqrt{\mathcal{F}}r)\right] - \mathcal{F}\frac{J_1(\sqrt{\mathcal{F}}r)}{\sqrt{\mathcal{F}}r}\right]\right\}}_{=0}\Bigg|_{r=\xi} + \\
&+ \mathcal{B}_4 \underbrace{\frac{d}{dr}\left\{\left[\frac{\mathcal{F}}{2}\left[Y_2(\sqrt{\mathcal{F}}r) + Y_0(\sqrt{\mathcal{F}}r)\right] - \mathcal{F}\frac{Y_1(\sqrt{\mathcal{F}}r)}{\sqrt{\mathcal{F}}r}\right]\right\}}_{=0}\Bigg|_{r=\xi} = \mathcal{B}_2\mathcal{F}\frac{1}{\xi}\,.
\end{aligned}$$

Consequently, discontinuity condition (8.96) leads to the equation

$$\mathcal{B}_2 = \frac{R_o^3}{I_1 E_1}\frac{1}{2\pi\xi R_o}\frac{1}{\mathcal{F}}\xi = \frac{R_o^2}{I_1 E_1 \frac{f_R R_o^2}{I_1 E_1}}\frac{1}{2\pi\xi}\xi = \frac{1}{2\pi f_R}\,. \tag{8.101d}$$

The last two equations for the integration constants are obtained from the boundary conditions (8.91):

$$\mathcal{B}_1 + \mathcal{B}_3 J_0(\sqrt{\mathcal{F}}) + \mathcal{B}_4 Y_0(\sqrt{\mathcal{F}}) = 0\,, \tag{8.102a}$$

$$\left(\frac{\mathrm{d}^2\mathcal{G}(r,\xi)}{\mathrm{d}r^2} + \frac{\nu}{r}\frac{\mathrm{d}\mathcal{G}(r,\xi)}{\mathrm{d}r}\right)\bigg|_{r=1} = -\,\mathcal{K}\frac{\mathrm{d}\mathcal{G}(r,\xi)}{\mathrm{d}r}\bigg|_{r=1}\,, \tag{8.102b}$$

where

$$\mathcal{K} = \frac{R_o k_\gamma}{I_1 E_1} \tag{8.103}$$

is the earlier mentioned dimensionless spring constant. The left side of boundary condition $(8.102\text{b})_2$ can be manipulated into a more suitable form if we take into account that

$$\frac{\mathrm{d}^2\mathcal{G}(r)}{\mathrm{d}r^2} + \frac{1}{r}\frac{\mathrm{d}\mathcal{G}(r)}{\mathrm{d}r} - (1-\nu)\frac{1}{r}\frac{\mathrm{d}\mathcal{G}(r)}{\mathrm{d}r} = \mathcal{B}_2\left\{-\frac{1}{r^2} + \frac{1}{r^2} - (1-\nu)\frac{1}{r^2}\right\} +$$
$$+\mathcal{B}_3\left\{\underbrace{\frac{\mathcal{F}}{2}\left[J_2(\sqrt{\mathcal{F}}r) - J_0(\sqrt{\mathcal{F}}r)\right] - \mathcal{F}\frac{J_1(\sqrt{\mathcal{F}}r)}{\sqrt{\mathcal{F}}r}}_{-\mathcal{F}J_0(\sqrt{\mathcal{F}}r)} + (1-\nu)\mathcal{F}\frac{J_1(\sqrt{\mathcal{F}}r)}{\sqrt{\mathcal{F}}r}\right\} +$$
$$+\mathcal{B}_4\left\{\underbrace{\frac{\mathcal{F}}{2}\left[Y_2(\sqrt{\mathcal{F}}r) - Y_0(\sqrt{\mathcal{F}}r)\right] - \mathcal{F}\frac{Y_1(\sqrt{\mathcal{F}}r)}{\sqrt{\mathcal{F}}r}}_{-\mathcal{F}Y_0(\sqrt{\mathcal{F}}r)} + (1-\nu)\mathcal{F}\frac{Y_1(\sqrt{\mathcal{F}}r)}{\sqrt{\mathcal{F}}r}\right\} =$$
$$= -\mathcal{B}_2(1-\nu)\frac{1}{r^2} + \mathcal{B}_3\left[(1-\nu)\mathcal{F}\frac{J_1(\sqrt{\mathcal{F}}r)}{\sqrt{\mathcal{F}}r} - \mathcal{F}J_0(\sqrt{\mathcal{F}}r)\right] +$$
$$+\mathcal{B}_4\left[(1-\nu)\mathcal{F}\frac{Y_1(\sqrt{\mathcal{F}}r)}{\sqrt{\mathcal{F}}r} - \mathcal{F}Y_0(\sqrt{\mathcal{F}}r)\right]$$

from where for $r = 1$ we get

$$\left.\frac{d^2\mathcal{G}(r,\xi)}{dr^2} + \nu\frac{1}{r}\frac{d\mathcal{G}(r,\xi)}{dr}\right|_{r=1} = -\mathcal{B}_2(1-\nu) +$$
$$+\mathcal{B}_3\left[(1-\nu)\mathcal{F}\frac{J_1(\sqrt{\mathcal{F}})}{\sqrt{\mathcal{F}}} - \mathcal{F}J_0(\sqrt{\mathcal{F}})\right] + \mathcal{B}_4\left[(1-\nu)\mathcal{F}\frac{Y_1(\sqrt{\mathcal{F}})}{\sqrt{\mathcal{F}}} - \mathcal{F}Y_0(\sqrt{\mathcal{F}})\right]. \tag{8.104}$$

As regards the right side of $(8.102\text{b})_2$, we have

$$-\left.\mathcal{K}\frac{d}{dr}\mathcal{G}(r,\xi)\right|_{r=1} = -\frac{\mathcal{K}}{1-\nu}\left(\mathcal{B}_2 - \mathcal{B}_3\sqrt{\mathcal{F}}J_1(\sqrt{\mathcal{F}}) - \mathcal{B}_4\sqrt{\mathcal{F}}Y_1(\sqrt{\mathcal{F}})\right). \tag{8.105}$$

If we equalize (8.104) and (8.105), we obtain

$$-\mathcal{B}_2\left(1 - \frac{\mathcal{K}}{1-\nu}\right) + \mathcal{B}_3\left[J_1(\sqrt{\mathcal{F}})\sqrt{\mathcal{F}}\left(1 - \frac{\mathcal{K}}{1-\nu}\right) - \frac{\mathcal{F}}{1-\nu}J_0(\sqrt{\mathcal{F}})\right] +$$
$$+\mathcal{B}_4\left[Y_1(\sqrt{\mathcal{F}})\sqrt{\mathcal{F}}\left(1 - \frac{\mathcal{K}}{1-\nu}\right) - \frac{\mathcal{F}}{1-\nu}Y_0(\sqrt{\mathcal{F}})\right] = 0. \tag{8.106}$$

Introducing the notations $J_{nr} = J_n(\sqrt{\mathcal{F}}r)$, $J_{n\xi} = J_n(\sqrt{\mathcal{F}}\xi)$, $J_{n1} = J_n(\sqrt{\mathcal{F}})$, $Y_{nr} = Y_n(\sqrt{\mathcal{F}}r)$, $Y_{n\xi} = Y_n(\sqrt{\mathcal{F}}\xi)$, $Y_{n1} = Y_n(\sqrt{\mathcal{F}})$ $(n = 0, 1)$ and

$$\underline{\mathcal{A}} = \begin{bmatrix} 1 & J_{0\xi} & -1 - \ln\xi & -J_{0\xi} & -Y_{0\xi} \\ 0 & -\sqrt{\mathcal{F}}J_{1\xi} & 0 & -\frac{1}{\xi} & \sqrt{\mathcal{F}}J_{1\xi} & \sqrt{\mathcal{F}}Y_{1\xi} \\ 0 & J_{0\xi} & 0 & 0 & -J_{0\xi} & -Y_{0\xi} \\ 0 & 0 & 0 & 1 & 0 & 0 \\ 0 & 0 & 1 & 0 & J_{01} & Y_{01} \\ 0 & 0 & 0 & -\mathcal{K}_\nu & J_{11}\sqrt{\mathcal{F}}\mathcal{K}_\nu - \mathcal{F}_\nu J_{01} & Y_{11}\sqrt{\mathcal{F}}\mathcal{K}_\nu - \mathcal{F}_\nu Y_{01} \end{bmatrix},$$

$$\begin{aligned} \underline{\mathcal{X}}^T &= \left[\mathcal{A}_1 \middle| \mathcal{A}_2 \middle| \mathcal{B}_1 \middle| \mathcal{B}_2 \middle| \mathcal{B}_3 \middle| \mathcal{B}_4\right], \\ \underline{\mathcal{F}}^T &= \left[0 | 0 | 0 | 1/2\pi f_R | 0 | 0\right], \\ \mathcal{F}_\nu &= \frac{\mathcal{F}}{1-\nu}, \qquad \mathcal{K}_\nu = 1 - \frac{\mathcal{K}}{1-\nu} \end{aligned} \tag{8.107}$$

we can rewrite Eqs. (8.101), (8.102a), and (8.106) in the following form:

$$\underline{\mathcal{A}}\,\underline{\mathcal{X}} = \underline{\mathcal{F}}. \tag{8.108}$$

The solutions are given by the following relations:

$$\begin{aligned} \mathcal{D}_c\mathcal{A}_1 = \xi\ln\xi\left[\sqrt{\mathcal{F}}\mathcal{K}_\nu J_{11}\left(J_{1\xi}Y_{0\xi} - Y_{1\xi}J_{0\xi}\right) - \mathcal{F}_\nu J_{01}\left(J_{1\xi}Y_{0\xi} - Y_{1\xi}J_{0\xi}\right)\right] - \\ -\mathcal{K}_\nu J_{01}\xi\left(J_{1\xi}Y_{0\xi} - Y_{1\xi}J_{0\xi}\right) + \mathcal{K}_\nu J_{0\xi}\left(J_{11}Y_{01} - J_{01}Y_{11}\right), \end{aligned} \tag{8.109a}$$

$$\begin{aligned} \sqrt{\mathcal{F}}\mathcal{D}_c\mathcal{A}_2 = J_{0\xi}\left(\sqrt{\mathcal{F}}\mathcal{K}_\nu Y_{11} - \mathcal{F}_\nu Y_{01}\right) + \mathcal{K}_\nu\sqrt{\mathcal{F}}\xi\left(J_{1\xi}Y_{0\xi} - J_{0\xi}Y_{1\xi}\right) + \\ +Y_{0\xi}\left(\mathcal{F}_\nu J_{01} - \sqrt{\mathcal{F}}\mathcal{K}_\nu J_{11}\right), \end{aligned} \tag{8.109b}$$

$$\mathcal{D}_c\mathcal{B}_1 = \mathcal{K}_\nu\left[J_{0\xi}\left(J_{11}Y_{01} - J_{01}Y_{11}\right) + J_{01}\xi\left(J_{0\xi}Y_{1\xi} - J_{1\xi}Y_{0\xi}\right)\right], \tag{8.109c}$$

$$\mathcal{B}_2 = \frac{1}{2\pi f_R}, \tag{8.109d}$$

$$\sqrt{\mathcal{F}}\mathcal{D}_c\mathcal{B}_3 = -J_{0\xi}\left(\sqrt{\mathcal{F}}\mathcal{K}_\nu Y_{11} - \mathcal{F}_\nu Y_{01}\right) + \mathcal{K}_\nu\sqrt{\mathcal{F}}\xi\left(J_{1\xi}Y_{0\xi} - J_{0\xi}Y_{1\xi}\right), \tag{8.109e}$$

$$\sqrt{\mathcal{F}}\mathcal{D}_c\mathcal{B}_4 = -J_{0\xi}\left(-\mathcal{F}_\nu J_{01} + \sqrt{\mathcal{F}}\mathcal{K}_\nu J_{11}\right), \tag{8.109f}$$

where

$$\mathcal{D}_c = 2\left(-J_{01}\mathcal{F}_\nu + \sqrt{\mathcal{F}}\mathcal{K}_\nu J_{11}\right)\left(J_{1\xi}Y_{0\xi} - J_{0\xi}Y_{1\xi}\right)\xi\pi f_R. \tag{8.110}$$

Let us substitute solutions (8.109), (8.110) into (8.94) and take the equations

$$J_{1\xi}Y_{0\xi} - J_{0\xi}Y_{1\xi} = \frac{2}{\pi\xi\sqrt{\mathcal{F}}} \qquad J_{11}Y_{01} - J_{01}Y_{11} = \frac{2}{\pi\sqrt{\mathcal{F}}},$$

which follow from (B.2.7e) into account. After some further manipulations, we get the sought Green function:

$$\boxed{\begin{aligned} \mathcal{G}(r,\xi) &= \frac{\mathcal{K}_\nu\left(J_{01} - J_{0\xi} - J_{0r}\right) + \frac{\pi}{2}J_{0r}J_{0\xi}\left(\mathcal{F}_\nu Y_{01} - \sqrt{\mathcal{F}}\mathcal{K}_\nu Y_{11}\right)}{2\pi f_R(\mathcal{F}_\nu J_{01} - \sqrt{\mathcal{F}}\mathcal{K}_\nu J_{11})} + \\ &\quad + \frac{\ln\xi - \frac{\pi}{2}J_{0r}Y_{0\xi}}{2\pi f_R}, \quad 0 \le r \le \xi, \\ \mathcal{G}(r,\xi) &= \frac{\mathcal{K}_\nu\left(J_{01} - J_{0r} - J_{0\xi}\right) + \frac{\pi}{2}J_{0\xi}J_{0r}\left(\mathcal{F}_\nu Y_{01} - \sqrt{\mathcal{F}}\mathcal{K}_\nu Y_{11}\right)}{2\pi f_R(\mathcal{F}_\nu J_{01} - \sqrt{\mathcal{F}}\mathcal{K}_\nu J_{11})} + \\ &\quad + \frac{\ln r - \frac{\pi}{2}J_{0\xi}Y_{0r}}{2\pi f_R}, \quad \xi \le r \le 1. \end{aligned}} \tag{8.111}$$

If $\mathcal{K} = R_o k_\gamma / I_1 E_1 \to 0$ in (8.111) we get the Green function of the simply supported plate:

$$\boxed{\begin{aligned} \mathcal{G}(r,\xi) &= \frac{J_{01} - J_{0\xi} - J_{0r} + \frac{\pi}{2}J_{0\xi}J_{0r}(\mathcal{F}_\nu Y_{01} - \sqrt{\mathcal{F}}Y_{11})}{2\pi f_R(\mathcal{F}_\nu J_{01} - \sqrt{\mathcal{F}}J_{11})} + \\ &\quad + \frac{\ln\xi - \frac{\pi}{2}J_{0r}Y_{0\xi}}{2\pi f_R}, \quad 0 \le r \le \xi, \\ \mathcal{G}(r,\xi) &= \frac{J_{01} - J_{0r} - J_{0\xi} + \frac{\pi}{2}J_{0r}J_{0\xi}\left(\mathcal{F}_\nu Y_{01} - \sqrt{\mathcal{F}}Y_{11}\right)}{2\pi f_R(\mathcal{F}_\nu J_{01} - \sqrt{\mathcal{F}}J_{11})} + \\ &\quad + \frac{\ln r - \frac{\pi}{2}J_{0\xi}Y_{0r}}{2\pi f}, \quad \xi \le r \le 1. \end{aligned}} \tag{8.112}$$

If $\mathcal{K} = R_o k_\gamma / I_1 E_1$ tends to infinity in (8.111) we get the Green function of the clamped plate:

$$\boxed{\begin{aligned} \mathcal{G}(r,\xi) &= \frac{J_{11}\ln\xi + \frac{1}{\sqrt{\mathcal{F}}}\left(J_{0r} + J_{0\xi} - J_{01}\right) + \frac{\pi}{2}J_{0r}\left(J_{0\xi}Y_{11} - Y_{0\xi}J_{11}\right)}{2\pi f_R J_{11}}, \\ &\qquad 0 \le r \le \xi, \\ \mathcal{G}(r,\xi) &= \frac{J_{11}\ln r + \frac{1}{\sqrt{\mathcal{F}}}\left(J_{0\xi} + J_{0r} - J_{01}\right) + \frac{\pi}{2}J_{0\xi}\left(J_{0r}Y_{11} - Y_{0r}J_{11}\right)}{2\pi f_R J_{11}}, \\ &\qquad \xi \le r \le 1. \end{aligned}} \tag{8.113}$$

8.11.3.4 Green Functions for Tensile f_R

If the in-plane load is a tensile one

$$w(r) = c_1 + c_2 \ln r + c_3 I_0\left(\sqrt{\mathcal{F}}r\right) + c_4 K_0\left(\sqrt{\mathcal{F}}r\right) \tag{8.114}$$

is the general solution we need when determining the Green function—c_1, c_2, c_3, and c_4 which are again undetermined constants of integration while $I_0\left(\sqrt{\mathcal{F}}r\right)$ and $K_0\left(\sqrt{\mathcal{F}}r\right)$ are Bessel functions. With the knowledge of the general solution, we shall assume that the Green function, which satisfies conditions (8.90), can be given in the following form:

$$G(r,\xi) = \mathcal{A}_1 + \mathcal{A}_3 I_0(\sqrt{\mathcal{F}}r)\,, \qquad r < \xi\,, \tag{8.115a}$$

$$G(r,\xi) = \mathcal{B}_1 + \mathcal{B}_2 \ln r + \mathcal{B}_3 I_0(\sqrt{\mathcal{F}}r) + \mathcal{B}_4 K_0(\sqrt{\mathcal{F}}r)\,, \qquad r > \xi \tag{8.115b}$$

in which the integration constants are denoted in the same as for the case of compressive f—however, this may not cause any misunderstanding. If we substitute the Green function (8.115) into continuity conditions (8.95), discontinuity condition (8.96), and boundary conditions (8.91), we obtain a system of linear equations for $\mathcal{A}_1$, $\mathcal{A}_3$, $\mathcal{B}_1, \ldots, \mathcal{B}_4$. Here, we present the result, i.e., the equation system only, the corresponding calculations are left for Problem 8.5. By introducing the notations $I_{nr} = I_n(\sqrt{\mathcal{F}}r)$, $I_{n\xi} = I_n(\sqrt{\mathcal{F}}\xi)$, $I_{n1} = I_n(\sqrt{\mathcal{F}})$, $K_{nr} = K_n(\sqrt{\mathcal{F}}r)$, $K_{n\xi} = K_n(\sqrt{\mathcal{F}}\xi)$, $K_{n1} = K_n(\sqrt{\mathcal{F}})$ $(n = 0, 1)$,

$$\underline{\mathcal{A}} = \begin{bmatrix} 1 & I_{0\xi} & -1 & -\ln\xi & -I_{0\xi} & -K_{0\xi} \\ 0 & \sqrt{\mathcal{F}}I_{1\xi} & 0 & -\frac{1}{\xi} & -\sqrt{\mathcal{F}}I_{1\xi} & \sqrt{\mathcal{F}}K_{1\xi} \\ 0 & I_{0\xi} & 0 & 0 & -I_{0\xi} & -K_{0\xi} \\ 0 & 0 & 0 & 1 & 0 & 0 \\ 0 & 0 & 1 & 0 & I_{01} & K_{01} \\ 0 & 0 & 0 & -\mathcal{K}_\nu & \mathcal{F}_\nu I_{01} - \sqrt{\mathcal{F}}I_{11}\mathcal{K}_\nu & \mathcal{F}_\nu K_{01} + \sqrt{\mathcal{F}}K_{11}\mathcal{K}_\nu \end{bmatrix},$$

and utilizing the notations given by Eqs. (8.107) we can give the equation system in the form equation system (8.108) has. The solutions are given by the following equations:

$$\begin{aligned} \mathcal{A}_1\mathcal{D}_t = {}& \xi\ln\xi\left[\sqrt{\mathcal{F}}\mathcal{K}_\nu I_{11}\left(I_{0\xi}K_{1\xi} + I_{1\xi}K_{0\xi}\right) - \mathcal{F}_\nu I_{01}\left(I_{0\xi}K_{1\xi} + I_{1\xi}K_{0\xi}\right)\right] + \\ & + \mathcal{K}_\nu I_{01}\xi\left(I_{0\xi}K_{1\xi} + I_{1\xi}K_{0\xi}\right) - \mathcal{K}_\nu I_{0\xi}\left(I_{01}K_{11} + K_{01}I_{11}\right)\,, \end{aligned} \tag{8.116a}$$

$$\begin{aligned} \sqrt{\mathcal{F}}\mathcal{D}_t\mathcal{A}_2 = {}& I_{0\xi}\left(\sqrt{\mathcal{F}}\mathcal{K}_\nu K_{11} + \mathcal{F}_\nu K_{01}\right) - \mathcal{K}_\nu\sqrt{\mathcal{F}}\xi\left(I_{0\xi}K_{1\xi} + I_{1\xi}K_{0\xi}\right) - \\ & - K_{0\xi}\left(\mathcal{F}_\nu I_{01} - \sqrt{\mathcal{F}}\mathcal{K}_\nu I_{11}\right)\,, \end{aligned} \tag{8.116b}$$

$$\mathcal{B}_1\mathcal{D}_t = \mathcal{K}_\nu \left[-I_{0\xi}\left(I_{11}K_{01} + I_{01}K_{11}\right) + I_{01}\xi\left(I_{0\xi}K_{1\xi} + I_{1\xi}K_{0\xi}\right)\right], \tag{8.116c}$$

$$\mathcal{B}_2 = \frac{1}{2\pi f_R}, \tag{8.116d}$$

$$\sqrt{\mathcal{F}}\mathcal{D}_t\mathcal{B}_3 = I_{0\xi}\left(\sqrt{\mathcal{F}}\mathcal{K}_\nu K_{11} + \mathcal{F}_\nu K_{01}\right) - \mathcal{K}_\nu\sqrt{\mathcal{F}}\xi\left(I_{0\xi}K_{1\xi} + I_{1\xi}K_{0\xi}\right), \tag{8.116e}$$

$$\sqrt{\mathcal{F}}\mathcal{D}_t\mathcal{B}_3 = I_{0\xi}\left(-\mathcal{F}_\nu I_{01} + \sqrt{\mathcal{F}}\mathcal{K}_\nu I_{11}\right), \tag{8.116f}$$

where

$$\mathcal{D}_t = 2\left(\sqrt{\mathcal{F}}\mathcal{K}_\nu I_{11} - \mathcal{F}_\nu I_{01}\right)\left(I_{0\xi}K_{1\xi} + I_{1\xi}K_{0\xi}\right)\xi\pi f\sqrt{\mathcal{F}}. \tag{8.117}$$

Let us substitute solutions (8.116), (8.117) into the definition of the Green function (8.115). If we take the relations

$$I_{0\xi}K_{1\xi} + I_{1\xi}K_{0\xi} = \frac{1}{\xi\sqrt{\mathcal{F}}} \qquad I_{11}K_{01} + I_{01}K_{11} = \frac{1}{\sqrt{\mathcal{F}}}$$

into account after some manipulations, we get

$$\begin{aligned}
\mathcal{G}(r,\xi) &= \frac{\mathcal{K}_\nu\left(I_{0\xi}+I_{0r}-I_{01}\right)-I_{0r}I_{0\xi}\left(\mathcal{F}_\nu K_{01}+\sqrt{\mathcal{F}}\mathcal{K}_\nu K_{11}\right)}{2\pi f_R(\mathcal{F}_\nu I_{01}-\sqrt{\mathcal{F}}\mathcal{K}_\nu I_{11})}+\\
&\quad+\frac{\ln\xi+I_{0r}K_{0\xi}}{2\pi f_R}, \quad 0<r\le\xi,\\
\mathcal{G}(\xi,r) &= \frac{\mathcal{K}_\nu(I_{0r}+I_{0\xi}-I_{01})-I_{0\xi}I_{0r}\left(\mathcal{F}_\nu K_{01}+\sqrt{\mathcal{F}}\mathcal{K}_\nu K_{11}\right)}{2\pi f_R(\mathcal{F}_\nu I_{01}-\sqrt{\mathcal{F}}\mathcal{K}_\nu I_{11})}+\\
&\quad+\frac{\ln r+K_{0r}I_{0\xi}}{2\pi f_R}, \quad \xi\le r\le 1.
\end{aligned} \tag{8.118}$$

If $\mathcal{K} = R_o k_\gamma / I_1 E_1 \to 0$ in (8.118), we get the Green function of the simply supported plate:

$$\begin{aligned}
\mathcal{G}(r,\xi) &= \frac{I_{0\xi}+I_{0r}-I_{01}-I_{0r}I_{0\xi}\left(\sqrt{\mathcal{F}}K_{11}+\mathcal{F}_\nu K_{01}\right)}{2f_R(\mathcal{F}_\nu I_{01}-\sqrt{\mathcal{F}}I_{11})}+\\
&\quad+\frac{\ln\xi+I_{0r}K_{0\xi}}{2f_R}, \quad 0<r\le\xi,\\
\mathcal{G}(r,\xi) &= \frac{I_{0r}+I_{0\xi}-I_{01}-I_{0\xi}I_{0r}\left(\sqrt{\mathcal{F}}K_{11}-\mathcal{F}_\nu K_{01}\right)}{2f_R(\mathcal{F}_\nu I_{01}-\sqrt{\mathcal{F}}I_{11})}+\\
&\quad+\frac{\ln r+I_{0\xi}K_{0r}}{2f_R}, \quad \xi\le r\le 1.
\end{aligned} \tag{8.119}$$

If $\mathcal{K} = R_o k_\gamma / I_1 E_1 \to 0$ Eq. (8.118) yields the Green function of the clamped plate:

$$
\boxed{
\begin{aligned}
G(r,\xi) &= \frac{I_{11}\ln\xi + \frac{1}{\sqrt{F}}\left(I_{01} - I_{0\xi} - I_{0r}\right) + I_{0r}\left(I_{0\xi}K_{11} + K_{0\xi}I_{11}\right)}{2I_{11}\pi f_R}\,, \\
&\qquad 0 < r \le \xi\,, \\
G(\xi,r) &= \frac{I_{11}\ln r + \frac{1}{\sqrt{F}}\left(I_{01} - I_{0r} - I_{0\xi}\right) + I_{0\xi}\left(I_{0r}K_{11} + K_{0r}I_{11}\right)}{2I_{11}\pi f_R}\,, \\
&\qquad \xi \le r \le 1\,.
\end{aligned}
}
\tag{8.120}
$$

8.11.3.5 Solutions of Statical Boundary Value Problems

Given the Green functions, the deflection due to an axisymmetric load $p_z(r)$ can always be calculated as

$$
w(\xi) = 2\pi R_o^2 \int_0^1 G(\xi,r)\,p_z(r)\,r\,\mathrm{d}r\,. \tag{8.121}
$$

8.12 Eigenvalues with Multiplicity

In the present section, it will be assumed that the eigenvalue problem (8.1), (8.2) is self-adjoint and positive definite. Then it holds for the eigenfunctions $y_i(x)$ $(i = 1, 2, 3, \ldots)$ that

$$
(y_i, y_i)_M > 0\,.
$$

The normalized eigenfunctions $\psi_i(x)$ are defined by the following equation:

$$
\psi_i = y_i / \sqrt{(y_i, y_i)_M}\,. \tag{8.122a}
$$

It is obvious that

$$
(\psi_i, \psi_i)_M = 1\,. \tag{8.122b}
$$

According to (8.17), the normalized eigenfunctions $\psi_i(x)$ also satisfy the following equation:

$$
(\psi_i, \psi_n)_K = \begin{cases} \lambda_n \text{ if } i = n, \\ 0 \text{ if } i \ne n. \end{cases} \qquad i, n = 1, 2, 3, \ldots \tag{8.123}
$$

The eigenvalue λ_L $(L \in 1, 2, 3, \ldots)$ of the eigenvalue problem (8.1), (8.2) has a multiplicity $r > 1$ if the number of those linearly independent eigenfunctions which belong to λ_L is r. Let us denote these normalized eigenfunctions by $\psi_{kL}(x)$ $(k \in 1, 2, \dots, r)$ where the second subscript identifies the eigenvalue λ_L. It is obvious that

$$K[\psi_{kL}] = \lambda_L M[\psi_{kL}], \qquad U_r[\psi_{kL}] = 0 \qquad (k = 1, 2, \dots, r). \tag{8.124}$$

The linear combinations

$$\hat{\psi}_{1L} = \sum_{k=1}^{r} c_{1k}\psi_{kL}, \quad \hat{\psi}_{2L} = \sum_{k=1}^{r} c_{2k}\psi_{kL}, \quad \dots \quad \hat{\psi}_{rL} = \sum_{k=1}^{r} c_{rk}\psi_{kL}, \tag{8.125}$$

in which the constants c_{1k}, c_{2k}, etc. are the weights or coefficients, are clearly eigenfunctions of the eigenvalue problem (8.1), (8.2). The normalized eigenfunctions $\psi_{kL}(x)$ constitute, however, such a set of functions in which the functions $\psi_{kL}(x)$ are not orthogonal to each other. In the sequel, we shall construct an orthogonal set by selecting the coefficients c_{1k}, c_{2k}, etc. appropriately. The procedure is basically the same as that in Sect. 5.2.4 and is, in fact, the Gram–Schmidt orthogonalization [12]. Let

$$\hat{\psi}_{1L} = \psi_{1L}. \tag{8.126a}$$

With $\hat{\psi}_{1L}$ we define $\hat{\psi}^*_{2L}$ by the equation

$$\hat{\psi}^*_{2L} = \psi_{2L} - \hat{\psi}_{1L}(\psi_{2L}, \hat{\psi}_{1L})_M \tag{8.126b}$$

where the term $\hat{\psi}_{1L}(\psi_{2L}, \hat{\psi}_{1L})_M$ is the projection of ψ_{2L} on $\hat{\psi}_{1L}$. Since

$$(\hat{\psi}^*_{2L}, \hat{\psi}_{1L})_M = (\psi_{2L}, \hat{\psi}_{1L})_M - \underbrace{(\hat{\psi}_{1L}, \hat{\psi}_{1L})_M}_{=1}(\psi_{2L}, \hat{\psi}_{1L})_M = 0, \tag{8.126c}$$

it follows that $\hat{\psi}^*_{2L}$ and

$$\hat{\psi}_{2L} = \hat{\psi}^*_{2L}/(\hat{\psi}^*_{2L}, \hat{\psi}^*_{2L})_M, \qquad \left[(\hat{\psi}_{2L}, \hat{\psi}_{2L})_M = 1\right] \tag{8.126d}$$

are both orthogonal to $\hat{\psi}_{1L}$. To proceed we define $\hat{\psi}^*_{3L}$ by the equation

$$\hat{\psi}^*_{3L} = \psi_{3L} - \hat{\psi}_{1L}(\psi_{3L}, \hat{\psi}_{1L})_M - \hat{\psi}_{2L}(\psi_{3L}, \hat{\psi}_{2L})_M, \tag{8.126e}$$

where the terms $\hat{\psi}_{1L}(\psi_{3L}, \hat{\psi}_{1L})_M$ and $\hat{\psi}_{2L}(\psi_{3L}, \hat{\psi}_{2L})_M$ are the projections of ψ_{3L} on $\hat{\psi}_{1L}$ and $\hat{\psi}_{2L}$. Since

$$\begin{aligned}(\hat{\psi}^*_{3L}, \hat{\psi}_{1L})_M = (\psi_{3L}, \hat{\psi}_{1L})_M &- \underbrace{(\hat{\psi}_{1L}, \hat{\psi}_{1L})_M}_{=1}(\psi_{3L}, \hat{\psi}_{1L})_M - \\ &- \underbrace{(\hat{\psi}_{1L}, \hat{\psi}_{2L})_M}_{=0}(\psi_{3L}, \hat{\psi}_{2L})_M = 0,\end{aligned} \tag{8.126f}$$

$$
\begin{aligned}
(\hat{\psi}_{3L}^{*}, \hat{\psi}_{2L})_M = (\psi_{3L}, \hat{\psi}_{2L})_M - \underbrace{(\hat{\psi}_{1L}, \hat{\psi}_{2L})_M}_{=0}(\psi_{3L}, \hat{\psi}_{1L})_M - \\
- \underbrace{(\hat{\psi}_{2L}, \hat{\psi}_{2L})_M}_{=1}(\psi_{3L}, \hat{\psi}_{2L})_M = 0,
\end{aligned} \tag{8.126g}
$$

it follows that $\hat{\psi}_{3L}^{*}$ and

$$
\hat{\psi}_{3L} = \hat{\psi}_{3L}^{*}/(\hat{\psi}_{3L}^{*}, \hat{\psi}_{3L}^{*})_M \,, \qquad \left[(\hat{\psi}_{3L}, \hat{\psi}_{3L})_M = 1\right] \tag{8.126h}
$$

are orthogonal to $\hat{\psi}_{1L}$ and $\hat{\psi}_{2L}$.

On the basis of all that has been said above

$$
\boxed{
\begin{aligned}
&\hat{\psi}_{1L} = \psi_{1L} \,, \\
&\hat{\psi}_{i+1,L}^{*} = \psi_{i+1,L} - \sum_{k=1}^{i} \hat{\psi}_{kL}(\psi_{i+1,L}, \hat{\psi}_{kL})_M \,, \quad i = 1, 2, \ldots, r-1 \\
&\hat{\psi}_{i+1,L} = \hat{\psi}_{i+1,L}^{*}/(\hat{\psi}_{i+1,L}^{*}, \hat{\psi}_{i+1,L}^{*})_M \,, \quad \left[(\hat{\psi}_{i+1,L}, \hat{\psi}_{i+1,L})_M = 1\right]
\end{aligned}
} \tag{8.127}
$$

is the orthogonalization procedure.

8.13 Properties of the Rayleigh Quotient

We shall assume that the eigenvalue problem (8.1), (8.2) is positive definite and simple. Then

$$
\lambda M\,[y] = \lambda g_0(x) y(x), \tag{8.128}
$$

where $g_0(x) > 0\ \forall x \in [a, b]$. Hence, it follows from (8.33) and (8.28a) that

$$
\begin{aligned}
&\mathcal{R}[u(x)] = \frac{(u, u)_K}{(u, u)_M} = \frac{\displaystyle\sum_{n=0}^{\kappa} \int_a^b u^{(n)}(x) f_n(x) u^{(n)}(x)\,\mathrm{d}x + K_0[u(x)]}{\displaystyle\int_a^b u(x) g_0(x) u(x)\,\mathrm{d}x} \,, \\
&K_0[u(x)] = \left[\sum_{n=0}^{\kappa} \sum_{r=0}^{n-1} (-1)^{(n+r)} u(x)^{(r)} \left[f_n(x)\, u^{(n)}(x)\right]^{(n-1-r)}\right]_a^b
\end{aligned} \tag{8.129}
$$

in which and in the sequel $u(x)$ is a comparison function which is not identically equal to zero if $x \in [a, b]$.

The Rayleigh quotient has the following properties:

1. Because $(u,u)_K > 0$ and $(u,u)_M > 0$ the Rayleigh quotient is a positive quantity.
2. Let

$$|u| = \sqrt{\int_a^b u^2(x)\,\mathrm{d}x}\,. \tag{8.130}$$

   The Rayleigh quotient is independent of $|u|$.
   Let ν be a non-zero real number. If we write $\nu u(x)$ for $u(x)$ in $(8.129)_1$, we get

$$\mathcal{R} = \frac{(\nu u, \nu u)_K}{(\nu u, \nu u)_M} = \frac{(u,u)_K\ \nu^2}{(u,u)_M\ \nu^2} = \frac{(u,u)_K}{(u,u)_M}, \tag{8.131}$$

   which shows that $\mathcal{R}$ is really independent of ν, i.e., the length of $u(x)$.
3. The Rayleigh quotient has a positive lower limit. This follows from the fact that $0 < (u,u)_K < \infty$ and $0 < (u,u)_M < \infty$.
4. The lower limit of the Rayleigh quotient is the smallest eigenvalue λ_1.
   The system of eigenfunctions ψ_ℓ $(\ell = 1, 2, \ldots, \infty)$ is complete. Hence, any comparison function $u(x)$ can be given as a linear combination of the eigenfunctions ψ_ℓ:

$$u(x) = c_1\psi_1 + c_2\psi_2 + c_3\psi_3 + \cdots + c_n\psi_n + \cdots = \sum_{\ell=1}^{\infty} c_\ell \psi_\ell \tag{8.132}$$

   in which the constant c_n is the weight of the eigenfunction ψ_ℓ. Since

$$(\psi_\ell, \psi_\ell)_M = 1, \qquad (\psi_k, \psi_\ell)_M = 0 \quad (k \neq \ell) \quad \text{and} \quad (\psi_\ell, \psi_\ell)_K = \lambda_\ell$$

   for the Rayleigh quotient, we get

$$\begin{aligned}
\mathcal{R} = \frac{(u,u)_K}{(u,u)_M} = \\
= \frac{(c_1\psi_1 + c_2\psi_2 + \cdots + c_n\psi_n + \cdots, c_1\psi_1 + c_2\psi_2 + \cdots + c_n\psi_n + \cdots)_K}{(c_1\psi_1 + c_2\psi_2 + \cdots + c_n\psi_n + \cdots, c_1\psi_1 + c_2\psi_2 + \cdots + c_n\psi_n + \cdots)_M} = \\
= \frac{c_1^2(\psi_1,\psi_1)_K + c_2^2(\psi_2,\psi_2)_K + \cdots + c_n^2(\psi_n,\psi_n)_K + \cdots}{c_1^2(\psi_1,\psi_1)_M + c_2^2(\psi_2,\psi_2)_M + \cdots + c_n^2(\psi_n,\psi_n)_M + \cdots} = \\
= \frac{\lambda_1 c_1^2 + \lambda_2 c_2^2 + \cdots + \lambda_n c_n^2 + \cdots}{c_1^2 + c_2^2 + \cdots + c_n^2 + \cdots}
\end{aligned}$$

   from where it follows that

$$\mathcal{R} = \lambda_1 \frac{c_1^2 + \frac{\lambda_2}{\lambda_1}c_2^2 + \cdots + \frac{\lambda_n}{\lambda_1}c_n^2 + \cdots}{c_1^2 + c_2^2 + \cdots + c_n^2 + \cdots}\,. \tag{8.133}$$

Hence,

$$\mathcal{R}_{\min} = \lambda_1 \,. \tag{8.134}$$

We remark that a more rigorous proof of the minimum property of the Rayleigh quotient can be found in Chap. 3 of [1].

Exercise 8.3 Assume that a fixed-fixed beam is preloaded by an axial force. Making use of the minimum property of the Rayleigh quotient, estimate the first natural frequency of the beam.

Vibration of a beam subjected to an axial force is governed by the differential equation (7.147). This means the Rayleigh quotient of the differential equation

$$\underbrace{\frac{\mathrm{d}^4 W(x)}{\mathrm{d}x^4} \mp \frac{N}{IE}\frac{\mathrm{d}^2 W(x)}{\mathrm{d}x^2}}_{K(W)} = \underbrace{\tilde{\lambda} W(x)}_{M(W)}, \qquad \tilde{\lambda} = \frac{\rho A \tilde{\omega}^2}{IE} \tag{8.135}$$

should be be determined in the first step of the solution. On the basis of (8.129) and the minimum property of the Rayleigh quotient, we may write

$$\mathcal{R}[u] = \frac{\displaystyle\int_0^\ell uK[u]dz}{\displaystyle\int_0^\ell uM[u]dz} = \frac{\displaystyle\int_0^\ell \left(\frac{\mathrm{d}^2 u}{\mathrm{d}x^2}\right)^2 \mathrm{d}x \pm \frac{N}{IE}\int_0^\ell \left(\frac{\mathrm{d}u}{\mathrm{d}x}\right)^2 \mathrm{d}x}{\displaystyle\int_0^\ell u^2\,\mathrm{d}x} \geq \tilde{\lambda}_1 \tag{8.136}$$

since the term $K_0[u(x)]$ in Eq. (8.129) is zero for fixed-fixed, simply supported, and fixed-pinned beams. By taking the minimum property of the Rayleigh quotient into account, it follows from here that

$$\frac{\rho A \tilde{\omega}_1^2}{I_x E} \leq \frac{\displaystyle\int_0^\ell \left(\frac{\mathrm{d}^2 u}{\mathrm{d}x^2}\right)^2 \mathrm{d}x}{\displaystyle\int_0^\ell u^2 \mathrm{d}x} \left\{ 1 \pm \frac{N}{IE\dfrac{\displaystyle\int_0^\ell \left(\frac{\mathrm{d}^2 u}{\mathrm{d}x^2}\right)^2 \mathrm{d}x}{\displaystyle\int_0^\ell \left(\frac{\mathrm{d}u}{\mathrm{d}x}\right)^2 \mathrm{d}x}} \right\} \tag{8.137}$$

in which the sign is positive if N is tensile and negative if N is compressive.

For $N = 0$, Eq. (8.135) yields the differential equation

$$\underbrace{\frac{\mathrm{d}^4 W(x)}{\mathrm{d}x^4}}_{K[W]} = \underbrace{\lambda\, W(x)}_{M[u]}, \quad \lambda = \frac{\rho A \omega^2}{IE}, \tag{8.138}$$

which describes the free transverse vibrations of the fixed-fixed, simply supported, and fixed-pinned beams with no preload. Then the Rayleigh quotient is of the form

$$\mathcal{R}[u] = \frac{\displaystyle\int_0^\ell \left(\frac{\mathrm{d}^2 u}{\mathrm{d}x^2}\right)^2 \mathrm{d}x}{\displaystyle\int_0^\ell u^2 \mathrm{d}x} \geq \lambda_1 . \tag{8.139}$$

Hence,

$$\lambda_1 = \frac{\rho A \omega_1^2}{I_x E} \leq \frac{\displaystyle\int_0^\ell \left(\frac{\mathrm{d}^2 u}{\mathrm{d}x^2}\right)^2 \mathrm{d}x}{\displaystyle\int_0^\ell u^2 \mathrm{d}x} = \frac{\rho A \omega_{1a}^2}{I_x E}, \tag{8.140}$$

where ω_{1a} is an approximation of the first eigenfrequency.

By setting $\tilde{\lambda}$ to zero and selecting the positive sign in (8.135)—then N is a compressive force—we obtain the differential equation valid for the stability problem of beams:

$$\underbrace{-\frac{\mathrm{d}^4 W(x)}{\mathrm{d}x^4}}_{K[W]} = \frac{N}{IE}\frac{\mathrm{d}^2 W(x)}{\mathrm{d}x^2} = \check{\lambda}\underbrace{\frac{\mathrm{d}^2 W(x)}{\mathrm{d}x^2}}_{M[W]}, \qquad \check{\lambda} = \frac{N}{IE}. \tag{8.141}$$

The Rayleigh quotient of this equation is given by

$$\mathcal{R}[u] = \frac{\displaystyle\int_0^\ell \left(\frac{\mathrm{d}^2 u}{\mathrm{d}x^2}\right)^2 \mathrm{d}x}{\displaystyle\int_0^\ell \left(\frac{\mathrm{d}u}{\mathrm{d}x}\right)^2 \mathrm{d}x} \geq \check{\lambda}_1 . \tag{8.142}$$

Thus

$$\check{\lambda}_1 = \frac{N_1}{IE} = \frac{N_{\mathrm{crit}}}{IE} \leq \mathcal{R}[u] = \frac{\displaystyle\int_0^\ell \left(\frac{\mathrm{d}^2 u}{\mathrm{d}x^2}\right)^2 \mathrm{d}x}{\displaystyle\int_0^\ell \left(\frac{\mathrm{d}u}{\mathrm{d}x}\right)^2 \mathrm{d}x} = \frac{N_{\mathrm{crit\,a}}}{IE}, \tag{8.143}$$

where $N_{\mathrm{crit\,a}}$ is an approximation of the smallest critical load.

Making use of Eqs. (8.140) and (8.143), the following estimation can be obtained for $\tilde{\omega}_1^2$:

$$\frac{\tilde{\omega}_1^2}{\omega_{1a}^2} \leq \left(1 \pm \frac{N}{N_{\mathrm{crit\,a}}}\right), \tag{8.144a}$$

where

$$\frac{\rho A \omega_{1a}^2}{IE} = \frac{\displaystyle\int_0^\ell \left(\frac{\mathrm{d}^2 u}{\mathrm{d}x^2}\right)^2 \mathrm{d}x}{\displaystyle\int_0^\ell u^2 \mathrm{d}x} \quad \text{and} \quad \frac{N_{\mathrm{crit\,a}}}{IE} = \frac{\displaystyle\int_0^\ell \left(\frac{\mathrm{d}^2 u}{\mathrm{d}x^2}\right)^2 \mathrm{d}x}{\displaystyle\int_0^\ell \left(\frac{\mathrm{d}u}{\mathrm{d}x}\right)^2 \mathrm{d}x}. \tag{8.144b}$$

It is worth mentioning two things here:

(a) The structure of estimation (8.144a) is the same as that of Eq. (7.160) derived for simply supported beams with a preload.
(b) Estimation (8.144a) is valid formally for simply supported, fixed-pinned, and fixed-fixed beams as well. In all three cases, we have to select an appropriate comparison function to make estimation (8.144a) applicable to calculating an upper limit of $\tilde{\omega}_1$.

Let

$$u = \sin^2 \frac{\pi x}{\ell} \tag{8.145}$$

be the comparison function for the fixed-fixed beam. Since

$$\frac{du}{dx} = \frac{\pi}{\ell} \sin \frac{2\pi x}{\ell} \tag{8.146}$$

it follows that this comparison function satisfies the boundary conditions

$$u(0) = \left.\frac{du}{dx}\right|_{x=0} = u(\ell) = \left.\frac{du}{dx}\right|_{x=\ell} = 0\,. \tag{8.147}$$

Substitute now the derivatives

$$\frac{d^2u}{dx^2} = 2\frac{\pi^2}{\ell^2} \cos \frac{2\pi x}{\ell}, \qquad \frac{d^4u}{dx^4} = -8\frac{\pi^4}{\ell^4} \cos \frac{2\pi x}{\ell} \tag{8.148}$$

into the differential equation of the stability problem. We get

$$\frac{d^4u}{dx^4} + \frac{N}{IE}\frac{d^2u}{dx^2} = 2\frac{\pi^2}{\ell^2}\left(\frac{N}{IE} - \frac{4\pi^2}{\ell^2}\right) \cos \frac{2\pi x}{\ell} = 0, \tag{8.149}$$

which shows that $u(x)$ is a solution of the stability problem if

$$N = N_{\text{crit}} = \frac{4\pi^2 IE}{\ell^2}, \tag{8.150}$$

where N_{crit} is the smallest critical load.

Using the integrals

$$\int_0^\ell (u)^2\, dx = \int_0^\ell \left(\sin^2 \frac{\pi x}{\ell}\right)^2 dx = \frac{3}{8}\ell\,, \tag{8.151a}$$

$$\int_0^\ell \left(\frac{du}{dx}\right)^2 dx = \left(\frac{\pi}{\ell}\right)^2 \int_0^\ell \left(\sin \frac{2\pi x}{\ell}\right)^2 dx = \frac{\pi^2}{2\ell}, \tag{8.151b}$$

$$\int_0^\ell \left(\frac{d^2u}{dx^2}\right)^2 dx = 4\frac{\pi^4}{\ell^4} \int_0^\ell \left(\cos \frac{2\pi x}{\ell}\right)^2 dx = 2\frac{\pi^4}{\ell^3} \tag{8.151c}$$

Eq. $(8.144\text{b})_2$ yields

$$N_{\text{crit a}} = IE \frac{\displaystyle\int_0^\ell \left(\frac{\mathrm{d}^2 u}{\mathrm{d}x^2}\right)^2 \mathrm{d}x}{\displaystyle\int_0^\ell \left(\frac{\mathrm{d}u}{\mathrm{d}x}\right)^2 \mathrm{d}x} = 4IE\frac{\pi^2}{\ell^2} = \frac{39.478IE}{\ell^2} = N_{\text{crit}}\,.$$

We get in the same way from $(8.144\text{b})_1$ that

$$\frac{\rho A \omega_{1a}^2}{IE} = \frac{\displaystyle\int_0^\ell \left(\frac{\mathrm{d}^2 u}{\mathrm{d}x^2}\right)^2 \mathrm{d}x}{\displaystyle\int_0^\ell u^2 \mathrm{d}x} = \frac{16}{3}\frac{\pi^4}{l^4} = \frac{519.515}{l^4}\,. \tag{8.152}$$

After substituting (8.3) and (8.152) into (8.144a), we obtain the following estimation:

$$\tilde{\omega}_1^2 \leq 519.515\frac{IE}{\rho A \ell^4}\left(1 \pm \frac{N}{N_{\text{crit}}}\right) = 519.515\frac{IE}{\rho A \ell^4}\left(1 \pm \frac{N\ell^2}{39.478IE}\right)\,. \tag{8.153}$$

Thus

$$\tilde{\omega}_1 \leq \frac{22.792}{\ell^2}\sqrt{\frac{IE}{\rho A}\left(1 \pm \frac{N\ell^2}{39.478IE}\right)} \tag{8.154}$$

is the estimation in its final form.

If $N = 0$ the exact value of $\tilde{\omega}_1$ should coincide with

$$\omega_1 = \frac{22.373}{\ell^2}\sqrt{\frac{IE}{\rho A}}\,. \tag{8.155}$$

It can be checked with ease that then the error is 1.87% only.

8.14 Some Concepts of Integral Equations

The equation

$$y(x) = f(x) + \lambda \int_a^b \mathcal{K}(x,\xi)\, y(\xi)\, \mathrm{d}x\,, \qquad (a < b,\ b - a = \ell) \tag{8.156}$$

is a Fredholm[4] integral equation of the second kind in which $y(x)$ is the unknown function, λ is a constant parameter, the symmetric function $\mathcal{K} = \mathcal{K}(\xi, x)$ is the kernel, and the given $f(x)$ is the inhomogeneity in Eq. (8.156) [13–15]. It is assumed that

[4]Eric Ivar Fredholm (1866–1927).

the kernel $\mathcal{K}$ is a continuous and differentiable function of the variables x and ξ $(x, \xi \in [a, b])$.

For $f(x) = 0$, integral equation (8.156) is homogeneous:

$$y(x) = \lambda \int_a^b \mathcal{K}(\xi, x)\, y(\xi)\, \mathrm{d}x\,. \tag{8.157}$$

It is obvious that $y(x) = 0$ satisfies this equation. This solution is the trivial one.

It is assumed that the kernel satisfies the condition

$$\mathcal{K}(\xi, x) \neq \sum_{\ell=1}^{m} a(x)\, b(\xi), \qquad m \geq 1, \tag{8.158}$$

where $a(x)$ and $b(\xi)$ are continuously differentiable functions in $[a, b]$.

The kernel is said to be positive definite if

$$\int_a^b \int_a^b v(x)\, \mathcal{K}(\xi, x)\, v(\xi)\, \mathrm{d}\xi\, \mathrm{d}x > 0 \tag{8.159}$$

for any $v(x)$ $(x \in [a, b])$ which are not identically equal to zero. In the sequel, it will be assumed that the kernel is positive definite.

Solutions different from the trivial one exist for the special values of the parameter λ only. These λ values are called eigenvalues and constitute a set with an infinite number of real elements if condition (8.158) is satisfied. The corresponding $y(x)$ solutions are the eigenfunctions.

Problem (8.157) governed by a homogeneous Fredholm integral equation is also referred to as eigenvalue problem.

The eigenfunctions $y_i(x)$ $(i = 1, 2, 3, \ldots)$ can be normalized:

$$\psi_i(x) = \frac{y_i(x)}{|y_i(x)|}, \quad |y_i(x)| = \sqrt{(y_i, y_i)_I}, \quad (y_i, y_i)_I = \int_a^b y_i(x) y_i(x)\, \mathrm{d}x, \tag{8.160}$$

where ψ_k is the normalized eigenfunction. The number of the eigenfunctions that belong to an eigenvalue is the multiplicity of the eigenvalue which we denote by r $(r = 1, 2, 3, \ldots)$.

If $r = 1$ the eigenvalue considered has only one eigenfunction.

Let λ_i and λ_ℓ be two not necessarily different eigenvalues each with only one eigenfunction. Then the orthogonality condition

$$(\psi_i, \psi_\ell)_I = \begin{cases} 1 & \text{if } i = \ell\,, \\ 0 & \text{if } i \neq \ell\,. \end{cases} \qquad i, \ell = 1, 2, 3, \ldots \tag{8.161}$$

is satisfied.

If $r > 1$ the number of linearly independent eigenfunctions which belong to the considered eigenvalue is r. These eigenfunctions are, in general, not orthogonal to each other. We can, however, orthogonalize them by using the Gram–Schmidt orthogonalization procedure presented in Sect. 8.12.

Assume that there exists such a continuous function $v(x)$ $(x \in [a, b])$ that the function $u(x)$ $(x \in [a, b])$ can be represented by the integral

$$u(x) = \int_a^b \mathcal{K}(x, \xi) v(\xi) \, \mathrm{d}x \, . \tag{8.162}$$

Then

$$u(x) = \sum_{\ell=1}^{\infty} c_\ell \psi_\ell(x), \tag{8.163a}$$

where

$$\int_a^b u(\xi)\psi_k(\xi) \, \mathrm{d}\xi = \sum_{\ell=1}^{\infty} c_\ell \int_a^b \psi_k(\xi)\psi_\ell(\xi) \, \mathrm{d}\xi = \sum_{\ell=1}^{\infty} c_\ell \, (\psi_k, \psi_\ell)_I \underset{(8.161)}{= \uparrow =} c_\ell \, . \tag{8.163b}$$

The normalized eigenfunction $\psi_\ell(x)$ satisfies the homogeneous integral equation:

$$\psi_\ell(x) = \lambda_\ell \int_a^b \mathcal{K}(x, \xi) \, \psi_\ell(\xi) \, \mathrm{d}\xi \, . \tag{8.164}$$

Multiply this equation by $\psi_\ell(x)$ and integrate the result with respect to x. We have

$$1 = \lambda_\ell \int_a^b \int_a^b \psi_\ell(x) \, \mathcal{K}(x, \xi) \, \psi_\ell(\xi) \, \mathrm{d}\xi \, \mathrm{d}x \, . \tag{8.165}$$

This equation is satisfied if

$$\mathcal{K}(x, \xi) = \sum_{\ell=1}^{\infty} \frac{1}{\lambda_\ell} \psi_\ell(x) \, \psi_\ell(\xi) \tag{8.166}$$

from where it follows with regard to (8.161) that

$$\int_a^b \mathcal{K}(\xi, \xi) \, \mathrm{d}\xi = \sum_{\ell=1}^{\infty} \frac{1}{\lambda_\ell} \, . \tag{8.167}$$

Equations (8.166) and (8.167) were established by Mercer[5] [16].

There is a way for estimating the eigenvalues [17]. Assume that the continuous function $v(x)$, $x \in [a, b]$ is normalized and orthogonal to the first $n - 1$ $(n \geq 2)$

[5]James Mercer (1883–1932).

eigenfunctions:

$$(v, v)_I = 1\,, \qquad (v, \psi_\ell)_I = 0\,, \quad (\ell = 1, 2, \ldots, n-1)\,. \tag{8.168}$$

Then

$$\frac{1}{\lambda_n} = \max \int_a^b \int_a^b v(x)\, \mathcal{K}(x, \xi)\, v(\xi)\, \mathrm{d}\xi\, \mathrm{d}x\,. \tag{8.169a}$$

If the orthogonalization conditions $(8.168)_2$ are not fulfilled, we get an estimation for the first eigenvalue:

$$\frac{1}{\lambda_1} = \max \int_a^b \int_a^b v(x)\, \mathcal{K}(x, \xi)\, v(\xi)\, \mathrm{d}\xi\, \mathrm{d}x\,. \tag{8.169b}$$

8.15 Transformation to an Algebraic Eigenvalue Problem

8.15.1 Classical Solution

Let

$$s = \frac{1}{b-a}\,(x-a) = \frac{1}{\ell}\,(x-a) \quad \text{and} \quad \sigma = \frac{1}{\ell}\,(\xi - a) \tag{8.170a}$$

be new independent variables. It is obvious that

$$\sigma = \begin{cases} 0 & \text{if } x = a\,, \\ 1 & \text{if } x = b \end{cases} \qquad \text{and} \qquad \mathrm{d}\xi = \ell\, \mathrm{d}\sigma\,. \tag{8.170b}$$

Using (8.170), we can rewrite eigenvalue problem (8.157) in the following form:

$$y(s) = \lambda \int_0^1 \hat{\mathcal{K}}(s, \sigma)\, y(\sigma)\, \mathrm{d}\sigma\,, \qquad \hat{\mathcal{K}}(s, \sigma) = \ell K(x, \xi)\,. \tag{8.171}$$

In the sequel, we shall transform eigenvalue problem (8.171) into an algebraic eigenvalue problem [18]. Consider the integral formula

$$J(\phi) = \int_0^1 \phi(\sigma)\mathrm{d}\sigma \approxeq \sum_{j=0}^{n} w_j \phi(\sigma_j)\,, \qquad \sigma_j \in [0, 1]\,, \quad j = 0, 1, \ldots, n \tag{8.172}$$

where $\phi(\sigma)$ is a scalar and the weights w_j are known. Making use of the above integral formula, the following equation can be obtained from (8.171):

$$\sum_{j=0}^{n} w_j\, \hat{\mathcal{K}}(s, \sigma_j)\, y(\sigma_j) = \chi y(s)\,, \qquad \chi = 1/\mathcal{A}\,, \quad s \in [0, 1]\,. \tag{8.173}$$

Its solution yields an approximate eigenvalue $\mathcal{A} = 1/\chi$ and an approximate eigenfunction $y(s)$. After setting s to $s_i = \sigma_i$ $(i = 0, 1, 2, \ldots, n)$, we have

$$\sum_{j=0}^{n} w_j \hat{\mathcal{K}}(s_i, \sigma_j)\, y\left(\sigma_j\right) = \chi y(s_i) \quad \chi = 1/\mathcal{A} \qquad s \in [0, 1]\,, \quad i = 0, 1, \ldots, n \tag{8.174}$$

or

$$\underline{\hat{\mathcal{K}}}\,\underline{\mathcal{D}}\,\underline{\tilde{\mathbf{y}}} = \chi\, \underline{\tilde{\mathbf{y}}}, \tag{8.175a}$$

where $\underline{\hat{\mathcal{K}}} = [\hat{\mathcal{K}}(s_i, \sigma_j)]$ is symmetric,

$$\underline{\mathcal{D}} = \text{diag}(w_0|\ldots|w_k|\ldots|w_n|) \quad \text{and} \quad \underline{\tilde{\mathbf{y}}}^T = [y(s_0)|y(s_1)|\ldots|y(s_n)]. \tag{8.175b}$$

The approximate eigenvalues χ_r and eigenvectors $\tilde{y}_r$ $(r = 0, 1, \ldots, n)$ are solutions of the algebraic eigenvalue problem (8.175). The corresponding eigenfunctions can be obtained by substituting back into Eq. (8.173):

$$y_r(s) = \frac{1}{\chi_r} \sum_{j=0}^{n} w_j \hat{\mathcal{K}}(s, \sigma_j)\, y_r(\sigma_j)\,. \tag{8.176}$$

Divide the interval $[0, 1]$ into equidistant subintervals of length $\mathfrak{L}$ and apply the integration formula to each subinterval. By repeating the line of thought leading to Eq. (8.175), one can show that the algebraic eigenvalue problem obtained in this way is of the same structure as that of Eq. (8.175).

8.15.2 *Solution Using the Boundary Element Technique*

In the previous subsection, a method for finding approximate solutions is presented. This section investigates what happens if the boundary element technique is applied to determine an approximate solution. The boundary element technique is a subinterval method [19] similar to the finite element method. This means that the interval $[a, b]$, in which the solution for $y(x)$ in the equation

$$\Lambda y(x) = \int_a^b \mathcal{K}(x, \xi)\, y(\xi)\, \mathrm{d}\xi\,, \qquad \Lambda\lambda = 1\,, \tag{8.177}$$

is sought, is divided into parts (subintervals) called elements. Figure 8.6 depicts the interval $[a, b]$ which is now divided into $n_e = 5$ elements. These elements and their lengths are denoted in the same way by $\mathfrak{L}_1, \ldots, \mathfrak{L}_{n_e}$. The points where the unknown function $y(x)$ is considered are the nodes (or nodal points) taken at the extremes (or ends) and the midpoint of an element. Elements with one node at the midpoint, or with two nodes at the end points, can also be applied. Since more nodes result in greater accuracy and slightly less elements, the three-node elements are preferred in the sequel.

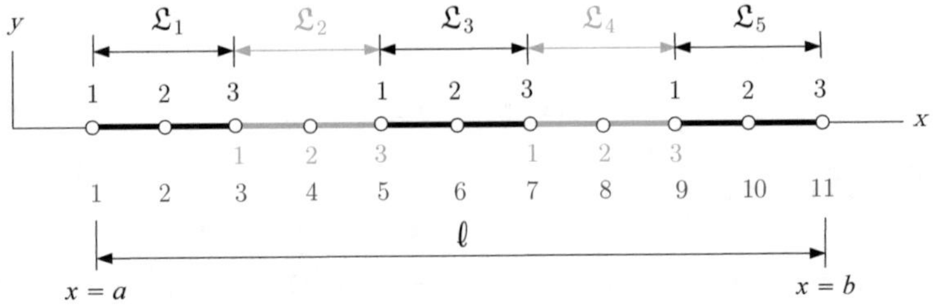

Fig. 8.6 Local and global node numbers

The nodal points are numbered, locally by 1, 2, and 3, whereas globally by $i = 1, 2, \ldots, 2n_e + 1$. Figure 8.6 shows both the local and the global node numbers.

The ξ coordinates of the local nodes and the values of the unknown function $y(x)$ at these nodes are the elements of the column matrices $\underline{\boldsymbol{\xi}}^e$ and $\underline{\mathbf{y}}^e$:

$$\underset{(3\times1)}{\underline{\boldsymbol{\xi}}^e} = \begin{bmatrix} \xi_1^e \\ \xi_2^e \\ \xi_3^e \end{bmatrix}, \qquad \underset{(3\times1)}{\underline{\mathbf{y}}^e} = \begin{bmatrix} y_1^e \\ y_2^e \\ y_3^e \end{bmatrix}, \qquad (e = 1, 2, \ldots, n_e)\,. \tag{8.178}$$

The interpolation functions N_1, N_2, and N_3 are defined by the following equations:

$$N_1(\eta) = \frac{1}{2}\eta(\eta - 1), \quad N_2(\eta) = 1 - \eta^2, \quad N_3(\eta) = \frac{1}{2}\eta(\eta + 1), \quad \eta \in [-1, 1]\,. \tag{8.179}$$

Table 8.4 Interpolation functions at the nodes

Node number	η	N_1	N_2	N_3
1	-1	1	0	0
2	0	0	1	0
3	1	0	0	1

Table 8.4 contains the values of the interpolation functions for $\eta = -1$, $\eta = 0$, and $\eta = 1$. Hence, the unknown function $y(x)$ can be approximated quadratically on the element $\mathfrak{L}_e$ by the equation

$$y^e(\eta) = \left[\, N_1(\eta) \middle| N_2(\eta) \middle| N_3(\eta) \,\right] \begin{bmatrix} y_1^e \\ y_2^e \\ y_3^e \end{bmatrix} = \underset{(1\times3)}{\underline{\mathbf{N}}}\;\underset{(3\times1)}{\underline{\mathbf{y}}^e}\,, \tag{8.180a}$$

where

$$\underset{(1\times 3)}{\mathbf{N}} = \left[N_1(\eta) \middle| N_2(\eta) \middle| N_3(\eta) \right]. \tag{8.180b}$$

As regards the geometry, we can write in the same manner that

$$\xi^e(\eta) = \left[N_1(\eta) \middle| N_2(\eta) \middle| N_3(\eta) \right] \begin{bmatrix} x_1^e \\ x_2^e \\ x_3^e \end{bmatrix} = \underset{(1\times 3)}{\mathbf{N}} \underset{(3\times 1)}{\mathbf{x}^e} =$$
$$= \frac{1}{2}\eta^2 \left(x_1^e - 2x_2^e + x_3^e\right) + \frac{1}{2}\eta \left(x_3^e - x_1^e\right) + x_2^e = \underset{x_2^e = \frac{1}{2}(x_1^e + x_3^e)}{\uparrow} = \frac{1}{2}\eta \left(x_3^e - x_1^e\right) + x_2^e. \tag{8.181}$$

Thus

$$\begin{aligned} \mathrm{d}\xi^e &= \left(\frac{\mathrm{d}N_1}{\mathrm{d}\eta} x_1^e + \frac{\mathrm{d}N_2}{\mathrm{d}\eta} x_2^e + \frac{\mathrm{d}N_1}{\mathrm{d}\eta} x_3^e \right) \mathrm{d}\eta = \frac{1}{2}\left(x_3^e - x_1^e\right) \mathrm{d}\eta = \mathcal{J} \mathrm{d}\eta, \\ \mathcal{J} &= \frac{1}{2}\left(x_3^e - x_1^e\right) = \frac{\mathfrak{L}_e}{2}. \end{aligned} \tag{8.182}$$

Integrating element by element in (8.177) and substituting then (8.180), (8.182), we get

$$\Lambda y(x) = \sum_{e=1}^{n_e} \int_{\mathfrak{L}_e} \mathcal{K}\left[x, \xi(\eta)\right] \left[N_1(\eta) | N_2(\eta) | N_3(\eta)\right] \mathcal{J}(\eta) \mathrm{d}\eta \begin{bmatrix} y_1^e \\ y_2^e \\ y_3^e \end{bmatrix} =$$
$$= \sum_{e=1}^{n_{be}} \int_{\eta=-1}^{\eta=1} \mathcal{K}\left[x, \xi(\eta)\right] \underset{(1\times 3)}{\mathbf{N}}(\eta) \mathcal{J}(\eta) \mathrm{d}\eta \underset{(3\times 1)}{\mathbf{y}^e}. \tag{8.183}$$

Note that this equation is the pair of Eq. (8.173) since the mathematical meaning of these two equations is basically the same.

Let us denote the coordinates of the nodes by x_i $(i = 1, 2, \ldots, 2n_e + 1)$. Then, it follows from Eq. (8.183) that

$$\Lambda y(x_i) = \sum_{e=1}^{n_e} \int_{\eta=-1}^{1} \mathcal{K}\left[x_i, \xi(\eta)\right] \left[N_1(\eta) | N_2(\eta) | N_3(\eta)\right] \mathcal{J}(\eta) \mathrm{d}\eta \begin{bmatrix} y_1^e \\ y_2^e \\ y_3^e \end{bmatrix},$$
$$(i = 1, 2, \ldots, 2n_e + 1). \tag{8.184}$$

For our later consideration, we shall introduce the following notations:

$$k_r^{ie} = \int_{\eta=-1}^{\eta=1} \mathcal{K}\left[x_i, \xi(\eta)\right] N_r(\eta) \mathcal{J}(\eta) \mathrm{d}\eta, \quad y(x_i) = y_i,$$
$$(i = 1, 2, 3, \ldots, 2n_e + 1; \; r = 1, 2, 3) \tag{8.185}$$

where i identifies the global number of the node, e is the number of the element over which the integral is taken, and r is the number of the interpolation function. Using these notations, Eq. (8.184) can be rewritten in the following form:

$$\left[k_1^{i1}\middle|k_2^{i1}\middle|k_3^{i1}+k_1^{i2}\middle|k_2^{i2}\middle|k_3^{i2}+k_1^{i3}\middle|k_1^{i3}\middle|\ldots\middle|k_3^{i,n_e-1}+k_1^{in_e}\middle|k_2^{in_e}\middle|k_3^{in_e}\right]\times$$

$$\times\begin{bmatrix} y_1 \\ y_2 \\ \cdots \\ y_{2n_e} \\ y_{2n_e+1} \end{bmatrix} - \Lambda y_i = 0, \qquad (i = 1, 2, 3, \ldots, 2n_e + 1)\,. \tag{8.186}$$

Here, we have as many equations as there are unknowns.

For the sake of making the structure of these equations more clear, it is worth introducing further notations:

$$k_{i1} = k_1^{i1}, \quad k_{i2} = k_2^{i1}, \quad k_{i3} = k_3^{i1}+k_1^{i2}, \quad k_{i4} = k_2^{i2}, \quad k_{i5} = k_3^{i2}+k_1^{i3}, \ldots,$$
$$k_{i,2n_e} = k_2^{in_e}, \quad k_{i,2n_e+1} = k_3^{in_e}\,. \tag{8.187}$$

If we use these notations and unite Eqs. (8.186), we arrive at the following matrix equation:

$$\left\{\begin{bmatrix} k_{11} & k_{12} & \cdots & k_{1,2n_e+1} \\ k_{21} & k_{22} & \cdots & k_{2,2n_e+1} \\ \cdots & \cdots & \cdots & \cdots \\ k_{n_e1} & k_{n_e2} & \cdots & k_{2n_e+1,2n_e+1} \end{bmatrix} - \Lambda \begin{bmatrix} 1 & 0 & \cdots & 0 \\ 0 & 1 & \cdots & 0 \\ & \cdots & & \\ 0 & 0 & \cdots & 1 \end{bmatrix}\right\} \begin{bmatrix} y_1 \\ y_2 \\ \cdots \\ y_{2n_e+1} \end{bmatrix} = \begin{bmatrix} 0 \\ 0 \\ \cdots \\ 0 \end{bmatrix}. \tag{8.188}$$

Equation (8.188) is an algebraic eigenvalue problem with $\Lambda = 1/\lambda$ as eigenvalue. After solving it numerically, we have both the eigenvalues $\Lambda_r = 1/\lambda_r$ and the corresponding eigenvectors

$$\underset{(1\times(2n_e+1))}{\mathbf{y}_r^T} = \left[y_{1r}\middle|y_{2r}\middle|y_{3r}\middle|\cdots\middle|y_{2n_e+1,r}\right], \qquad (r = 1, 2, 3, \ldots, 2n_e + 1).$$

The last issue we have to deal with is that of the numerical integration. Introduce the notation

$$\phi(\eta) = \mathcal{K}\left[x_i, \xi(\eta)\right] N_r(\eta)\mathcal{J}(\eta)\,. \tag{8.189}$$

With (8.189) we can rewrite integral (8.185) in the following form:

$$k_r^{ie} = \int_{\eta=-1}^{\eta=1} \mathcal{K}\left[x_i, \xi(\eta)\right] N_r(\eta)\mathcal{J}(\eta)\,d\eta = \int_{\eta=-1}^{\eta=1} \phi(\eta)\,d\eta \tag{8.190}$$

which can be computed by applying one of the Gaussian quadrature rules presented in Sect. B.3.

8.16 Vibration of the Preloaded Beams

8.16.1 Formulation of the Governing Equations

Let us assume that the preloaded pinned-pinned, fixed-pinned, or fixed-fixed beam vibrates. According to Eq. (7.147), the amplitude $W(x)$ of the vibrations should satisfy the differential equation

$$\frac{\mathrm{d}^4 W}{\mathrm{d}x^4} \mp p^2 \frac{\mathrm{d}^2 W}{\mathrm{d}x^2} = \tilde{\lambda} W, \tag{8.191}$$

where

$$p = \sqrt{\frac{N}{IE}} \quad \text{and} \quad \tilde{\lambda} = \frac{\rho A \tilde{\omega}^2}{IE}. \tag{8.192}$$

Differential equation (8.191) and the boundary conditions shown in Tables 8.2 and 8.3 for fixed-pinned and fixed-fixed beams define eigenvalue problems in which the eigenvalue $\tilde{\lambda}$ is a function of the parameter p (of the axial force N).

Let $r(x)$ $(x \in [0, \ell])$ be a continuous function. Further, let $\mathcal{K}(x, \xi)$ be the Green function of the differential equation

$$\frac{\mathrm{d}^4 W}{\mathrm{d}x^4} \mp p^2 \frac{\mathrm{d}^2 W}{\mathrm{d}x^2} = r \tag{8.193}$$

under the boundary condition detailed in Tables 8.2 and 8.3. It is obvious on the basis of Sect. 8.11.2 that

$$\mathcal{K}(x, \xi) = IE\, \mathcal{G}(x, \xi), \tag{8.194}$$

where the functions $\mathcal{G}(x, \xi)$ are presented in the third columns of Tables 8.2 and 8.3. On the basis of Eq. (8.44), solutions to the boundary value problems determined by differential equation (8.191) and the boundary conditions in Tables 8.2 and 8.3 can be given in the form

$$W(x) = \int_0^\ell \mathcal{K}(x, \xi) r(\xi)\, \mathrm{d}\xi. \tag{8.195}$$

If $r(\xi) = \tilde{\lambda} W(\xi)$, a homogeneous integral equation is obtained:

$$W(x) = \tilde{\lambda} \int_0^\ell \mathcal{K}(x, \xi) W(\xi)\, \mathrm{d}\xi\,. \tag{8.196}$$

The eigenvalue problem determined by the homogeneous integral equation (8.196) can be solved for $\tilde{\lambda}$ numerically by applying the algorithm that is detailed in Sect. 8.15.2.

8.17 Computational Results for Fixed-Pinned and Fixed-Fixed Beams

8.17.1 Data for the Computations

It follows from Eq. $(7.82)_2$ and Table 7.5 that

$$\lambda_1 = \frac{\rho A \omega^2}{IE} = \beta_1^4 = \frac{3.926602^4}{\ell^4} \tag{8.197a}$$

is the first eigenvalue for the free vibrations of unloaded and fixed-pinned beams. A comparison of the above equation and the fourth row in Table 7.5 yields the first eigenvalue for the unloaded and fixed-fixed beams:

$$\lambda_1 = \frac{4.730041^4}{\ell^4} . \tag{8.197b}$$

As regards the smallest critical load,

$$N_{1\,\mathrm{crit}} = \frac{IE}{\ell^2} \frac{\pi^2}{0.699^2} \tag{8.198a}$$

for fixed-pinned beams and

$$N_{1\,\mathrm{crit}} = \frac{IE}{\ell^2} \frac{\pi^2}{(0.5)^2} \tag{8.198b}$$

for fixed-fixed beams. Hence,

$$p_{1\,\mathrm{crit}}^2 = \frac{N_{1\,\mathrm{crit}}}{IE} = \frac{\pi^2}{0.699^2 \ell^2} \quad \text{and} \quad p_{1\,\mathrm{crit}}^2 = \frac{\pi^2}{0.5^2 \ell^2} . \tag{8.199}$$

8.17.2 Computational Results

A program in Fortran 90 has been developed for solving the eigenvalue problem given by the homogeneous integral equation (8.196) numerically. The solution algorithm is based on Sect. 8.15.2. The computational results for fixed-pinned beams are shown in Table 8.5 which presents the quotient $\tilde{\omega}_1^2/\omega_1^2 = \tilde{\lambda}_1/\lambda_1$ for the uniform steps of the non-dimensional load parameter $N/N_{1\,\mathrm{crit}}$. The columns' difference contains the difference between the results obtained for two consecutive iteration steps.

Table 8.5 Computational results for fixed-pinned beam

		Fixed-pinned beam			
		Compression		Tension	
Load step	$N/N_{1\,\text{crit}}$ $(p^2/p^2_{1\,\text{crit}})$	$\tilde{\omega}_1^2/\omega_1^2$ $(\tilde{\lambda}_1/\lambda_1)$	Difference	$\tilde{\omega}_1^2/\omega_1^2$ $(\tilde{\lambda}_1/\lambda_1)$	Difference
1	0.0000	1.00000000		1.00000000	
2	0.1250	0.87761168	-0.12238832	1.12223613	0.12223613
3	0.2500	0.75444648	-0.12316520	1.24378388	0.12154776
4	0.3750	0.63064337	-0.12380311	1.36487259	0.12108871
5	0.5000	0.50614099	-0.12450238	1.48553498	0.12066239
6	0.6250	0.38086865	-0.12527234	1.60580030	0.12026532
7	0.7500	0.25474443	-0.12612422	1.72569481	0.11989451
8	0.8750	0.12767281	-0.12707162	1.84524215	0.11954734
9	1.0000	-0.00045841	-0.12813122	1.96446368	0.11922153

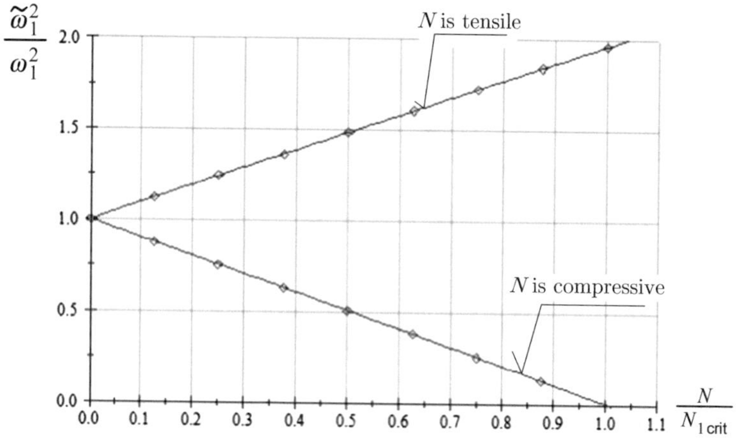

Fig. 8.7 The effect of the axial force on the first circular frequencies for a fixed-pinned beam

The quadratic polynomials

$$\frac{\tilde{\omega}_1^2}{\omega_1^2} = \frac{\tilde{\lambda}_1}{\lambda_1} = 0.9999 - 0.9755\frac{N}{N_{1\text{crit}}} - 2.4305 \times 10^{-2}\left(\frac{N}{N_{1\,\text{crit}}}\right)^2, \tag{8.200a}$$

$$\frac{\tilde{\omega}_1^2}{\omega_1^2} = \frac{\tilde{\lambda}_1}{\lambda_1} = 1.0002 + 0.9774\frac{N}{N_{1\,\text{crit}}} - 1.3204 \times 10^{-2}\left(\frac{N}{N_{1\,\text{crit}}}\right)^2 \tag{8.200b}$$

are fitted onto the computational results both for compression and tension. Their graphs are shown in Fig. 8.7 which represents the computational results by diamonds.

The solutions computed for fixed-fixed beams are included in Table 8.6 which shows the results in the same manner as Table 8.5.

Table 8.6 Computational results for fixed-fixed beam

		Fixed-fixed beam			
		Compression		Tension	
Load step	$N/N_{1\,\mathrm{crit}}$ $(p^2/p^2_{1\,\mathrm{crit}})$	$\tilde{\omega}_1^2/\omega_1^2$ $(\tilde{\lambda}_1/\lambda_1)$	Difference	$\tilde{\omega}_1^2/\omega_1^2$ $(\tilde{\lambda}_1/\lambda_1)$	Difference
1	0.0000	1.00000000		1.00000000	
2	0.1250	0.87869836	-0.12238832	1.12139263	0.12139263
3	0.2500	0.75619154	-0.12250682	1.24169360	0.12030097
4	0.3750	0.63282404	-0.12336750	1.36136285	0.11966925
5	0.5000	0.50851948	-0.12430455	1.48044338	0.11908054
6	0.6250	0.38319119	-0.12532830	1.59897397	0.11853058
7	0.7500	0.25674027	-0.12645092	1.71698962	0.11801566
8	0.8750	0.12905339	-0.12768688	1.83452212	0.11753249
9	1.0000	0.00000000	-0.12905339	1.95160032	0.11707820

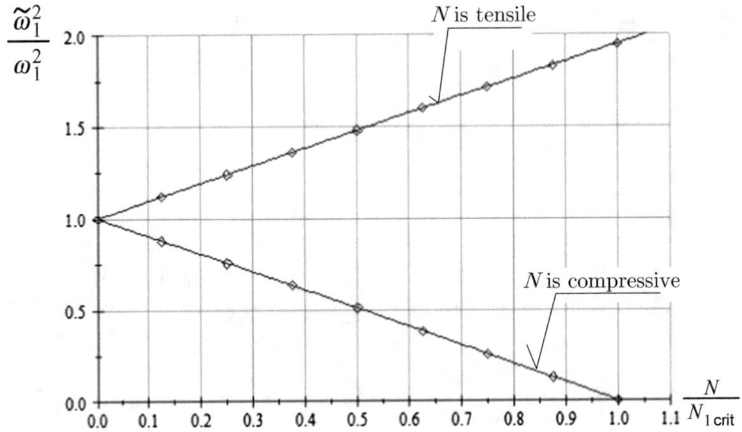

Fig. 8.8 The effect of the axial force on the first circular frequencies for a pinned-pinned beam

Quadratic polynomials are fitted onto the results again. They are given by the equations

$$\frac{\tilde{\omega}_1^2}{\omega_1^2} = \frac{\tilde{\lambda}_1}{\lambda_1} = 0.9999 - 0.9656\frac{N}{N_{1\,\mathrm{crit}}} - 3.4067 \times 10^{-2}\left(\frac{N}{N_{1\,\mathrm{crit}}}\right)^2, \tag{8.201a}$$

$$\frac{\tilde{\omega}_1^2}{\omega_1^2} = \frac{\tilde{\lambda}_1}{\lambda_1} = 1.0002 + 0.9698\frac{N}{N_{1\,\mathrm{crit}}} - 1.8552 \times 10^{-2}\left(\frac{N}{N_{1\,\mathrm{crit}}}\right)^2. \tag{8.201b}$$

The graphs, which represent the polynomials fitted onto the solutions, are shown in Fig. 8.8. The points that represent the numerical solutions are denoted by diamonds.

Exercise 8.4 Let us assume that $N/N_{1\,\text{crit}} = 0.5$. Compare the estimation given by Eq. (8.153) to the result given in Table 8.6.

It follows from Eqs. (8.197b) and $(7.82)_2$ that

$$\lambda_1 = \frac{\rho A \omega_1^2}{IE} = \frac{(4.730041)^4}{\ell^4} = \frac{500.564\,009742}{\ell^4}.$$

Hence

$$\omega_1^2 = \frac{500.564\,009742 IE}{\rho A \ell^4}$$

is the square of the first circular frequency of the unloaded fixed-fixed beam. Recalling Eq. (8.153), the quotient $\tilde{\omega}_1^2/\omega_1^2$ can be estimated as

$$\frac{\tilde{\omega}_1^2}{\omega_1^2} \leq \frac{519.515}{500.564\,009742}\left(1 \pm \frac{N}{N_{1\,\text{crit}}}\right) = 1.037\,859\left(1 \pm \frac{N}{N_{1\,\text{crit}}}\right)$$

from where

$$\left.\frac{\tilde{\omega}_1^2}{\omega_1^2}\right|_{\frac{N}{N_{1\,\text{crit}}}=0.5} \leq 1.037\,859\ (1 - 0.5) = 0.518\,929\,5.$$

With the exact value

$$0.50851948$$

taken from Table 8.6, the relative error is

$$\delta = \frac{0.518\,929\,5}{0.50851948} \times 100 - 100 = 2.047\%,$$

which shows that the estimation (8.153) is sufficiently accurate.

8.18 Vibration of Preloaded Circular Plates

8.18.1 Equation of Motion

If the circular plate preloaded in its plane vibrates, the transverse load is the force of inertia:

$$p_z = -\rho \frac{\partial^2 w}{\partial t^2}, \tag{8.202}$$

where ρ is the plate density for a unit area on the middle surface. Substitution of (8.202) into the equilibrium condition (8.86) yields the equation of motion:

$$\left(\frac{\partial^2}{\partial r^2} + \frac{1}{r}\frac{\partial}{\partial r}\right)\left[\left(\frac{\partial^2}{\partial r^2} + \frac{1}{r}\frac{\partial}{\partial r}\right) w \pm \mathcal{F} w\right] = \frac{R_o^4}{I_1 E_1}\rho \frac{\partial^2 w}{\partial t^2}. \tag{8.203}$$

The solution is sought in the form

$$w(r,t) = W(r)\gamma(t), \tag{8.204}$$

where $W(r)$ is the amplitude of the vibrations and $\gamma(t)$ is an unknown function. Substituting this solution into (8.203) results in the equation

$$\frac{I_1 E_1}{R_o^4 \rho}\frac{1}{W}\left(\tilde{\Delta}\tilde{\Delta}W \pm \mathcal{F}\tilde{\Delta}W\right) = -\frac{1}{\gamma(t)}\frac{\mathrm{d}^2\gamma(t)}{\mathrm{d}t^2} = \omega^2\,, \quad \tilde{\Delta} = \frac{\mathrm{d}^2}{\mathrm{d}r^2} + \frac{1}{r}\frac{\mathrm{d}}{\mathrm{d}r}\,. \tag{8.205}$$

Since the left side of this equation is independent of t whereas the middle part is independent of r, it follows that they have the same constant value denoted by ω^2 (ω is the eigenfrequency of the vibrations). Hence, $W(r)$ and $\gamma(t)$ should satisfy the following equations:

$$\left(\frac{\mathrm{d}^2}{\mathrm{d}r^2} + \frac{1}{r}\frac{\mathrm{d}}{\mathrm{d}r}\right)\left[\left(\frac{\mathrm{d}^2}{\mathrm{d}r^2} + \frac{1}{r}\frac{\mathrm{d}}{\mathrm{d}r}\right)W \pm \mathcal{F}W\right] = \frac{R_o^4}{I_1 E_1}\rho\omega^2 W\,, \tag{8.206}$$

$$\frac{\mathrm{d}^2\gamma(t)}{\mathrm{d}t^2} + \omega^2\gamma(t) = 0\,. \tag{8.207}$$

Equation (8.206) and boundary conditions (8.89)–(8.91) define three self-adjoint eigenvalue problems.

8.18.2 Integral Equation of the Eigenvalue Problem

Since $\rho\omega^2 W$ corresponds to p_z in Eq. (8.86), it follows from Eq. (8.121) that the amplitude $W(r)$ should satisfy the integral equation

$$W(r) = \lambda \int_0^1 \hat{G}(r,\xi)\, W(\xi)\, \xi\, \mathrm{d}\xi \quad \text{where} \quad \lambda = \rho\omega^2\,, \qquad \hat{G}(r,\xi) = 2\pi R_o^2 G(r,\xi)\,, \tag{8.208}$$

which can be manipulated into the following form:

$$W(r) = \rho\omega^2 \int_0^1 \frac{R_o^2}{f_R}\tilde{G}(r,\xi)\, W(\xi)\, \xi\, \mathrm{d}\xi = \underbrace{\rho\frac{\omega^2 R_o^4}{I_1 E_1}}_{\mathcal{A}} \int_0^1 \underbrace{\frac{1}{\frac{f_R R_o^2}{I_1 E_1}}}_{\mathcal{F}} \tilde{G}(r,\xi)\, W(\xi)\, \xi\, \mathrm{d}\xi\,. \tag{8.209}$$

Here, $\mathcal{A}$ and $\mathcal{F}$ are dimensionless quantities: $\mathcal{A}$ is proportional to the square of an eigenfrequency and $\mathcal{F}$ is proportional to the in-plane load. By introducing the new unknown function

$$y(r) = \sqrt{r}W(r), \tag{8.210}$$

we have

$$\underbrace{\sqrt{r}W(r)}_{y(r)} = \mathcal{A}\int_0^1 \underbrace{\sqrt{r}\frac{\tilde{G}(r,\xi)}{\mathcal{F}}\sqrt{(\xi)}}_{\mathcal{G}(r,\xi)}\,\underbrace{\sqrt{\xi}W(\xi)}_{y(\xi)}\,\mathrm{d}\xi\,, \tag{8.211}$$

that is,

$$y(r) = \mathcal{A}\int_0^1 \mathcal{G}(r,\xi)y(\xi)\mathrm{d}\xi\,. \tag{8.212}$$

The above equation is a homogeneous Fredholm integral equation with a symmetric kernel. At the same time, this equation is an eigenvalue problem with $\mathcal{A}$ as an eigenvalue, which is a function of the dimensionless in-plane load $\mathcal{F}$.

8.19 Computational Results

8.19.1 Data for the Computations

The numerical solution is based on Sect. 8.15.2. On the basis of (8.209), it is worth introducing the following dimensionless quantities:

$$\mathcal{A}_{oi} = \rho\frac{R_o^2}{I_1E_1}\omega_{oi}^2\,, \qquad \mathcal{A}_i = \rho\frac{R_o^2}{I_1E_1}\omega_i^2 \quad \text{and} \quad \mathcal{F}_{o1} = \frac{R_o^2}{I_1E_1}f_{R1}\,, \tag{8.213}$$

where ω_{oi} and ω_i are the i-th ($i = 1, 2, \ldots$) natural frequencies of the unloaded and loaded plates, while f_{R1} is the first critical load.

The dimensionless unknown $\mathcal{A}(\mathcal{F})$ and the parameter $\mathcal{F}$ in Eq. (8.212)—$\mathcal{G}(r,\xi)$ depends on $\mathcal{F}$—are formally independent of the geometrical data and for the clamped plates of the material properties of the plate. For the simple and spring supported plates, they depend on ν. It is obvious that the previous statement is valid for $\mathcal{A}_{o1}$ and $\mathcal{F}_{o1}$ as well. For computational purposes, the following parameters are selected: $R_o = 100$ mm, $b = 5$ mm, $E = 2.1 \times 10^5$, $\nu = 0.33333$—the last two values are typical for a mild steel.

8.19.2 Clamped Plate

For the clamped plate $\mathcal{A}_{o1} = 104.85$, $\mathcal{F}_{o1} = 14.682$. The computational results for $i = 1$ are presented in Table 8.7.

Table 8.7 Computational results for clamped plate

$\mathcal{F}/\mathcal{F}_{o1}$	0.068	0.136	0.204	0.272	0.341	0.409	0.477
$\mathcal{A}_1/\mathcal{A}_{o1}$ – compression	0.929	0.863	0.796	0.730	0.663	0.595	0.528
$\mathcal{A}_1/\mathcal{A}_{o1}$ – tension	1.061	1.127	1.193	1.258	1.324	1.389	1.454
$\mathcal{F}/\mathcal{F}_{o1}$	0.545	0.613	0.681	0.749	0.817	0.886	0.954
$\mathcal{A}_1/\mathcal{A}_{o1}$ – compression	0.460	0.392	0.324	0.255	0.186	0.117	0.048
$\mathcal{A}_1/\mathcal{A}_{o1}$ – tension	1.5189	1.584	1.648	1.713	1.777	1.841	1.906

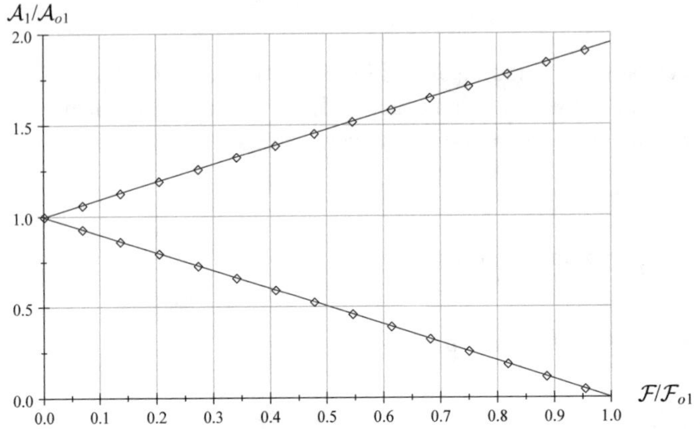

Fig. 8.9 The effect of the constant in plane load on the first circular frequencies for clamped plate

Making use of the computational results, the following polynomials are fitted onto the discrete points—these are denoted by diamonds in Fig. 8.9:

$$\frac{\mathcal{A}_1}{\mathcal{A}_{o1}} = 0.994\,99 - 0.965\,94\frac{\mathcal{F}}{\mathcal{F}_{o1}} - 2.851\,4 \times 10^{-2}\left(\frac{\mathcal{F}}{\mathcal{F}_{o1}}\right)^2, \tag{8.214a}$$

$$\frac{\mathcal{A}_1}{\mathcal{A}_{o1}} = 0.995\,61 + 0.968\,14\frac{\mathcal{F}}{\mathcal{F}_{o1}} - 1.474\,2 \times 10^{-2}\left(\frac{\mathcal{F}}{\mathcal{F}_{o1}}\right)^2. \tag{8.214b}$$

Observe that the approximate solutions (8.214) are practically linear in the interval $\mathcal{F}/\mathcal{F}_{o1} \in [0, 1]$.

8.19.3 Simply Supported Plate

For a simply supported plate, the computational results are presented in Table 8.8 under the assumption that $i = 1$. These are denoted by diamonds in Fig. 8.10.

Table 8.8 Computational results for simply supported plate

$\mathcal{F}/\mathcal{F}_{o1}$	0.070	0.140	0.210	0.280	0.350	0.420	0.490
$\mathcal{A}_1/\mathcal{A}_{o1}$ – compression	0.930	0.860	0.790	0.720	0.650	0.580	0.510
$\mathcal{A}_1/\mathcal{A}_{o1}$ – tension	1.071	1.141	1.211	1.281	1.351	1.421	1.491
$\mathcal{F}/\mathcal{F}_{o1}$	0.560	0.630	0.700	0.770	0.817	0.840	0.910
$\mathcal{A}_1/\mathcal{A}_{o1}$ – compression	0.440	0.370	0.300	0.230	0.159	0.089	0.019
$\mathcal{A}_1/\mathcal{A}_{o1}$ – tension	1.561	1.631	1.701	1.771	1.841	1.911	1.981

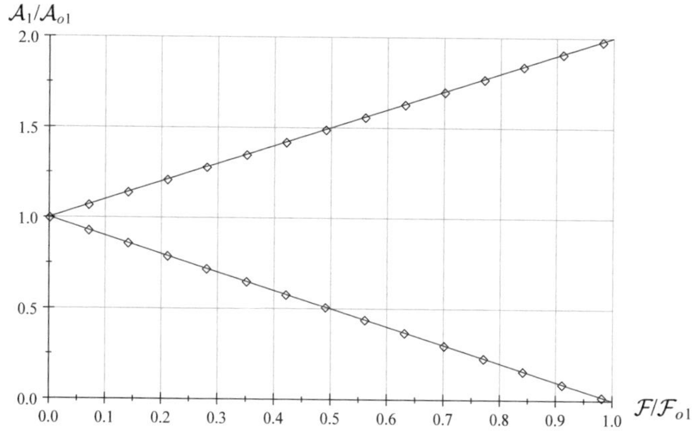

Fig. 8.10 The effect of the constant in plane load on the first circular frequencies for simply supported plate

The approximate solutions obtained again by fitting polynomials onto the computational results are practically linear functions. These are also shown in Fig. 8.10 ($\mathcal{A}_{o1} = 24.838$, $\mathcal{F}_{o1} = 4.285$):

$$\frac{\mathcal{A}_1}{\mathcal{A}_{o1}} = 1.0004 - 1.0007\frac{\mathcal{F}}{\mathcal{F}_{o1}} \simeq 1.000 - 1.000\frac{\mathcal{F}}{\mathcal{F}_{o1}}, \tag{8.215a}$$

$$\frac{\mathcal{A}_1}{\mathcal{A}_{o1}} = 1.0004 + 1.0007\frac{\mathcal{F}}{\mathcal{F}_{o1}} \simeq 1.000 + 1.000\frac{\mathcal{F}}{\mathcal{F}_{o1}}. \tag{8.215b}$$

The square of the first natural frequency of the plate when unloaded and the first critical load can be calculated by using Eq. (8.215).

8.19.4 Spring Supported Plate

For a spring supported plate, the results depend on the dimensionless spring constant $\mathcal{K}$ as well. If $i = 1$, Table 8.9 shows the results we have obtained for $\mathcal{A}_{o1}(\mathcal{K})$ and $\mathcal{F}_{o1}(\mathcal{K})$.

Table 8.9 Computational results for $\mathcal{A}_{o1}\mathcal{K}$ and $\mathcal{F}_{o1}\mathcal{K}$

$\mathcal{K}$	0.0	1.00	5.00	10.00	50.00	100.00	1000.00	$\mathcal{K}\to\infty$
$\mathcal{A}_{o1}$	24.838	37.36	63.118	76.677	96.791	100.41	103.97	104.85
$\mathcal{F}_{o1}$	4.285	6.498	10.481	12.182	14.115	14.397	14.657	14.68

Observe that for $\mathcal{K}=0$ and $\mathcal{K}\to\infty$ Table 8.9 contains the values valid for simply supported and clamped plates. For a compressive f_R, the following functions can be fitted onto the results obtained:

$$\mathcal{K}=1.00 \qquad \frac{\mathcal{A}_1}{\mathcal{A}_{o1}}=0.9937-1.0067\frac{\mathcal{F}}{\mathcal{F}_{o1}}, \tag{8.216a}$$

$$\mathcal{K}=5.00 \qquad \frac{\mathcal{A}_1}{\mathcal{A}_{o1}}=1.000-0.99119\frac{\mathcal{F}}{\mathcal{F}_{o1}}-8.7863\times10^{-3}\left(\frac{\mathcal{F}}{\mathcal{F}_{o1}}\right)^2, \tag{8.216b}$$

$$\mathcal{K}=10.00 \qquad \frac{\mathcal{A}_1}{\mathcal{A}_{o1}}=1.000-0.98223\frac{\mathcal{F}}{\mathcal{F}_{o1}}-1.7763\times10^{-2}\left(\frac{\mathcal{F}}{\mathcal{F}_{o1}}\right)^2, \tag{8.216c}$$

$$\mathcal{K}=50.00 \qquad \frac{\mathcal{A}_1}{\mathcal{A}_{o1}}=1.000-0.9721\frac{\mathcal{F}}{\mathcal{F}_{o1}}-2.7839\times10^{-2}\left(\frac{\mathcal{F}}{\mathcal{F}_{o1}}\right)^2, \tag{8.216d}$$

$$\mathcal{K}=100.00 \qquad \frac{\mathcal{A}_1}{\mathcal{A}_{o1}}=1.000-0.9712\frac{\mathcal{F}}{\mathcal{F}_{o1}}-2.82\times10^{-2}\left(\frac{\mathcal{F}}{\mathcal{F}_{o1}}\right)^2, \tag{8.216e}$$

$$\mathcal{K}=1000.00 \qquad \frac{\mathcal{A}_1}{\mathcal{A}_{o1}}=0.99997-0.97108\frac{\mathcal{F}}{\mathcal{F}_{o1}}-2.8827\times10^{-2}\left(\frac{\mathcal{F}}{\mathcal{F}_{o1}}\right)^2. \tag{8.216f}$$

For a tensile f_r, Eqs. (8.217) are the polynomials we have fitted onto the computational results.

$$\mathcal{K}=1.00 \qquad \frac{\mathcal{A}_1}{\mathcal{A}_{o1}}=0.99364+1.0068\frac{\mathcal{F}}{\mathcal{F}_{o1}}, \tag{8.217a}$$

$$\mathcal{K}=5.00 \qquad \frac{\mathcal{A}_1}{\mathcal{A}_{o1}}=1.0002+0.99203\frac{\mathcal{F}}{\mathcal{F}_{o1}}-4.511\times10^{-3}\left(\frac{\mathcal{F}}{\mathcal{F}_{o1}}\right)^2, \tag{8.217b}$$

$$\mathcal{K}=10.00 \qquad \frac{\mathcal{A}_1}{\mathcal{A}_{o1}}=1.0004+0.98359\frac{\mathcal{F}}{\mathcal{F}_{o1}}-8.6524\times10^{-3}\left(\frac{\mathcal{F}}{\mathcal{F}_{o1}}\right)^2, \tag{8.217c}$$

$$\mathcal{K}=50.00 \qquad \frac{\mathcal{A}_1}{\mathcal{A}_{o1}}=1.0006+0.97411\frac{\mathcal{F}}{\mathcal{F}_{o1}}-1.4003\times10^{-2}\left(\frac{\mathcal{F}}{\mathcal{F}_{o1}}\right)^2, \tag{8.217d}$$

$$\mathcal{K}=100.00 \qquad \frac{\mathcal{A}_1}{\mathcal{A}_{o1}}=1.0006+0.97360\frac{\mathcal{F}}{\mathcal{F}_{o1}}-1.4546\times10^{-2}\left(\frac{\mathcal{F}}{\mathcal{F}_{o1}}\right)^2, \tag{8.217e}$$

$$\mathcal{K}=1000.00 \qquad \frac{\mathcal{A}_1}{\mathcal{A}_{o1}}=1.0047+0.94510\frac{\mathcal{F}}{\mathcal{F}_{o1}}+1.8687\times10^{-2}\left(\frac{\mathcal{F}}{\mathcal{F}_{o1}}\right)^2. \tag{8.217f}$$

Observe that functions (8.216) and (8.217) are almost linear functions. For this reason, Table 8.10 presents the computational results for $\mathcal{K}=100.0$ only.

Table 8.10 Computational results for spring supported plate if $\mathcal{K} = 100.0$

$\mathcal{F}/\mathcal{F}_{o1}$	0.069	0.139	0.208	0.278	0.347	0.417	0.486
$\mathcal{A}_1/\mathcal{A}_{o1}$ – compression	0.933	0.865	0.796	0.729	0.659	0.590	0.521
$\mathcal{A}_1/\mathcal{A}_{o1}$ – tension	1.068	1.136	1.203	1.270	1.337	1.404	1.471
$\mathcal{F}/\mathcal{F}_{o1}$	0.556	0.625	0.695	0.764	0.834	0.903	0.972
$\mathcal{A}_1/\mathcal{A}_{o1}$ – compression	0.452	0.382	0.312	0.241	0.171	0.100	0.028
$\mathcal{A}_1/\mathcal{A}_{o1}$ – tension	1.537	1.604	1.670	1.736	1.802	1.868	1.934

Figure 8.11 shows function (8.216e) and (8.217e) fitted onto the above results.

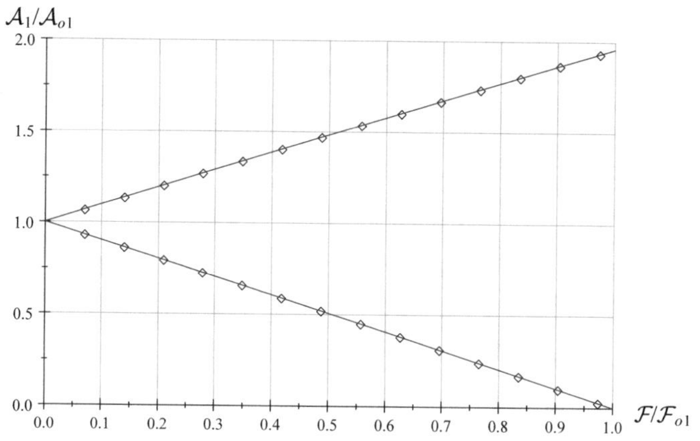

Fig. 8.11 The effect of the constant in plane load on the first circular frequencies for spring supported plate if $\kappa = 100.0$

8.20 Problems

Problem 8.1 Prove that the last column in Table 8.1 contains the correct form of the Green functions for fixed-pinned, fixed-fixed, and fixed-free beams.

Problem 8.2 Solution of Problem 7.5 is based on a given flexibility matrix. Check the elements of the flexibility matrix. (Hint: use the Green function of the simply supported beam.)

Problem 8.3 Determine the Green function for preloaded fixed-fixed beams.

Problem 8.4 Determine the Green function for preloaded fixed-pinned beams.

Problem 8.5 Derive equation system (8.108) for tensile preload with unknowns $\mathcal{A}_1$, $\mathcal{A}_3$, $\mathcal{B}_1, \ldots, \mathcal{B}_4$ using continuity conditions (8.95), discontinuity condition (8.96), and boundary conditions (8.91).

References

1. L. Collatz, *Eigenwertaufgaben mit Technischen Anwendungen*. Russian edn. in 1968 (Akademische Verlagsgesellschaft Geest & Portig K.G., Leipzig, 1963)
2. E. Kamke, Über die Definiten Selbstadjungierten Eigenwertaufgaben bei gewöhnlichen Differentialgleichungen IV. Math. Z. **48**, 67–100 (1942)
3. L. Collatz, *The Numerical Treatment of Differential Equations*, 3rd edn. (Springer, Berlin, 1966)
4. N.M. Matveev, *Analytic Theory of Differential Equations* [in Russian] (Leningrad Pedagogic University, St. Petersburg, 1989)
5. M. Bocher, Some theorems concerning linear differential equations of the second order. Bull. Am. Math. Soc. **6**(7), 279–280 (1901)
6. M. Bocher, Boundary problems and Green's functions for linear differential and difference equations. Ann. Math. **13.1/4**, 71–88 (1911–1912)
7. E.L. Ince, *Ordinary Differential Equations. (a)* (Longmans, Green and Co., London, 1926), pp. 256–258
8. J.M. Höené-Wronski, *Réfutation de la Théorie des Fonctions Analytiques de Lagrange* (Blankenstein, Paris, 1812)
9. N. Szűcs, Vibrations of circular plates subjected to an in-plane load. GÉP LVIII.5-6 (2007 (in Hungarian)), pp. 41–47
10. N. Szűcs, G. Szeidl, Vibration of circular plates subjected to constant radial load in their plane. J. Comput. Appl. Mech. **12**(1), 57–76 (2017). https://doi.org/10.32973/jcam.2017.004
11. I. Kozák, *Strength of Materials V. - Thin Walled Structures and the Theory of Plates and Shells* (in Hungarian). Tankönyvkiadó (Publisher of Textbooks) (Budapest, Hungary, 1967), pp. 287–291
12. G.B. Arken, H.J. Weber, *Mathematical Methods for Physicists*, 7th edn. (Elsevier/Academic, Amsterdam, 2005)
13. E.I. Fredholm, On a new method for solving the Dirichlet problem. Sur une nouvelle méthode pour la résolution du probléme de Dirichlet. Stockh. Öfv. **57**, 39–46 (1900)
14. E.I. Fredholm, Sur une classe d'équations fonctionnelles. Acta Math. **27**, 365–390 (1903). https://doi.org/10.1007/bf02421317
15. W.V. Lovitt, *Linear Integral Equations* (McGraw-Hill, New York, 1924)
16. J. Mercer, Functions of positive and negative type and their connection with the theory of integral equations. Philos. Trans. R. Soc. A **209**, 441–458 (1909). https://doi.org/10.1098/rsta.1909.0016
17. R. Courant, D. Hilbert, *Methods of Mathematical Physics*, vol. 1 (Interscience Publishers, New York, 1953), pp. 148–150
18. C.T.H. Baker, in *The Numerical Treatment of Integral Equations - Monographs on Numerical Analysis*, ed. by L. Fox, J. Walsh (Clarendon Press, Oxford, 1977)
19. C.A. Brebbia, I. Dominguez, *Boundary Elements, an Introductory Course*, 2nd edn. (1992)

Chapter 9
Eigenvalue Problems of Ordinary Differential Equation Systems

9.1 Vibrations Described by Differential Equation Systems

9.1.1 *Engesser–Timoshenko Beam*

9.1.1.1 Kinematical Hypothesis and Inner Forces

In this subsection, the kinematical hypothesis of the Engesser–Timoshenko beam theory is presented. The theory that can be applied to take the effect of shear into account was proposed by Engesser and Timoshenko [1, 2]. The assumptions detailed in Sect. 7.5.1 remain valid except the following: (vii) the distributed moment load $\boldsymbol{\mu} = \mu_y(x)\mathbf{i}_y$ is zero and (ix) the displacement field of the beam is of the form

$$\mathbf{u} = u_x(x, z)\,\mathbf{i}_x + u_z(x)\,\mathbf{i}_z = (u(x) + \psi_y(x)z)\,\mathbf{i}_x + w(x)\,\mathbf{i}_z\,, \qquad u_y = 0, \tag{9.1}$$

where $\psi_y(x)$ is the angle of rotation of the centerline. This quantity is independent of $w(x)$ and it is, therefore, a new kinematic variable.

For the displacement field given by Eq. (9.1)

$$\varepsilon_{xx} = \frac{\partial u_x}{\partial x} = \frac{\mathrm{d}u}{\mathrm{d}x} + \frac{\mathrm{d}\psi_y}{\mathrm{d}x}z = \varepsilon_o + \frac{\mathrm{d}\psi_y}{\mathrm{d}x}z\,, \quad \varepsilon_{yy} = \frac{\partial u_y}{\partial y} = 0\,, \quad \varepsilon_{zz} = \frac{\partial u_z(x)}{\partial z} = 0 \tag{9.2a}$$

and

$$\gamma_{xy} = \gamma_{yx} = \frac{1}{2}\left(\frac{\partial u_x}{\partial y} + \frac{\partial u_y}{\partial x}\right) = 0\,, \quad \gamma_{yz} = \gamma_{zy} = \frac{1}{2}\left(\frac{\partial u_y}{\partial z} + \frac{\partial u_z}{\partial y}\right) = 0\,,$$
$$\gamma_{zx} = \gamma_{zx} = \frac{1}{2}\left(\frac{\partial u_z}{\partial x} + \frac{\partial u_x}{\partial z}\right) = \frac{\mathrm{d}w}{\mathrm{d}x} + \psi_y \tag{9.2b}$$

are the components of the strain tensor. According to Hooke's law

G. Szeidl and L. P. Kiss, *Mechanical Vibrations*, Foundations of Engineering Mechanics,
https://doi.org/10.1007/978-3-030-45074-8_9

$$\sigma_{xx} = E\varepsilon_{xx} = E\varepsilon_o + E\frac{\mathrm{d}\psi_y}{\mathrm{d}x}z\,, \qquad \tau_{xz} = G\gamma_{xz} = G\left(\frac{\mathrm{d}w}{\mathrm{d}x} + \psi_y\right) \tag{9.3}$$

are the non-zero stress components where

$$G = \frac{E}{2(1+\nu)} \tag{9.4}$$

in which ν is the Poisson number. Hence

$$N = \int_A \sigma_{xx}\,\mathrm{d}A = \int_A \left[E\varepsilon_o + E\frac{\mathrm{d}\psi_y}{\mathrm{d}x}z\right]\mathrm{d}A = AE\varepsilon_o + E\frac{\mathrm{d}\psi_y}{\mathrm{d}x}\underbrace{\int_A z\,\mathrm{d}A}_{Q_y=0} = AE\varepsilon_o \tag{9.5a}$$

is the axial force and

$$M_y = \int_A z\sigma_{xx}\,\mathrm{d}A = \int_A \left[E\varepsilon_o z + \frac{\mathrm{d}\psi_y}{\mathrm{d}x}z^2\right]\mathrm{d}A =$$
$$= AE\varepsilon_o \underbrace{\int_A z\,\mathrm{d}A}_{Q_y=0} + E\frac{\mathrm{d}\psi_y}{\mathrm{d}x}\underbrace{\int_A z^2\,\mathrm{d}A}_{=I} = IE\frac{\mathrm{d}\psi_y}{\mathrm{d}x} \tag{9.5b}$$

where Q_y is the first moment of the cross section with respect to a principal axis of the cross section and it is therefore equal to zero, while $I_y = I$ is the moment of inertia.

9.1.1.2 Shear Correction Factor

As regards the shear force, we have

$$V_z = -\int_A \tau_{xz}\,\mathrm{d}A = -\int_A G\gamma_{xz}\,\mathrm{d}A = -AG\gamma_{xz} = -AG\left(\frac{\mathrm{d}w}{\mathrm{d}x} + \psi_y\right). \tag{9.5c}$$

Since γ_{xz} is constant neither τ_{xz} nor V_z are accurate in the previous equation: they should, therefore, be modified in such a way that the strain energy due to shearing be the same as that obtained from a good approximation of the shear stress. The latter is calculated by using equilibrium methods. The first moment of the shaded area of the cross section (Fig. 9.1) with respect to the centroidal axis y is given by the relation

$$Q'_y(z) = \int_{A'} \zeta\,\mathrm{d}A = \int_{A'} \zeta\, t(\zeta)\mathrm{d}\zeta. \tag{9.6}$$

Formula

$$\tau_{zx} = -\frac{V_z\, Q'_y(z)}{It(z)} \tag{9.7}$$

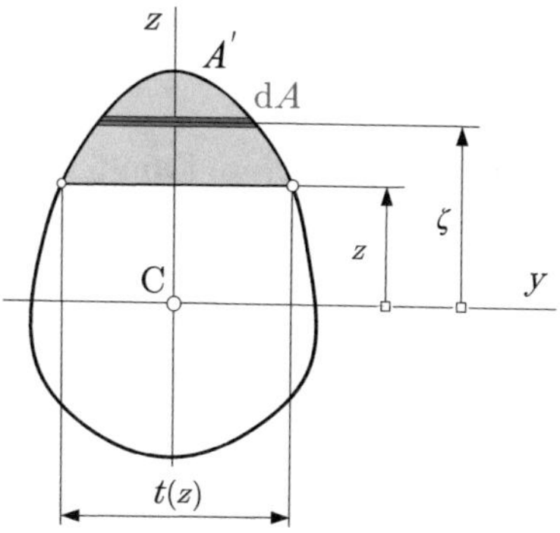

Fig. 9.1 Cross section A with its parts A' and $\mathrm{d}A$

for the shear stress can be proven, as has already been mentioned above, by using equilibrium methods.[1]. Making use of (9.7), the strain energy for a unit length of the beam due to shear can be calculated in the following manner:

$$U_\tau = \frac{1}{2G}\int_{A'} \tau_{xz}^2(z)\,\mathrm{d}A = \frac{1}{2G}\int_{A'}\left[\frac{V_z\,Q'_y(z)}{It(z)}\right]^2 \mathrm{d}A = \frac{1}{2G}\int_{A'} \frac{V_z^2\,(Q'_y(z))^2}{I^2 t(z)}\,\mathrm{d}z\,. \tag{9.8}$$

It also holds that

$$U_{\tau\gamma} = \frac{1}{2}\int_{A'} \tau_{xz}\gamma_{xz}\,\mathrm{d}A = \frac{1}{2}\int_{A'} \tau_{xz}\,\mathrm{d}A\,\gamma_{xz} = -\frac{1}{2}V_z\,\gamma_{xz}, \tag{9.9}$$

where it has been taken into account that according to Eq. $(9.2\text{b})_3$ γ_{xz} is independent of the coordinates y, z. It is obvious that $U_\tau \neq U_{\tau\gamma}$, though, in principle, they should have the same value. However, manipulating Eq. (9.5c) into the form

$$V_z = -AG\kappa_y\,\gamma_{xz}\,, \qquad \kappa_y = \frac{I^2}{A}\frac{1}{\displaystyle\int_A \frac{(Q'_y(z))^2}{t(z)}\,\mathrm{d}z}\,, \tag{9.10}$$

in which κ_y is the shear correction factor, ensures that $U_{\tau\gamma}$ will be equal to U_τ—substitute γ_{xz} from $(9.10)_1$ into (9.9). The shear correction factors for some simple cross sections can be found in [3, 4]. A few typical values are presented in Appendix B.4.

Remark 9.1 Equation $(9.10)_2$ is the special case of equation

$$\kappa_y = \frac{V_z^2}{A\displaystyle\int_A \left(\tau_{zx}^2 + \tau_{zy}^2\right)\,\mathrm{d}A} \tag{9.11}$$

[1] See for instance http://www.bu.edu/moss/mechanics-of-materials-bending-shear-stress

from where it can be obtained by substituting (9.7) for τ_{zx} and setting τ_{zy} to zero [5].

9.1.1.3 Equilibrium Equations in Terms of Displacement Components

The following equations

$$N = AE\varepsilon_o\,, \qquad M_y = IE\frac{\mathrm{d}\psi_y}{\mathrm{d}x}\,, \qquad V_z = -AG\kappa_y\left(\frac{\mathrm{d}w}{\mathrm{d}x} + \psi_y\right) \tag{9.12}$$

are the constitutive relations for the inner forces and the bending moment within the framework of the Timoshenko beam theory.

For static problems, there are two equilibrium conditions to be satisfied. The first one is Eq. (7.69), repeated here for completeness:

$$-\frac{\mathrm{d}V_z}{\mathrm{d}x} + f_z = 0\,. \tag{9.13a}$$

The second is obtained from (7.70) by taking $(7.71)_{1,2}$ into account and assuming that the distributed moment acting on the centerline is not zero, i.e., $\boldsymbol{\mu} = \mu_y\mathbf{i}_y$:

$$\frac{\mathrm{d}M_y}{\mathrm{d}x} + N\frac{\mathrm{d}w}{\mathrm{d}x} + V_z + \mu_y = 0\,. \tag{9.13b}$$

Note that this equation coincides with Eq. (7.72) if $\mu_y = 0$.

Substitution of the constitutive relations $(9.12)_{2,3}$ into the equilibrium conditions (9.13) yields

$$\begin{gathered}\frac{\mathrm{d}}{\mathrm{d}x}\left[AG\kappa_y\left(\frac{\mathrm{d}w}{\mathrm{d}x} + \psi_y\right)\right] + f_z = 0\,,\\ \frac{\mathrm{d}}{\mathrm{d}x}\left(IE\frac{\mathrm{d}\psi_y}{\mathrm{d}x}\right) \pm N\frac{\mathrm{d}w}{\mathrm{d}x} - AG\kappa_y\left(\frac{\mathrm{d}w}{\mathrm{d}x} + \psi_y\right) + \mu_y = 0,\end{gathered} \tag{9.14}$$

where (and in the sequel) the (upper)[lower] sign of N belongs to a (tensile) [compressive] axial force—from now on N is regarded as if it were a positive quantity and the applied sign reflects if it is tensile or compressive force. If the beam is uniform—this is an assumption—Eq. (9.14) is simplified:

$$\boxed{\begin{gathered}AG\kappa_y\left(\frac{\mathrm{d}^2 w}{\mathrm{d}x^2} + \frac{\mathrm{d}\psi_y}{\mathrm{d}x}\right) + f_z = 0\,,\\ IE\frac{\mathrm{d}^2\psi_y}{\mathrm{d}x^2} \pm N\frac{\mathrm{d}w}{\mathrm{d}x} - AG\kappa_y\left(\frac{\mathrm{d}w}{\mathrm{d}x} + \psi_y\right) + \mu_y = 0\,.\end{gathered}} \tag{9.15}$$

These equations can be rewritten in matrix form as well:

$$\begin{bmatrix}\hat{\chi} & 0\\ 0 & 1\end{bmatrix}\begin{bmatrix}w^{(2)}\\ \psi^{(2)}\end{bmatrix}+\begin{bmatrix}0 & \hat{\chi}\\ -\hat{\chi}\pm\hat{N} & 0\end{bmatrix}\begin{bmatrix}w^{(1)}\\ \psi^{(1)}\end{bmatrix}+ \\ +\begin{bmatrix}0 & 0\\ 0 & -\hat{\chi}\end{bmatrix}\begin{bmatrix}w\\ \psi\end{bmatrix}+\begin{bmatrix}\hat{f}\\ \hat{\mu}\end{bmatrix}=\begin{bmatrix}0\\ 0\end{bmatrix}, \tag{9.16a}$$

where

$$\hat{\chi}=\frac{AG\kappa_y}{IE},\quad \hat{N}=\frac{N}{IE},\quad \hat{f}=\frac{f_z}{IE},\quad \hat{\mu}=\frac{\mu_z}{IE}\quad\text{and}\quad \psi=\psi_y. \tag{9.16b}$$

Substituting $w^{(2)}+\psi_y^{(1)}$ and then $w^{(2)}$ form Eq. (9.15)$_1$ into the derivative of Eq. (9.15)$_2$ yields an equation for the unknown ψ_y:

$$\frac{\mathrm{d}^3\psi_y}{\mathrm{d}x^3}\mp\frac{N}{IE}\frac{\mathrm{d}\psi_y}{\mathrm{d}x}+f_z\left(1\mp\frac{N}{IE\,AG\kappa_y}\right)+\frac{\mathrm{d}\mu_y}{\mathrm{d}x}=0. \tag{9.17}$$

Special cases:

(a) There is no axial force, i.e., $N=0$. Then

$$\begin{aligned}AG\kappa_y\left(\frac{\mathrm{d}^2w}{\mathrm{d}x^2}+\frac{\mathrm{d}\psi_y}{\mathrm{d}x}\right)+f_z=0,\\ IE\frac{\mathrm{d}^2\psi_y}{\mathrm{d}x^2}-AG\kappa_y\left(\frac{\mathrm{d}w}{\mathrm{d}x}+\psi_y\right)+\mu_y=0\end{aligned} \tag{9.18}$$

are the governing equations.

(b) If in addition to this there is no distributed moment load, it follows from Eq. (9.18) that

$$\frac{\mathrm{d}^3\psi_y}{\mathrm{d}x^3}+f_z=0\quad\text{and}\quad IE\frac{\mathrm{d}^4w}{\mathrm{d}x^4}+\frac{1}{AG\kappa_y}\frac{\mathrm{d}^2f_z}{\mathrm{d}x^2}=f_z \tag{9.19}$$

if ψ_y is given in terms of w.

For our later considerations, we present some solutions which can be used for determining the Green function matrices:

- If the axial force is compressive, the homogeneous differential equation system

$$\begin{bmatrix}\hat{\chi} & 0\\ 0 & 1\end{bmatrix}\begin{bmatrix}w^{(2)}\\ \psi^{(2)}\end{bmatrix}+\begin{bmatrix}0 & \hat{\chi}\\ -\hat{\chi}-\hat{N} & 0\end{bmatrix}\begin{bmatrix}w^{(1)}\\ \psi^{(1)}\end{bmatrix}+\begin{bmatrix}0 & 0\\ 0 & -\hat{\chi}\end{bmatrix}\begin{bmatrix}w\\ \psi\end{bmatrix}=\begin{bmatrix}0\\ 0\end{bmatrix} \tag{9.20}$$

has the following solutions:

$$\boxed{\begin{aligned} w &= C_1 + C_2 x + \frac{C_3}{\hat{n}} \cos \hat{n}x - \frac{C_4}{\hat{n}} \sin \hat{n}x \,, \\ \psi &= -C_2 \frac{1}{\hat{\chi}} \left(\hat{\chi} + \hat{N} \right) + C_3 \sin \hat{n}x + C_4 \cos \hat{n}x \,. \end{aligned}} \tag{9.21}$$

Here and in the sequel $C_1, \ldots C_4$ are undermined integration constants which differ from each other in the considered cases and

$$\hat{n} = \sqrt{N/IE} \,. \tag{9.22}$$

- If the axial force is tensile solution of the homogeneous differential equation system

$$\begin{bmatrix} \hat{\chi} & 0 \\ 0 & 1 \end{bmatrix} \begin{bmatrix} w^{(2)} \\ \psi^{(2)} \end{bmatrix} + \begin{bmatrix} 0 & \hat{\chi} \\ -\hat{\chi} + \hat{N} & 0 \end{bmatrix} \begin{bmatrix} w^{(1)} \\ \psi^{(1)} \end{bmatrix} + \begin{bmatrix} 0 & 0 \\ 0 & -\hat{\chi} \end{bmatrix} \begin{bmatrix} w \\ \psi \end{bmatrix} = \begin{bmatrix} 0 \\ 0 \end{bmatrix} \tag{9.23}$$

is given by the following equations:

$$\boxed{\begin{aligned} w &= C_1 + C_2 x - \frac{C_3}{\hat{n}} \cosh \hat{n}x - \frac{C_4}{\hat{n}} \sinh \hat{n}x \,, \\ \psi &= -C_2 \frac{1}{\hat{\chi}} \left(\hat{\chi} - \hat{N} \right) + C_3 \sinh \hat{n}x + C_4 \cosh \hat{n}x \,. \end{aligned}} \tag{9.24}$$

Exercise 9.1 Find the solution of the homogeneous differential equation systems (9.20) and (9.23) if the axial force $N = 0$.

Note that the two homogeneous differential equations coincide with each other. It can be checked with ease that the solutions of the differential equations

$$\psi_y^{(3)} = 0 \,, \qquad w^{(2)} + \psi_y^{(1)} = 0$$

should also satisfy the differential equation

$$\psi_y^{(2)} - \hat{\chi} \left(w^{(2)} + \psi_y \right) = 0 \,,$$

which yields a requirement for the integration constants. Hence,

$$\boxed{\begin{aligned} w &= C_1 + C_2 x - \frac{1}{2} C_3 x^2 - \frac{1}{3} C_4 x^3 \,, \\ \psi &= -C_2 + C_3 x + C_4 \left(\frac{2}{\hat{\chi}} + x^2 \right) . \end{aligned}} \tag{9.25}$$

9.1.1.4 Boundary Conditions

Equilibrium equations (9.15) should be supplemented with appropriate boundary conditions. Comparison of Eqs. (9.13b), $(9.12)_2$, and Table 7.2 leads to the boundary conditions presented again in a table form:

Table 9.1 Boundary conditions for Timoshenko beams

	Support arrangements	Boundary conditions	
1.	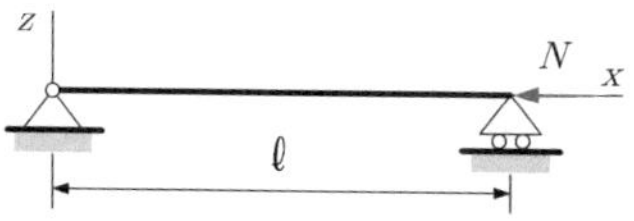	$\omega(0)=0$ $\psi_y^{(1)}(0)=0$	$\omega(\ell)=0$ $\psi_y^{(1)}(\ell)=0$
2.	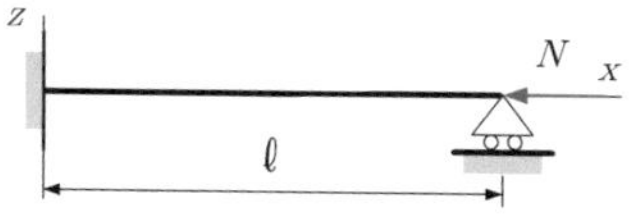	$\omega(0)=0$ $\psi_y(0)=0$	$\omega(\ell)=0$ $\psi_y^{(1)}(\ell)=0$
3.	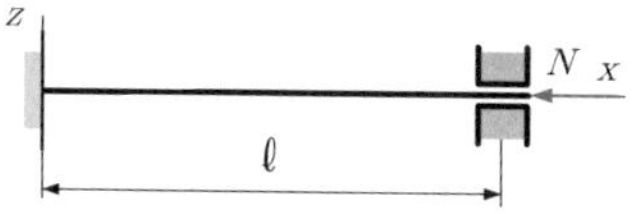	$\omega(0)=0$ $\psi_y(0)=0$	$\omega(\ell)=0$ $\psi_y(\ell)=0$
4.		$\omega(0)=0$ $\psi_y(0)=0$	$\psi_y^{(1)}(\ell)=0$ $IE\psi_y^{(2)}(\ell)\pm N\omega^{(1)}(\ell)=0$

If the left or right endpoint of the beam is restrained (vertically) (rotationally) by a (spring) [torsional spring]—see Fig. 7.9 for details—then the boundary conditions are given by the following equations:

$$-IE\psi_y^{(2)} \mp Nw^{(1)} + k_{z\ell}w = 0\,, \qquad -IE\psi_y^{(2)} \mp Nw^{(1)} - k_{zr}w = 0\,; \tag{9.26a}$$

$$IE\psi_y^{(1)} - k_{\gamma\ell}\psi_y = 0\,, \qquad IE\psi_y^{(1)} + k_{\gamma r}\psi_y = 0\,. \tag{9.26b}$$

9.1.2 Equations of Motion

For the vibrating axially loaded Timoshenko beam

$$f_z = -\rho A\frac{\partial^2 w}{\partial t^2} \quad \text{and} \quad \mu_z = -\rho I\frac{\partial^2 \psi_y}{\partial t^2} \tag{9.27}$$

are the intensities of distributed and moment loads on the axis x. Substituting (9.27) into the equilibrium conditions (9.15) results in the equations of motion:

$$
\begin{aligned}
AG\kappa_y\left(\frac{\partial^2 w}{\partial x^2}+\frac{\partial\psi_y}{\partial x}\right)&=\rho A\frac{\partial^2 w}{\partial t^2}\,,\\
IE\frac{\partial^2\psi_y}{\partial x^2}\pm N\frac{\partial w}{\partial x}-AG\kappa_y\left(\frac{\partial w}{\partial x}+\psi_y\right)&=\rho I\frac{\partial^2\psi_y}{\partial t^2}\,.
\end{aligned}
\tag{9.28}
$$

For harmonic vibrations we may assume that

$$
\begin{bmatrix} w(x,t)\\ \psi(x,t)\end{bmatrix}=\begin{bmatrix} W(x)\\ \Psi(x)\end{bmatrix}\sin\omega t,
\tag{9.29}
$$

where $W(X)$ and $\Psi(x)$ are the unknown amplitudes of the motion and ω is the unknown circular frequency. Substituting (9.29) into (9.28) and dropping $\sin\omega t$ yields

$$
\begin{aligned}
AG\kappa_y\left(W^{(2)}+\Psi_y^{(1)}\right)&=-\rho\omega^2 AW\,,\\
IE\Psi_y^{(2)}\pm NW^{(1)}-AG\kappa_y\left(W^{(1)}+\Psi_y\right)&=-\rho\omega^2 I\Psi_y\,.
\end{aligned}
\tag{9.30}
$$

Equation (9.30) can be rewritten in matrix form as well:

$$
\begin{aligned}
&\begin{bmatrix}\hat{\chi} & 0\\ 0 & 1\end{bmatrix}\begin{bmatrix}W^{(2)}\\ \Psi^{(2)}\end{bmatrix}+\begin{bmatrix}0 & \hat{\chi}\\ -\hat{\chi}\pm\hat{N} & 0\end{bmatrix}\begin{bmatrix}W^{(1)}\\ \Psi^{(1)}\end{bmatrix}+\\
&\qquad+\begin{bmatrix}0 & 0\\ 0 & -\hat{\chi}\end{bmatrix}\begin{bmatrix}W\\ \Psi\end{bmatrix}=-\lambda\begin{bmatrix}1 & 0\\ 0 & r_y^2\end{bmatrix}\begin{bmatrix}W\\ \Psi\end{bmatrix},
\end{aligned}
\tag{9.31}
$$

where

$$
\lambda=\frac{\rho A\omega^2}{IE},\quad r_y^2=\frac{I}{A}\quad\text{and}\quad\hat{\chi}=\frac{AG\kappa_y}{IE},\quad\hat{N}=\frac{N}{IE}\,.
\tag{9.32}
$$

After eliminating Ψ_y in Eq. $(9.30)_2$, we get an equation for W:

$$
\frac{IE}{\rho A}W^{(4)}+\frac{I}{A}\omega^2\left(1+\frac{E}{G\kappa_y}\right)W^{(2)}\mp\frac{N}{\rho A}W^{(2)}+\frac{I}{A}\omega^2\frac{\rho\omega^2}{G\kappa_y}W-\omega^2 W=0
$$

or

$$
W^{(4)}+\lambda r_y^2\left(1+\frac{1}{\hat{\kappa}r_y^2}\right)W^{(2)}\mp\hat{N}\,W^{(2)}-\lambda W+\lambda^2\frac{r_y^2}{\hat{\kappa}}W=0\,.
\tag{9.33}
$$

Special Cases:

1. If the axial force N is zero we get

$$
\begin{aligned}
&\begin{bmatrix}\hat{\chi} & 0\\ 0 & 1\end{bmatrix}\begin{bmatrix}W^{(2)}\\ \Psi^{(2)}\end{bmatrix}+\begin{bmatrix}0 & \hat{\chi}\\ -\hat{\chi} & 0\end{bmatrix}\begin{bmatrix}W^{(1)}\\ \Psi^{(1)}\end{bmatrix}+\\
&\qquad+\begin{bmatrix}0 & 0\\ 0 & -\hat{\chi}\end{bmatrix}\begin{bmatrix}W\\ \Psi\end{bmatrix}=-\lambda\begin{bmatrix}1 & 0\\ 0 & r_y^2\end{bmatrix}\begin{bmatrix}W\\ \Psi\end{bmatrix}
\end{aligned}
\tag{9.34a}
$$

or

$$W^{(4)} + \lambda r_y^2 \left(1 + \frac{1}{\hat{\kappa} r_y^2}\right) W^{(2)} - \lambda W + \lambda^2 \frac{r_y^2}{\hat{\kappa}} W = 0 . \tag{9.34b}$$

2. If the effect of the shear stresses is neglected, i.e., the terms that contain $\hat{\kappa}$ are dropped, we obtain from Eq. (9.34b):

$$W^{(4)} + \lambda r_y^2 W^{(2)} - \lambda W = 0 . \tag{9.35}$$

3. If the effect of the rotary inertia is neglected, i.e., the term that includes λ^2 and the term $\lambda r_y^2 W^{(2)}$—both in Eq. (9.34b)—are dropped we have

$$W^{(4)} + \frac{\lambda}{\hat{\kappa}} W^{(2)} - \lambda W = 0 . \tag{9.36}$$

4. If neither the rotary inertia nor the effect of the shear are taken into account, the result is the classical differential equation for the amplitude of the vibration

$$W^{(4)} = \lambda W . \tag{9.37}$$

5. If the Timoshenko beam does not vibrate, i.e., $\lambda = 0$, the equation describes the stability problem of axially loaded beams—the force N is a compressive one:

$$\begin{bmatrix} \hat{\chi} & 0 \\ 0 & 1 \end{bmatrix} \begin{bmatrix} W^{(2)} \\ \Psi^{(2)} \end{bmatrix} + \begin{bmatrix} 0 & \hat{\chi} \\ -\hat{\chi} - \hat{N} & 0 \end{bmatrix} \begin{bmatrix} W^{(1)} \\ \Psi^{(1)} \end{bmatrix} + \\ + \begin{bmatrix} 0 & 0 \\ 0 & -\hat{\chi} \end{bmatrix} \begin{bmatrix} W \\ \Psi \end{bmatrix} = \begin{bmatrix} 0 \\ 0 \end{bmatrix} . \tag{9.38}$$

Exercise 9.2 Find λ for a simply supported beam if the axial force $N = 0$.
The amplitude functions

$$W = C \sin \frac{k\pi x}{\ell}, \qquad \Psi_y = D \cos \frac{k\pi x}{\ell} \qquad k = 1, 2, 3, \dots \tag{9.39}$$

satisfy the boundary conditions shown in Row 1 of Table 9.1. Substitution of the above solutions into the differential equation system (9.34a) results in a homogeneous linear equation system for the undetermined integration constants C and D:

$$C \left(\lambda - \hat{\chi} \left(\frac{k\pi}{\ell}\right)^2\right) - D \hat{\chi} \frac{k\pi}{\ell} = 0 , \tag{9.40a}$$

$$-C \hat{\chi} \left(\frac{k\pi}{\ell}\right) + D \left[- \left(\frac{k\pi}{\ell}\right)^2 - \hat{\chi} + \lambda r_y^2 \right] = 0 . \tag{9.40b}$$

Solutions for C and D different from the trivial one exist if and only if the determinant of the coefficient matrix is zero:

$$\begin{vmatrix} \lambda - \hat{\chi}\left(\frac{k\pi}{\ell}\right)^2 & -\hat{\chi}\frac{k\pi}{\ell} \\ -\hat{\chi}\left(\frac{k\pi}{\ell}\right) & -\left(\frac{k\pi}{\ell}\right)^2 - \hat{\chi} + \lambda r_y^2 \end{vmatrix} = \tag{9.41}$$

$$\lambda^2 - \lambda\frac{\hat{\chi}}{r_y^2}\left[1 + r_y^2\left(1 + \frac{1}{\hat{\chi} r_y^2}\right)\left(\frac{k\pi}{\ell}\right)^2\right] + \frac{\hat{\chi}}{r_y^2}\left(\frac{k\pi}{\ell}\right)^4 = 0 \tag{9.42}$$

from where

$$\lambda = \frac{\hat{\chi}}{2r_y^2}\left[1 + r_y^2\left(1 + \frac{1}{\hat{\chi} r_y^2}\right)\left(\frac{k\pi}{\ell}\right)^2\right] \pm$$
$$\pm\sqrt{\left(\frac{\hat{\chi}}{2r_y^2}\right)^2\left[1 + r_y^2\left(1 + \frac{1}{\hat{\chi} r_y^2}\right)\left(\frac{k\pi}{\ell}\right)^2\right]^2 - \frac{\hat{\chi}}{r_y^2}\left(\frac{k\pi}{\ell}\right)^4} \tag{9.43}$$

are the two roots. The discriminant $\mathcal{D}$ that is the expression under the square root sign can be manipulated into the following form:

$$\mathcal{D} = \frac{1}{4\ell^4 r_y^4}\left[\pi^4 k^4\left(\hat{\chi} r_y^2 - 1\right)^2 + \hat{\chi}\ell^2\left(\hat{\chi}\ell^2 + 2\pi^2 k^2 + 2\pi^2 k^2 \hat{\chi} r_y^2\right)\right] > 0\,. \tag{9.44}$$

Hence, it is a positive quantity. Since the first term on the right side of (9.43) is always greater than the second one, it follows that $\lambda > 0$ if $\hat{\chi}$ is positive. The smaller eigenvalue belongs to the bending vibration mode of the beam, while the greater one corresponds to the vibration mode characterized by shear [6].

9.2 Eigenvalue Problem for a Class of Differential Equations

9.2.1 The Differential Equation System

Determination of the eigenfrequencies of a vibrating Timoshenko beam has led to an eigenvalue problem governed by the homogeneous differential equation system (9.34a) and the homogeneous boundary conditions presented in Row 1 of Table 9.1.

Consider now the class of those eigenvalue problems which is governed by the homogeneous ordinary differential equation system (referred to as eigenvalue problem of ODES)

$$\underline{\mathbf{K}}[\,\underline{\mathbf{y}}\,] = \lambda \underline{\mathbf{M}}[\,\underline{\mathbf{y}}\,] \tag{9.45a}$$

and the homogeneous boundary conditions

$$\underline{\mathbf{U}}_r[\,\underline{\mathbf{y}}\,] = \underline{\mathbf{0}}\,, \qquad r = 1, 2, \ldots 2\kappa \tag{9.45b}$$

where $\underline{\mathbf{y}}^T(x) = [y_1(x)|y_2(x)|\ldots|y_n(x)]$; $n \geq 2$ is the unknown function vector, 2κ and 2μ ($\kappa > \mu$) are the order of the differential operators $\underline{\mathbf{K}}$ and $\underline{\mathbf{M}}$—see Eq. (9.46) below—while λ is a parameter (the eigenvalue sought).

Let

$$\underset{(n\times n)}{\underline{\mathbf{K}}_\nu(x)}, \quad (\nu = 0, 1, ..., 2\kappa) \qquad \text{and} \qquad \underset{(n\times n)}{\underline{\mathbf{M}}_\nu(x)}, \quad (\nu = 0, 1, ..., 2\mu,\ \mu < \kappa)$$

be continuous otherwise arbitrary quadratic matrices in the interval $x \in [a, b]$, $a < b$. It is assumed that the matrices $\underline{\mathbf{K}}_{2\kappa}(x)$ and $\underline{\mathbf{M}}_{2\mu}(x)$ have inverse for all $x \in [a, b]$.

The differential operators $\underline{\mathbf{K}}[\,\underline{\mathbf{y}}\,]$ and $\underline{\mathbf{M}}[\,\underline{\mathbf{y}}\,]$ are defined by the following relationships:

$$\begin{aligned} \underline{\mathbf{K}}[\,\underline{\mathbf{y}}\,] &= \sum_{\nu=0}^{2\kappa} \underline{\mathbf{K}}_\nu(x)\underline{\mathbf{y}}^{(\nu)}(x)\,, \qquad \frac{\mathrm{d}^\nu(\ldots)}{\mathrm{d}x^\nu} = (\ldots)^{(\nu)}\,; \\ \underline{\mathbf{M}}[\,\underline{\mathbf{y}}\,] &= \sum_{\nu=0}^{2\mu} \underline{\mathbf{M}}_\nu(x)\underline{\mathbf{y}}^{(\nu)}(x)\,. \end{aligned} \tag{9.46}$$

Let

$$\underset{(n\times n)}{\underline{\boldsymbol{A}}_{\nu r}} \qquad \text{and} \qquad \underset{(n\times n)}{\underline{\boldsymbol{B}}_{\nu r}} \qquad \nu, r = 1, 2, \ldots, 2\kappa$$

be constant quadratic matrices. The homogeneous boundary conditions prescribed by (9.45b) can be given in the following form:

$$\underline{\mathbf{U}}_r[\,\underline{\mathbf{y}}\,] = \sum_{\nu=1}^{2\kappa}\left[\underline{\boldsymbol{A}}_{\nu r}\,\underline{\mathbf{y}}^{(\nu-1)}(a) + \underline{\boldsymbol{B}}_{\nu r}\,\underline{\mathbf{y}}^{(\nu-1)}(b)\right] = 0\,, \quad r = 1, 2, \ldots 2\kappa \tag{9.47}$$

where the rows of the matrix $\underline{\mathbf{U}}_r$ should be linearly independent of each other.

Eigenvalue problem (9.45) is said to be simple if

$$\underline{\mathbf{M}}[\,\underline{\mathbf{y}}\,] = \underline{\mathbf{M}}_0(x)\underline{\mathbf{y}}(x)\,. \tag{9.48}$$

The function vectors (or column matrices) $\underline{\mathbf{u}}(x)$ and $\underline{\mathbf{v}}(x)$ ($|\underline{\mathbf{u}}(x)|$ and $|\underline{\mathbf{v}}(x)|$ are not identically equal to zero if $x \in [a, b]$) are called (a) comparison function vectors if they satisfy boundary conditions (9.45b) and (b) eigenfunction vectors (or simply eigenfunctions) if they satisfy differential equation (9.45a) and boundary conditions (9.45b).

The integrals

$$\left(\underline{\mathbf{u}}, \underline{\mathbf{v}}\right)_K = \int_a^b \underline{\mathbf{u}}^T(x)\,\underline{\mathbf{K}}[\,\underline{\mathbf{v}}(x)]\,\mathrm{d}x\,, \quad \left(\underline{\mathbf{u}}, \underline{\mathbf{v}}\right)_M = \int_a^b \underline{\mathbf{u}}^T(x)\,\underline{\mathbf{M}}[\,\underline{\mathbf{v}}(x)]\,\mathrm{d}x \tag{9.49}$$

taken on the set of the comparison function vectors $\underline{\mathbf{u}}(x)$, $\underline{\mathbf{v}}(x)$ are products defined on the operators $\underline{\mathbf{K}}$ and $\underline{\mathbf{M}}$.

Eigenvalue problem of ordinary differential equations (9.45) is said to be self-adjoint if the products (9.49) are commutative, i.e., if it holds that

$$\left(\underline{\mathbf{u}},\underline{\mathbf{v}}\right)_K=\left(\underline{\mathbf{v}},\underline{\mathbf{u}}\right)_K \quad \text{and} \quad \left(\underline{\mathbf{u}},\underline{\mathbf{v}}\right)_M=\left(\underline{\mathbf{v}},\underline{\mathbf{u}}\right)_M . \tag{9.50}$$

Exercise 9.3 Show that the eigenvalue problem defined by Eq. (9.34a) and the boundary conditions in Row 1 of Table 9.1 is self-adjoint.

It is obvious that this eigenvalue problem is governed by the differential equation

$$\begin{aligned}\underline{\mathbf{K}}[\underline{\mathbf{y}}]&=\sum_{\nu=0}^{2}\underline{\mathbf{K}}_\nu(x)\,\underline{\mathbf{y}}^{(\nu)}(x)=\\&=\underbrace{\begin{bmatrix}\hat{\chi} & 0\\ 0 & 1\end{bmatrix}}_{\underline{\mathbf{K}}_2}\begin{bmatrix}y_1\\ y_2\end{bmatrix}^{(2)}+\underbrace{\begin{bmatrix}0 & \hat{\chi}\\ -\hat{\chi} & 0\end{bmatrix}}_{\underline{\mathbf{K}}_1}\begin{bmatrix}y_1\\ y_2\end{bmatrix}^{(1)}+\underbrace{\begin{bmatrix}0 & 0\\ 0 & -\hat{\chi}\end{bmatrix}}_{\underline{\mathbf{K}}_0}\begin{bmatrix}y_1\\ y_2\end{bmatrix}=\\&\qquad=\lambda\underline{\mathbf{M}}[\underline{\mathbf{y}}]=\lambda\underbrace{\begin{bmatrix}-1 & 0\\ 0 & -r_y^2\end{bmatrix}}_{\underline{\mathbf{M}}_0}\begin{bmatrix}y_1\\ y_2\end{bmatrix}\end{aligned} \tag{9.51a}$$

associated with the boundary conditions

$$y_1(0)=y_1(\ell)=0 \quad \text{and} \quad y_2(0)=y_2(\ell)=0 . \tag{9.51b}$$

Hence,

$$\begin{aligned}\left(\underline{\mathbf{u}},\underline{\mathbf{v}}\right)_K&=\int_0^\ell \underline{\mathbf{u}}^T\underline{\mathbf{K}}_2\underline{\mathbf{v}}^{(2)}\mathrm{d}x+\int_0^\ell \underline{\mathbf{u}}^T\underline{\mathbf{K}}_1\underline{\mathbf{v}}^{(1)}\mathrm{d}x+\int_0^\ell \underline{\mathbf{u}}^T\underline{\mathbf{K}}_0\underline{\mathbf{v}}\,\mathrm{d}x=\\&=\underline{\mathbf{u}}^T\underline{\mathbf{K}}_2\underline{\mathbf{v}}^{(1)}\Big|_0^\ell-\int_0^\ell\left(\underline{\mathbf{u}}^T\right)^{(1)}\underline{\mathbf{K}}_2\underline{\mathbf{v}}^{(1)}\mathrm{d}x+\int_0^\ell \underline{\mathbf{u}}^T\underline{\mathbf{K}}_1\underline{\mathbf{v}}^{(1)}\mathrm{d}x+\\&+\int_0^\ell \underline{\mathbf{u}}^T\underline{\mathbf{K}}_0\underline{\mathbf{v}}\mathrm{d}x=\underbrace{\left(\hat{\chi}v_1u_1^{(1)}+v_2u_2^{(1)}\right)\Big|_0^\ell-\hat{\chi}u_1v_2\Big|_0^\ell}_{=0}-\\&-\hat{\chi}\int_0^\ell\left(u_1^{(1)}v_2+v_1^{(1)}u_2\right)\mathrm{d}x-\int_0^\ell\left(\hat{\chi}v_1^{(1)}u_1^{(1)}+v_2^{(1)}u_2^{(1)}\right)\mathrm{d}x-\\&-\int_0^\ell\hat{\chi}v_2u_2\mathrm{d}x=\left(\underline{\mathbf{v}},\underline{\mathbf{u}}\right)_K\end{aligned} \tag{9.52a}$$

and

$$\left(\underline{\mathbf{u}},\underline{\mathbf{v}}\right)_M=\int_0^\ell \underline{\mathbf{u}}^T\underline{\mathbf{M}}_0\underline{\mathbf{v}}\,\mathrm{d}x=-\int_0^\ell\left(v_1u_1+r_y^2v_2u_2\right)\mathrm{d}x=\left(\underline{\mathbf{v}},\underline{\mathbf{u}}\right)_M , \tag{9.52b}$$

which means that the boundary value problem considered is really self-adjoint.

It can be proved that the eigenvalue problems defined by differential equation (9.34a) and the boundary conditions in Row 2, 3 and 4 of Table 9.1 are all self-adjoint. The proof is left for Problem 9.1.

9.2.2 Orthogonality of the Eigenfunction Vectors

The line of thought in the present section is basically the same as that of Sect. 8.3. It is assumed that the eigenvalue problem (9.45) is self-adjoint. Let $\underline{\mathbf{y}}_k$ and $\underline{\mathbf{y}}_\ell$ be two different eigenvector functions ($k, \ell = 1, 2, 3, \ldots$), which therefore satisfy the differential equation systems

$$\underline{\mathbf{K}}[\,\underline{\mathbf{y}}_k\,] = \lambda_k\,\underline{\mathbf{M}}[\,\underline{\mathbf{y}}_k\,], \qquad \underline{\mathbf{K}}[\,\underline{\mathbf{y}}_\ell\,] = \lambda_\ell\,\underline{\mathbf{M}}[\,\underline{\mathbf{y}}_\ell\,] \tag{9.53}$$

where λ_k and λ_ℓ are the eigenvalues that belong to the eigenfunction vectors $\underline{\mathbf{y}}_k$ and $\underline{\mathbf{y}}_\ell$—$\lambda_k \neq \lambda_\ell$.

Multiply Eq. $(9.53)_1$ by $\underline{\mathbf{y}}_\ell(x)$, and Eq. $(9.53)_2$ by $\underline{\mathbf{y}}_k(x)$. Integrating the result over the interval $[a, b]$ yields

$$\left(\underline{\mathbf{y}}_\ell, \underline{\mathbf{y}}_k\right)_K = \lambda_k \left(\underline{\mathbf{y}}_\ell, \underline{\mathbf{y}}_k\right)_M, \qquad \left(\underline{\mathbf{y}}_k, \underline{\mathbf{y}}_\ell\right)_K = \lambda_\ell \left(\underline{\mathbf{y}}_k, \underline{\mathbf{y}}_\ell\right)_M. \tag{9.54}$$

Since the products $\left(\underline{\mathbf{y}}_\ell, \underline{\mathbf{y}}_k\right)_K$ and $\left(\underline{\mathbf{y}}_\ell, \underline{\mathbf{y}}_k\right)_M$ are commutative, the difference $(9.54)_1$-$(9.54)_2$ is of the form

$$(\lambda_k - \lambda_\ell)\left(\underline{\mathbf{y}}_k, \underline{\mathbf{y}}_\ell\right)_M = 0, \tag{9.55}$$

where $\lambda_k - \lambda_\ell \neq 0$. Hence

$$\left(\underline{\mathbf{y}}_k, \underline{\mathbf{y}}_\ell\right)_M = 0 \qquad k \neq \ell\,. \tag{9.56a}$$

In view of this equation, it follows from (8.12) that

$$\left(\underline{\mathbf{y}}_k, \underline{\mathbf{y}}_\ell\right)_K = 0 \qquad k \neq \ell\,. \tag{9.56b}$$

Fulfillment of Eq. (9.56) proves that the eigenvector functions $\underline{\mathbf{y}}_k(x)$ and $\underline{\mathbf{y}}_\ell(x)$ are orthogonal to each other in general sense.

For $\underline{\mathbf{M}}[\,\underline{\mathbf{y}}\,] = \underline{\mathbf{1}}\,\underline{\mathbf{y}}$ where $\underline{\mathbf{1}}$ is the $n \times n$ unit matrix, the eigenvector functions are orthogonal in traditional sense.

With regard to (9.54), it follows on the basis of (9.56) that

$$\left(\underline{\mathbf{y}}_k, \underline{\mathbf{y}}_\ell\right)_K = \begin{cases} \lambda_\ell \left(\underline{\mathbf{y}}_k, \underline{\mathbf{y}}_\ell\right)_M & \text{if } k = \ell, \\ 0 & \text{if } k \neq \ell. \end{cases} \qquad k, \ell = 1, 2, 3 \ldots \tag{9.57}$$

9.2.3 On the Reality of the Eigenvalues for Eigenvalue Problem (9.45)

We shall prove that the eigenvalues are real numbers if the eigenvalue problem of ODES (9.45) is self-adjoint. Assume first that the eigenvalue is not real but complex.

If the eigenvalue λ_ℓ is complex, then so is the corresponding eigenvector function which can, therefore, be given in the following way:

$$\underline{\mathbf{y}}_\ell(x) = \underline{\mathbf{u}}_\ell(x) + i\underline{\mathbf{v}}_\ell(x)\,, \qquad \lambda_\ell = \mathfrak{a} + i\mathfrak{b},$$

where $\underline{\mathbf{u}}_\ell(x)$ and $\underline{\mathbf{v}}_\ell(x)$ are real vector functions, while $\mathfrak{a}$ and $\mathfrak{b}$ are real constants. It is also obvious that the complex conjugates $\bar{\underline{\mathbf{y}}}_\ell$ and $\bar{\lambda}$ satisfy differential equation (9.45a). On the basis of (9.54) we can write

$$\left(\bar{\underline{\mathbf{y}}}_\ell, \underline{\mathbf{y}}_\ell\right)_K = \lambda_\ell \left(\bar{\underline{\mathbf{y}}}_\ell, \underline{\mathbf{y}}_\ell\right)_M \qquad \left(\underline{\mathbf{y}}_\ell, \bar{\underline{\mathbf{y}}}_\ell\right)_K = \bar{\lambda} \left(\underline{\mathbf{y}}_\ell, \bar{\underline{\mathbf{y}}}_\ell\right)_M . \tag{9.58}$$

Subtract $(9.58)_2$ from $(9.58)_1$ by taking into account that the products $\left(\bar{\underline{\mathbf{y}}}_\ell, \underline{\mathbf{y}}_\ell\right)_K$ and $\left(\bar{\underline{\mathbf{y}}}_\ell, \underline{\mathbf{y}}_\ell\right)_M$ are commutative. We have

$$2i\mathfrak{b}\left(\underline{\mathbf{y}}_\ell, \bar{\underline{\mathbf{y}}}_\ell\right)_M = 2i\mathfrak{b}\left((\underline{\mathbf{u}}_\ell, \underline{\mathbf{u}}_\ell)_M + (\underline{\mathbf{v}}_\ell, \underline{\mathbf{v}}_\ell)_M\right) = 0\,.$$

This means that $\mathfrak{b}=0$, i.e., the eigenvalue λ_ℓ is real if $(\underline{\mathbf{u}}_\ell, \underline{\mathbf{u}}_\ell)_M + (\underline{\mathbf{v}}_\ell, \underline{\mathbf{v}}_\ell)_M \neq 0$.

9.2.4 Rayleigh Quotient

For $\ell = k$ Eq. (9.54) yields

$$\left(\underline{\mathbf{y}}_\ell, \underline{\mathbf{y}}_\ell\right)_K = \lambda_\ell \left(\underline{\mathbf{y}}_\ell, \underline{\mathbf{y}}_\ell\right)_M \tag{9.59}$$

from where it follows that

$$\lambda_\ell = \frac{\left(\underline{\mathbf{y}}_\ell, \underline{\mathbf{y}}_\ell\right)_K}{\left(\underline{\mathbf{y}}_\ell, \underline{\mathbf{y}}_\ell\right)_M}. \tag{9.60}$$

On the basis of Eq. (9.60), the Rayleigh quotient is defined on the set of comparative function vectors as

$$\mathcal{R}[\underline{\mathbf{u}}(x)] = \frac{(\underline{\mathbf{u}}, \underline{\mathbf{u}})_K}{(\underline{\mathbf{u}}, \underline{\mathbf{u}})_M} = \frac{\displaystyle\int_a^b \underline{\mathbf{u}}^T(x)\,\underline{\mathbf{K}}\,[\underline{\mathbf{u}}(x)]\,\mathrm{d}x}{\displaystyle\int_a^b \underline{\mathbf{u}}^T(x)\,\underline{\mathbf{M}}\,[\underline{\mathbf{u}}(x)]\,\mathrm{d}x}. \tag{9.61}$$

The eigenvalue problem (9.45) is positive definite if the eigenvalues are positive, positive semidefinite if one eigenvalue is zero and the other eigenvalues are all positive, negative semidefinite if one eigenvalue is zero and the other eigenvalues are all negative, and finally negative definite if the eigenvalues are negative.

Assume that

$$\left(\underline{\mathbf{u}},\underline{\mathbf{u}}\right)_K>0\,,\quad\text{and}\quad\left(\underline{\mathbf{u}},\underline{\mathbf{u}}\right)_M>0 \tag{9.62}$$

for any comparison function vector $\underline{\mathbf{u}}(x)$. Then the eigenvalue problem (9.45) is positive definite (or full definite).

Exercise 9.4 Find the Rayleigh quotient for the eigenvalue problem defined by Eq. (9.34a) and the boundary conditions in Row 1 of Table 9.1.

Comparison of (9.61) and (9.52) yields

$$\mathcal{R}[\underline{\mathbf{u}}(x)]=\frac{\left(\underline{\mathbf{u}},\underline{\mathbf{u}}\right)_K}{\left(\underline{\mathbf{u}},\underline{\mathbf{u}}\right)_M}=\frac{\displaystyle\int_0^\ell\left\{\hat{\chi}\left[(u_1^{(1)})^2+(u_2)^2\right]+(u_2^{(1)})^2+2\hat{\chi}u_1^{(1)}u_2\right\}\mathrm{d}x}{\displaystyle\int_0^\ell\left[(u_1)^2+r_y^2(u_2)^2\right]\mathrm{d}x}\,. \tag{9.63}$$

9.2.5 *Determination of Eigenvalues*

The general solution of the differential equation $\underline{\mathbf{K}}[\,\underline{\mathbf{y}}\,]=\lambda\,\underline{\mathbf{M}}[\,\underline{\mathbf{y}}\,]$ can be given in the following form:

$$\underline{\mathbf{y}}(x)=\sum_{\ell=1}^{2\kappa\times n}\underline{\mathbf{z}}_\ell(x,\lambda)\mathcal{A}_\ell, \tag{9.64}$$

where $\underline{\mathbf{z}}_\ell(x,\lambda)$ stand for the linearly independent particular solutions while the undetermined integration constants are denoted by $\mathcal{A}_\ell$. The latter constants can be obtained from the boundary conditions:

$$\sum_{\ell=1}^{2\kappa\times n}\underline{\mathbf{U}}_r[\underline{\mathbf{z}}_\ell(x,\lambda)]\,\mathcal{A}_\ell=0\,,\qquad r=1,2,\dots 2\kappa\,. \tag{9.65}$$

Since these equations constitute a homogeneous linear equation system for the unknown $\mathcal{A}_\ell$ solutions different from the trivial one (nontrivial solutions) exist if and only if the determinant of the system is zero:

$$\Delta(\lambda)=\det\bigl[\underline{\mathbf{U}}_r[\underline{\mathbf{z}}_\ell(x,\lambda)]\bigr]=0\,. \tag{9.66}$$

This is the equation which should be solved to find the eigenvalues λ. We remark that $\Delta(\lambda)$ is often referred to as characteristic determinant.

9.2.6 The Green Function Matrix

Consider the inhomogeneous ordinary differential equation system

$$\underline{\mathbf{L}}[\underline{\mathbf{y}}(x)] = \underline{\mathbf{r}}(x), \tag{9.67a}$$

where the differential operator of order κ is given by the equation

$$\underline{\mathbf{L}}[\underline{\mathbf{y}}] = \sum_{\nu=0}^{\kappa} \underline{\mathbf{P}}_{\nu}(x)\underline{\mathbf{y}}^{(\nu)}(x), \tag{9.67b}$$

in which $\underline{\mathbf{y}}^T(x) = [y_1(x)|y_2(x)|\ldots|y_n(x)]$; $n \geq 2$ is the unknown function vector, κ is the order of the differential operator, $\underline{\mathbf{r}}(x) = [r_1(x)|r_2(x)|\ldots|r_n(x)]$ is a known inhomogeneity and the coefficients

$$\underset{(n\times n)}{\underline{\mathbf{P}}_{\nu}(x)}, \quad \nu = 0, 1, ..., \kappa, \quad x \in [a, b], \ a < b$$

are continuous otherwise arbitrary quadratic matrices.

It is assumed that $\underline{\mathbf{P}}_{\kappa}(x)$ has an inverse for all $x \in [a, b]$. If there exist no inverse, the differential equation system is called degenerated [7]. Let

$$\underset{(n\times n)}{\underline{\mathcal{A}}_{\nu r}} \quad \text{and} \quad \underset{(n\times n)}{\underline{\mathcal{B}}_{\nu r}} \quad \nu, r = 1, 2, \ldots, \kappa$$

be constant quadratic matrices.

The differential equation system (9.67) is associated with the following homogeneous boundary conditions:

$$\underline{\mathbf{U}}_r[\underline{\mathbf{y}}] = \sum_{\nu=1}^{\kappa} \left[\underline{\mathcal{A}}_{\nu r}\underline{\mathbf{y}}^{(\nu-1)}(a) + \underline{\mathcal{B}}_{\nu r}\underline{\mathbf{y}}^{(\nu-1)}(b)\right] = 0, \quad r = 1, 2, \ldots \kappa \tag{9.68}$$

which are basically the same as the boundary conditions given by Eq. (9.47).

Solution of the boundary value problem (9.67), (9.68) can be given in the form

$$\underline{\mathbf{y}}(x) = \int_a^b \underline{\mathbf{G}}(x, \xi)\underline{\mathbf{r}}(\xi)\, \mathrm{d}\xi, \tag{9.69}$$

where $\underline{\mathbf{G}}(x, \xi)$ is the Green function matrix [8] defined by the following four properties:

1. The Green function matrix is a continuous function of x and ξ in each of the triangles $a \leq x \leq \xi \leq b$ and $a \leq \xi \leq x \leq b$. In addition, it is κ times differentiable with respect to x and the derivatives

$$\frac{\partial^n \underline{\mathbf{G}}(x,\xi)}{\partial x^n} = \underline{\mathbf{G}}^{(n)}(x,\xi)\,, \qquad (n = 1, 2, \ldots, \kappa)$$

are also continuous functions of x and ξ in the triangles $a \le x \le \xi \le b$ and $a \le \xi \le x \le b$.

2. Let ξ be fixed in $[a, b]$. The Green function matrix and its derivatives

$$\underline{\mathbf{G}}^{(n)}(x,\xi) = \frac{\partial^n \underline{\mathbf{G}}(x,\xi)}{\partial x^n}\,, \qquad (n = 1, 2, \ldots, \kappa - 2) \tag{9.70}$$

should be continuous for $x = \xi$:

$$\lim_{\varepsilon \to 0} \left[\underline{\mathbf{G}}^{(n)}(\xi+\varepsilon,\xi) - \underline{\mathbf{G}}^{(n)}(\xi-\varepsilon,\xi)\right] =$$
$$= \left[\underline{\mathbf{G}}^{(n)}(\xi+0,\xi) - \underline{\mathbf{G}}^{(n)}(\xi-0,\xi)\right] = 0\,, \quad (n = 0, 1, 2, \ldots \kappa - 2). \tag{9.71a}$$

The derivative $\underline{\mathbf{G}}^{(\kappa-1)}(x,\xi)$ should, however, have a jump

$$\lim_{\varepsilon \to 0} \left[\underline{\mathbf{G}}^{(\kappa-1)}(\xi+\varepsilon,\xi) - \underline{\mathbf{G}}^{(\kappa-1)}(\xi-\varepsilon,\xi)\right] =$$
$$= \left[\underline{\mathbf{G}}^{(\kappa-1)}(\xi+0,\xi) - \underline{\mathbf{G}}^{(\kappa-1)}(\xi-0,\xi)\right] = \underline{\mathbf{P}}_\kappa^{-1}(\xi) \tag{9.71b}$$

if $x = \xi$.

3. Let $\underline{\boldsymbol{\alpha}}^T = [\alpha_1|\alpha_2|\ldots|\alpha_n]\,, \alpha_\nu \ne 0\,(\nu = 1, 2, \ldots, n)$ be an arbitrary constant matrix with finite elements. For a fixed $\xi \in [a, b]$, the product $\underline{\mathbf{G}}(x,\xi)\underline{\boldsymbol{\alpha}}$ as a function of x $(x \ne \xi)$ should satisfy the homogeneous differential equation

$$\underline{\mathbf{L}}\left[\underline{\mathbf{G}}(x,\xi)\underline{\boldsymbol{\alpha}}\right] = \underline{\mathbf{0}}\,.$$

4. The product $\underline{\mathbf{G}}(x,\xi)\underline{\boldsymbol{\alpha}}$ as a function of x should satisfy the boundary conditions

$$\underline{\mathbf{U}}_r\left[\underline{\mathbf{G}}(x,\xi)\underline{\boldsymbol{\alpha}}\right] = \underline{\mathbf{0}}\,, \qquad (r = 1, 2, 3, \ldots, \kappa). \tag{9.72}$$

9.2.7 *Calculation of the Green Function Matrix*

The definition of the Green function matrix given in the previous subsection is a constructive one which means that it provides the means that are needed to calculate the Green function matrix.

The general solution of the homogeneous differential equation system

$$\underline{\mathbf{L}}[\,\underline{\mathbf{y}}(x)] = \underline{\mathbf{r}}(x) \tag{9.73}$$

can be given in the form

$$\underline{\mathbf{y}} = \left[\sum_{\ell=1}^{\kappa} \underset{(n\times n)}{\underline{\mathbf{Y}}_\ell(x)} \underset{(n\times n)}{\underline{\mathbf{C}}_\ell} \right] \underset{(n\times 1)}{\underline{\mathbf{e}}} \,, \tag{9.74}$$

where each column of the matrices $\underline{\mathbf{Y}}_\ell$ satisfies the homogeneous differential equation system (9.73), $\underline{\mathbf{C}}_\ell$ is a constant quadratic matrix, while $\underline{\mathbf{e}}$ is a constant column matrix (a constant vector).

Recalling the third property of the definition and the structure of the general solution (9.73) the Green function matrix can be expressed in the following manner [8]:

$$\underset{(n\times n)}{\underline{\mathbf{G}}(x,\xi)} = \sum_{\ell=1}^{\kappa} \underline{\mathbf{Y}}_\ell(x) \left[\underline{\mathbf{A}}_\ell(\xi) \pm \underline{\mathbf{B}}_\ell(\xi) \right] \,, \tag{9.75}$$

where the sign is [positive](negative) if $[x \le \xi](\xi \le x)$.

The above choice automatically assures the fulfillment of the first property of the definition.

On the basis of the second property, the following equations can be set up for calculating the elements of the matrices $\underline{\mathbf{B}}_\ell(\xi)$:

$$\left. \begin{aligned} \sum_{\ell=1}^{\kappa} \underline{\mathbf{Y}}_\ell(\xi)\, \underline{\mathbf{B}}_\ell(\xi) &= \underline{\mathbf{0}} \,, \\ \sum_{\ell=1}^{\kappa} \underline{\mathbf{Y}}_\ell^{(1)}(\xi)\, \underline{\mathbf{B}}_\ell(\xi) &= \underline{\mathbf{0}} \,, \\ &\dots \\ \sum_{\ell=1}^{\kappa} \underline{\mathbf{Y}}_\ell^{(\kappa-2)}(\xi)\, \underline{\mathbf{B}}_\ell(\xi) &= \underline{\mathbf{0}} \,, \\ \sum_{\ell=1}^{\kappa} \underline{\mathbf{Y}}_\ell^{(\kappa-1)}(\xi)\, \underline{\mathbf{B}}_\ell(\xi) &= -\frac{1}{2} \underline{\mathbf{P}}_\kappa^{-1}(\xi) \,, \end{aligned} \right\} \tag{9.76}$$

where $\underline{\mathbf{0}}$ is the $n \times n$ zero matrix. Let us denote the ν-th column ($\nu = 1, 2, \dots, n$) of the matrix $\underline{\mathbf{B}}_\ell$ by $\underset{(n\times 1)}{\underline{\mathbf{B}}_{\ell\nu}}$:

$$\underset{(n\times n)}{\underline{\mathbf{B}}_\ell} = [\, \underset{(n\times 1)}{\underline{\mathbf{B}}_{\ell 1}} \mid \underset{(n\times 1)}{\underline{\mathbf{B}}_{\ell 2}} \mid \cdots \mid \underset{(n\times 1)}{\underline{\mathbf{B}}_{\ell n}} \,]. \tag{9.77}$$

The matrix

$$\underset{(1\times(\kappa\times n))}{\underline{\boldsymbol{\mathcal{B}}}_\nu^T} = \left[\underset{(1\times n)}{\underline{\mathbf{B}}_{1\nu}^T} | \dots | \underset{(1\times n)}{\underline{\mathbf{B}}_{i\nu}^T} | \dots | \underset{(1\times n)}{\underline{\mathbf{B}}_{\kappa\nu}^T} \right], \qquad i = 2, 3, \dots, \kappa - 1 \tag{9.78}$$

is that of the unknowns for a fixed ν.

The i-th column ($i = 1, 2, \dots, n$) of the matrix $\underline{\mathbf{Y}}_\ell$ is denoted by $\underline{\boldsymbol{\eta}}_{\ell i}$:

$$\underset{(n\times n)}{\underline{\mathbf{Y}}_\ell} = [\ \underset{(n\times 1)}{\underline{\boldsymbol{\eta}}_{\ell 1}} \mid \underset{(n\times 1)}{\underline{\boldsymbol{\eta}}_{\ell 2}} \mid \cdots \mid \underset{(n\times 1)}{\underline{\boldsymbol{\eta}}_{\ell n}}\ ]. \tag{9.79}$$

The matrix $\underline{\boldsymbol{\mathcal{B}}}_\nu$ is the solution of a linear equation system of the form

$$\underline{\mathbf{W}}\,\underline{\boldsymbol{\mathcal{B}}}_\nu = \underline{\boldsymbol{\mathcal{P}}}_\nu, \tag{9.80a}$$

where

$$\underline{\mathbf{W}} = \begin{bmatrix} \underline{\boldsymbol{\eta}}_{11} & \cdots & \underline{\boldsymbol{\eta}}_{1n} & \cdots & \underline{\boldsymbol{\eta}}_{\ell 1} & \cdots & \underline{\boldsymbol{\eta}}_{\ell n} & \cdots & \underline{\boldsymbol{\eta}}_{\kappa 1} & \cdots & \underline{\boldsymbol{\eta}}_{\kappa n} \\ \underline{\boldsymbol{\eta}}^{(1)}_{11} & \cdots & \underline{\boldsymbol{\eta}}^{(1)}_{1n} & \cdots & \underline{\boldsymbol{\eta}}^{(1)}_{\ell 1} & \cdots & \underline{\boldsymbol{\eta}}^{(1)}_{\ell n} & \cdots & \underline{\boldsymbol{\eta}}^{(1)}_{\kappa 1} & \cdots & \underline{\boldsymbol{\eta}}^{(1)}_{\kappa n} \\ & & & & \cdots & \cdots & \cdots & & & & \\ \underline{\boldsymbol{\eta}}^{(\kappa-1)}_{11} & \cdots & \underline{\boldsymbol{\eta}}^{(\kappa-1)}_{1n} & \cdots & \underline{\boldsymbol{\eta}}^{(\kappa-1)}_{\ell 1} & \cdots & \underline{\boldsymbol{\eta}}^{(\kappa-1)}_{\ell n} & \cdots & \underline{\boldsymbol{\eta}}^{(\kappa-1)}_{\kappa 1} & \cdots & \underline{\boldsymbol{\eta}}^{(\kappa-1)}_{\kappa n} \end{bmatrix} \tag{9.80b}$$

and $\underline{\boldsymbol{\mathcal{P}}}_\nu$ is the transpose of the ν-th row in the matrix:

$$\big[\underset{1}{\underline{\mathbf{0}}} \,|\, \underset{2}{\underline{\mathbf{0}}} \,|\cdots|\, \underset{\kappa-1}{\underline{\mathbf{0}}} \,|\, \underset{\kappa}{-\tfrac{1}{2}\underline{\mathbf{P}}_\kappa^{-1}}\big]\,. \tag{9.80c}$$

Note that the coefficient matrix $\underline{\mathbf{W}}$ is the same for each $\underline{\boldsymbol{\mathcal{B}}}_\nu$.

It is also worth mentioning that the first three properties of the definition have all been satisfied.

After having determined the matrices $\underline{\mathbf{B}}_\ell$, we can proceed with the calculation of the matrices $\underline{\mathbf{A}}_\ell$ by using the fourth property of the definition. Let $\underline{\boldsymbol{\alpha}}$ be the ν-th unit vector in the $n \times n$ space:

$$\underline{\boldsymbol{\alpha}}^T = [\ \underset{1}{0} | \underset{2}{0} | \cdots | \underset{\nu}{1} | \cdots | \underset{n}{0}\ ]\,. \tag{9.81}$$

With the above $\underline{\boldsymbol{\alpha}}^T$, the fourth property leads to the following equations:

$$\underline{\mathbf{U}}_r\left[\sum_{\ell=1}^{n}\underline{\mathbf{Y}}_\ell(x)\underline{\mathbf{A}}_\ell(\xi)\underline{\boldsymbol{\alpha}}\right] = \mp\underline{\mathbf{U}}_r\left[\sum_{\ell=1}^{n}\underline{\mathbf{Y}}_\ell(x)\underline{\mathbf{B}}_\ell(\xi)\underline{\boldsymbol{\alpha}}\right] \quad r = 1, 2, \ldots, \kappa\,. \tag{9.82}$$

The matrix

$$\underset{(1\times(\kappa\times n))}{\underline{\boldsymbol{\mathcal{A}}}^T_\nu} = \left[\underset{(1\times n)}{\underline{\mathbf{A}}^T_{1\nu}} | \ldots | \underset{(1\times n)}{\underline{\mathbf{A}}^T_{i\nu}} | \ldots | \underset{(1\times n)}{\underline{\mathbf{A}}^T_{\kappa\nu}}\right], \qquad i = 2, 3, \ldots, \kappa-1 \tag{9.83}$$

is that of the unknowns for a fixed ν in $\underline{\boldsymbol{\alpha}}$. Making use of (9.78) and (9.82), the equation system to be solved is of the form

$$\underline{\mathbf{U}}_r[\underline{\mathbf{Y}}_1, \underline{\mathbf{Y}}_2, \ldots, \underline{\mathbf{Y}}_n]\underline{\boldsymbol{\mathcal{A}}}_\nu = \mp\underline{\mathbf{U}}_r[\underline{\mathbf{Y}}_1, \underline{\mathbf{Y}}_2, \ldots, \underline{\mathbf{Y}}_n]\underline{\boldsymbol{\mathcal{B}}}_\nu \quad r = 1, 2, 3, \ldots, \kappa \tag{9.84}$$

where $\underline{\mathbf{U}}_r\left[\underline{\mathbf{Y}}_1, \underline{\mathbf{Y}}_2, \ldots, \underline{\mathbf{Y}}_n\right]$ is a matrix with size $n \times (\kappa \times n)$. If the boundary conditions are linearly independent, the determinant of equation system (9.84) is different from zero. Then equation system (9.84) is solvable, i.e., there is a unique solution for $\underline{\boldsymbol{\mathcal{A}}}_\nu$ (for the matrices $\underline{\mathbf{A}}_\ell$). This means that there exists the Green function matrix of the boundary value problem (9.67), (9.68).

9.2.8 Symmetry of the Green Function Matrix

The present section is a formal generalization of Sect. 8.10. Consider the following two inhomogeneous boundary value problems:

$$\underline{\mathbf{L}}\left[\underline{\mathbf{u}}(x)\right] = \underline{\mathbf{r}}(x)\,, \qquad \underline{\mathbf{U}}_r[\underline{\mathbf{u}}(x)] = 0\,; \tag{9.85a}$$

$$\underline{\mathbf{L}}\left[\underline{\mathbf{v}}(x)\right] = \underline{\mathbf{s}}(x)\,, \qquad \underline{\mathbf{U}}_r[\underline{\mathbf{v}}(x)] = 0, \tag{9.85b}$$

where $\underline{\mathbf{L}}$ is defined by Eq. (9.67b), $\underline{\mathbf{u}}(x)$ and $\underline{\mathbf{v}}(x)$ stand for the unknown functions, while $\underline{\mathbf{r}}(x)$ and $\underline{\mathbf{s}}(x)$ are continuous inhomogeneities in the interval $x \in [a, b]$. It is assumed that the boundary value problems (9.85a) and (9.85b) are self-adjoint. Hence,

$$(\underline{\mathbf{u}}, \underline{\mathbf{v}})_L - (\underline{\mathbf{v}}, \underline{\mathbf{u}})_L = 0,$$

in which on the basis of Eq. (9.69) it holds that

$$\underline{\mathbf{u}}(x) = \int_a^b \underline{\mathbf{G}}(x, \xi)\underline{\mathbf{r}}(\xi)\,\mathrm{d}\xi \quad \text{and} \quad \underline{\mathbf{v}}(x) = \int_a^b \underline{\mathbf{G}}(x, \xi)\underline{\mathbf{s}}(\xi)\,\mathrm{d}\xi\,.$$

Consequently,

$$(\underline{\mathbf{u}}, \underline{\mathbf{v}})_L - (\underline{\mathbf{v}}, \underline{\mathbf{u}})_L = \int_a^b \left(\underline{\mathbf{u}}^T\,\underline{\mathbf{L}}[\underline{\mathbf{v}}] - \underline{\mathbf{v}}^T\,\underline{\mathbf{L}}[\underline{\mathbf{u}}]\right)\mathrm{d}x =$$

$$= \int_a^b \int_a^b \underline{\mathbf{s}}^T(\xi)\left[\underline{\mathbf{G}}^T(x, \xi) - \underline{\mathbf{G}}(\xi, x)\right]\underline{\mathbf{r}}(\xi)\,\mathrm{d}\xi\mathrm{d}x = 0, \tag{9.86}$$

where both $\underline{\mathbf{r}}(x)$ and $\underline{\mathbf{s}}(x)$ are arbitrary continuous and non-zero vector functions in the interval $[a, b]$. Thus

$$\boxed{\underline{\mathbf{G}}^T(x, \xi) = \underline{\mathbf{G}}(\xi, x)\,.} \tag{9.87}$$

If this equation is satisfied, the Green function matrix is called cross-symmetric.

9.2.9 Green Function Matrices for Some Beam Problems

9.2.9.1 Pinned-Pinned Timoshenko Beam

Find the Green function matrix for the boundary value problem governed by the differential equation system

$$AG\kappa_y\left(\frac{\mathrm{d}^2 w}{\mathrm{d}\hat{x}^2}+\frac{\mathrm{d}\psi_y}{\mathrm{d}\hat{x}}\right)+f_z=0\,,\quad IE\frac{\mathrm{d}^2\psi_y}{\mathrm{d}\hat{x}^2}-AG\kappa_y\left(\frac{\mathrm{d}w}{\mathrm{d}\hat{x}}+\psi_y\right)+\mu_y=0 \tag{9.88a}$$

and the boundary conditions

$$w(0)=\psi_y^{(1)}(0)=0,\qquad w(\ell)=\psi_y^{(1)}(\ell)=0, \tag{9.88b}$$

where, for the sake of our later considerations, the earlier independent variable x is denoted by $\hat{x}$. According to (9.15) and Table 9.1, these equations describe the mechanical behavior of a pinned-pinned Timoshenko beam if the axial force N is zero.

Introducing the dimensionless variables

$$\hat{x}=x\ell,\quad y_1=\frac{w}{\ell},\quad y_2=\psi_y,\quad \chi=\frac{AG\kappa_y\ell^2}{IE}=\hat{\chi}\ell^2, \tag{9.89}$$

where x is a new dimensionless independent variable differential equations (9.88a) and boundary conditions (9.88b) can be rewritten in the following form:

$$\chi\left(\frac{\mathrm{d}^2 y_1}{\mathrm{d}x^2}+\frac{\mathrm{d}y_2}{\mathrm{d}x}\right)=-\frac{\ell^3}{IE}f_z=r_1,\,,\quad \frac{\mathrm{d}^2 y_2}{\mathrm{d}x^2}-\chi\left(\frac{\mathrm{d}y_1}{\mathrm{d}x}+y_2\right)=-\frac{\ell^2}{IE}\mu_y=r_2 \tag{9.90a}$$

and

$$y_1(0)=y_2^{(1)}(0)=0,\qquad y_1(1)=y_2^{(1)}(1)=0\,. \tag{9.90b}$$

By using matrix notation, differential equations (9.90a) can be rewritten in the following form:

$$\underbrace{\begin{bmatrix}\chi & 0\\ 0 & 1\end{bmatrix}}_{\mathbf{\underline{K}}_2}\begin{bmatrix}y_1\\ y_2\end{bmatrix}^{(2)}+\underbrace{\begin{bmatrix}0 & \chi\\ -\chi & 0\end{bmatrix}}_{\mathbf{\underline{K}}_1}\begin{bmatrix}y_1\\ y_2\end{bmatrix}^{(1)}+\underbrace{\begin{bmatrix}0 & 0\\ 0 & -\chi\end{bmatrix}}_{\mathbf{\underline{K}}_0}\begin{bmatrix}y_1\\ y_2\end{bmatrix}=\underbrace{\begin{bmatrix}r_1\\ r_2\end{bmatrix}}_{\mathbf{\underline{r}}}. \tag{9.91}$$

On the basis of Eq. (9.25), the general solution of the homogeneous equation assumes the following form:

$$\begin{bmatrix}y_1\\ y_2\end{bmatrix}=\underbrace{\begin{bmatrix}1 & x\\ 0 & -1\end{bmatrix}}_{\mathbf{\underline{Y}}_1(x)}\begin{bmatrix}C_1\\ C_2\end{bmatrix}+\underbrace{\begin{bmatrix}-\frac{1}{2}x^2 & -\frac{1}{3}x^3\\ x & \frac{2}{\chi}+x^2\end{bmatrix}}_{\mathbf{\underline{Y}}_2(x)}\begin{bmatrix}C_3\\ C_4\end{bmatrix}. \tag{9.92}$$

The Green function matrix of differential equation (9.91) can be expressed as

$$\underset{(2\times 2)}{\underline{\mathbf{G}}(x,\xi)} = \sum_{\ell=1}^{2} \underline{\mathbf{Y}}_{\ell}(x) \left[\underline{\mathbf{A}}_{\ell}(\xi) \pm \underline{\mathbf{B}}_{\ell}(\xi)\right], \tag{9.93}$$

where the sign is [positive](negative) if $[x \leq \xi](\xi \leq x)$. For the sake of the further calculations, it is worth partitioning the matrices $\underline{\mathbf{Y}}_{\ell}$, $\underline{\mathbf{B}}_{\ell}$, and $\underline{\mathbf{A}}_{\ell}$:

$$\underline{\mathbf{Y}}_{\ell} = \begin{bmatrix} \overset{\ell}{Y}_{11} & \overset{\ell}{Y}_{12} \\ \overset{\ell}{Y}_{21} & \overset{\ell}{Y}_{22} \end{bmatrix} = \begin{matrix} 1\,\{ \\ 1\,\{ \end{matrix} \underbrace{\begin{bmatrix} \underline{\mathbf{Y}}_{\ell 1} \\ \underline{\mathbf{Y}}_{\ell 2} \end{bmatrix}}_{(2\times 2)}, \tag{9.94a}$$

$$\underline{\mathbf{A}}_{\ell} = \begin{bmatrix} \overset{\ell}{A}_{11} & \overset{\ell}{A}_{12} \\ \overset{\ell}{A}_{21} & \overset{\ell}{A}_{22} \end{bmatrix} = \begin{bmatrix} \underset{(2\times 1)}{\underline{\mathbf{A}}_{\ell 1}} & \underset{(2\times 1)}{\underline{\mathbf{A}}_{\ell 2}} \end{bmatrix}, \tag{9.94b}$$

$$\underline{\mathbf{B}}_{\ell} = \begin{bmatrix} \overset{\ell}{B}_{11} & \overset{\ell}{B}_{12} \\ \overset{\ell}{B}_{21} & \overset{\ell}{B}_{22} \end{bmatrix} = \begin{bmatrix} \underset{(2\times 1)}{\underline{\mathbf{B}}_{\ell 1}} & \underset{(2\times 1)}{\underline{\mathbf{B}}_{\ell 2}} \end{bmatrix}. \tag{9.94c}$$

It follows from Eq. (9.76) that the matrices $\underline{\mathbf{B}}_{\ell}$ should satisfy the following equations:

$$\sum_{\ell=1}^{2} \underline{\mathbf{Y}}_{\ell}(\xi)\, \underline{\mathbf{B}}_{\ell}(\xi) = \begin{bmatrix} 1 & \xi \\ 0 & -1 \end{bmatrix} \begin{bmatrix} \overset{1}{B}_{11} & \overset{1}{B}_{12} \\ \overset{1}{B}_{21} & \overset{1}{B}_{22} \end{bmatrix} + \\ + \begin{bmatrix} -\frac{1}{2}\xi^2 & -\frac{1}{3}\xi^3 \\ \xi & \frac{2}{\chi} + \xi^2 \end{bmatrix} \begin{bmatrix} \overset{2}{B}_{11} & \overset{2}{B}_{12} \\ \overset{2}{B}_{21} & \overset{2}{B}_{22} \end{bmatrix} = \begin{bmatrix} 0 & 0 \\ 0 & 0 \end{bmatrix}, \tag{9.95a}$$

$$\sum_{\ell=1}^{2} \underline{\mathbf{Y}}_{\ell}^{(1)}(\xi)\, \underline{\mathbf{B}}_{\ell}(\xi) = \begin{bmatrix} 0 & 1 \\ 0 & 0 \end{bmatrix} \begin{bmatrix} \overset{1}{B}_{11} & \overset{1}{B}_{12} \\ \overset{1}{B}_{21} & \overset{1}{B}_{22} \end{bmatrix} + \\ + \begin{bmatrix} -\xi & -\xi^2 \\ 1 & 2\xi \end{bmatrix} \begin{bmatrix} \overset{2}{B}_{11} & \overset{2}{B}_{12} \\ \overset{2}{B}_{21} & \overset{2}{B}_{22} \end{bmatrix} = -\frac{1}{2} \begin{bmatrix} \frac{1}{\chi} & 0 \\ 0 & 1 \end{bmatrix}. \tag{9.95b}$$

By introducing the new variables

$$a = \overset{1}{B}_{1i}, \quad b = \overset{1}{B}_{2i} \quad c = \overset{2}{B}_{1i} \quad d = \overset{2}{B}_{2i} \tag{9.96}$$

for $i = 1$ we get the equation system

$$\begin{bmatrix} 1 & \xi & -\frac{1}{2}\xi^2 & -\frac{1}{3}\xi^3 \\ 0 & -1 & \xi & \frac{2}{\chi}+\xi^2 \\ 0 & 1 & -\xi & -\xi^2 \\ 0 & 0 & 1 & 2\xi \end{bmatrix} \begin{bmatrix} a \\ b \\ c \\ d \end{bmatrix} = \begin{bmatrix} 0 \\ 0 \\ -\frac{1}{2\chi} \\ 0 \end{bmatrix}$$

from where

$$\begin{aligned} a &= \overset{1}{B}_{11} = \frac{1}{2\chi}\xi - \frac{1}{12}\xi^3, & b &= \overset{1}{B}_{21} = \frac{1}{4}\xi^2 - \frac{1}{2\chi}, \\ c &= \overset{2}{B}_{11} = \frac{1}{2}\xi, & d &= \overset{2}{B}_{21} = -\frac{1}{4}. \end{aligned} \tag{9.97a}$$

If $i = 2$ we have in a similar way:

$$\begin{bmatrix} 1 & \xi & -\frac{1}{2}\xi^2 & -\frac{1}{3}\xi^3 \\ 0 & -1 & \xi & \frac{2}{\chi}+\xi^2 \\ 0 & 1 & -\xi & -\xi^2 \\ 0 & 0 & 1 & 2\xi \end{bmatrix} \begin{bmatrix} a \\ b \\ c \\ d \end{bmatrix} = \begin{bmatrix} 0 \\ 0 \\ 0 \\ -\frac{1}{2} \end{bmatrix}$$

from where

$$\begin{aligned} a &= \overset{1}{B}_{12} = \frac{1}{4}\xi^2, & b &= \overset{1}{B}_{22} = -\frac{1}{2}\xi, \\ c &= \overset{2}{B}_{12} = -\frac{1}{2}, & d &= \overset{2}{B}_{22} = 0\,. \end{aligned} \tag{9.97b}$$

Note that the matrices $\underline{\mathbf{B}}_\ell$ are independent of the boundary conditions.

According to Eq. (9.72) the product $\underline{\mathbf{G}}(x,\xi)\underline{\boldsymbol{\alpha}}$ should satisfy boundary conditions (9.90b).

If $\underline{\boldsymbol{\alpha}}^T = [1|0]$, the boundary conditions yield the following equation system:

$$\begin{aligned} \sum_{\ell=1}^{2} \underline{\mathbf{Y}}_{\ell 1}(x)\big|_{x=0}\, \underline{\mathbf{A}}_{\ell 1}(\xi) &= -\sum_{\ell=1}^{2} \underline{\mathbf{Y}}_{\ell 1}(x)\big|_{x=0}\, \underline{\mathbf{B}}_{\ell 1}(\xi), \\ \sum_{\ell=1}^{2} \underline{\mathbf{Y}}_{\ell 1}(x)\big|_{x=1}\, \underline{\mathbf{A}}_{\ell 1}(\xi) &= +\sum_{\ell=1}^{2} \underline{\mathbf{Y}}_{\ell 1}(x)\big|_{x=1}\, \underline{\mathbf{B}}_{\ell 1}(\xi), \\ \sum_{\ell=1}^{2} \underline{\mathbf{Y}}^{(1)}_{\ell 2}(x)\Big|_{x=0}\, \underline{\mathbf{A}}_{\ell 1}(\xi) &= -\sum_{\ell=1}^{2} \underline{\mathbf{Y}}^{(1)}_{\ell 2}(x)\Big|_{x=0}\, \underline{\mathbf{B}}_{\ell 1}(\xi), \\ \sum_{\ell=1}^{2} \underline{\mathbf{Y}}^{(1)}_{\ell 2}(x)\Big|_{x=1}\, \underline{\mathbf{A}}_{\ell 1}(\xi) &= +\sum_{\ell=1}^{2} \underline{\mathbf{Y}}^{(1)}_{\ell 2}(x)\Big|_{x=1}\, \underline{\mathbf{B}}_{\ell 1}(\xi). \end{aligned} \tag{9.98}$$

If $\underline{\boldsymbol{\alpha}}^T = [0|1]$, the boundary conditions result in the equation system

$$
\begin{aligned}
&\sum_{\ell=1}^{2} \underline{\mathbf{Y}}_{\ell 1}(x)\big|_{x=0} \underline{\mathbf{A}}_{\ell 2}(\xi) = -\sum_{\ell=1}^{2} \underline{\mathbf{Y}}_{\ell 1}(x)\big|_{x=0} \underline{\mathbf{B}}_{\ell 2}(\xi),\\
&\sum_{\ell=1}^{2} \underline{\mathbf{Y}}_{\ell 1}(x)\big|_{x=1} \underline{\mathbf{A}}_{\ell 2}(\xi) = +\sum_{\ell=1}^{2} \underline{\mathbf{Y}}_{\ell 1}(x)\big|_{x=1} \underline{\mathbf{B}}_{\ell 2}(\xi),\\
&\sum_{\ell=1}^{2} \underline{\mathbf{Y}}^{(1)}_{\ell 2}(x)\Big|_{x=0} \underline{\mathbf{A}}_{\ell 2}(\xi) = -\sum_{\ell=1}^{2} \underline{\mathbf{Y}}^{(1)}_{\ell 2}(x)\Big|_{x=0} \underline{\mathbf{B}}_{\ell 2}(\xi),\\
&\sum_{\ell=1}^{2} \underline{\mathbf{Y}}^{(1)}_{\ell 2}(x)\Big|_{x=1} \underline{\mathbf{A}}_{\ell 2}(\xi) = +\sum_{\ell=1}^{2} \underline{\mathbf{Y}}^{(1)}_{\ell 2}(x)\Big|_{x=1} \underline{\mathbf{B}}_{\ell 2}(\xi).
\end{aligned}
\tag{9.99}
$$

Since the coefficient matrices on the left sides are the same in (9.98) and (9.99), the two equations can be united:

$$
\begin{bmatrix} 1 & 0 & 0 & 0 \\ 1 & 1 & -\frac{1}{2} & -\frac{1}{3} \\ 0 & 0 & 1 & 0 \\ 0 & 0 & 1 & 2 \end{bmatrix}
\begin{bmatrix} \overset{1}{A}_{1i} \\ \overset{1}{A}_{2i} \\ \overset{2}{A}_{i1} \\ \overset{2}{A}_{2i} \end{bmatrix}
=
\begin{bmatrix} -a \\ a+b-\frac{1}{2}c-\frac{1}{3}d \\ -c \\ c+2d \end{bmatrix}
\qquad i = 1, 2\,.
$$

The solutions are as follows:

$$
\overset{1}{A}_{1i} = -a\,, \qquad \overset{1}{A}_{2i} = 2a + b - \frac{2}{3}c\,,
$$
$$
\overset{2}{A}_{1i} = -c\,, \qquad \overset{2}{A}_{2i} = c + d
$$

from where by substituting (9.97a) and (9.97b) we obtain

$$
\begin{aligned}
&\overset{1}{A}_{11} = -\frac{1}{2\chi}\xi + \frac{1}{12}\xi^3\,, &&\overset{1}{A}_{21} = -\frac{1}{2\chi} + \frac{\xi}{\chi} - \frac{\xi}{3} + \frac{1}{4}\xi^2 - \frac{1}{6}\xi^3\\
&\overset{2}{A}_{11} = -\frac{1}{2}\xi\,, &&\overset{2}{A}_{21} = \frac{1}{2}\xi - \frac{1}{4}
\end{aligned}
\tag{9.100a}
$$

and

$$
\begin{aligned}
&\overset{1}{A}_{12} = -\frac{1}{4}\xi^2\,, \quad &&\overset{1}{A}_{22} = \frac{1}{2}\xi^2 - \frac{1}{2}\xi + \frac{1}{3}\,,\\
&\overset{2}{A}_{11} = \frac{1}{2}\,, &&\overset{2}{A}_{22} = -\frac{1}{2}\,.
\end{aligned}
\tag{9.100b}
$$

Hence,

$$
\begin{aligned}
\underline{\mathbf{G}}(x,\xi) &= \underline{\mathbf{Y}}_1(x)\left[\underline{\mathbf{A}}_1(\xi) \pm \underline{\mathbf{B}}_1(\xi)\right] + \underline{\mathbf{Y}}_2(x)\left[\underline{\mathbf{A}}_2(\xi) \pm \underline{\mathbf{B}}_2(\xi)\right] = \\
&= \begin{bmatrix} 1 & x \\ 0 & -1 \end{bmatrix} \left\{ \begin{bmatrix} -\frac{1}{2\chi}\xi + \frac{1}{12}\xi^3 & -\frac{1}{4}\xi^2 \\ -\frac{1}{2\chi} + \frac{\xi}{\chi} - \frac{\xi}{3} + \frac{1}{4}\xi^2 - \frac{1}{6}\xi^3 & \frac{1}{2}\xi^2 - \frac{1}{2}\xi + \frac{1}{3} \end{bmatrix} \pm \right.
\end{aligned}
$$

$$\left.\left.\pm\begin{bmatrix}\frac{1}{2\chi}\xi-\frac{1}{12}\xi^3 & \frac{1}{4}\xi^2\\ \frac{1}{4}\xi^2-\frac{1}{2\chi} & -\frac{1}{2}\xi\end{bmatrix}\right]\right\}+$$

$$+\begin{bmatrix}-\frac{1}{2}x^2 & -\frac{1}{3}x^3\\ x & \frac{2}{\chi}+x^2\end{bmatrix}\left\{\begin{bmatrix}-\frac{1}{2}\xi & \frac{1}{2}\\ \frac{1}{2}\xi-\frac{1}{4} & -\frac{1}{2}\end{bmatrix}\pm\begin{bmatrix}\frac{1}{2}\xi & -\frac{1}{2}\\ -\frac{1}{4} & 0\end{bmatrix}\right\} \tag{9.101}$$

is the Green function matrix. Since the boundary value problem defined by differential equation (9.91) and boundary conditions (9.90b) is self-adjoint —we remind the reader of Exercise 9.3—it follows that the above Green function matrix is cross-symmetric.

Exercise 9.5 A concentrated mass is attached to the simply supported beam shown in Fig. 9.2. It is assumed that the mass m is much greater than that of the beam. The effect of the beam mass can, therefore, be neglected. Find the circular frequency of the mass if it vibrates vertically given that the effect of the rotation can be left out of consideration.

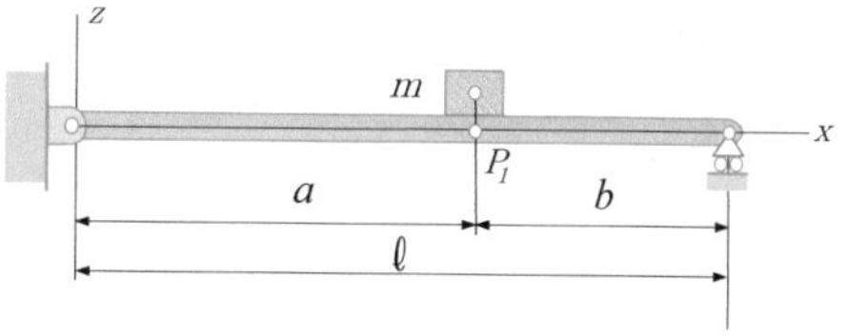

Fig. 9.2 Simply supported beam with a concentrated mass

If we use the Euler–Bernoulli beam theory, the flexibility at the point $x = a$ can be determined using the Green function given for simply supported beams in Table 8.1:

$$f_{11} = \int_0^\ell \mathcal{G}(a,\xi)\,\delta(a-\xi)\,\mathrm{d}\xi = \int_0^\ell \frac{a}{6IE\ell}(\ell-\xi)\left(2\ell\xi-\xi^2-a^2\right)\delta(a-\xi)\,\mathrm{d}\xi =$$

$$= \frac{ab}{6IE\ell}\left(2\ell a-2a^2\right) = \frac{a^2b^2}{3IE\ell}\,.$$

Hence,

$$k_{11} = \frac{1}{f_{11}} = \frac{3IE\ell}{a^2b^2}$$

is the corresponding rigidity and

$$\omega^2 = \frac{k_{11}}{m} = \frac{3IE\ell}{a^2b^2m} \tag{9.102}$$

is the square of the circular frequency sought.

The solution steps are the same within the framework of the Timoshenko beam theory. The flexibility can be calculated utilizing the Green function matrix (9.101) and taking into account the fact what form the right side of the differential equations

(9.90a) have. Recalling that ξ and x are now dimensionless coordinates and using the relations $x_a = a/\ell, r_1(\xi) = -\delta(x_a - \xi)\,\ell^3/IE, r_2(\xi) = 0, \hat{\xi} = \ell\xi$ we have

$$\begin{bmatrix} y_1(x_a) \\ y_2(x_a) \end{bmatrix} = \int_0^1 \begin{bmatrix} G_{11}(x_a,\xi) & G_{12}(x_a,\xi) \\ G_{21}(x_a,\xi) & G_{22}(x_a,\xi) \end{bmatrix} \begin{bmatrix} r_1(\xi) \\ r_2(\xi) \end{bmatrix} \mathrm{d}\xi =$$

$$= -\frac{\ell^3}{IE}\int_0^\ell \begin{bmatrix} G_{11}\left(x_a,\frac{\hat{\xi}}{\ell}\right) & G_{12}\left(x_a,\frac{\hat{\xi}}{\ell}\right) \\ G_{21}\left(x_a,\frac{\hat{\xi}}{\ell}\right) & G_{22}\left(x_a,\frac{\hat{\xi}}{\ell}\right) \end{bmatrix} \begin{bmatrix} \delta\left(x_a - \frac{\hat{\xi}}{\ell}\right) \\ 0 \end{bmatrix} \frac{\mathrm{d}\hat{\xi}}{\ell}$$

from where

$$y_1(x_a) = \frac{w(x_a)}{\ell} = \frac{1}{\ell} f_{11} = -\frac{\ell^3}{IE}\int_0^\ell G_{11}\left(x_a,\frac{\hat{\xi}}{\ell}\right)\delta\left(x_a - \frac{\hat{\xi}}{\ell}\right)\frac{\mathrm{d}\hat{\xi}}{\ell} =$$

$$= -\frac{1}{\ell}\frac{\ell^3}{IE}\left\{\begin{bmatrix} 1 & x_a \end{bmatrix}\begin{bmatrix} -\frac{1}{\chi}x_a + \frac{1}{6}x_a^3 \\ \frac{x_a}{\chi} - \frac{x_a}{3} - \frac{1}{6}x_a^3 \end{bmatrix} + \begin{bmatrix} -\frac{1}{2}x_a^2 & -\frac{1}{3}x_a^3 \end{bmatrix}\begin{bmatrix} -x_a \\ \frac{1}{2}x_a \end{bmatrix}\right\} =$$

$$= \frac{1}{\ell}\frac{\ell^3}{3IE}\frac{x_a}{\chi}(1 - x_a)\left(-\chi x_a^2 + \chi x_a + 3\right) = \frac{1}{\ell}\frac{1}{3IE\ell}\frac{ab}{\chi}\left(\chi ab + 3\ell^2\right) =$$

$$= \frac{1}{\ell}\frac{ab}{3IE\ell}\left(ab + 3\frac{\ell^2}{\chi}\right).$$

Hence,

$$f_{11} = \frac{ab}{3IE\ell}\left(ab + 3\frac{\ell^2}{\chi}\right) \quad \text{and} \quad k_{11} = \frac{1}{f_{11}} = \frac{3IE\ell}{ab\left(ab + 3\frac{\ell^2}{\chi}\right)}.$$

Consequently,

$$\omega^2 = \frac{k_{11}}{m} = \frac{3IE\ell}{mab\left(ab + 3\frac{\ell^2}{\chi}\right)} = \underset{\chi=\frac{AG\kappa_y\ell^2}{IE}}{\uparrow} \quad \underset{E=2G(1+\nu)}{\uparrow} \quad \underset{r_y^2=\frac{I}{A}}{\uparrow} = \frac{3IE\ell}{mab\left(ab + \frac{6r_y^2(1+\nu)}{\kappa_y}\right)}. \tag{9.103}$$

9.2.9.2 Fixed-Pinned Timoshenko Beam

According to Table 9.1 differential equation (9.91) is associated with the following boundary conditions:

$$y_1(0) = y_2(0) = 0, \qquad y_1(1) = y_2^{(1)}(1) = 0\,. \tag{9.104}$$

When calculating the Green function matrix, it is sufficient to determine the matrices $\underline{\mathbf{A}}_1$ and $\underline{\mathbf{A}}_2$ since the matrices $\underline{\mathbf{B}}_1$ and $\underline{\mathbf{B}}_2$ are independent of the boundary conditions. Without entering into details, we have

$$\underline{\mathbf{G}}(x,\xi) = \underline{\mathbf{Y}}_1(x)\left[\underline{\mathbf{A}}_1(\xi) \pm \underline{\mathbf{B}}_1(\xi)\right] + \underline{\mathbf{Y}}_2(x)\left[\underline{\mathbf{A}}_2(\xi) \pm \underline{\mathbf{B}}_2(\xi)\right] = \begin{bmatrix} 1 & x \\ 0 & -1 \end{bmatrix} \times$$
$$\times \left\{ \left[\begin{matrix} -\frac{1}{2\chi}\xi + \frac{1}{12}\xi^3 & \frac{1}{4}\xi^2 \\ -\frac{1}{4\chi(\chi+3)}\left(2\xi^3\chi + \xi^2\chi^2 - 3\xi^2\chi - 12\xi + 2\chi + 6\right) & \frac{1}{2}\frac{\xi}{\chi+3}\left(3\xi + \chi - 3\right) \end{matrix} \right] \pm \right.$$
$$\left. \pm \begin{bmatrix} \frac{1}{2\chi}\xi - \frac{1}{12}\xi^3 & \frac{1}{4}\xi^2 \\ \frac{1}{4}\xi^2 - \frac{1}{2\chi} & -\frac{1}{2}\xi \end{bmatrix} \right\} + \begin{bmatrix} -\frac{1}{2}x^2 & -\frac{1}{3}x^3 \\ x & \frac{2}{\chi} + x^2 \end{bmatrix} \times$$
$$\times \left\{ \frac{1}{4(\chi+3)} \begin{bmatrix} -2\left(3\xi - \xi\chi + 3\xi^2\chi - \xi^3\chi\right) & -2\left(3\chi\xi^2 - 6\chi\xi + \chi + 3\right) \\ -\chi\xi^3 + 3\chi\xi^2 + 6\xi - \chi - 3 & 3\chi\xi\left(\xi - 2\right) \end{bmatrix} \pm \right.$$
$$\left. \pm \begin{bmatrix} \frac{1}{2}\xi & -\frac{1}{2} \\ -\frac{1}{4} & 0 \end{bmatrix} \right\}. \tag{9.105}$$

9.2.9.3 Fixed-Fixed Timoshenko Beam

The Green function matrix is as follows:

$$\underline{\mathbf{G}}(x,\xi) = \underline{\mathbf{Y}}_1(x)\left[\underline{\mathbf{A}}_1(\xi) \pm \underline{\mathbf{B}}_1(\xi)\right] + \underline{\mathbf{Y}}_2(x)\left[\underline{\mathbf{A}}_2(\xi) \pm \underline{\mathbf{B}}_2(\xi)\right] =$$
$$= \begin{bmatrix} 1 & x \\ 0 & -1 \end{bmatrix} \left\{ \left[\begin{matrix} -\frac{1}{2\chi}\xi + \frac{1}{12}\xi^3 & -\frac{1}{4}\xi^2 \\ -\frac{1}{4\chi(\chi+12)}\left(8\xi^3\chi + \xi^2\chi^2 - 48\xi + 2\chi + 24\right) & \frac{\xi}{2(\chi+12)}\left(12\xi + \chi\right) \end{matrix} \right] \pm \right.$$
$$\left. \pm \begin{bmatrix} \frac{1}{2\chi}\xi - \frac{1}{12}\xi^3 & \frac{1}{4}\xi^2 \\ \frac{1}{4}\xi^2 - \frac{1}{2\chi} & -\frac{1}{2}\xi \end{bmatrix} \right\} + \begin{bmatrix} -\frac{1}{2}x^2 & -\frac{1}{3}x^3 \\ x & \frac{2}{\chi} + x^2 \end{bmatrix} \times$$
$$\times \left\{ \frac{1}{\chi+12} \begin{bmatrix} \frac{1}{2}\left(\xi\chi - 12\xi^2 - 4\xi^2\chi + 2\xi^3\chi\right) & -\frac{1}{2}\left(\chi - 24\xi - 8\xi\chi + 6\xi^2\chi + 12\right) \\ -\frac{1}{4}\left(4\chi\xi^3 - 6\chi\xi^2 - 24\xi + \chi + 12\right) & 3\xi\chi\left(\xi - 1\right) \end{bmatrix} \pm \right.$$
$$\left. \pm \begin{bmatrix} \frac{1}{2}\xi & -\frac{1}{2} \\ -\frac{1}{4} & 0 \end{bmatrix} \right\}. \tag{9.106}$$

The paper-and-pencil calculations that lead to the above result are again omitted.

9.2.9.4 Fixed-Free Timoshenko Beam

For completeness, the Green function matrix for the fixed-free Timoshenko beam is also presented here:

$$\underline{\mathbf{G}}(x,\xi) = \underline{\mathbf{Y}}_1(x)\left[\underline{\mathbf{A}}_1(\xi) \pm \underline{\mathbf{B}}_1(\xi)\right] + \underline{\mathbf{Y}}_2(x)\left[\underline{\mathbf{A}}_2(\xi) \pm \underline{\mathbf{B}}_2(\xi)\right] =$$
$$= \begin{bmatrix} 1 & x \\ 0 & -1 \end{bmatrix} \left\{ \begin{bmatrix} -\frac{1}{2\chi}\xi + \frac{1}{12}\xi^3 & -\frac{1}{4}\xi^2 \\ -\frac{1}{4\chi}\left(\chi\xi^2 + 2\right) & \frac{1}{2}\xi \end{bmatrix} \pm \begin{bmatrix} \frac{1}{2\chi}\xi - \frac{1}{12}\xi^3 & \frac{1}{4}\xi^2 \\ \frac{1}{4}\xi^2 - \frac{1}{2\chi} & -\frac{1}{2}\xi \end{bmatrix} \right\} +$$
$$+ \begin{bmatrix} -\frac{1}{2}x^2 & -\frac{1}{3}x^3 \\ x & \frac{2}{\chi} + x^2 \end{bmatrix} \left\{ \begin{bmatrix} \frac{1}{2}\xi & -\frac{1}{2} \\ -\frac{1}{4} & 0 \end{bmatrix} \pm \begin{bmatrix} \frac{1}{2}\xi & -\frac{1}{2} \\ -\frac{1}{4} & 0 \end{bmatrix} \right\}. \tag{9.107}$$

9.2.9.5 Axially Loaded and Pinned-Pinned Timoshenko Beam

Find the Green function matrix if the pinned-pinned Timoshenko beam is axially loaded. It is obvious on the basis of Eq. (9.15) and Table 9.1 that the Green function matrix in question belongs to the boundary value problem governed by the differential equations

$$\begin{aligned} AG\kappa_y\left(\frac{\mathrm{d}^2 w}{\mathrm{d}x^2}+\frac{\mathrm{d}\psi_y}{\mathrm{d}x}\right)+f_z=0\,, \\ IE\frac{\mathrm{d}^2\psi_y}{\mathrm{d}x^2}\pm N\frac{\mathrm{d}w}{\mathrm{d}x}-AG\kappa_y\left(\frac{\mathrm{d}w}{\mathrm{d}x}+\psi_y\right)+\mu_y=0, \end{aligned} \tag{9.108a}$$

which are associated with the boundary conditions

$$w(0)=\psi_y^{(1)}(0)=0, \qquad w(\ell)=\psi_y^{(1)}(\ell)=0. \tag{9.108b}$$

Note that Eq. (9.108a) differs from Eq. (9.88a) in one term only while boundary conditions (9.88b) and (9.108b) coincide with each other. Recalling definition (9.90a) of r_1, r_2 and introducing dimensionless variables by the use of Eq. (9.89) differential equations (9.108a) can be rewritten in the following form:

$$\underbrace{\begin{bmatrix}\chi & 0\\ 0 & 1\end{bmatrix}}_{\mathbf{\underline{K}}_2}\begin{bmatrix}y_1\\ y_2\end{bmatrix}^{(2)}+\underbrace{\begin{bmatrix}0 & \chi\\ -\chi\pm\mathcal{N} & 0\end{bmatrix}}_{\mathbf{\underline{K}}_1}\begin{bmatrix}y_1\\ y_2\end{bmatrix}^{(1)}+\underbrace{\begin{bmatrix}0 & 0\\ 0 & -\chi\end{bmatrix}}_{\mathbf{\underline{K}}_0}\begin{bmatrix}y_1\\ y_2\end{bmatrix}=\begin{bmatrix}r_1\\ r_2\end{bmatrix}, \tag{9.109}$$

where

$$\mathcal{N}=n^2=\frac{N\ell^2}{IE}\,. \tag{9.110}$$

(a) Green function matrix if the axial force is a compressive one.
If the axial force is compressive, the sign of $\mathcal{N}$ is negative in differential equation (9.109). Recalling (9.20) and (9.21), it is obvious that the general solution of the homogeneous equation is given by the following equation:

$$\begin{bmatrix}y_1\\ y_2\end{bmatrix}=\underbrace{\begin{bmatrix}1 & x\\ 0 & -\frac{1}{\chi}(\chi+\mathcal{N})\end{bmatrix}}_{\mathbf{\underline{Y}}_1(x)}\begin{bmatrix}C_1\\ C_2\end{bmatrix}+\underbrace{\begin{bmatrix}\frac{1}{n}\cos nx & -\frac{1}{n}\sin nx\\ \sin nx & \cos nx\end{bmatrix}}_{\mathbf{\underline{Y}}_2(x)}\begin{bmatrix}C_3\\ C_4\end{bmatrix}. \tag{9.111}$$

The relatively long calculation of the Green function matrix is left for Problem 9.3. Here the final result is presented only:

$$\begin{aligned}\mathbf{\underline{G}}(x,\xi)&=\mathbf{\underline{Y}}_1(x)\left[\mathbf{\underline{A}}_1(\xi)\pm\mathbf{\underline{B}}_1(\xi)\right]+\mathbf{\underline{Y}}_2(x)\left[\mathbf{\underline{A}}_2(\xi)\pm\mathbf{\underline{B}}_2(\xi)\right]=\\ &=\begin{bmatrix}1 & x\\ 0 & -\kappa\end{bmatrix}\left\{\begin{bmatrix}\frac{\xi}{2\chi(\kappa-1)} & -\frac{1}{2n^2}\\ -\frac{2\xi-1}{2\chi(\kappa-1)} & \frac{1}{n^2}\end{bmatrix}\pm\begin{bmatrix}-\frac{\xi}{\chi(\kappa-1)} & \frac{1}{n^2}\\ \frac{1}{\chi(\kappa-1)} & 0\end{bmatrix}\right\}+\end{aligned}$$

$$+\frac{1}{2}\begin{bmatrix}\frac{1}{n}\cos nx & -\frac{1}{n}\sin nx\\ \sin nx & \cos nx\end{bmatrix}\left\{\begin{bmatrix}-\frac{\kappa\sin n\xi}{2\chi(\kappa-1)} & \frac{\cos n\xi}{2n}\\ \frac{\kappa(2\sin n\xi\cos n+\cos n\xi\sin n)}{2\chi(\kappa-1)\sin n} & \frac{2\cos n\cos n\xi+\sin n\sin n\xi}{2n\sin n}\end{bmatrix}\pm\right.$$
$$\left.\pm\begin{bmatrix}\frac{\kappa\sin n\xi}{2\chi(\kappa-1)} & -\frac{\cos n\xi}{2n}\\ \frac{\kappa\cos n\xi}{2\chi(\kappa-1)} & \frac{\sin n\xi}{2n}\end{bmatrix}\right\}, \tag{9.112a}$$

where

$$\kappa=\frac{1}{\chi}(\chi+\mathcal{N}). \tag{9.112b}$$

(b) Green function matrix if the axial force is a tensile one.

Then the sign of $\mathcal{N}$ is positive in differential equation (9.109). On the basis of Eqs. (9.23) and (9.24), it follows that the general solution of the homogeneous equation assumes the following form:

$$\begin{bmatrix}y_1\\ y_2\end{bmatrix}=\underbrace{\begin{bmatrix}1 & x\\ 0 & -\frac{1}{\chi}(\chi-\mathcal{N})\end{bmatrix}}_{\underline{\mathbf{Y}}_1(x)}\begin{bmatrix}C_1\\ C_2\end{bmatrix}+\underbrace{\begin{bmatrix}-\frac{1}{n}\cosh nx & -\frac{1}{n}\sinh nx\\ \sinh nx & \cosh nx\end{bmatrix}}_{\underline{\mathbf{Y}}_2(x)}\begin{bmatrix}C_3\\ C_4\end{bmatrix}. \tag{9.113}$$

Calculation of the Green function matrix is detailed in the solution of Problem 9.3. The result is as follows:

$$\underline{\mathbf{G}}(x,\xi)=\underline{\mathbf{Y}}_1(x)\left[\underline{\mathbf{A}}_1(\xi)\pm\underline{\mathbf{B}}_1(\xi)\right]+\underline{\mathbf{Y}}_2(x)\left[\underline{\mathbf{A}}_2(\xi)\pm\underline{\mathbf{B}}_2(\xi)\right]=$$
$$=\begin{bmatrix}1 & x\\ 0 & -\kappa\end{bmatrix}\left\{\begin{bmatrix}\frac{\xi}{2\chi(\kappa-1)} & \frac{1}{2n^2}\\ -\frac{2\xi-1}{2\chi(\kappa-1)} & -\frac{1}{n^2}\end{bmatrix}\pm\begin{bmatrix}-\frac{\xi}{2\chi(\kappa-1)} & -\frac{1}{n^2}\\ \frac{1}{2\chi(\kappa-1)} & 0\end{bmatrix}\right\}+$$
$$+\begin{bmatrix}-\frac{\cosh nx}{n} & -\frac{\sinh nx}{n}\\ \sinh nx & \cosh nx\end{bmatrix}\times$$
$$\times\left\{\begin{bmatrix}\frac{\kappa\sinh n\xi}{2\chi(\kappa-1)} & \frac{\cosh n\xi}{2n}\\ \frac{\kappa(\cosh n\xi\sinh n-2\sinh n\xi\cosh n)}{2\chi(\kappa-1)\sinh n} & \frac{\sinh n\xi\sinh n-2\cosh n\xi\cosh n}{n\sinh n}\end{bmatrix}\pm\right.$$
$$\left.\pm\begin{bmatrix}-\frac{\kappa\sinh n\xi}{2\chi(\kappa-1)} & -\frac{\cosh n\xi}{2n}\\ \frac{\kappa\cosh n\xi}{2\chi(\kappa-1)} & \frac{\sinh n\xi}{2n}\end{bmatrix}\right\}, \tag{9.114}$$

where

$$\kappa=\frac{1}{\chi}(\chi-\mathcal{N}). \tag{9.115}$$

9.2.10 *Solution of Static Boundary Value Problems*

Let us assume that the Green function matrices are known for the boundary value problems related to pinned-pinned, fixed-pinned, fixed-fixed, and fixed-free Timoshenko beams provided that there are no axial force acting on the beams. Assume further that the Green function matrix of the axially loaded and pinned-pinned Timoshenko beam is also known. If the dimensionless distributed load i.e., $\underline{\mathbf{r}}(x)$, which is the right side of Eq. (9.91) (or Eq. (9.109)), is known then, on the basis of Eq. (9.69), the solution for $\underline{\mathbf{y}}$ can be given in a closed form:

$$\underline{\mathbf{y}}(x) = \int_0^1 \underline{\mathbf{G}}(x,\xi)\underline{\mathbf{r}}(\xi)\,\mathrm{d}\xi\,. \tag{9.116}$$

9.2.11 Multiple Eigenvalues of ODES

Let the eigenvalue problem (9.45) of ODES be self-adjoint and positive definite. Then the eigenfunction vectors $\underline{\mathbf{y}}_i(x)$ $(i = 1, 2, 3, \ldots)$ fulfill the condition

$$\left(\underline{\mathbf{y}}_i, \underline{\mathbf{y}}_i\right)_M > 0\,.$$

It is clear that the normalized eigenfunction vectors

$$\underline{\psi}_i = \underline{\mathbf{y}}_i / \sqrt{\left(\underline{\mathbf{y}}_i, \underline{\mathbf{y}}_i\right)_M} \tag{9.117a}$$

satisfy the equation

$$\left(\underline{\psi}_i, \underline{\psi}_i\right)_M = 1\,. \tag{9.117b}$$

Comparison of Eqs. (9.54)$_1$, (9.56), and (9.117) shows that the normalized eigenfunction vectors $\underline{\psi}_k(x)$ also satisfy the equation

$$\left(\underline{\psi}_i, \underline{\psi}_n\right)_K = \begin{cases} \lambda_n & \text{if } i = n, \\ 0 & \text{if } i \neq n. \end{cases} \qquad i, n = 1, 2, 3, \ldots \tag{9.118}$$

The eigenvalue λ_L $(L \in 1, 2, 3, \ldots)$ of the eigenvalue problem (9.45) is a multiple one with multiplicity $r > 1$ if the number of the linearly independent eigenfunction vectors that belong to λ_L is r.

The normalized eigenfunction vectors with multiplicity r are denoted by $\underline{\psi}_{kL}(x)$ $(k \in 1, 2, \ldots, r)$ where the second subscript is that of the eigenvalue λ_L. It is obvious that

$$\underline{\mathbf{K}}[\,\underline{\psi}_{kL}] = \lambda_L\,\underline{\mathbf{M}}[\,\underline{\psi}_{kL}]\,, \quad \underline{\mathbf{U}}_r[\,\underline{\psi}_{kL}] = \underline{\mathbf{0}}\,, \qquad (k = 1, 2, \ldots, r). \tag{9.119}$$

Let c_{1k}, c_{2k} $(k = 1, 2, \ldots, r)$, etc. be given weights or coefficients. The linear combinations

$$\hat{\underline{\psi}}_{1L} = \sum_{k=1}^{r} c_{1k}\underline{\psi}_{kL}, \quad \hat{\underline{\psi}}_{2L} = \sum_{k=1}^{r} c_{2k}\underline{\psi}_{kL}, \quad \ldots \quad \hat{\underline{\psi}}_{rL} = \sum_{k=1}^{r} c_{rk}\underline{\psi}_{kL} \tag{9.120}$$

are also eigenfunction vectors of the eigenvalue problem. The normalized eigenfunctions $\underline{\psi}_{kL}(x)$ are, however, not orthogonal to each other. Recalling the Gram–Schmidt orthogonalization presented in Sect. 8.12, it can be seen that the procedure

$$\boxed{\begin{aligned} &\hat{\underline{\psi}}_{1L} = \underline{\psi}_{1L}\,, \\ &\hat{\underline{\psi}}^*_{i+1,L} = \underline{\psi}_{i+1,L} - \sum_{k=1}^{i} \hat{\underline{\psi}}_{kL}\big(\underline{\psi}_{i+1,L}, \hat{\underline{\psi}}_{kL}\big)_M\,, \qquad i = 1, 2, \ldots, r-1 \\ &\hat{\underline{\psi}}_{i+1,L} = \hat{\underline{\psi}}^*_{i+1,L}/\big(\hat{\underline{\psi}}^*_{i+1,L}, \hat{\underline{\psi}}^*_{i+1,L}\big)_M\,, \qquad \Big[\big(\hat{\underline{\psi}}_{i+1,L}, \hat{\underline{\psi}}_{i+1,L}\big)_M = 1\Big] \end{aligned}} \tag{9.121}$$

yields orthogonal eigenfunction vectors.

9.2.12 Properties of the Rayleigh Quotient of ODES

We shall assume that the eigenvalue problem of ODES (9.45) is positive definite and simple. Then it holds that

$$\underline{\mathbf{M}}\Big[\underline{\mathbf{y}}\Big] = \underline{\mathbf{M}}_0(x)\underline{\mathbf{y}}(x), \tag{9.122}$$

where the products $\big(\underline{\mathbf{u}}, \underline{\mathbf{u}}\big)_K$, $\big(\underline{\mathbf{u}}, \underline{\mathbf{u}}\big)_M$ regarded on the set of the comparison vectors $\mathbf{u}$ are positive.

It can be proved—we remind the reader of the proofs given in Sect. 8.13 and in the book [9]—that the Rayleigh quotient

$$\mathcal{R}[u(x)] = \frac{\big(\underline{\mathbf{u}}, \underline{\mathbf{u}}\big)_K}{\big(\underline{\mathbf{u}}, \underline{\mathbf{u}}\big)_M} \tag{9.123}$$

has the following properties:

1. The Rayleigh quotient $\mathcal{R}[u(x)]$ is a positive quantity.
2. Let

$$|u| = \sqrt{\int_a^b \underline{\mathbf{u}}^T \underline{\mathbf{u}}\, \mathrm{d}x}\,. \tag{9.124}$$

   The Rayleigh quotient is independent of $|u|$.
3. The Rayleigh quotient has a positive lower limit.
4. The lower limit of the Rayleigh quotient is the smallest eigenvalue λ_1.

Exercise 9.6 Estimate the first eigenvalue for the vibrating pinned-pinned Timoshenko beam.

As is well known, the Rayleigh quotient provides an upper limit for λ_1. Hence, (9.63) gives the estimation

$$\mathcal{R}[\underline{\mathbf{u}}(x)] = \frac{\displaystyle\int_0^\ell \left\{\hat{\chi}\left[(u_1^{(1)})^2 + (u_2)^2\right] + (u_2^{(1)})^2 + 2\hat{\chi}u_1^{(1)}u_2\right\} \mathrm{d}x}{\displaystyle\int_0^\ell \left[(u_1)^2 + r_y^2(u_2)^2\right] \mathrm{d}x} > \lambda_1, \quad (9.125)$$

in which u_1 and u_2 should be comparative functions. It can be checked with ease that the functions

$$u_1 = \sin\frac{\pi x}{\ell}, \qquad u_2 = -\frac{\pi}{\ell}\cos\frac{\pi x}{\ell}$$

are comparative ones.

Note that u_1 is the first eigenfunction of the vibrating pinned-pinned Euler–Bernoulli beam, while u_2 is the corresponding rotation.

Substitution of u_1 and u_2 into (9.125) yields

$$\int_0^\ell \left\{\hat{\chi}\left[(u_1^{(1)})^2 + (u_2)^2\right] + (u_2^{(1)})^2 + 2\hat{\chi}u_1^{(1)}u_2\right\} \mathrm{d}x = \left(\frac{\pi}{\ell}\right)^4 \int_0^\ell \sin^2\frac{\pi x}{\ell}\,\mathrm{d}x = \frac{\pi^4}{2\ell^3}$$

for the numerator and

$$\int_0^\ell \left[(u_1)^2 + r_y^2(u_2)^2\right] \mathrm{d}x = \int_0^\ell \sin^2\frac{\pi x}{\ell}\,\mathrm{d}x + r_y^2\left(\frac{\pi}{\ell}\right)^2 \int_0^\ell \cos^2\frac{\pi x}{\ell}\,\mathrm{d}x = \frac{1}{2\ell}\left(\ell^2 + \pi^2 r_y^2\right)$$

for the denominator of the Rayleigh quotient. Thus

$$\mathcal{R}[\underline{\mathbf{u}}(x)] = \frac{\pi^4}{\ell^2\left(\ell^2 + \pi^2 r_y^2\right)} > \lambda_1 \quad (9.126)$$

is the estimation for the first eigenvalue. It is worth mentioning that estimation (9.126) (a) is independent of κ_y and (b) coincides with the first eigenvalue of the vibrating pinned-pinned Euler–Bernoulli beam if r_y^2 is set to zero.

9.2.13 Eigenvalue Problems Governed by a System of Fredholm Integral Equations

Equation

$$\underset{(2\times1)}{\underline{\mathbf{y}}(x)} = \lambda \int_a^b \underset{(2\times2)}{\underline{\mathcal{K}}(x,\xi)}\, \underset{(2\times1)}{\underline{\mathbf{y}}(\xi)}\, \mathrm{d}\xi \quad (9.127)$$

is a Fredholm integral equation system in which $x, \xi \in [a, b]$, $b > a$, $\underline{\mathbf{y}}(x)$ is the unknown function vector (unknown function matrix), $\underline{\mathcal{K}}(x, \xi)$ is the given kernel function matrix (or simply kernel):

$$\underline{\mathcal{K}} = \begin{bmatrix} \mathcal{K}_{11}(x,\xi) & \mathcal{K}_{12}(x,\xi) \\ \mathcal{K}_{21}(x,\xi) & \mathcal{K}_{22}(x,\xi) \end{bmatrix}, \quad (9.128)$$

while λ is a parameter (the eigenvalue sought). It will be assumed that (a) the kernel is differentiable with respect to x and ξ, (b) the kernel satisfies the inequality

$$\underline{\mathcal{K}}(x,\xi) \neq \sum_{\ell=1}^{m} \underset{(2\times 1)}{\underline{\mathbf{a}}(x)}\, \underset{(1\times 2)}{\underline{\mathbf{b}}(\xi)}, \qquad m \geq 1 \tag{9.129}$$

in which $\underline{\mathbf{a}}(x)$ and $\underline{\mathbf{b}}(\xi)$ are differentiable, and (c) the kernel is cross-symmetric:

$$\underline{\mathcal{K}}(x,\xi) = \underline{\mathcal{K}}^T(\xi, x)\,. \tag{9.130}$$

Under these conditions, (a) the number of different eigenvalues is infinite, (b) the eigenvalues are real numbers, and (c) the number of solutions for $\underline{\mathbf{y}}(x)$—they are called eigenfunction vectors (or simply eigenfunctions)—that belong to an eigenvalue is finite (this number is the multiplicity of the eigenvalue and is denoted by r).

The kernel is positive definite if for any continuous and non-zero $\underline{\mathbf{v}}(x)$ it holds that

$$\int_a^b\int_a^b \underset{(1\times 2)}{\underline{\mathbf{v}}^T(x)}\, \underset{(2\times 1)}{\underline{\mathcal{K}}(x,\xi)}\, \underset{(2\times 1)}{\underline{\mathbf{v}}(\xi)}\, \mathrm{d}\xi\, \mathrm{d}x > 0\,. \tag{9.131}$$

If the kernel is positive definite, the eigenvalues are positive numbers:

$$0 < \lambda_1 < \lambda_2 < \lambda_3 < \cdots$$

The eigenfunction vector that belongs to λ_L $(L = 1, 2, 3, \ldots)$ is denoted by $\underline{\mathbf{y}}_L$ if the multiplicity of λ_L is one, by $\underline{\mathbf{y}}_{kL}$ if the multiplicity of λ_L is r $(k = 1, 2, \ldots, r)$.

The eigenfunction vectors can be normalized:

$$\begin{aligned} \underline{\psi}_i(x) &= \frac{\underline{\mathbf{y}}_i(x)}{|\underline{\mathbf{y}}_i(x)|}, \qquad |\underline{\mathbf{y}}_i(x)| = \sqrt{\left(\underline{\mathbf{y}}_i, \underline{\mathbf{y}}_i\right)_I}, \\ \left(\underline{\mathbf{y}}_i, \underline{\mathbf{y}}_i\right)_I &= \int_a^b \underline{\mathbf{y}}_i^T(x)\, \underline{\mathbf{y}}_i(x)\, \mathrm{d}x, \quad i = 1, 2, 3, \ldots \end{aligned} \tag{9.132}$$

where $\underline{\psi}_i$ is the normalized eigenfunction vector. Let λ_i and λ_ℓ be two not necessarily different eigenvalues each with only one eigenfunction vector. It can be proved that the orthogonality condition

$$\int_a^b \underline{\psi}_i^T(x)\, \underline{\psi}_\ell(x)\, \mathrm{d}x = \left(\underline{\psi}_i, \underline{\psi}_\ell\right)_I = \begin{cases} 1 & \text{if } i = \ell\,, \\ 0 & \text{if } i \neq \ell\,. \end{cases} \quad i, \ell = 1, 2, 3, \ldots \tag{9.133}$$

holds. If $r > 1$ the number of linearly independent eigenfunction vectors which belong to the considered eigenvalue is r. These eigenfunctions are, in general, not orthogonal to each other. They can, however, be made orthogonal by applying the Gram–Schmidt orthogonalization procedure presented in Sect. 9.2.11.

9.2.14 Calculation of Eigenvalues

The boundary element technique can also be applied to reducing the eigenvalue problem governed by the Fredholm integral equation system

$$\Lambda \underset{(2\times1)}{\underline{\mathbf{y}}(x)} = \int_a^b \underset{(2\times1)}{\underline{\mathcal{K}}(x,\xi)} \, \underset{(2\times2)}{\underline{\mathbf{y}}(\xi)} \, \mathrm{d}\xi \,, \qquad \Lambda\lambda = 1 \,, \tag{9.134}$$

to an algebraic eigenvalue problem. The procedure detailed below is based on Sect. 8.15.2.

The interval $[a, b]$ is the same as the one shown in Fig. 8.6. The number of elements in this interval is $n_e = 5$. The elements and their lengths are also denoted in the same way as earlier, i.e., by $\mathfrak{L}_1, \ldots, \mathfrak{L}_{n_e}$. The nodal points where the unknown function vector $\underline{\mathbf{y}}(x)$ is considered are taken at the extremes (or ends) and the midpoint of an element. The nodal point numbering is shown in Fig. 8.6—the numbering scheme has remained unchanged. Let $\underline{\mathbf{x}}^e$ and $\underline{\mathbf{y}}_k^e$ be the matrices of the nodal coordinates and the unknown function vectors at the nodes of the element $\mathfrak{L}_e$:

$$\left(\underline{\mathbf{x}}^e\right)^T = \left[x_1^e \; x_2^e \; x_3^e\right], \quad \left(\underline{\mathbf{y}}_k^e\right)^T = \left[y_{k1}^e \; y_{k2}^e\right], \quad k = 1, 2, 3 \tag{9.135}$$

It is assumed that the approximation over the elements is quadratic. If this is the case, the shape functions $N_k(\eta)$ are given by Eq. (8.179). Making use of the notations

$$\underset{(6\times1)}{\underline{\mathbf{y}}^e} = \begin{bmatrix} \underline{\mathbf{y}}_1^e \\ \underline{\mathbf{y}}_2^e \\ \underline{\mathbf{y}}_3^e \end{bmatrix} \quad \text{and} \quad \underset{(2\times2)}{\underline{\mathbf{N}}_k(\eta)} = \begin{bmatrix} N_k(\eta) & 0 \\ 0 & N_k(\eta) \end{bmatrix}, \tag{9.136}$$

the unknown function matrix over the element $\mathfrak{L}_e$ can be approximated by the equation

$$\underset{(2\times1)}{\underline{\mathbf{y}}}\left[\xi^e(\eta)\right] = \left[\underline{\mathbf{N}}_1(\eta) \middle| \underline{\mathbf{N}}_2(\eta) \middle| \underline{\mathbf{N}}_3(\eta)\right] \begin{bmatrix} \underline{\mathbf{y}}_1^e \\ \underline{\mathbf{y}}_2^e \\ \underline{\mathbf{y}}_3^e \end{bmatrix} = \underset{(2\times6)}{\underline{\mathbf{N}}} \; \underset{(6\times1)}{\underline{\mathbf{y}}^e} \,, \tag{9.137a}$$

where

$$\xi^e(\eta) = \left[N_1(\eta) \middle| N_2(\eta) \middle| N_3(\eta)\right] \begin{bmatrix} x_1^e \\ x_2^e \\ x_3^e \end{bmatrix} = \frac{1}{2}\eta\left(x_3^e - x_1^e\right) + x_2^e \,. \tag{9.137b}$$

Note that the derivation of the latter equation is omitted since it should obviously coincide with Eq. (8.181). According to relations (8.182), it also holds that

$$\mathrm{d}\xi^e = \mathcal{J}\mathrm{d}\eta \,, \qquad \mathcal{J} = \frac{1}{2}\left(x_3^e - x_1^e\right) = \frac{\mathfrak{L}_e}{2} \,. \tag{9.138}$$

Substituting (9.137a) and (9.138) into (9.134) and integrating, then element by element yields

$$\Lambda \underline{\mathbf{y}}(x) = \sum_{e=1}^{n_e} \int_{\mathfrak{L}_e} \underline{\mathcal{K}}[x, \xi(\eta)] \left[\underline{\mathbf{N}}_1(\eta) | \underline{\mathbf{N}}_2(\eta) | \underline{\mathbf{N}}_3(\eta) \right] \mathcal{J}(\eta) \mathrm{d}\eta \begin{bmatrix} \underline{\mathbf{y}}_1^e \\ \underline{\mathbf{y}}_2^e \\ \underline{\mathbf{y}}_3^e \end{bmatrix} =$$

$$= \sum_{e=1}^{n_{be}} \int_{\eta=-1}^{\eta=1} \underset{(2\times 2)}{\underline{\mathcal{K}}} [x, \xi(\eta)] \underset{(2\times 6)}{\underline{\mathbf{N}}} (\eta) \mathcal{J}(\eta) \mathrm{d}\eta \underset{(6\times 1)}{\underline{\mathbf{y}}^e} . \qquad (9.139)$$

If the node coordinates are denoted by x_i $(i = 1, 2, \ldots, 2n_e + 1)$ and the previous equation is taken at these points we obtain

$$\Lambda \underline{\mathbf{y}}(x_i) = \sum_{e=1}^{n_e} \int_{\eta=-1}^{\eta=1} \underline{\mathcal{K}}[x_i, \xi(\eta)] \left[\underline{\mathbf{N}}_1(\eta) | \underline{\mathbf{N}}_2(\eta) | \underline{\mathbf{N}}_3(\eta) \right] \mathcal{J}(\eta) \mathrm{d}\eta \begin{bmatrix} \underline{\mathbf{y}}_1^e \\ \underline{\mathbf{y}}_2^e \\ \underline{\mathbf{y}}_3^e \end{bmatrix},$$

$$(i = 1, 2, \ldots, 2n_e + 1) . \qquad (9.140)$$

For the sake of simplicity it is worth introducing further notations:

$$\underset{(2\times 2)}{\underline{\mathbf{k}}_r^{ie}} = \int_{\eta=-1}^{\eta=1} \underset{(2\times 2)}{\underline{\mathcal{K}}} [x_i, \xi(\eta)] \underset{(2\times 2)}{\underline{\mathbf{N}}_r} (\eta) \mathcal{J}(\eta) \mathrm{d}\eta , \qquad \underset{(2\times 1)}{\underline{\mathbf{y}}} (x_i) = \underset{(2\times 1)}{\underline{\mathbf{y}}_i} ,$$

$$(i = 1, 2, 3, \ldots, 2n_e + 1;\ r = 1, 2, 3), \qquad (9.141)$$

where i identifies the global number of the node, e is the number of the element over which the integral is taken, and r is the number of the interpolation function matrix. With notations (9.141), Eq. (9.140) assumes the following form:

$$\left[\underline{\mathbf{k}}_1^{i1} | \underline{\mathbf{k}}_2^{i1} | \underline{\mathbf{k}}_3^{i1} + \underline{\mathbf{k}}_1^{i2} | \underline{\mathbf{k}}_2^{i2} | \underline{\mathbf{k}}_3^{i2} + \underline{\mathbf{k}}_1^{i3} | \underline{\mathbf{k}}_1^{i3} | \ldots | \underline{\mathbf{k}}_3^{i,n_e-1} + \underline{\mathbf{k}}_1^{in_e} | \underline{\mathbf{k}}_2^{in_e} | \underline{\mathbf{k}}_3^{in_e} \right] \times$$

$$\times \begin{bmatrix} \underline{\mathbf{y}}_1 \\ \underline{\mathbf{y}}_2 \\ \cdots \\ \underline{\mathbf{y}}_{2n_e} \\ \underline{\mathbf{y}}_{2n_e+1} \end{bmatrix} - \Lambda \underline{\mathbf{y}}_i = \begin{bmatrix} 0 \\ 0 \end{bmatrix}, \qquad (i = 1, 2, 3, \ldots, 2n_e + 1), \qquad (9.142)$$

in which the column matrices $\underline{\mathbf{y}}_i$ are the unknowns. Note that the number of Eq. (9.142) coincides with the number of unknowns. By introducing the notations

$$\underline{\mathbf{k}}_{i1} = \underline{\mathbf{k}}_1^{i1},\ \ \underline{\mathbf{k}}_{i2} = \underline{\mathbf{k}}_2^{i1},\ \ \underline{\mathbf{k}}_{i3} = \underline{\mathbf{k}}_3^{i1} + \underline{\mathbf{k}}_1^{i2},\ \ \underline{\mathbf{k}}_{i4} = \underline{\mathbf{k}}_2^{i2},\ \ \underline{\mathbf{k}}_{i5} = \underline{\mathbf{k}}_3^{i2} + \underline{\mathbf{k}}_1^{i3}, \ldots,$$

$$\underline{\mathbf{k}}_{i,2n_e} = \underline{\mathbf{k}}_2^{in_e},\ \ \underline{\mathbf{k}}_{i,2n_e+1} = \underline{\mathbf{k}}_3^{in_e} \qquad (9.143a)$$

and

$$\underset{(2\times2)}{\underline{\mathbf{1}}} = \begin{bmatrix} 1 & 0 \\ 0 & 1 \end{bmatrix}, \quad \underset{(2\times2)}{\underline{\mathbf{0}}} = \begin{bmatrix} 0 & 0 \\ 0 & 0 \end{bmatrix}, \quad \underset{(2\times1)}{\underline{\hat{\mathbf{0}}}} = \begin{bmatrix} 0 \\ 0 \end{bmatrix}, \tag{9.143b}$$

the structure of equation system (9.142) becomes easy to survey:

$$\left\{ \begin{bmatrix} \underline{\mathbf{k}}_{11} & \underline{\mathbf{k}}_{12} & \cdots & \underline{\mathbf{k}}_{1,2n_e+1} \\ \underline{\mathbf{k}}_{21} & \underline{\mathbf{k}}_{22} & \cdots & \underline{\mathbf{k}}_{2,2n_e+1} \\ & \cdots\cdots\cdots\cdots & & \\ \underline{\mathbf{k}}_{n_e1} & \underline{\mathbf{k}}_{n_e2} & \cdots & \underline{\mathbf{k}}_{2n_e+1,2n_e+1} \end{bmatrix} - \Lambda \begin{bmatrix} \underline{\mathbf{1}} & \underline{\mathbf{0}} & \cdots & \underline{\mathbf{0}} \\ \underline{\mathbf{0}} & \underline{\mathbf{1}} & \cdots & \underline{\mathbf{0}} \\ & \cdots\cdots & & \\ \underline{\mathbf{0}} & \underline{\mathbf{0}} & \cdots & \underline{\mathbf{1}} \end{bmatrix} \right\} \begin{bmatrix} \underline{\mathbf{y}}_1 \\ \underline{\mathbf{y}}_2 \\ \cdots \\ \underline{\mathbf{y}}_{2n_e+1} \end{bmatrix} = \\ = \begin{bmatrix} \underline{\hat{\mathbf{0}}} \\ \underline{\hat{\mathbf{0}}} \\ \cdots \\ \underline{\hat{\mathbf{0}}} \end{bmatrix}. \tag{9.144}$$

Equation (9.144) is again an algebraic eigenvalue problem—similar to Eq. (8.188)—with $\Lambda = 1/\lambda$ as eigenvalue. After solving it numerically, we have both the eigenvalues $\Lambda_r = 1/\lambda_r$ and the corresponding eigenvectors

$$\underset{(1\times(4n_e+2))}{\underline{\mathbf{y}}_r^T} = \left[\underline{\mathbf{y}}_{1r}^T \middle| \underline{\mathbf{y}}_{2r}^T \middle| \underline{\mathbf{y}}_{3r}^T \middle| \cdots \middle| \underline{\mathbf{y}}_{2n_e+1,r}^T \right], \quad (r = 1, 2, 3, \ldots, 2n_e + 1).$$

As regards the issue of numerical integration, it is worthy introducing the notation

$$\underline{\phi}(\eta) = \underline{\mathcal{K}}\left[x_i, \xi(\eta)\right] \underline{\mathbf{N}}_r(\eta) \mathcal{J}(\eta). \tag{9.145}$$

With (9.145) one can rewrite integral (9.141) in the following form:

$$\underline{\mathbf{k}}_r^{ie} = \int_{\eta=-1}^{\eta=1} \underline{\mathcal{K}}\left[x_i, \xi(\eta)\right] \underline{\mathbf{N}}_r(\eta) \mathcal{J}(\eta)\, \mathrm{d}\eta = \int_{\eta=-1}^{\eta=1} \underline{\phi}(\eta)\, \mathrm{d}\eta \tag{9.146}$$

for which it is easy to apply the Gaussian quadrature rules presented in Sect. B.3.

9.3 Free Vibrations of Timoshenko Beams

By the use of dimensionless variables on the base of Eq. (9.89), the motion equations (9.29) can be rewritten in the following form:

$$\begin{aligned} \frac{AG\kappa_y\ell^2}{IE}\left(\frac{\mathrm{d}^2 y_1}{\mathrm{d}x^2} + \frac{\mathrm{d}y_2}{\mathrm{d}x}\right) &= -\frac{\rho A\omega^2}{IE}\ell^4 y_1, \\ \frac{\mathrm{d}^2 y_2}{\mathrm{d}x^2} \pm \frac{N\ell^2}{IE}\frac{\mathrm{d}y_1}{\mathrm{d}x} - \frac{AG\kappa_y\ell^2}{IE}\left(\frac{\mathrm{d}y_1}{\mathrm{d}x} + y_2\right) &= -\frac{\rho A\omega^2}{IE}\ell^4\frac{I}{A\ell^2}y_2 \end{aligned} \tag{9.147}$$

or

$$\chi\left(\frac{\mathrm{d}^2 y_1}{\mathrm{d}x^2}+\frac{\mathrm{d}y_2}{\mathrm{d}x}\right)=-\lambda y_1\,,$$
$$\frac{\mathrm{d}^2 y_2}{\mathrm{d}x^2}\pm\mathcal{N}\frac{\mathrm{d}y_1}{\mathrm{d}x}-\chi\left(\frac{\mathrm{d}y_1}{\mathrm{d}x}+y_2\right)=-\lambda\mathfrak{r}^2 y_2, \tag{9.148a}$$

where y_1 and y_2 are dimensionless amplitude functions, x is a dimensionless coordinate and

$$\chi=\frac{AG\kappa_y\ell^2}{IE},\quad \mathcal{N}=\frac{N\ell^2}{IE},\quad \lambda=\frac{\rho A\omega^2}{IE}\ell^4,\quad \mathfrak{r}^2=\frac{I}{A\ell^2}. \tag{9.148b}$$

If there is no axial force, the amplitude function should satisfy the equations

$$\chi\left(\frac{\mathrm{d}^2 y_1}{\mathrm{d}x^2}+\frac{\mathrm{d}y_2}{\mathrm{d}x}\right)=-\lambda y_1\qquad \frac{\mathrm{d}^2 y_2}{\mathrm{d}x^2}-\chi\left(\frac{\mathrm{d}y_1}{\mathrm{d}x}+y_2\right)=-\lambda\mathfrak{r}^2 y_2, \tag{9.149}$$

which can be rewritten in matrix form as well:

$$\underbrace{\begin{bmatrix}\chi & 0\\ 0 & 1\end{bmatrix}}_{\underline{\mathbf{K}}_2}\begin{bmatrix}y_1\\ y_2\end{bmatrix}^{(2)}+\underbrace{\begin{bmatrix}0 & \chi\\ -\chi & 0\end{bmatrix}}_{\underline{\mathbf{K}}_1}\begin{bmatrix}y_1\\ y_2\end{bmatrix}^{(1)}+\underbrace{\begin{bmatrix}0 & 0\\ 0 & -\chi\end{bmatrix}}_{\underline{\mathbf{K}}_0}\begin{bmatrix}y_1\\ y_2\end{bmatrix}=\lambda\underbrace{\begin{bmatrix}-1 & 0\\ 0 & -\mathfrak{r}^2\end{bmatrix}}_{\underline{\mathbf{M}}_0}\begin{bmatrix}y_1\\ y_2\end{bmatrix}. \tag{9.150}$$

It is clear that the left side of the above equation coincides with the left side of Eq. (9.91). The expression

$$\lambda\begin{bmatrix}-1 & 0\\ 0 & -\mathfrak{r}^2\end{bmatrix}\begin{bmatrix}y_1\\ y_2\end{bmatrix}$$

on the right side corresponds to $\underline{\mathbf{r}}$ in the solution (9.116). Hence,

$$\begin{bmatrix}y_1(x)\\ y_2(x)\end{bmatrix}=\lambda\int_0^1\begin{bmatrix}G_{11}(x,\xi) & G_{12}(x,\xi)\\ G_{21}(x,\xi) & G_{22}(x,\xi)\end{bmatrix}\begin{bmatrix}-1 & 0\\ 0 & -\mathfrak{r}^2\end{bmatrix}\begin{bmatrix}y_1\\ y_2\end{bmatrix}\mathrm{d}\xi\,,$$

where for pinned-pinned, fixed-pinned, fixed-fixed, and fixed-free beams the Green function matrices are given by Eqs. (9.101), (9.105), (9.106), and (9.107), respectively. Introduction of the new variables $\underline{\mathcal{Y}}$ and $\underline{\mathcal{K}}$ as shown by the equation

$$\underbrace{\begin{bmatrix}1 & 0\\ 0 & \mathfrak{r}\end{bmatrix}\begin{bmatrix}y_1(x)\\ y_2(x)\end{bmatrix}}_{\underline{\mathcal{Y}}(x)}=\lambda\int_0^1\underbrace{\begin{bmatrix}1 & 0\\ 0 & \mathfrak{r}\end{bmatrix}\begin{bmatrix}-G_{11}(x,\xi) & -G_{12}(x,\xi)\\ -G_{21}(x,\xi) & -G_{22}(x,\xi)\end{bmatrix}\begin{bmatrix}1 & 0\\ 0 & \mathfrak{r}\end{bmatrix}}_{\underline{\mathcal{K}}(x,\xi)}\underbrace{\begin{bmatrix}1 & 0\\ 0 & \mathfrak{r}\end{bmatrix}\begin{bmatrix}y_1\\ y_2\end{bmatrix}}_{\underline{\mathcal{Y}}(\xi)}\mathrm{d}\xi$$

results in an eigenvalue problem governed by the Fredholm integral equation

$$\underline{\mathcal{Y}}(x)=\lambda\int_0^1\underline{\mathcal{K}}(x,\xi)\,\underline{\mathcal{Y}}(\xi)\,\mathrm{d}\xi\,. \tag{9.151}$$

The eigenvalue problem governed by the homogeneous integral equation (9.151) can be solved numerically by applying the solution procedure presented in Sect. 9.2.14. A FORTRAN 90 program was written by applying the solution procedure mentioned. Some results are presented graphically in Figs. 9.3, 9.4, 9.5, and 9.6.

Let $\breve{\omega}_k$ $(k = 1, 2)$ be the first two natural circular frequencies of the Timoshenko beam considered. The first natural frequency of the same beam within the framework of the Euler–Bernoulli beam theory is denoted by ω_1.

These figures show the quotient $\hat{\omega}_k/\omega_k$ against $\mathfrak{r} = r_y/\ell$, where r_y is the radius of gyration and ℓ is the length of the beam. A further parameter for the graphs shown in Figs. 9.3, 9.4, 9.5, and 9.6 is the quotient $\mathfrak{E} = E/G\kappa_y$.

It follows from Eq. (9.148b) which define the dimensionless parameters χ and $\mathfrak{r}^2$ that $\chi = 1/\mathfrak{E}\mathfrak{r}^2$. If $\mathfrak{r} = 0$, the solution for $\hat{\omega}$ coincides with that of the Euler beam theory. It follows from this fact by taking the data of Table 7.5 and Eq. $(7.86\text{a})_5$ into account that the solution curves in Figs. 9.3, 9.4, 9.5, and 9.6 start from the ordinates (points)

$$[0, 1],\ [0, 4]\ \text{(Pinned-Pinned Beam)};$$
$$[0, 1],\ [0, (7.0685/3.9266)^2 = 3.2406]\ \text{(Fixed-Pinned Beam)};$$
$$[0, 1],\ [0, (7.8532/4.7300)^2 = 2.7565]\ \text{(Fixed-Fixed Beam)};$$
$$[0, 1],\ [0, (4.6941/1.8751)^2 = 6.2668]\ \text{(Fixed-Free Beam)}.$$

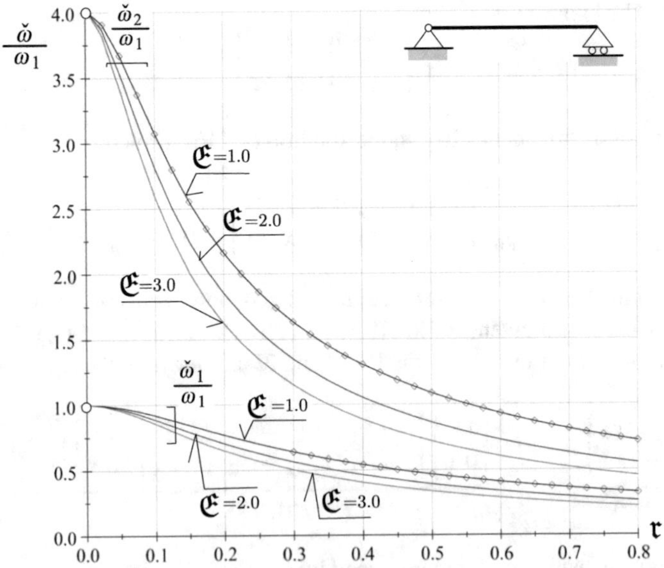

Fig. 9.3 Quotient $\hat{\omega}_k/\omega_k$ against $\mathfrak{r} = r_y/\ell$ for simply supported Timoshenko beam

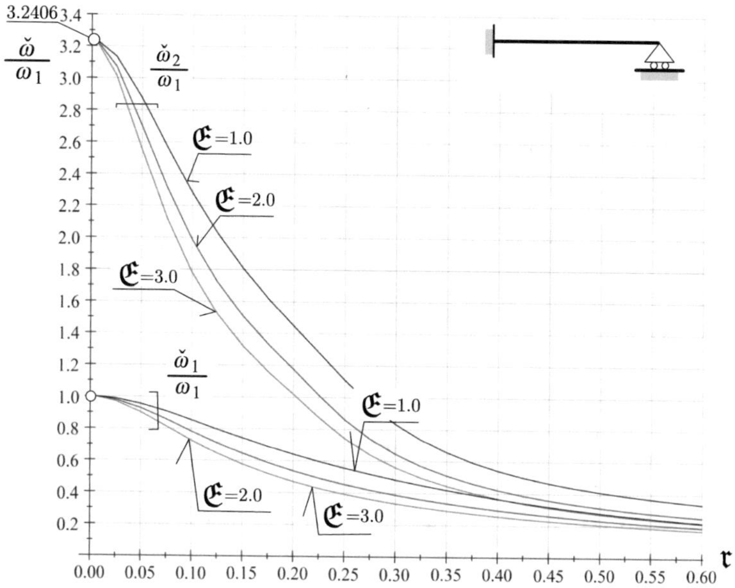

Fig. 9.4 Quotient $\hat{\omega}_k/\omega_k$ against $\mathfrak{r} = r_y/\ell$ for fixed-pinned Timoshenko beam

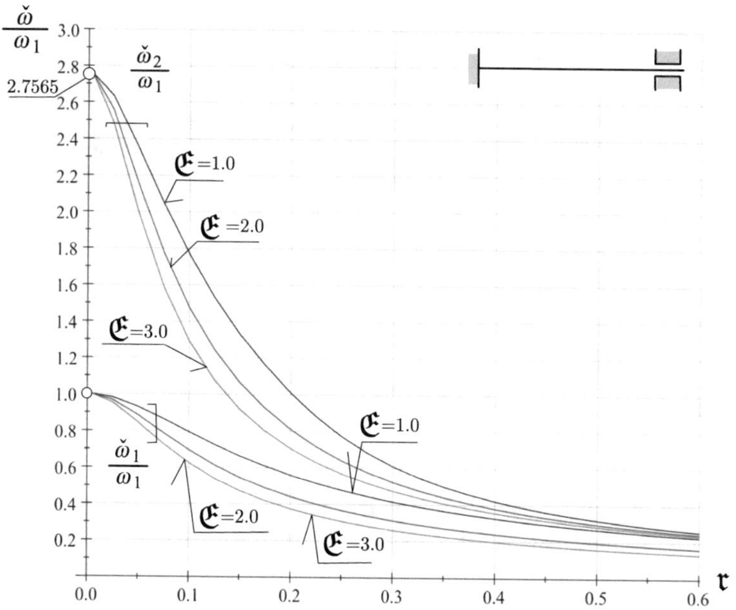

Fig. 9.5 Quotient $\hat{\omega}_k/\omega_k$ against $\mathfrak{r} = r_y/\ell$ for fixed-fixed Timoshenko beam

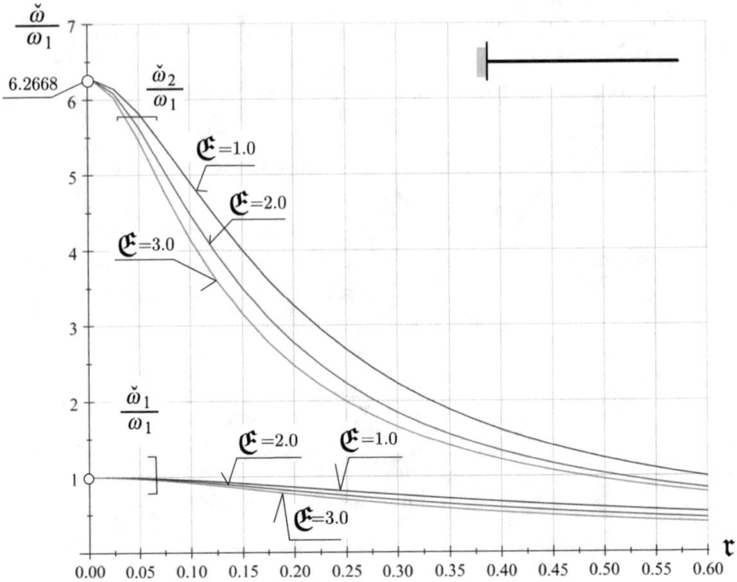

Fig. 9.6 Quotient $\hat{\omega}_k/\omega_k$ against $\mathfrak{r} = r_y/\ell$ for fixed-free Timoshenko beam

The graphs shown are drown by joining the points that represent the numerical solutions by short straight lines. There is a closed-form solution for the eigenvalue λ for pinned-pinned beams—see Problem 9.4 and Eq. (C.9.48) in its solution. The points depicted by diamonds in Fig. 9.3 represent the analytical solution. They fit onto the numerical solution with a very good accuracy.

9.4 Problems

Problem 9.1 Prove that the eigenvalue problems defined by differential equation (9.34a) and the boundary conditions shown in Rows 2, 3, and 4 of Table 9.1 are all self-adjoint.

Problem 9.2 A concentrated mass is attached to the right end of the beam shown in Fig. 9.7. It is assumed that the mass m is much greater than that of the beam. Assume

Fig. 9.7 Fixed beam with a concentrated mass at the right end of the beam

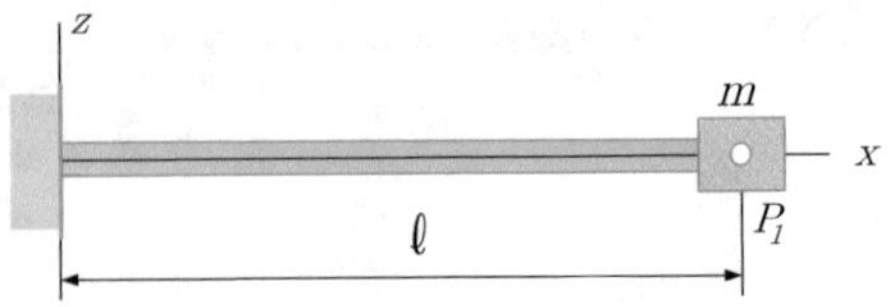

that the mass vibrates vertically. Prove that the square of its circular frequency is given by

$$\omega^2 = \frac{k_{11}}{m} = \frac{3IE}{\ell^3 m} \qquad \text{(Euler–Bernoulli beam theory)}$$

or by

$$\omega^2 = \frac{3IE}{\ell^3\left(1 + \frac{2r_y^2}{\ell^2}\frac{(1+\nu)}{\kappa_y}\right)} \qquad \text{(Timoshenko beam theory)} \tag{9.152}$$

Problem 9.3 Detail the calculation of the Green function matrix for axially loaded and pinned-pinned Timoshenko beams in order to prove the validity of the solutions (9.112) and (9.114).

Problem 9.4 Find the frequency equations for the free vibrations of pinned-pinned Timoshenko beams.

References

1. S.P. Timoshenko, On the correction for shear of the differential equation for transverse vibrations of prismatic bars. Lond. Edinb. Dublin Philos. Mag. J. Sci. **41**(245), 744–746 (1921). https://doi.org/10.1080/14786442108636264
2. F. Engesser, Die knickfestigkeit grader stëabe. Central Bauvervaltung **11**, 483–486 (1981)
3. G.R. Cowper, The shear coefficient in Timoshenko's beam theory. J. Appl. Mech. **33**(2), 335–340. (1966). https://doi.org/10.1115/1.3625046
4. G.R. Cowper, On the accuracy of Timoshenko's beam theory. ASCE, J. Eng. Mech. Div. **94**(6), 1447–1453 (1968)
5. A. Marinetti, G. Oliveto, *On the Evaluation of the Shear Correction Factors: a Boundary Element Approach* (2009)
6. M. Levinson, D.W. Cooke, On the frequency spectra of Timohsenko beams. J. Sound Vib. **84**(3), 319–324 (1982). https://doi.org/10.1016/0022-460X(82)90480-1
7. G. Szeidl, Effect of Change in Length on the Natural Frequencies and Stability of Circular Beams. in Hungarian. Ph.D. thesis. Department of Mechanics, University of Miskolc, Hungary (1975), 158 pp
8. J.G. Obádovics, On the Boundary and Initial Value Problems of Differential Equation Systems. Ph.D. thesis. Hungarian Academy of Sciences (1967), (in Hungarian)
9. L. Collatz, *Eigenwertaufgaben mit Technischen Anwendungen*. Russian Edition in 1968 (Akademische Verlagsgesellschaft Geest & Portig K.G., 1963)

Chapter 10
Eigenvalue Problems Described by Degenerated Systems of Ordinary Differential Equations

10.1 Vibration of Heterogeneous Curved Beams

10.1.1 Introductory Remarks

As regards the vibration problem of curved beams, it is worthwhile to mention that Den Hartog [1, 2] is known to be the first to deal with the free vibrations of circular beams. Other early but relevant results, considering the inextensibility of the centerline, were achieved in [3, 4]. A more recent research by Qatu and Elsharkawy [5] presents an exact model and numerical solutions for the free vibrations of laminated arches. With the differential quadrature method, Kang et al. [6] determine the eigenfrequencies for the in- and out-of-plane vibrations of circular Timoshenko arches with rotatory inertia and shear deformations included. Tüfekçi and Arpaci [7] present exact solutions for the differential equations which describe the in-plane free harmonic vibrations of extensible curved beams. Krishnan and Suresh [8] tackle the very same issue with a shear-deformable finite element (FE) model. Paper [9] by Ecsedi and Dluhi analyzes some dynamic features of non-homogeneous curved beams and closed rings assuming cross-sectional heterogeneity. Elastic foundation is taken into account in [10]. Survey paper [11] by Hajianmaleki and Qatu collects a bunch of references up until the early 2010s in the topic investigated. Kovács [12] considers layered arches with both perfect and even imperfect bonding between any two adjacent layers.

The present chapter is aimed at clarifying the vibratory properties of circular beams made of heterogeneous material. The differential equation system that describes these vibrations is a degenerated one. The numerical solution will be based on the use of the Green function matrices which should be redefined since the differential equation system is, as has just been mentioned, a degenerated one.

G. Szeidl and L. P. Kiss, *Mechanical Vibrations*, Foundations of Engineering Mechanics,
https://doi.org/10.1007/978-3-030-45074-8_10

10.1.2 Equilibrium Equations

10.1.2.1 Fundamental Assumptions

The following assumptions will be applied when the equations that govern the vibration problem of heterogeneous curved beams are derived:

(i) the beam is uniform, i.e., the cross section is constant;
(ii) the radius of the centerline is constant and the cross section is symmetric with respect to the plane of the centerline;
(iii) the material of the beam is heterogeneous which means that the modulus of elasticity depends on the cross-sectional coordinates only; thus, it is independent of the coordinate measured on the centerline of the beam (cross-sectional heterogeneity)—since the beam is heterogeneous, a precise definition for the centerline will be given later;
(iv) the displacements and deformations are small;
(v) the centerline of the beam keeps its own plane during the deformations (vibrations);
(vi) the normal stress parallel to the centerline satisfies the relation $\sigma_\xi \gg \sigma_\eta, \sigma_\zeta$—see Fig. 10.1 for the coordinate directions;
(vi) the work done by the shear stresses τ can be neglected when it is compared to that of the normal stress σ_ξ;
(vii) the beam is subjected to a distributed load lying in the plane of the centerline and/or to concentrated forces.

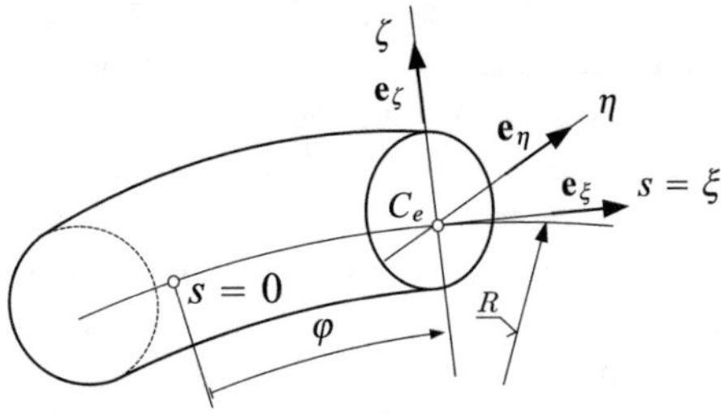

Fig. 10.1 Curved beam element and the selected coordinate systems

The coordinate system selected is the polar one shown in Fig. 10.1. The unit vector $\mathbf{e}_\xi$ is tangent to the centerline, $\mathbf{e}_\eta$ is perpendicular to the plane of the centerline while the unit vectors $\mathbf{e}_\xi$, $\mathbf{e}_\eta$, and $\mathbf{e}_\zeta$ form a right-hand triad, hence $\mathbf{e}_\zeta = \mathbf{e}_\xi \times \mathbf{e}_\eta$. Consequently, e_ζ is perpendicular to the centerline in its plane. It is clear from Fig. 10.1 that R is the radius of the centerline, s is the arc coordinate on the centerline, and φ is the polar angle. It also holds that

$$\frac{\mathrm{d}\mathbf{e}_\xi}{\mathrm{d}s} = -\frac{1}{R}\mathbf{e}_\zeta\,; \qquad \frac{\mathrm{d}\mathbf{e}_\zeta}{\mathrm{d}s} = \frac{1}{R}\mathbf{e}_\xi\,. \tag{10.1}$$

In accordance with assumption (iii), the modulus of elasticity should satisfy the relation $E(\eta, \zeta) = E(-\eta, \zeta)$ which means that E is an even function of the η-coordinate. The E-weighted first moment of the cross section with respect to the axis η is defined by the equation

$$Q_{e\eta} = \int_A E(\eta, \zeta)\zeta \mathrm{d}A. \tag{10.2}$$

The location of the centerline—that can also be called E-weighted centerline—is determined uniquely by the requirement $Q_{e\eta} = 0$. The E-weighted centerline intersects the cross section at the point C_e called E-weighted center.

For our later considerations, it is worth introducing the following three quantities [13, 14]:

$$A_{eR} = \int_A \frac{R}{R+\zeta} E(\eta, \zeta)\mathrm{d}A = \int_A \left(1 - \frac{\zeta}{R} + \frac{\zeta^2}{R^2} - \ldots\right) E(\eta, \zeta)\mathrm{d}A \cong$$
$$\cong \int_A E(\eta, \zeta)\mathrm{d}A \simeq A_e, \tag{10.3a}$$

$$Q_{eR} = \int_A \frac{R}{R+\zeta} E(\eta, \zeta)\zeta\mathrm{d}A = \int_A \left(1 - \frac{\zeta}{R} + \frac{\zeta^2}{R^2} - \ldots\right) \zeta E(\eta, \zeta)\mathrm{d}A \cong$$
$$\cong \underbrace{\int_A \zeta E(\eta, \zeta)\mathrm{d}A}_{Q_{e\eta}=0} - \frac{1}{R}\underbrace{\int_A \zeta^2 E(\eta, \zeta)\mathrm{d}A}_{I_{e\eta}} \tag{10.3b}$$

and

$$I_{eR} = \int_A \frac{R}{R+\zeta} E(\eta, \zeta)\zeta^2\mathrm{d}A = \int_A \left(1 - \frac{\zeta}{R} + \frac{\zeta^2}{R^2} - \ldots\right) \zeta^2 E(\eta, \zeta)\mathrm{d}A \cong$$
$$\cong \int_A \zeta^2 E(\eta, \zeta)\mathrm{d}A = I_{e\eta}, \tag{10.3c}$$

where A_e and $I_{e\eta}$ are the E-weighted cross-sectional area and the E-weighted moment of inertia. In the sequel, it will be assumed that $A_{eR} = A_e$, $Q_{eR} = -I_{e\eta}/R$ and $I_{eR} = I_{e\eta}$. Note that these quantities characterize the geometry of the cross section and the material distribution over the cross section.

The nabla operator in the considered polar coordinate system has the following form:

$$\nabla = \frac{\partial}{\partial s}\frac{1}{1+\frac{\zeta}{R}}\mathbf{e}_\xi + \frac{\partial}{\partial \eta}\mathbf{e}_\eta + \frac{\partial}{\partial \zeta}\mathbf{e}_\zeta\,. \tag{10.4}$$

10.1.2.2 Strain State

As regards the deformations, it is assumed that the cross section rotates as if it were a rigid body—Euler–Bernoulli hypothesis. Then

$$\mathbf{u} = [u_o(s) + \psi_{o\eta}(s)\zeta]\mathbf{e}_\xi + w_o(s)\mathbf{e}_\zeta \tag{10.5}$$

is the displacement field where u_o and w_o are the tangential and normal displacement components on the centerline, while $\psi_{o\eta}$ is the rotation of the cross section about the η-axis. As is well known

$$\boldsymbol{\psi} = -\frac{1}{2}(\mathbf{u} \times \nabla)$$

is the rotation field in the beam. Hence

$$\boldsymbol{\psi}|_{\zeta=0} = -\frac{1}{2}(\mathbf{u} \times \nabla)\bigg|_{\zeta=0} . \tag{10.6}$$

Substituting (10.4), (10.1) and taking the relation $\mathbf{e}_\zeta \times \mathbf{e}_\xi = \mathbf{e}_\eta$ into account yields

$$\frac{1}{2}\frac{u_o}{R}\mathbf{e}_\eta - \frac{1}{2}\frac{\mathrm{d}w_o}{\mathrm{d}s}\mathbf{e}_\eta + \frac{1}{2}\psi_{o\eta}\mathbf{e}_\eta$$

for the right side of (10.6). As regards the left side, we have

$$\boldsymbol{\psi}|_{\zeta=0} = \psi_{o\eta}\mathbf{e}_\eta .$$

Thus

$$\psi_{o\eta} = \frac{u_o}{R} - \frac{\mathrm{d}w_o}{\mathrm{d}s} \tag{10.7}$$

is the rotation on the centerline. The strain tensor is given by the equation

$$\boldsymbol{\varepsilon} = \frac{1}{2}(\mathbf{u} \circ \nabla + \nabla \circ \mathbf{u})$$

from where

$$\varepsilon_\xi = \mathbf{e}_\xi \cdot \frac{1}{2}(\mathbf{u} \circ \nabla + \nabla \circ \mathbf{u}) \cdot \mathbf{e}_\xi = \frac{1}{2}\mathbf{e}_\xi \cdot \frac{\partial \mathbf{u}}{\partial s}\frac{1}{1+\frac{\zeta}{R}} + \frac{1}{2}\frac{1}{1+\frac{\zeta}{R}}\frac{\partial \mathbf{u}}{\partial s} \cdot \mathbf{e}_\xi =$$
$$= \frac{1}{1+\frac{\zeta}{R}}\frac{\partial \mathbf{u}}{\partial s} \cdot \mathbf{e}_\xi = \frac{R}{R+\zeta}\left(\frac{\mathrm{d}u_o}{\mathrm{d}s} + \frac{w_o}{R} + \frac{\mathrm{d}\psi_{o\eta}}{\mathrm{d}s}\zeta\right)$$

is the axial strain in the coordinate direction ξ. Since $\zeta = 0$ on the η-axis

$$\varepsilon_{o\xi} = \frac{\mathrm{d}u_o}{\mathrm{d}s} + \frac{w_o}{R} \tag{10.8}$$

is the axial strain on the centerline. The curvature κ_o of the centerline is defined by the equation

$$\kappa_o = \frac{\mathrm{d}\psi_{o\eta}}{\mathrm{d}s} = -\frac{\mathrm{d}}{\mathrm{d}s}\left(\frac{\mathrm{d}w_o}{\mathrm{d}s} - \frac{u_o}{R}\right). \tag{10.9}$$

Consequently,

$$\varepsilon_\xi = \frac{1}{1+\frac{\zeta}{R}}\left[\frac{\mathrm{d}u_o}{\mathrm{d}s} + \frac{w_o}{R} - \zeta\frac{\mathrm{d}}{\mathrm{d}s}\left(\frac{\mathrm{d}w_o}{\mathrm{d}s} - \frac{u_o}{R}\right)\right] = \frac{1}{1+\frac{\zeta}{R}}\left(\varepsilon_{o\xi} + \zeta\kappa_o\right). \tag{10.10}$$

10.1.2.3 Inner Forces

In accordance with Fig. 10.2, the positive inner forces, i.e., the axial force N, the shear force V, and the bending moment M acting on the cross section, are shown in Fig. 10.2. Since $\sigma_\xi \gg \sigma_\eta, \sigma_\zeta$, the simple Hooke's law is applicable, i.e., $\sigma_\xi = E(\eta,\zeta)\varepsilon_\xi$. With (10.10)

$$N = \int_A \sigma_\xi \mathrm{d}A = \varepsilon_{o\xi}\int_A \frac{R}{R+\zeta}E(\eta,\zeta)\mathrm{d}A + \kappa_o \int_A \frac{R}{R+\zeta}E(\eta,\zeta)\zeta\mathrm{d}A \tag{10.11}$$

is the axial force. It is clear that $\mathrm{d}s = R\mathrm{d}\varphi$. With regard to this relation, it holds

$$\frac{\mathrm{d}^k}{\mathrm{d}s^k}(\ldots) = \frac{1}{R^k}\frac{\mathrm{d}^k}{\mathrm{d}\varphi^k}(\ldots) = \frac{1}{R^k}(\ldots)^{(k)} \qquad k = 1, 2, 3, \ldots \tag{10.12}$$

By substituting (10.3) for the integrals and changing over to derivatives taken with respect to the polar angle Eq. (10.11) for the axial force can be manipulated into a more suitable form. After performing some rearrangement, we have

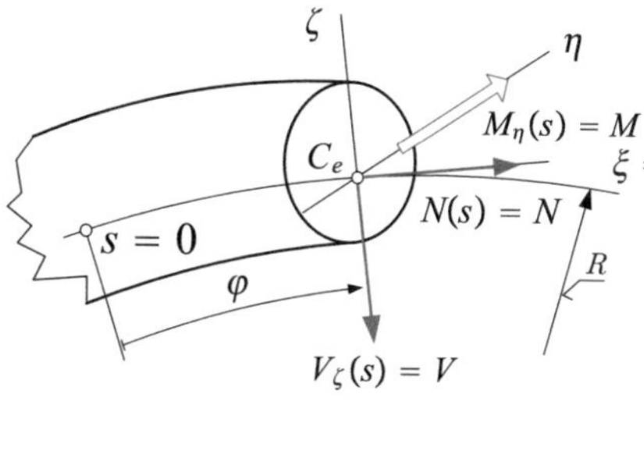

Fig. 10.2 Inner forces acting on the cross section

$$N = \frac{I_{e\eta}}{R^3}\left[\left(\frac{A_e R^2}{I_{e\eta}} - 1\right)\left(u_o^{(1)} + w_o\right) + w_o^{(2)} + w_o\right] =$$
$$= \frac{I_{e\eta}}{R^3}\left[m\left(u_o^{(1)} + w_o\right) + w_o^{(2)} + w_o\right], \tag{10.13}$$

where

$$m = \frac{A_e R^2}{I_{e\eta}} - 1\,. \tag{10.14}$$

As regards the bending moment, the following equation can be deduced:

$$M = \int_A \zeta \sigma_\xi \mathrm{d}A = \varepsilon_{o\xi} \int_A \frac{R}{R+\zeta} E(\eta,\zeta)\zeta \mathrm{d}A + \kappa_o \int_A \frac{R}{R+\zeta} E(\eta,\zeta)\zeta^2 \mathrm{d}A\,. \tag{10.15}$$

Making use of (10.3) and then (10.8), (10.9) formula (10.15) for the bending moment can be rewritten in the following form:

$$M = -\frac{I_{e\eta}}{R^2}\left(w_o^{(2)} + w_o\right). \tag{10.16}$$

10.1.2.4 Equilibrium Conditions

It follows from equilibrium equations (7.2a) and (7.2b) that the resultant of the inner forces $\mathbf{F}_C = N\mathbf{e}_\xi - V\mathbf{e}_\zeta$ and the moment resultant of the inner forces $\mathbf{M} = M\mathbf{e}_\eta$ should satisfy the following equations:

$$\left(\frac{\mathrm{d}N}{\mathrm{d}s} + f_\xi\right)\mathbf{e}_\xi + N\frac{\mathrm{d}\mathbf{e}_\xi}{\mathrm{d}s} - \left(\frac{\mathrm{d}V}{\mathrm{d}s} - f_\zeta\right)\mathbf{e}_\zeta - V\frac{\mathrm{d}\mathbf{e}_\zeta}{\mathrm{d}s} = \mathbf{0}\,, \tag{10.17a}$$

$$\frac{\mathrm{d}M}{\mathrm{d}s}\mathbf{e}_\eta + V\mathbf{e}_\eta + \left[\frac{\mathrm{d}}{\mathrm{d}s}(u_o\mathbf{e}_\xi + w_o\mathbf{e}_\zeta)\right] \times (N\mathbf{e}_\xi - V\mathbf{e}_\zeta) + \mu\mathbf{e}_\eta = \mathbf{0}, \tag{10.17b}$$

where the intensity of the distributed load acting on the centerline is denoted by f_ξ and f_ζ, while μ is the intensity of the moment load distribution. Substitution of the derivatives given by Eqs. (10.1) into (10.17a) results in two scalar equations:

$$\frac{\mathrm{d}V}{\mathrm{d}s} + \frac{N}{R} - f_\zeta = 0\,, \qquad \frac{\mathrm{d}N}{\mathrm{d}s} - \frac{V}{R} + f_\xi = 0\,. \tag{10.18}$$

Using again (10.1) for the derivatives in (10.17b) and performing then the cross product yields a scalar equation if the result is dot multiplied by $\mathbf{e}_\eta$:

$$\frac{\mathrm{d}M}{\mathrm{d}s}+V+V\underbrace{\left(\frac{\mathrm{d}u_o}{\mathrm{d}s}+\frac{w_o}{R}\right)}_{\varepsilon_{o\xi}}+N\underbrace{\left(\frac{\mathrm{d}w_o}{\mathrm{d}s}-\frac{u_o}{R}\right)}_{-\psi_{o\eta}}+\mu=0\,. \tag{10.19}$$

Within the framework of the linear theory, the quadratic terms can be dropped in (10.19). In the sequel, it will be assumed that the intensity of the moment load distribution μ is zero. Under these conditions, Eq. (10.19) becomes much simpler:

$$\frac{\mathrm{d}M}{\mathrm{d}s}+V=0\,. \tag{10.20}$$

This equation makes possible the elimination of the shear force V in Eq. (10.18). It can be checked with ease that eliminating V in (10.18) results in the following two equations:

$$\frac{\mathrm{d}N}{\mathrm{d}s}+\frac{1}{R}\frac{\mathrm{d}M}{\mathrm{d}s}+f_\xi=0\,,\qquad \frac{\mathrm{d}^2M}{\mathrm{d}s^2}-\frac{N}{R}+f_\zeta=0\,. \tag{10.21}$$

Equations for the dimensionless displacement components $U_o=u_o/R$ and $W_o=w_o/R$ can be obtained by substituting (10.13) for N and (10.16) for M in (10.21). After changing the derivatives with respect to s to those with respect to φ and performing some rearrangement, the following differential equation system (presented here in matrix form) is obtained:

$$\begin{bmatrix}0&0\\0&1\end{bmatrix}\begin{bmatrix}U_o\\W_o\end{bmatrix}^{(4)}+\begin{bmatrix}-m&0\\0&2\end{bmatrix}\begin{bmatrix}U_o\\W_o\end{bmatrix}^{(2)}+\begin{bmatrix}0&-m\\m&0\end{bmatrix}\begin{bmatrix}U_o\\W_o\end{bmatrix}^{(1)}+$$
$$+\begin{bmatrix}0&0\\0&m+1\end{bmatrix}\begin{bmatrix}U_o\\W_o\end{bmatrix}^{(0)}=\frac{R^3}{I_{e\eta}}\begin{bmatrix}f_\xi\\f_\zeta\end{bmatrix}. \tag{10.22}$$

Equation (10.22) should be supplemented with boundary conditions. If the support is pinned-pinned, the displacement and the bending moment are zero. If the supports are fixed, the displacement and the rotation are zero. Table 10.1 shows the three support arrangements for which the determination of the first four natural frequencies is one of our main objectives in the present chapter. Table 10.1 contains the boundary conditions too; they are presented in terms of the displacement components here by using Eqs. (10.7) and (10.16). Note that the central angle of the beam is $\bar{\vartheta}=2\vartheta$.

Table 10.1 Boundary conditions for curved beams

	Support arrangements	Boundary conditions	
1.	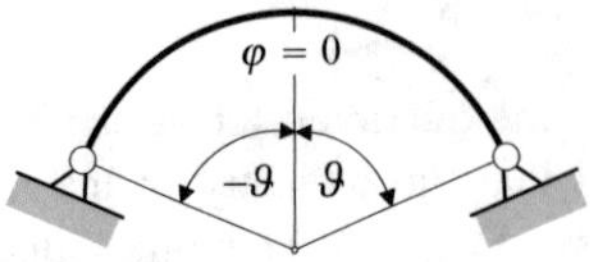	$U_o(-\vartheta) = 0$ $W_o(-\vartheta) = 0$ $W_o^{(2)}(-\vartheta) = 0$	$U_o(\vartheta) = 0$ $W_o(\vartheta) = 0$ $W_o^{(2)}(\vartheta) = 0$
2.	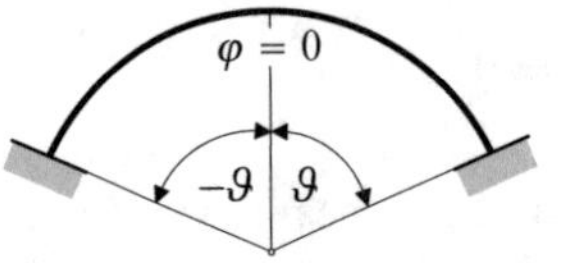	$U_o(-\vartheta) = 0$ $W_o(-\vartheta) = 0$ $W_o^{(1)}(-\vartheta) = 0$	$U_o(\vartheta) = 0$ $W_o(\vartheta) = 0$ $W_o^{(1)}(\vartheta) = 0$
3.	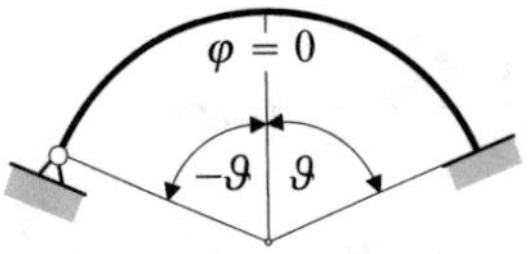	$U_o(-\vartheta) = 0$ $W_o(-\vartheta) = 0$ $W_o^{(2)}(-\vartheta) = 0$	$U_o(\vartheta) = 0$ $W_o(\vartheta) = 0$ $W_o^{(1)}(\vartheta) = 0$

Remark 10.1 Differential equation system (10.22) can be rewritten in the following matrix form:

$$\underline{\mathbf{K}}[\,\underline{\mathbf{y}}(x)] = \underline{\mathbf{r}}(x), \tag{10.23}$$

where

$$\underline{\mathbf{K}}[\,\underline{\mathbf{y}}\,] = \sum_{\nu=0}^{4} \underline{\mathbf{K}}_{\nu}(x)\underline{\mathbf{y}}^{(\nu)}(x)\,, \qquad \underline{\mathbf{K}}_3(x) = \underline{\mathbf{0}}\,. \tag{10.24}$$

On the basis of (10.22) and (10.24), differential equation system (10.23) can be detailed as

$$\underbrace{\begin{bmatrix} 0 & 0 \\ 0 & 1 \end{bmatrix}}_{\underline{\mathbf{K}}_4} \begin{bmatrix} y_1 \\ y_2 \end{bmatrix}^{(4)} + \underbrace{\begin{bmatrix} -m & 0 \\ 0 & 2 \end{bmatrix}}_{\underline{\mathbf{K}}_2} \begin{bmatrix} y_1 \\ y_2 \end{bmatrix}^{(2)} + \underbrace{\begin{bmatrix} 0 & -m \\ m & 0 \end{bmatrix}}_{\underline{\mathbf{K}}_1} \begin{bmatrix} y_1 \\ y_2 \end{bmatrix}^{(1)} + \\ + \underbrace{\begin{bmatrix} 0 & 0 \\ 0 & m+1 \end{bmatrix}}_{\underline{\mathbf{K}}_0} \begin{bmatrix} y_1 \\ y_2 \end{bmatrix}^{(0)} = \begin{bmatrix} r_1 \\ r_2 \end{bmatrix} \tag{10.25}$$

in which $x = \varphi$ and

$$\begin{bmatrix} y_1 \\ y_2 \end{bmatrix} = \begin{bmatrix} U_o \\ W_o \end{bmatrix}, \qquad \begin{bmatrix} r_1 \\ r_2 \end{bmatrix} = \frac{R^3}{I_{e\eta}} \begin{bmatrix} f_\xi \\ f_\zeta \end{bmatrix}. \tag{10.26}$$

It is worth mentioning that $\underline{\mathbf{r}}$ is a dimensionless distributed load. Since the matrix $\underline{\mathbf{K}}_4$ has no inverse differential equation system (10.25) is degenerated.

10.1.3 Equations of Motion

For the problem of free vibrations the distributed loads are forces of inertia. Thus

$$f_\xi = -\rho_a A \frac{\partial^2 u_o}{\partial t^2} = -R\rho_a A \frac{\partial^2 U_o}{\partial t^2} \qquad f_\zeta = -\rho_a A \frac{\partial^2 w_o}{\partial t^2} = -R\rho_a A \frac{\partial^2 W_o}{\partial t^2}, \tag{10.27}$$

where

$$\rho_a = \int_A \rho(\eta, \zeta)\mathrm{d}A\,, \qquad \rho(\eta, \zeta) = \rho(-\eta, \zeta) \tag{10.28}$$

is the average of the density ρ over the cross section of area A and t denotes time. Note that now the displacement components are functions of φ and t.

With (10.27) equation (10.22) assumes the following form:

$$\begin{bmatrix} 0 & 0 \\ 0 & 1 \end{bmatrix} \begin{bmatrix} U_o \\ W_o \end{bmatrix}^{(4)} + \begin{bmatrix} -m & 0 \\ 0 & 2 \end{bmatrix} \begin{bmatrix} U_o \\ W_o \end{bmatrix}^{(2)} + \begin{bmatrix} 0 & -m \\ m & 0 \end{bmatrix} \begin{bmatrix} U_o \\ W_o \end{bmatrix}^{(1)} + \\ + \begin{bmatrix} 0 & 0 \\ 0 & m+1 \end{bmatrix} \begin{bmatrix} U_o \\ W_o \end{bmatrix}^{(0)} = -R^4 \frac{\rho_a A}{I_{e\eta}} \frac{\partial^2}{\partial t^2} \begin{bmatrix} U_o \\ W_o \end{bmatrix}. \tag{10.29}$$

For harmonic vibrations, it is assumed that the solutions are

$$U_o(\varphi, t) = U(\varphi) \sin \omega t\,, \qquad W_o(\varphi, t) = W(\varphi) \sin \omega t, \tag{10.30}$$

where $U(\varphi)$ and $W(\varphi)$ are the amplitude functions and ω is the circular frequency of the vibrations. Substituting solution (10.30) into (10.29) yields the following differential equation system for $U(\varphi)$ and $W(\varphi)$:

$$\begin{bmatrix} 0 & 0 \\ 0 & 1 \end{bmatrix} \begin{bmatrix} U \\ W \end{bmatrix}^{(4)} + \begin{bmatrix} -m & 0 \\ 0 & 2 \end{bmatrix} \begin{bmatrix} U \\ W \end{bmatrix}^{(2)} + \begin{bmatrix} 0 & -m \\ m & 0 \end{bmatrix} \begin{bmatrix} U \\ W \end{bmatrix}^{(1)} + \\ + \begin{bmatrix} 0 & 0 \\ 0 & m+1 \end{bmatrix} \begin{bmatrix} U \\ W \end{bmatrix}^{(0)} = \lambda \begin{bmatrix} 1 & 0 \\ 0 & 1 \end{bmatrix} \begin{bmatrix} U \\ W \end{bmatrix}, \tag{10.31a}$$

$$\lambda = R^4 \frac{\rho_a A \omega^2}{I_{e\eta}}, \tag{10.31b}$$

where λ is the dimensionless eigenvalue. It is clear from Table 10.1 that the amplitude functions should also satisfy the following boundary conditions:

Pinned-pinned beam:

$$U(-\vartheta) = W(-\vartheta) = W^{(2)}(-\vartheta) = 0, \qquad U(\vartheta) = W(\vartheta) = W^{(2)}(\vartheta) = 0. \tag{10.32}$$

Fixed-fixed beam:

$$U(-\vartheta) = W(-\vartheta) = W^{(1)}(-\vartheta) = 0, \qquad U(\vartheta) = W(\vartheta) = W^{(1)}(\vartheta) = 0. \quad (10.33)$$

Pinned-fixed beam:

$$U(-\vartheta) = W(-\vartheta) = W^{(2)}(-\vartheta) = 0, \qquad U(\vartheta) = W(\vartheta) = W^{(1)}(\vartheta) = 0. \quad (10.34)$$

The homogeneous differential equation (10.31) and the also homogeneous boundary conditions (10.32), (10.33), and (10.34) determine three eigenvalue problems with λ as the eigenvalue. All that has been said about eigenvalue problems governed by homogeneous differential equation systems and boundary conditions in Sects. 9.2.1, 9.2.2, 9.2.3, and 9.2.4 remain valid for the three eigenvalue problems determined by equations (10.31) and (10.32), (10.33), and (10.34). There is, however, a difference since the governing differential equation (10.32) is degenerated—the coefficient matrix $\underline{\mathbf{P}}_4$ in (10.25) has no inverse in contrast to $\underline{\mathbf{P}}_\kappa$ in (9.67b) for which the existence of the inverse is a requirement. For this reason, the definition of the Green function matrix should be rethought.

Remark 10.2 Using the notations introduced by Eqs. (10.23), (10.24), and (10.25), the eigenvalue problems defined by differential equation (10.32) and boundary conditions (10.32), (10.33), and (10.34) can be rewritten as follows:
Differential equation:

$$\underline{\mathbf{K}}[\underline{\mathbf{y}}(x)] = \sum_{\nu=0}^{4} \underline{\mathbf{K}}_\nu(x)\underline{\mathbf{y}}^{(\nu)}(x) = \lambda \underline{\mathbf{M}}_0(x)\underline{\mathbf{y}}(x),$$

$$\underline{\mathbf{M}}_0(x) = \begin{bmatrix} 1 & 0 \\ 0 & 1 \end{bmatrix}, \quad \underline{\mathbf{y}} = \begin{bmatrix} y_1 \\ y_2 \end{bmatrix} = \begin{bmatrix} U \\ W \end{bmatrix}, \quad (10.35)$$

where $x = \varphi$ is the dimensionless independent variable. Differential equation (10.35) is associated with the following boundary conditions:

Pinned-pinned beam:

$$y_1(x)|_{x=\pm\vartheta} = y_2(x)|_{x=\pm\vartheta} = y_2^{(2)}(x)\Big|_{x=\pm\vartheta} = 0. \quad (10.36)$$

Fixed-fixed beam:

$$y_1(x)|_{x=\pm\vartheta} = y_2(x)|_{x=\pm\vartheta} = y_2^{(1)}(x)\Big|_{x=\pm\vartheta} = 0. \quad (10.37)$$

Pinned-fixed beam:

$$y_1(x)|_{x=\pm\vartheta} = y_2(x)|_{x=\pm\vartheta} = y_2^{(2)}(x)\Big|_{x=-\vartheta} = y_2^{(1)}(x)\Big|_{x=\vartheta} = 0. \quad (10.38)$$

10.2 Green Function Matrix for Degenerated Differential Equation Systems

10.2.1 Definition

Consider the degenerated system of ordinary differential equations

$$\underline{\mathbf{L}}[\underline{\mathbf{y}}(x)] = \sum_{\nu=0}^{\kappa} \overset{\nu}{\underline{\mathbf{P}}}(x)\underline{\mathbf{y}}^{(\nu)}(x) = \begin{bmatrix} 0 & 0 \\ 0 & \overset{\kappa}{\underline{\mathbf{P}}}_{22} \end{bmatrix} \begin{bmatrix} \underline{\mathbf{y}}_1 \\ \underline{\mathbf{y}}_2 \end{bmatrix}^{(\kappa)} + \cdots +$$

$$+ \begin{bmatrix} 0 & 0 \\ 0 & \overset{k+1}{\underline{\mathbf{P}}}_{22} \end{bmatrix} \begin{bmatrix} \underline{\mathbf{y}}_1 \\ \underline{\mathbf{y}}_2 \end{bmatrix}^{(k+1)} + \begin{bmatrix} \overset{k}{\underline{\mathbf{P}}}_{11} & \overset{k}{\underline{\mathbf{P}}}_{12} \\ 0 & \overset{k}{\underline{\mathbf{P}}}_{22} \end{bmatrix} \begin{bmatrix} \underline{\mathbf{y}}_1 \\ \underline{\mathbf{y}}_2 \end{bmatrix}^{(k)} + \cdots +$$

$$+ \begin{bmatrix} \overset{s}{\underline{\mathbf{P}}}_{11} & \overset{s}{\underline{\mathbf{P}}}_{12} \\ \overset{s}{\underline{\mathbf{P}}}_{21} & \overset{s}{\underline{\mathbf{P}}}_{22} \end{bmatrix} \begin{bmatrix} \underline{\mathbf{y}}_1 \\ \underline{\mathbf{y}}_2 \end{bmatrix}^{(s)} + \cdots = \begin{bmatrix} \underline{\mathbf{r}}_1 \\ \underline{\mathbf{r}}_2 \end{bmatrix}, \tag{10.39}$$

where $\kappa > k > s > 0$ and n is the number of unknown functions (or the size of $\underline{\mathbf{y}}$), j is the size of $\underline{\mathbf{y}}_2$, and the matrices $\overset{\nu}{\underline{\mathbf{P}}}$ and $\underline{\mathbf{r}}$ are continuous for $x \in [a, b]$; $a < b$.

It will be assumed that

- $\overset{\kappa}{\underline{\mathbf{P}}}_{22}$ and $\overset{k}{\underline{\mathbf{P}}}_{11}$ are invertible if $x \in [a, b]$
- the system of differential equations (10.39) is associated with linear homogeneous boundary conditions of the form

$$\underline{\mathbf{U}}_{\mu}[\underline{\mathbf{y}}] = \sum_{\nu=1}^{\kappa} \left[\underline{\mathcal{A}}_{\nu\mu} \underline{\mathbf{y}}^{(\nu-1)}(a) + \underline{\mathcal{B}}_{\nu\mu} \underline{\mathbf{y}}^{(\nu-1)}(b) \right] =$$

$$= \sum_{\nu=1}^{\kappa} \left\{ \begin{bmatrix} \overset{11}{\underline{\mathcal{A}}}_{\nu\mu} & \overset{12}{\underline{\mathcal{A}}}_{\nu\mu} \\ \overset{21}{\underline{\mathcal{A}}}_{\nu\mu} & \overset{22}{\underline{\mathcal{A}}}_{\nu\mu} \end{bmatrix} \begin{bmatrix} \underline{\mathbf{y}}_1(a) \\ \underline{\mathbf{y}}_2(a) \end{bmatrix}^{(\nu-1)} + \begin{bmatrix} \overset{11}{\underline{\mathcal{B}}}_{\nu\mu} & \overset{12}{\underline{\mathcal{B}}}_{\nu\mu} \\ \overset{21}{\underline{\mathcal{B}}}_{\nu\mu} & \overset{22}{\underline{\mathcal{B}}}_{\nu\mu} \end{bmatrix} \begin{bmatrix} \underline{\mathbf{y}}_1(b) \\ \underline{\mathbf{y}}_2(b) \end{bmatrix}^{(\nu-1)} \right\} =$$

$$= \begin{bmatrix} \underset{((n-j)\times 1)}{\underline{\mathbf{0}}} \\ \underset{(j\times 1)}{\underline{\mathbf{0}}} \end{bmatrix}, \tag{10.40a}$$

where $\mu = 1, \ldots, \kappa$ and for $\nu > k$ the constant matrices $\underline{\mathcal{A}}_{\nu\mu}$ and $\underline{\mathcal{B}}_{\nu\mu}$ fulfill the conditions

$$\overset{11}{\underline{\mathcal{A}}}_{\nu\mu} = \overset{21}{\underline{\mathcal{A}}}_{\nu\mu} = \overset{11}{\underline{\mathcal{B}}}_{\nu\mu} = \overset{21}{\underline{\mathcal{B}}}_{\nu\mu} = \underline{\mathbf{0}}. \tag{10.40b}$$

- $\bar{l} = \kappa n - [(n-j)k + \kappa j]$ rows are identically zero in the hypermatrix

$$\underline{\mathbf{P}}_f = \begin{bmatrix} \underline{\mathcal{A}}_{11} & \ldots & \underline{\mathcal{A}}_{\kappa 1} & \underline{\mathcal{B}}_{11} & \ldots & \underline{\mathcal{B}}_{\kappa 1} \\ \cdots & \cdots & \cdots & \cdots & \cdots & \cdots \\ \underline{\mathcal{A}}_{1\kappa} & \ldots & \underline{\mathcal{A}}_{\kappa\kappa} & \underline{\mathcal{B}}_{1\kappa} & \ldots & \underline{\mathcal{B}}_{\kappa\kappa} \end{bmatrix}. \tag{10.40c}$$

By introducing appropriate new variables, differential equation system (10.39) can be replaced by $(n-j)k+\kappa j$ differential equations of order one for which it is possible to construct the corresponding Green function. In the present case, however, there is no need for this transformation since the definition presented here is based on the original equation [15].

Condition (10.40b) expresses that $\underline{\mathbf{y}}_1^{(\nu)}$ cannot appear in the boundary conditions if $\nu \geq k$. Condition (10.40c) is a restriction on the number of boundary conditions.

Solution to the degenerated boundary value problems (10.39), (10.40a) is sought in the same way as for the non-degenerated boundary value problem defined by Eqs. (9.67), (9.68), i.e., in the form of an integral:

$$\underline{\mathbf{y}}(x)=\int_a^b \underline{\mathbf{G}}(x,\xi)\underline{\mathbf{r}}(\xi)\,\mathrm{d}\xi =$$

$$=\int_a^b \begin{bmatrix} \underset{((l-j)\times(l-j))}{\underline{\mathbf{G}}_{11}(x,\xi)} & \underset{((l-j)\times j)}{\underline{\mathbf{G}}_{12}(x,\xi)} \\ \underset{(j\times(l-j))}{\underline{\mathbf{G}}_{21}(x,\xi)} & \underset{(j\times j)}{\underline{\mathbf{G}}_{22}(x,\xi)} \end{bmatrix} \begin{bmatrix} \underset{((l-j)\times 1)}{\underline{\mathbf{r}}_1(\xi)} \\ \underset{(j\times 1)}{\underline{\mathbf{r}}_2(\xi)} \end{bmatrix} \mathrm{d}\xi, \tag{10.41}$$

where $\underline{\mathbf{G}}(x,\xi)$ is again a Green function matrix defined by the following four properties:

1. The Green matrix function is a continuous function of x and ξ in each of the triangles $a \leq x \leq \xi \leq b$ and $a \leq \xi \leq x \leq b$. The functions

$$\left(\underline{\mathbf{G}}_{11}(x,\xi),\underline{\mathbf{G}}_{12}(x,\xi)\right)\ \left[\underline{\mathbf{G}}_{21}(x,\xi),\underline{\mathbf{G}}_{22}(x,\xi)\right]$$

are (k times) [κ times] differentiable with respect to x and the derivatives

$$\frac{\partial^\nu \underline{\mathbf{G}}(x,\xi)}{\partial x^\nu}=\underline{\mathbf{G}}^{(\nu)}(x,\xi), \qquad (\nu=1,2,\ldots,k),$$

$$\frac{\partial^\nu \underline{\mathbf{G}}_{2i}(x,\xi)}{\partial x^\nu}=\underline{\mathbf{G}}_{2i}^{(\nu)}(x,\xi), \qquad (\nu=1,2,\ldots,\kappa;\ i=1,2)$$

are continuous functions of x and ξ.

2. Let ξ be fixed in $[a,b]$. Though the derivatives

$$\underline{\mathbf{G}}_{11}^{(\nu)}(x,\xi)\ \ (\nu=1,2,\ldots,k-2),\quad \underline{\mathbf{G}}_{12}^{(\nu)}(x,\xi)\ \ (\nu=1,2,\ldots,k-1),$$

$$\underline{\mathbf{G}}_{21}^{(\nu)}(x,\xi)\ \ (\nu=1,2,\ldots,\kappa-1),\quad \underline{\mathbf{G}}_{22}^{(\nu)}(x,\xi)\ \ (\nu=1,2,\ldots,\kappa-2)$$

are continuous for $x=\xi$, the higher derivatives, i.e.,

$$\underline{\mathbf{G}}_{11}^{(k-1)}(x,\xi)\ \text{ and }\ \underline{\mathbf{G}}_{22}^{(\kappa-1)}(x,\xi)$$

have a jump on the diagonal given by the following two equations:

$$\lim_{\varepsilon\to 0}\left[\underline{\mathbf{G}}_{11}^{(k-1)}(\xi+\varepsilon,\xi)-\underline{\mathbf{G}}_{11}^{(k-1)}(\xi-\varepsilon,\xi)\right]=\overset{k}{\underline{\mathbf{P}}}{}_{11}^{-1}(\xi),$$

$$\lim_{\varepsilon\to 0}\left[\underline{\mathbf{G}}_{22}^{(\kappa-1)}(\xi+\varepsilon,\xi)-\underline{\mathbf{G}}_{22}^{(\kappa-1)}(\xi-\varepsilon,\xi)\right]=\overset{\kappa}{\underline{\mathbf{P}}}{}_{22}^{-1}(\xi).$$

3. Let $\underline{\boldsymbol{\alpha}}$ be an arbitrary otherwise constant vector. For a fixed $\xi\in[a,b]$ the vector $\underline{\mathbf{G}}(x,\xi)\underline{\boldsymbol{\alpha}}$ as a function of x $(x\neq\xi)$ should satisfy the homogeneous differential equation
$$\underline{\mathbf{L}}\left[\underline{\mathbf{G}}(x,\xi)\underline{\boldsymbol{\alpha}}\right]=\underline{\mathbf{0}}\,.$$

4. The vector $\underline{\mathbf{G}}(x,\xi)\underline{\boldsymbol{\alpha}}$ as a function of x should satisfy the boundary conditions
$$\underline{\mathbf{U}}_{\mu}\left[\underline{\mathbf{G}}(x,\xi)\underline{\boldsymbol{\alpha}}\right]=\underline{\mathbf{0}},\quad(\mu=1,\ldots,n)\,.$$

Remark 10.3 It is worth comparing the definition presented in Sect. 9.2.6 to the above definition. The difference appears in Properties 1 and 2; Properties 3 and 4 are basically the same as earlier.

Remark 10.4 If there exists the Green matrix function defined above for the boundary value problem (10.39), (10.40a) then the vector (10.41) satisfies differential equation (10.39) and the boundary conditions (10.40a).

The part of the statement concerning the boundary conditions follows immediately from the comparison of the fourth property of the definition. As regards the second part of our statement, substitute the representation (10.41) into (10.39) by utilizing that the matrices $\underline{\mathbf{G}}_{11}^{(k-1)}$ and $\underline{\mathbf{G}}_{22}^{(\kappa-1)}$ are discontinuous if $x=\xi$. In this way, we have

$$\begin{aligned}
\underline{\mathbf{L}}(\underline{\mathbf{y}})&=\sum_{\nu=0}^{\kappa}\overset{\nu}{\underline{\mathbf{P}}}(x)\underline{\mathbf{y}}^{(\nu)}(x)=\\
&=\begin{bmatrix}0&0\\0&\overset{\kappa}{\underline{\mathbf{P}}}_{22}(x)\end{bmatrix}\int_a^b\begin{bmatrix}0&0\\\underline{\mathbf{G}}_{21}^{(\kappa)}(x,\xi)&\underline{\mathbf{G}}_{22}^{(\kappa)}(x,\xi)\end{bmatrix}\begin{bmatrix}\underline{\mathbf{r}}_1(\xi)\\\underline{\mathbf{r}}_2(\xi)\end{bmatrix}\mathrm{d}\xi+\\
&+\begin{bmatrix}0&0\\0&\overset{\kappa}{\underline{\mathbf{P}}}_{22}(x)\left\{\underline{\mathbf{G}}_{22}^{(\kappa-1)}(x,x-0)-\underline{\mathbf{G}}_{22}^{(\kappa-1)}(x,x+0)\right\}\end{bmatrix}\begin{bmatrix}\underline{\mathbf{r}}_1(x)\\\underline{\mathbf{r}}_2(x)\end{bmatrix}+\\
&+\begin{bmatrix}0&0\\0&\overset{k+1}{\underline{\mathbf{P}}}_{22}(x)\end{bmatrix}\int_a^b\begin{bmatrix}0&0\\\underline{\mathbf{G}}_{21}^{(k+1)}(x,\xi)&\underline{\mathbf{G}}_{22}^{(k+1)}(x,\xi)\end{bmatrix}\begin{bmatrix}\underline{\mathbf{r}}_1(\xi)\\\underline{\mathbf{r}}_2(\xi)\end{bmatrix}\mathrm{d}\xi+\\
&+\begin{bmatrix}\overset{k}{\underline{\mathbf{P}}}_{11}(x)&\overset{k}{\underline{\mathbf{P}}}_{12}(x)\\\overset{k}{\underline{\mathbf{P}}}_{22}(x)&\overset{k}{\underline{\mathbf{P}}}_{22}(x)\end{bmatrix}\int_a^b\begin{bmatrix}\underline{\mathbf{G}}_{11}^{(k)}(x,\xi)&\underline{\mathbf{G}}_{12}^{(k)}(x,\xi)\\\underline{\mathbf{G}}_{21}^{(k)}(x,\xi)&\underline{\mathbf{G}}_{22}^{(k)}(x,\xi)\end{bmatrix}\begin{bmatrix}\underline{\mathbf{r}}_1(\xi)\\\underline{\mathbf{r}}_2(\xi)\end{bmatrix}\mathrm{d}\xi+\\
&+\begin{bmatrix}\overset{n}{\underline{\mathbf{P}}}_{11}(x)\left\{\underline{\mathbf{G}}_{11}^{(k-1)}(x,x-0)-\underline{\mathbf{G}}_{11}^{(k-1)}(x,x+0)\right\}&0\\0&0\end{bmatrix}\begin{bmatrix}\underline{\mathbf{r}}_1(x)\\\underline{\mathbf{r}}_2(x)\end{bmatrix}+\\
&+\ldots
\end{aligned}\tag{10.42}$$

Taking into account Property 1 of the definition and the fact that due to Property 2 the sum of the integrals on the right side of (10.42) vanishes, substitute the value of the jump. This transformation results in the right side of (10.39), i.e., the representation (10.41) of the solution really satisfies the differential equation (10.39).

Let $\underline{\mathbf{r}}(\xi) = \underline{\mathbf{e}}\delta(\xi - \xi_f)$ where $\underline{\mathbf{e}}$ is a constant vector attached mentally to the fixed point $\xi_f \in [a, b]$ and $\delta(\xi - \xi_f)$ is the Dirac delta function. Assume that each element of $\underline{\mathbf{e}}$ is equal to 1. It follows from (10.41) that

$$\underline{\mathbf{y}}(x) = \int_a^b \underline{\mathbf{G}}(x, \xi)\delta(\xi - \xi_f)\,\underline{\mathbf{e}}\,\mathrm{d}\xi = \underline{\mathbf{G}}(x, \xi_f)\,\underline{\mathbf{e}}\,. \tag{10.43}$$

This means that the i-th column in the matrix $\underline{\mathbf{G}}(x, \xi_f)$ is the solution that belongs to $\delta(\xi - \xi_f)\underline{\mathbf{e}}_i$, where $\underline{\mathbf{e}}_i$ is the column matrix in which the i-th element is 1 while the other elements are all equal to zero.

10.2.2 Existence and Calculation of the Green Matrix Function

In accordance with (9.74), the general solution of differential equation $\underline{\mathbf{L}}(\underline{\mathbf{y}}) = \underline{\mathbf{0}}$ can be given again in the form

$$\underline{\mathbf{y}} = \left[\sum_{\ell=1}^{\kappa} \underset{(n\times n)}{\underline{\mathbf{Y}}_\ell(x)}\ \underset{(n\times n)}{\underline{\mathbf{C}}_\ell}\right] \underset{(n\times 1)}{\underline{\mathbf{e}}}\,, \tag{10.44}$$

where $\underline{\mathbf{C}}_i$ is a constant non-singular matrix and $\underline{\mathbf{e}}$ is a constant vector.

For the sake of distinguishing the columns in the matrices $\underline{\mathbf{Y}}_i$, they are denoted by $\underline{\boldsymbol{\eta}}_{i\nu}$ $(\nu = 1, \ldots, l)$. In contrast to (9.74) $\bar{l} = \kappa n - [(n - j)k + \kappa j]$ columns are, however, identically equal to zero in the matrices $\underline{\mathbf{Y}}_i$. Consider now the hypermatrix

$$\underline{\mathbf{D}} = \begin{bmatrix} \underline{\mathrm{U}}_1[\underline{\boldsymbol{\eta}}_{11}] & \cdots & \underline{\mathrm{U}}_1[\underline{\boldsymbol{\eta}}_{1n}] & \cdots & \underline{\mathrm{U}}_1[\underline{\boldsymbol{\eta}}_{i\nu}] & \cdots & \underline{\mathrm{U}}_1[\underline{\boldsymbol{\eta}}_{\kappa n}] \\ \cdots\cdots & \cdots\cdots & \cdots\cdots & \cdots\cdots & \cdots\cdots & \cdots\cdots & \cdots\cdots \\ \underline{\mathrm{U}}_\kappa[\underline{\boldsymbol{\eta}}_{11}] & \cdots & \underline{\mathrm{U}}_\kappa[\underline{\boldsymbol{\eta}}_{1n}] & \cdots & \underline{\mathrm{U}}_\kappa[\underline{\boldsymbol{\eta}}_{i\nu}] & \cdots & \underline{\mathrm{U}}_\kappa[\underline{\boldsymbol{\eta}}_{\kappa n}] \end{bmatrix}. \tag{10.45}$$

Taking into account what has been said about the boundary conditions—here we think of the structure of the matrix $\underline{\mathbf{P}}_f$ –and recalling that $\bar{l}$ columns are identically zero—one can come to the conclusion that $\bar{l}$ rows and columns are identically zero in the hypermatrix $\underline{\mathbf{D}}$. Let $\check{\underline{\mathbf{D}}}$ be the matrix obtained by removing the zero rows and columns. Then we can establish the following statement:

If $\det(\check{\underline{\mathbf{D}}}) \neq 0$ then there exists a uniquely determined Green matrix function which meets Properties 1. to 4. of the definition. In addition, the solution given by (10.41) is the only solution of the boundary value problem (10.39), (10.40a) for an arbitrary right side $\underline{\mathbf{r}}$.

The proof of our statement is similar to that given in [16].

To prove the statement, we have to detail the calculation of the Green matrix function. The line of thought presented in the sequel is basically the same as the one presented in Sect. 9.2.7. The notations we are going to use are more or less also the same as those in the subsection mentioned.

With regard to the third property of the definition, the Green function matrix $\underline{\mathbf{G}}(x,\xi)$ is sought in the following form:

$$\underline{\mathbf{G}}(x,\xi)=\sum_{\ell=1}^{n}\underline{\mathbf{Y}}_\ell(x)\left[\underline{\mathbf{A}}_\ell(\xi)\pm\underline{\mathbf{B}}_\ell(\xi)\right], \tag{10.46}$$

where $\underline{\mathbf{A}}_\ell(\xi)$ and $\underline{\mathbf{B}}_\ell(\xi)$ are $n\times n$ matrices and the sign is [positive](negative) if $[x\leq\xi](\xi\leq x)$. As regards the structure of (10.46), it coincides formally with Eq. (9.75).

After partitioning the matrices $\underline{\mathbf{Y}}_\ell$ and $\underline{\mathbf{B}}_\ell$

$$\underbrace{\begin{matrix} n-j\,\{\\ j\,\{\end{matrix}\begin{bmatrix}\underline{\mathbf{Y}}_{\ell 1}\\ \underline{\mathbf{Y}}_{\ell 2}\end{bmatrix}}_{n\times n}\qquad \underbrace{\Big[\overbrace{\underline{\mathbf{B}}_{\ell 1}}^{n-j}\ \overbrace{\underline{\mathbf{B}}_{\ell 2}}^{j}\Big]}_{n\times n},$$

the following equation system can be obtained from Property 2 of the definition:

$$\begin{bmatrix}\sum_\ell\underline{\mathbf{Y}}_{\ell 1}\underline{\mathbf{B}}_{\ell 1} & \sum_\ell\underline{\mathbf{Y}}_{\ell 1}\underline{\mathbf{B}}_{\ell 2}\\ \sum_\ell\underline{\mathbf{Y}}_{\ell 2}\underline{\mathbf{B}}_{\ell 1} & \sum_\ell\underline{\mathbf{Y}}_{\ell 2}\underline{\mathbf{B}}_{\ell 2}\end{bmatrix}=\begin{bmatrix}\underline{\mathbf{0}} & \underline{\mathbf{0}}\\ \underline{\mathbf{0}} & \underline{\mathbf{0}}\end{bmatrix}, \tag{10.47a}$$

$$\begin{bmatrix}\sum_\ell\underline{\mathbf{Y}}^{(1)}_{\ell 1}\underline{\mathbf{B}}_{\ell 1} & \sum_\ell\underline{\mathbf{Y}}^{(1)}_{\ell 1}\underline{\mathbf{B}}_{\ell 2}\\ \sum_\ell\underline{\mathbf{Y}}^{(1)}_{\ell 2}\underline{\mathbf{B}}_{\ell 1} & \sum_\ell\underline{\mathbf{Y}}^{(1)}_{\ell 2}\underline{\mathbf{B}}_{\ell 2}\end{bmatrix}=\begin{bmatrix}\underline{\mathbf{0}} & \underline{\mathbf{0}}\\ \underline{\mathbf{0}} & \underline{\mathbf{0}}\end{bmatrix}, \tag{10.47b}$$

$$\cdots=\cdots$$

$$\begin{bmatrix}\sum_\ell\underline{\mathbf{Y}}^{(k-1)}_{\ell 1}\underline{\mathbf{B}}_{\ell 1} & \sum_\ell\underline{\mathbf{Y}}^{(k-1)}_{\ell 1}\underline{\mathbf{B}}_{\ell 2}\\ \sum_\ell\underline{\mathbf{Y}}^{(k-1)}_{\ell 2}\underline{\mathbf{B}}_{\ell 1} & \sum_\ell\underline{\mathbf{Y}}^{(k-1)}_{\ell 2}\underline{\mathbf{B}}_{\ell 2}\end{bmatrix}=\begin{bmatrix}-\frac{1}{2}\overset{k}{\underline{\mathbf{P}}}{}^{-1}_{11} & \underline{\mathbf{0}}\\ \underline{\mathbf{0}} & \underline{\mathbf{0}}\end{bmatrix}, \tag{10.47c}$$

$$\left[\sum_\ell\underline{\mathbf{Y}}^{(k)}_{\ell 2}\underline{\mathbf{B}}_{\ell 1}\ \sum_\ell\underline{\mathbf{Y}}^{(k)}_{\ell 2}\underline{\mathbf{B}}_{\ell 2}\right]=\left[\underline{\mathbf{0}}\ \underline{\mathbf{0}}\right], \tag{10.47d}$$

$$\cdots=\cdots$$

$$\left[\sum_\ell\underline{\mathbf{Y}}^{(\kappa-1)}_{\ell 2}\underline{\mathbf{B}}_{\ell 1}\ \sum_\ell\underline{\mathbf{Y}}^{(\kappa-1)}_{\ell 2}\underline{\mathbf{B}}_{\ell 2}\right]=\left[\underline{\mathbf{0}}\ -\frac{1}{2}\overset{n}{\underline{\mathbf{P}}}{}^{-1}_{22}\right], \tag{10.47e}$$

where $(\ell=1,2,\ldots,\kappa)$. Note that the sizes of the zero sub-matrices on the right sides of the above equations are the same as those of the corresponding sub-matrices on the left sides.

Let the ν-th column vector of $\underline{\mathbf{B}}_\ell$ be denoted by $\underline{\mathbf{B}}_{\ell\nu}$.

For a given ν, the matrix

$$\underset{(1\times(\kappa\times n))}{\underline{\boldsymbol{\mathcal{B}}}^T_\nu}=\left[\underset{(1\times n)}{\underline{\mathbf{B}}^T_{1\nu}}|\ldots|\underset{(1\times n)}{\underline{\mathbf{B}}^T_{i\nu}}|\ldots|\underset{(1\times n)}{\underline{\mathbf{B}}^T_{\kappa\nu}}\right],\quad i=2,3,\ldots,\kappa-1 \tag{10.48}$$

is that of the unknowns.

The system of equations (10.47) has the same coefficient matrix for each unknown vector $\underline{\mathcal{B}}_\nu$:

$$\underline{\mathbf{W}}\,\underline{\mathcal{B}}_\nu = \underline{\mathcal{P}}_\nu, \tag{10.49}$$

where

$$\underline{\mathbf{W}} = \begin{bmatrix} \underline{\eta}_{11\,1} & \cdots & \underline{\eta}_{1n\,1} & \cdots & \underline{\eta}_{i\nu\,1} & \cdots & \underline{\eta}_{\kappa 1\,1} & \cdots & \underline{\eta}_{\kappa n\,1} \\ \underline{\eta}_{11\,2} & \cdots & \underline{\eta}_{1n\,2} & \cdots & \underline{\eta}_{i\nu\,2} & \cdots & \underline{\eta}_{\kappa 1\,2} & \cdots & \underline{\eta}_{\kappa n\,2} \\ \cdots & \cdots & \cdots & \cdots & \cdots & \cdots & \cdots & \cdots & \cdots \\ \underline{\eta}^{(k-1)}_{11\,1} & \cdots & \underline{\eta}^{(k-1)}_{1n\,1} & \cdots & \underline{\eta}^{(k-1)}_{i\nu\,1} & \cdots & \underline{\eta}^{(k-1)}_{\kappa 1\,1} & \cdots & \underline{\eta}^{(k-1)}_{\kappa n\,1} \\ \underline{\eta}^{(k-1)}_{11\,2} & \cdots & \underline{\eta}^{(k-1)}_{1l\,2} & \cdots & \underline{\eta}^{(k-1)}_{i\nu\,2} & \cdots & \underline{\eta}^{(k-1)}_{\kappa 1\,2} & \cdots & \underline{\eta}^{(k-1)}_{\kappa n\,2} \\ \underline{\eta}^{(k)}_{11\,1} & \cdots & \underline{\eta}^{(k)}_{1n\,1} & \cdots & \underline{\eta}^{(k)}_{i\nu\,1} & \cdots & \underline{\eta}^{(k)}_{\kappa 1\,1} & \cdots & \underline{\eta}^{(k)}_{\kappa n\,1} \\ \cdots & \cdots & \cdots & \cdots & \cdots & \cdots & \cdots & \cdots & \cdots \\ \underline{\eta}^{(\kappa-1)}_{11\,2} & \cdots & \underline{\eta}^{(\kappa-1)}_{1n\,2} & \cdots & \underline{\eta}^{(\kappa-1)}_{i\nu\,2} & \cdots & \underline{\eta}^{(\kappa-1)}_{\kappa 1\,2} & \cdots & \underline{\eta}^{(\kappa-1)}_{\kappa n\,2} \end{bmatrix}$$

in which

$$\underline{\eta}_{i\nu} = \begin{bmatrix} \underline{\eta}_{i\nu\,1} \\ \underline{\eta}_{i\nu\,2} \end{bmatrix} \begin{matrix} \}\, n-j \\ \}\, j \end{matrix}$$

and $\underline{\mathcal{P}}_\nu$ is the ν-th column in the transpose of the matrix formulated from the right sides of Eqs. (10.47):

columns in the blocks

$$\begin{matrix} n & & n & n-j & j & j & & j & j & \\ \left[\begin{array}{c|c|c|c|c|c|c|c|c} 0 & \cdots & 0 & -\frac{1}{2}(\overset{k}{\underline{\mathbf{P}}_{11}^{-1}})^T & 0 & 0 & \cdots & 0 & 0 \\ \hline 0 & \cdots & 0 & 0 & 0 & 0 & \cdots & 0 & -\frac{1}{2}(\overset{n}{\underline{\mathbf{P}}_{22}^{-1}})^T \end{array}\right] & & & & & & & & & \begin{matrix} n\text{-}j \\ j \end{matrix} \\ 1 & & k-1 & k & & k+1 & & \kappa-1 & \kappa & \end{matrix}$$

block number

Note the size of the coefficient matrix $\underline{\mathbf{W}}$ is

$$((kn + (\kappa - k)j) \times \kappa n).$$

It also holds that $\bar{l} = \kappa n - [(n-j)k + \kappa j]$ columns in the matrix $\underline{\mathbf{W}}$ are identically equal to zero. If we remove the zero columns, the column number will be equal to the row number, i.e., a square matrix is obtained which will be denoted by $\check{\underline{\mathbf{W}}}$. Consequently, the elements in the unknown matrices $\underline{\mathcal{B}}_\nu$ with the same index are also set to zero. The matrix obtained is denoted by $\check{\underline{\mathcal{B}}}_\nu$, and it is the solution of the equation system

$$\underline{\check{\mathbf{W}}}\,\underline{\check{\boldsymbol{\mathcal{B}}}}_\nu = \underline{\boldsymbol{\mathcal{P}}}_\nu. \tag{10.50}$$

Introducing new variables, the system of ordinary differential equation $\underline{\mathbf{L}}[\underline{\mathbf{y}}]$ can be replaced by a system of ordinary differential equations of order one. The new variables can always be chosen in such a way that $\det(\underline{\check{\mathbf{W}}})$ coincides with the Wronsky determinant of the system of differential equations of order one. Since the Wronsky determinant differs from zero and the vector $\underline{\boldsymbol{\mathcal{P}}}_\nu$ has at least one non-zero element, the linear equation system (10.50) is soluble, i.e., there exists a solution different from the trivial one.

If the matrices $\underline{\mathbf{B}}_\ell$ have been determined, the functions $\underline{\mathbf{A}}_\ell$ can be calculated from Property 4 of the definition. The calculation steps are similar to those presented at the end of Sect. 9.2.7.

Let $\boldsymbol{\alpha}$ be the ν-th unit vector in the space of size $n \times n$. Then, it follows from the fourth property that

$$\underline{\mathbf{U}}_\mu\left[\sum_{\ell=1}^{n}\underline{\mathbf{Y}}_\ell(x)\underline{\mathbf{A}}_\ell(\xi)\boldsymbol{\alpha}\right] = \mp\underline{\mathbf{U}}_\mu\left[\sum_{\ell=1}^{n}\underline{\mathbf{Y}}_\ell(x)\underline{\mathbf{B}}_\ell(\xi)\boldsymbol{\alpha}\right], \quad \mu = 1, 2, \ldots, \kappa\,. \tag{10.51}$$

Let $\underline{\mathbf{A}}_{i\nu}$ be the ν-th column vector in $\underline{\mathbf{A}}_i$. The matrix defined by the equation

$$\underline{\boldsymbol{\mathcal{A}}}_\nu^T = \left[\underline{\mathbf{A}}_{1\nu}^T|\ldots|\underline{\mathbf{A}}_{i\nu}^T|\ldots|\underline{\mathbf{A}}_{n\nu}^T\right]$$

is, in fact, that of the unknowns which belong to the index ν. Due to the choice of $\boldsymbol{\alpha}$, Eq. (10.51) can be rewritten in the following form:

$$\underline{\mathbf{U}}_\mu\left[\underline{\mathbf{Y}}_1(x)|\ldots|\underline{\mathbf{Y}}_n(x)\right]\underline{\boldsymbol{\mathcal{A}}}_\nu = \mp\underline{\mathbf{U}}_\mu\left[\underline{\mathbf{Y}}_1(x)|\ldots|\underline{\mathbf{Y}}_n(x)\right]\underline{\boldsymbol{\mathcal{B}}}_\nu \quad \mu = 1, \ldots, \kappa. \tag{10.52}$$

Taking into account the linearity of $\underline{\mathbf{U}}_\mu$ and the structure of the column vectors of $\underline{\mathbf{Y}}_i$, it can readily be seen that the coefficient matrix on the right side of the equation system (10.52) coincides with the matrix $\underline{\mathbf{D}}$ given by Eq. (10.45).

For the same reasons as before (for obtaining a solvable linear equation system) the elements of $\underline{\boldsymbol{\mathcal{A}}}_\nu$ with the same index as the elements of $\underline{\boldsymbol{\mathcal{B}}}_\nu$ set to zero have, are also set to zero. The result is denoted by $\underline{\check{\boldsymbol{\mathcal{A}}}}_\nu$.

Taking into account the structure of $\underline{\mathbf{D}}$, we can remove the identically zero columns and rows in the left side of equation—this results in the matrix $\underline{\check{\mathbf{D}}}$ multiplied by $\underline{\check{\boldsymbol{\mathcal{A}}}}_\nu$. As regards the right side, a similar reasoning shows that the right side, which is a known quantity, has a similar form with the left side: $\underline{\check{\mathbf{D}}}\,\underline{\check{\boldsymbol{\mathcal{B}}}}_\nu$. This means that we have to solve the equation system

$$\underline{\check{\mathbf{D}}}\,\underline{\check{\boldsymbol{\mathcal{A}}}}_\nu = \mp\underline{\check{\mathbf{D}}}\,\underline{\check{\boldsymbol{\mathcal{B}}}}_\nu \tag{10.53}$$

for $\underline{\check{\boldsymbol{\mathcal{A}}}}_\nu$. The equation system (10.53) is solvable if $\det\underline{\check{\mathbf{D}}}$ is not equal to zero. If this condition is satisfied, then there exists the Green function matrix.

10.2.3 Green Function Matrices for Some Curved Beam Problems

10.2.3.1 Pinned-Pinned Curved Beam

In this subsection, it is our aim to determine the Green function matrix for pinned-pinned curved beams. Differential equation (10.25) and boundary conditions (10.36) constitute our point of departure. This means that we want determine the Green function matrix for the boundary value problem governed by the differential equation

$$\underbrace{\begin{bmatrix} 0 & 0 \\ 0 & 1 \end{bmatrix}}_{\underset{\underline{\mathbf{P}}}{4}=\underline{\mathbf{K}}_4}\begin{bmatrix} y_1 \\ y_2 \end{bmatrix}^{(4)} + \underbrace{\begin{bmatrix} -m & 0 \\ 0 & 2 \end{bmatrix}}_{\underset{\underline{\mathbf{P}}}{2}=\underline{\mathbf{K}}_2}\begin{bmatrix} y_1 \\ y_2 \end{bmatrix}^{(2)} + \underbrace{\begin{bmatrix} 0 & -m \\ m & 0 \end{bmatrix}}_{\underset{\underline{\mathbf{P}}}{1}=\underline{\mathbf{K}}_1}\begin{bmatrix} y_1 \\ y_2 \end{bmatrix}^{(1)} +$$

$$+\underbrace{\begin{bmatrix} 0 & 0 \\ 0 & m+1 \end{bmatrix}}_{\underset{\underline{\mathbf{P}}}{0}=\underline{\mathbf{K}}_0}\begin{bmatrix} y_1 \\ y_2 \end{bmatrix}^{(0)} = \begin{bmatrix} r_1 \\ r_2 \end{bmatrix} \tag{10.54}$$

associated with the boundary conditions

$$y_1(x)|_{x=\pm\vartheta} = y_2(x)|_{x=\pm\vartheta} = y_2^{(2)}(x)\Big|_{x=\pm\vartheta} = 0. \tag{10.55}$$

The general solution to the homogeneous part of Eq. (10.54) is given by the equation

$$\underline{\mathbf{y}} = \left[\sum_{i=1}^{4} \underset{(2\times2)}{\underline{\mathbf{Y}}_i}\ \underset{(2\times2)}{\underline{\mathbf{C}}_i}\right] \underset{(2\times1)}{\underline{\mathbf{e}}}\ , \tag{10.56}$$

where $\underline{\mathbf{C}}_i$ is a constant non-singular matrix, $\underline{\mathbf{e}}$ is a constant column matrix and

$$\underline{\mathbf{Y}}_1 = \begin{bmatrix} \cos x & 0 \\ \sin x & 0 \end{bmatrix}, \qquad \underline{\mathbf{Y}}_2 = \begin{bmatrix} -\sin x & 0 \\ \cos x & 0 \end{bmatrix}, \tag{10.57a}$$

$$\underline{\mathbf{Y}}_3 = \begin{bmatrix} -\sin x + x\cos x & (m+1)x \\ x\sin x & -m \end{bmatrix}, \quad \underline{\mathbf{Y}}_4 = \begin{bmatrix} -\cos x - x\sin x & 1 \\ x\cos x & 0 \end{bmatrix}. \tag{10.57b}$$

It follows from Eq. (10.46) that the Green function matrix can be given in the following form:

$$\underbrace{\underline{\mathbf{G}}(x,\xi)}_{(2\times 2)} = \sum_{\ell=1}^{4} \underline{\mathbf{Y}}_\ell(x)\left[\underline{\mathbf{A}}_\ell(\xi) \pm \underline{\mathbf{B}}_\ell(\xi)\right] =$$

$$= \begin{bmatrix} \cos x & 0 \\ \sin x & 0 \end{bmatrix} \left\{ \begin{bmatrix} \overset{1}{A}_{11} & \overset{1}{A}_{12} \\ 0 & 0 \end{bmatrix} \pm \begin{bmatrix} \overset{1}{B}_{11} & \overset{1}{B}_{12} \\ 0 & 0 \end{bmatrix} \right\} +$$

$$+ \begin{bmatrix} -\sin x & 0 \\ \cos x & 0 \end{bmatrix} \left\{ \begin{bmatrix} \overset{2}{A}_{11} & \overset{2}{A}_{12} \\ 0 & 0 \end{bmatrix} \pm \begin{bmatrix} \overset{2}{B}_{11} & \overset{2}{B}_{12} \\ 0 & 0 \end{bmatrix} \right\} +$$

$$+ \begin{bmatrix} -\sin x + x\cos x & (m+1)x \\ x\sin x & -m \end{bmatrix} \left\{ \begin{bmatrix} \overset{3}{A}_{11} & \overset{3}{A}_{12} \\ \overset{2}{A}_{21} & \overset{2}{A}_{22} \end{bmatrix} \pm \begin{bmatrix} \overset{3}{B}_{11} & \overset{3}{B}_{12} \\ \overset{3}{B}_{21} & \overset{3}{B}_{22} \end{bmatrix} \right\} +$$

$$+ \begin{bmatrix} -\cos x - x\sin x & 1 \\ x\cos x & 0 \end{bmatrix} \left\{ \begin{bmatrix} \overset{3}{A}_{11} & \overset{3}{A}_{12} \\ \overset{3}{A}_{21} & \overset{3}{A}_{22} \end{bmatrix} \pm \begin{bmatrix} \overset{3}{B}_{11} & \overset{3}{B}_{12} \\ \overset{3}{B}_{21} & \overset{3}{B}_{22} \end{bmatrix} \right\}. \qquad (10.58)$$

The sign is [positive](negative) if [$x \le \xi$] ($\xi \le x$). In accordance with (10.47), Property 2 of the definition leads to the following equation system for the non-zero elements of the matrices $\underline{\mathbf{B}}_\ell$:

$$\begin{bmatrix} \cos\xi & -\sin\xi & -\sin\xi + \xi\cos\xi & (1+m)\xi & -\cos\xi - \xi\sin\xi & 1 \\ \sin\xi & \cos\xi & \xi\sin\xi & -m & \xi\cos\xi & 0 \\ -\sin\xi & -\cos\xi & -\xi\sin\xi & 1+m & -\xi\cos\xi & 0 \\ \cos\xi & -\sin\xi & \xi\cos\xi + \sin\xi & 0 & -\xi\sin\xi + \cos\xi & 0 \\ -\sin\xi & -\cos\xi & -\xi\sin\xi + 2\cos\xi & 0 & -\xi\cos\xi - 2\sin\xi & 0 \\ -\cos\xi & \sin\xi & -\xi\cos\xi - 3\sin\xi & 0 & \xi\sin\xi - 3\cos\xi & 0 \end{bmatrix} \begin{bmatrix} \overset{1}{B}_{11} & \overset{1}{B}_{12} \\ \overset{2}{B}_{11} & \overset{2}{B}_{12} \\ \overset{3}{B}_{11} & \overset{3}{B}_{12} \\ \overset{3}{B}_{21} & \overset{3}{B}_{22} \\ \overset{4}{B}_{11} & \overset{4}{B}_{12} \\ \overset{4}{B}_{21} & \overset{4}{B}_{22} \end{bmatrix} =$$

$$= \begin{bmatrix} 0 & 0 \\ 0 & 0 \\ \dfrac{1}{2m} & 0 \\ 0 & 0 \\ 0 & 0 \\ 0 & -\dfrac{1}{2} \end{bmatrix}. \qquad (10.59)$$

Here, the coefficient matrix corresponds to $\check{\underline{\mathbf{W}}}$, the first column in the matrix of unknowns to $\check{\underline{\boldsymbol{\mathcal{B}}}}_1$ and the second to $\check{\underline{\boldsymbol{\mathcal{B}}}}_2$ in (10.50). As regards the right side, the first column corresponds to $\underline{\mathcal{P}}_1$ and the second to $\underline{\mathcal{P}}_2$ in (10.50).

Now the matrices $\overset{2}{\underline{\mathbf{P}}}_{11}$ and $\overset{4}{\underline{\mathbf{P}}}_{22}$ are scalars, and their values are $-m$ and 1. The solutions are given by the following equations:

$$
\begin{aligned}
&\overset{1}{B}_{11} = \frac{1}{2}\sin\xi - \frac{1}{4}\xi\cos\xi\,, &\quad& \overset{1}{B}_{12} = -\frac{1}{4}\cos\xi - \frac{1}{4}\xi\sin\xi\,,\\
&\overset{2}{B}_{11} = \frac{1}{4}\xi\sin\xi + \frac{1}{2}\cos\xi\,, && \overset{2}{B}_{12} = \frac{1}{4}\sin\xi - \frac{1}{4}\xi\cos\xi\,,\\
&\overset{3}{B}_{11} = \frac{1}{4}\cos\xi\,, && \overset{3}{B}_{12} = \frac{1}{4}\sin\xi\,,\\
&\overset{3}{B}_{21} = \frac{1}{2m}\,, && \overset{3}{B}_{22} = 0\,,\\
&\overset{4}{B}_{11} = -\frac{1}{4}\sin\xi\,, && \overset{4}{B}_{12} = \frac{1}{4}\cos\xi\,,\\
&\overset{4}{B}_{21} = -\frac{1}{2}(m+1)\frac{\xi}{m}\,, && \overset{4}{B}_{21} = \frac{1}{2}.
\end{aligned}
\tag{10.60}
$$

Note that the solutions obtained for the non-zero elements of the matrices $\underline{\mathbf{B}}_\ell$ are independent of the boundary conditions.

For the sake of a simplification in writing equations, it is worth introducing the following notations:

$$
a = \overset{1}{B}_{1i};\quad b = \overset{2}{B}_{1i};\quad c = \overset{3}{B}_{1i};\quad d = \overset{3}{B}_{2i};\quad e = \overset{4}{B}_{1i};\quad f = \overset{4}{B}_{2i},\quad (i = 1, 2).
\tag{10.61}
$$

By substituting the Green function matrix (10.58) into boundary conditions (10.55) and taking the above notation convention also into account, the following equation system is obtained for calculating the non-zero elements of the matrices $\underline{\mathbf{A}}_\ell$:

$$
\begin{bmatrix}
\cos\vartheta & \sin\vartheta & \sin\vartheta - \vartheta\cos\vartheta & -(m+1)\vartheta & -\cos\vartheta - \vartheta\sin\vartheta & 1\\
\cos\vartheta & -\sin\vartheta & -\sin\vartheta + \vartheta\cos\vartheta & (m+1)\vartheta & -\cos\vartheta - \vartheta\sin\vartheta & 1\\
-\sin\vartheta & \cos\vartheta & \vartheta\sin\vartheta & -m & -\vartheta\cos\vartheta & 0\\
\sin\vartheta & \cos\vartheta & \vartheta\sin\vartheta & -m & \vartheta\cos\vartheta & 0\\
\sin\vartheta & -\cos\vartheta & 2\cos\vartheta - \vartheta\sin\vartheta & 0 & 2\sin\vartheta + \vartheta\cos\vartheta & 0\\
-\sin\vartheta & -\cos\vartheta & 2\cos\vartheta - \vartheta\sin\vartheta & 0 & -2\sin\vartheta - \vartheta\cos\vartheta & 0
\end{bmatrix}
\begin{bmatrix}
\overset{1}{A}_{1i}\\ \overset{2}{A}_{1i}\\ \overset{3}{A}_{1i}\\ \overset{3}{A}_{2i}\\ \overset{4}{A}_{1i}\\ \overset{4}{A}_{2i}
\end{bmatrix}
=
$$

$$
=
\begin{bmatrix}
-a\cos\vartheta - b\sin\vartheta - c\,(\sin\vartheta - \vartheta\cos\vartheta) + d(m+1)\vartheta + e\,(\cos\vartheta + \vartheta\sin\vartheta) - f\\
a\cos\vartheta - b\sin\vartheta + c\,(-\sin\vartheta + \vartheta\cos\vartheta) + d(m+1)\vartheta - e\,(\cos\vartheta + \vartheta\sin\vartheta) + f\\
a\sin\vartheta - b\cos\vartheta - c\vartheta\sin\vartheta + dm + e\vartheta\cos\vartheta\\
a\sin\vartheta + b\cos\vartheta + c\vartheta\sin\vartheta - dm + e\vartheta\cos\vartheta\\
-a\sin\vartheta + b\cos\vartheta - c\,(2\cos\vartheta - \vartheta\sin\vartheta) - e\,(2\sin\vartheta + \vartheta\cos\vartheta)\\
-a\sin\vartheta - b\cos\vartheta + c\,(2\cos\vartheta - \vartheta\sin\vartheta) - e\,(2\sin\vartheta + \vartheta\cos\vartheta)
\end{bmatrix}.
\tag{10.62}
$$

Here, the first two equations are obtained from the boundary conditions $y_1(x)|_{x=\pm\vartheta} = 0$, the second two equations from the boundary conditions $y_2(x)|_{x=\pm\vartheta} = 0$, while the last two equations from the boundary conditions $y_2^{(2)}(x)|_{x=\pm\vartheta} = 0$.

Note that the coefficient matrix in (10.62) corresponds to $\underline{\check{\mathbf{D}}}$, the matrix of unknowns to $\underline{\check{\mathcal{A}}}_\nu$, while the right side to the product $\underline{\check{\mathbf{D}}}\,\underline{\check{\mathcal{B}}}_\nu$ in (10.53).

With D_1 and D_2 defined by the equation

$$D_1 = \sin^2 \vartheta, \qquad D_2 = m\vartheta + 2\,(m+1)\,\vartheta \cos^2 \vartheta - 3m \sin \vartheta \cos \vartheta, \tag{10.63}$$

the solutions for the non-zero elements of the matrices $\underline{\mathbf{A}}_\ell$ are

$$\overset{1}{A}_{1i} = \frac{1}{2D_1}\left[2\overset{2}{B}_{1i} \sin \vartheta \cos \vartheta + 2\overset{3}{B}_{1i}\vartheta - \overset{3}{B}_{2i} m\,(2 \sin \vartheta + \vartheta \cos \vartheta)\right], \tag{10.64a}$$

$$\overset{2}{A}_{1i} = \frac{1}{D_2}\left[\overset{1}{B}_{1i}\left(2(1+m)\vartheta \sin \vartheta \cos \vartheta - m \sin^2 \vartheta + 2m \cos^2 \vartheta\right) + \right.$$
$$\left. + \overset{4}{B}_{1i}\left(3m\vartheta^2 + 2\vartheta^2 - 2m\right) - \overset{4}{B}_{2i} m\,(\vartheta \sin \vartheta - 2 \cos \vartheta)\right], \tag{10.64b}$$

$$\overset{3}{A}_{1i} = \frac{1}{D_2}\left(\overset{1}{B}_{1i} m - \overset{4}{B}_{1i}\left(m \cos^2 \vartheta - 2m \sin^2 \vartheta + 2(1+m)\vartheta \sin \vartheta \cos \vartheta\right) + \right.$$
$$\left. + \overset{4}{B}_{2i} m \cos \vartheta\right), \tag{10.64c}$$

$$\overset{3}{A}_{2i} = \frac{2}{D_2}\left[\overset{1}{B}_{1i} \cos \vartheta + \overset{4}{B}_{1i}\,(\vartheta \sin \vartheta - \cos \vartheta) + \overset{4}{B}_{2i} \cos^2 \vartheta\right], \tag{10.64d}$$

$$\overset{4}{A}_{1i} = \frac{1}{2D_1}\left(-2\overset{3}{B}_{1i} \sin \vartheta \cos \vartheta + \overset{3}{B}_{2i} m \sin \vartheta\right), \tag{10.64e}$$

$$\overset{4}{A}_{2i} = \frac{1}{2D_1}\left[-2\overset{2}{B}_{1i} \sin \vartheta - 2\overset{3}{B}_{1i}\,(\sin \vartheta + \vartheta \cos \vartheta) + \right.$$
$$\left. + \overset{3}{B}_{2i}\left(m\vartheta \cos^2 \vartheta + 3m \sin \vartheta\,(\cos \vartheta + \vartheta \sin \vartheta) + 2\vartheta \sin^2 \vartheta\right)\right]. \tag{10.64f}$$

Here, the solutions are presented in terms of the non-zero elements of the matrices $\underline{\mathbf{B}}_\ell$ by using definition (10.61) of the quantities $a, \ldots, f$.

With regard to the fact that the solutions (10.60) and (10.64) obtained for the non-zero elements of the matrices $\underline{\mathbf{B}}_\ell$ and $\underline{\mathbf{A}}_\ell$ are relatively long formulas, we do not present the whole Green function matrix here.

10.2.3.2 Fixed-Fixed Curved Beam

The solution steps are the same as those applied to calculating the elements of the Green function matrix for pinned-pinned beams. Since the matrices $\underline{\mathbf{B}}_\ell$ are boundary condition-independent quantities, the solutions given by Eq. (10.60) are valid for fixed-fixed beams too. It is therefore sufficient to deal with the calculation of the non-zero elements of the matrices $\underline{\mathbf{A}}_{\ell}$.

For fixed-fixed beam, the first four boundary conditions are the same as those for the pinned-pinned beam. The last two boundary conditions are, however, different: $y_2^{(1)}(x)|_{x=\pm\vartheta} = 0$. Hence, the first four equations in (10.62) will not change. After utilizing the last to boundary conditions, we arrive at the following equation system:

$$\begin{bmatrix} \cos\vartheta & \sin\vartheta & \sin\vartheta-\vartheta\cos\vartheta & -(m+1)\vartheta & -\cos\vartheta-\vartheta\sin\vartheta & 1 \\ \cos\vartheta & -\sin\vartheta & -\sin\vartheta+\vartheta\cos\vartheta & (m+1)\vartheta & -\cos\vartheta-\vartheta\sin\vartheta & 1 \\ -\sin\vartheta & \cos\vartheta & \vartheta\sin\vartheta & -m & -\vartheta\cos\vartheta & 0 \\ \sin\vartheta & \cos\vartheta & \vartheta\sin\vartheta & -m & \vartheta\cos\vartheta & 0 \\ \cos\vartheta & \sin\vartheta & -\sin\vartheta-\vartheta\cos\vartheta & 0 & \cos\vartheta-\vartheta\sin\vartheta & 0 \\ \cos\vartheta & -\sin\vartheta & \sin\vartheta+\vartheta\cos\vartheta & 0 & \cos\vartheta-\vartheta\sin\vartheta & 0 \end{bmatrix} \begin{bmatrix} \overset{1}{A}_{1i} \\ \overset{2}{A}_{1i} \\ \overset{3}{A}_{1i} \\ \overset{3}{A}_{2i} \\ \overset{4}{A}_{1i} \\ \overset{4}{A}_{2i} \end{bmatrix} =$$

$$= \begin{bmatrix} -a\cos\vartheta - b\sin\vartheta - c\,(\sin\vartheta-\vartheta\cos\vartheta) + d(m+1)\vartheta + e\,(\cos\vartheta+\vartheta\sin\vartheta) - f \\ a\cos\vartheta - b\sin\vartheta + c\,(-\sin\vartheta+\vartheta\cos\vartheta) + d(m+1)\vartheta - e\,(\cos\vartheta+\vartheta\sin\vartheta) + f \\ a\sin\vartheta - b\cos\vartheta - c\vartheta\sin\vartheta + dm + e\vartheta\cos\vartheta \\ a\sin\vartheta + b\cos\vartheta + c\vartheta\sin\vartheta - dm + e\vartheta\cos\vartheta \\ -a\cos\vartheta - b\sin\vartheta + c\,(\sin\vartheta+\vartheta\cos\vartheta) - e\,(\cos\vartheta-\vartheta\sin\vartheta) \\ a\cos\vartheta - b\sin\vartheta + c\,(\sin\vartheta+\vartheta\cos\vartheta) + e\,(\cos\vartheta-\vartheta\sin\vartheta) \end{bmatrix}. \tag{10.65}$$

By using the notations

$$D_1 = \vartheta - \sin\vartheta\cos\vartheta\,, \tag{10.66a}$$

$$D_2 = m\sin\vartheta\,(\vartheta\cos\vartheta - 2\sin\vartheta) + (m+1)\,\vartheta^2 + \vartheta\cos\vartheta\sin\vartheta, \tag{10.66b}$$

the solutions can be given in the following forms:

$$\overset{1}{A}_{1i} = \frac{1}{D_1}\left[-\overset{2}{B}_{1i}\cos^2\vartheta + \overset{3}{B}_{1i}\vartheta^2 + \overset{3}{B}_{2i}m\,(\cos\vartheta - \vartheta\sin\vartheta)\right], \tag{10.67a}$$

$$\overset{2}{A}_{1i} = \frac{1}{D_2}\Bigg[\overset{1}{B}_{1i}\,(m+1)\,\vartheta\sin^2\vartheta + 2\overset{1}{B}_{1i}m\sin\vartheta\cos\vartheta - 2\overset{4}{B}_{1i}m\vartheta + \\ + \overset{4}{B}_{1i}\,(m+1)\,\vartheta^3 + \overset{4}{B}_{2i}m\,(\vartheta\cos\vartheta + \sin\vartheta)\Bigg], \tag{10.67b}$$

$$\overset{3}{A}_{1i} = \frac{1}{D_2}\Big[\overset{1}{B}_{1i}\,(m+1)\,\vartheta + \overset{4}{B}_{1i}\,(m+1)\,\vartheta\cos^2\vartheta - \\ -2\overset{4}{B}_{1i}m\sin\vartheta\cos\vartheta + \overset{4}{B}_{2i}m\sin\vartheta\Big], \tag{10.67c}$$

$$\overset{3}{A}_{2i} = \frac{1}{D_2}\left[2\overset{1}{B}_{1i}\sin\vartheta - 2\overset{4}{B}_{1i}\vartheta\cos\vartheta + \overset{4}{B}_{2i}\left(\vartheta + \sin\vartheta\cos\vartheta\right)\right], \tag{10.67d}$$

$$\overset{4}{A}_{1i} = \frac{1}{D_1}\left(\overset{2}{B}_{1i} - \overset{3}{B}_{1i}\sin^2\vartheta - \overset{3}{B}_{2i}m\cos\vartheta\right), \tag{10.67e}$$

$$\overset{4}{A}_{2i} = \frac{1}{D_1}\left[2\overset{2}{B}_{1i}\cos\vartheta - 2\overset{3}{B}_{1i}\vartheta\sin\vartheta + \overset{3}{B}_{2i}\left(m+1\right)\vartheta^2 - \right.$$
$$\left. - \overset{3}{B}_{2i}\left(m+1\right)\vartheta\sin\vartheta\cos\vartheta - 2\overset{3}{B}_{2i}m\cos^2\vartheta\right]. \tag{10.67f}$$

Note that the solutions are given again in terms of the non-zero elements of the matrices $\underline{\mathbf{B}}_\ell$ by utilizing relations (10.61). With (10.60) and (10.67), the Green function matrix for fixed-fixed beams can be given in a closed form.

10.2.3.3 Pinned-Fixed Curved Beam

It is obvious that the non-zero elements of the matrices $\underline{\mathbf{B}}_\ell$ can be obtained from Eq. (10.60). As regards the non-zero elements of the matrices $\underline{\mathbf{A}}_\ell$, the reader is referred to Problem 10.3. Its solution in Sect. C.10 contains the formulas for the non-zero elements of the matrices $\underline{\mathbf{A}}_\ell$—see Eq. (C.10.56) for details. With (10.60) and (C.10.56), the Green function matrix can also be given in a closed form for pinned-fixed curved beams.

10.2.3.4 Closing Remarks Concerning the Calculation of the Green Function Matrices

It is proven in the solution of Problem 10.1 that the differential equation (10.54), which coincides with left side of the differential equation (10.35), is self-adjoint. Hence, the Green function matrices we have determined for the pinned-pinned, fixed-fixed, and pinned-fixed curved beams should be cross-symmetric, i.e., they should satisfy the requirement

$$\mathbf{G}^T(x,\xi) = \mathbf{G}(\xi, x)\,.$$

Fulfillment of this equation is checked numerically—a Fortran 90 program is written for the computation of the three Green function matrices and the computational results prove that the Green function matrices mentioned are really cross-symmetric.

Let us assume that the dimensionless load is either a tangential or a normal unit load applied to the curved beam at the point ξ, i.e.,

$$\underline{\mathbf{r}} = \begin{bmatrix}\delta(\check{\xi}-\xi)\\ 0\end{bmatrix} \quad \text{or} \quad \underline{\mathbf{r}} = \begin{bmatrix}0\\ \delta(\check{\xi}-\xi)\end{bmatrix}, \tag{10.68}$$

where δ is the Dirac function, $\check{\xi}$ is a parameter ($\check{\xi} \in [-\vartheta, \vartheta]$), and ξ is fixed. Then, it follows from (10.41) that

$$\int_a^b \begin{bmatrix} G_{11}(x,\check{\xi}) & G_{12}(x,\check{\xi}) \\ G_{21}(x,\check{\xi}) & G_{22}(x,\check{\xi}) \end{bmatrix} \begin{bmatrix} \delta(\check{\xi}-\xi) \\ 0 \end{bmatrix} d\check{\xi} = \begin{bmatrix} G_{11}(x,\xi) \\ G_{21}(x,\xi) \end{bmatrix} \tag{10.69a}$$

or

$$\int_a^b \begin{bmatrix} G_{11}(x,\check{\xi}) & G_{12}(x,\check{\xi}) \\ G_{21}(x,\check{\xi}) & G_{22}(x,\check{\xi}) \end{bmatrix} \begin{bmatrix} 0 \\ \delta(\check{\xi}-\xi) \end{bmatrix} d\check{\xi} = \begin{bmatrix} G_{12}(x,\xi) \\ G_{22}(x,\xi) \end{bmatrix}. \tag{10.69b}$$

In words: (a) $G_{11}(x,\xi)$ and $G_{21}(x,\xi)$ are dimensionless tangential and normal displacements at the point x due to a tangential dimensionless unit load applied at the point ξ; (b) $G_{12}(x,\xi)$ and $G_{22}(x,\xi)$ are dimensionless tangential and normal displacements at the point x due to a normal dimensionless unit load applied at the point ξ. This is the geometrical meaning of the elements in the Green function matrix. For a pinned-fixed curved beam, Fig. 10.3 depicts graphically the four dimensionless displacement components and the dimensionless unit load if $x < 0$ and $\xi > 0$.

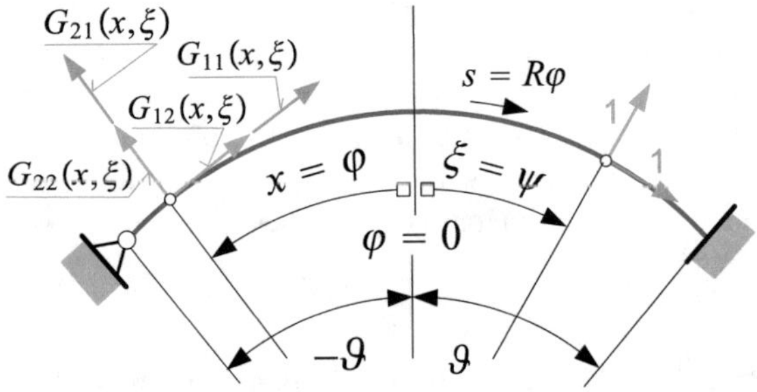

Fig. 10.3 Displacement components due to unit forces

10.2.4 Free Vibrations of Curved Beams

The eigenvalue problems determined by differential equation system (10.35)—the left side $\underline{\mathbf{K}}\left[\underline{\mathbf{y}}\right]$ of this equation is given by Eq. (10.24)—and boundary conditions (10.36), (10.37), and (10.38) describe the free vibrations of pinned-pinned, fixed-fixed, and pinned-fixed curved beams. This means that the eigenvalues—if they are known—make it possible to calculate the circular frequencies by the use of Eq. (10.31b), while the corresponding eigenfunctions coincide with the mode shapes of the free vibrations at the considered circular frequency.

According to Eq. (10.41), the solutions of the boundary value problems determined by the differential equation system

$$\underline{\mathbf{K}}[\underline{\mathbf{y}}] = \underline{\mathbf{r}}$$

repeated here on the basis of (10.24) for completeness, and boundary conditions (10.36), (10.37), and (10.38) can be given in closed forms provided that we know the Green function matrices for each of the mentioned boundary conditions:

$$\underline{\mathbf{y}}(x) = \int_{-\vartheta}^{\vartheta} \underline{\mathbf{G}}(x,\xi)\underline{\mathbf{r}}(\xi)\,\mathrm{d}\xi\,. \tag{10.70}$$

Here $\underline{\mathbf{r}}$ is the dimensionless load. It follows from (10.35) that

$$\underline{\mathbf{r}}(\xi) = \lambda\underline{\mathbf{M}}_0(x)\,\underline{\mathbf{y}}(x) = \lambda\underline{\mathbf{y}}(x)$$

is the dimensionless load for the vibration problems of curved beams. By substituting $\lambda\underline{\mathbf{y}}(x)$ for $\underline{\mathbf{r}}$ in (10.70), a homogeneous Fredholm integral equation system is obtained

$$\begin{bmatrix} y_1(x) \\ y_2(x) \end{bmatrix} = \lambda \int_{-\vartheta}^{\vartheta} \begin{bmatrix} G_{11}(x,\xi) & G_{12}(x,\xi) \\ G_{21}(x,\xi) & G_{22}(x,\xi) \end{bmatrix} \begin{bmatrix} y_1(\xi) \\ y_2(\xi) \end{bmatrix} \mathrm{d}\xi, \tag{10.71}$$

which can be solved numerically by applying the solution algorithm presented in Sect. 9.2.13—now $\underline{\mathbf{G}}(x,\xi)$ is the kernel function.

Remark 10.5 Consider a straight beam with the same length ℓ as that of the curved beam. It is also assumed that material distribution over the cross sections of the straight beam is the same as that of the curved beam which means that the straight beam is also heterogeneous. The free vibrations of such beams are described by the differential equation

$$\frac{\mathrm{d}^4 W}{\mathrm{d}x^4} = \lambda W\,, \qquad \lambda = \frac{\rho_a A\omega^2}{I_{ey}}, \tag{10.72}$$

where W is the amplitude of the deflection, ρ_a is the average density on the cross section, and

$$I_{ey} = \int_A z^2 E(y,z)\mathrm{d}A\,, \qquad E(y,z) = E(-y,z)\,. \tag{10.73}$$

We remark that the derivation of these equations is left for Problem 10.5.

Note that differential equation (10.72) which describes the free vibration of heterogeneous straight beams coincides with differential equation (7.82) which describes the free vibrations of homogeneous straight beams. In addition, the boundary conditions for pinned-pinned, fixed-fixed, and pinned-fixed beams are also the same independent of the fact whether the straight beam is homogeneous or not. Hence, the results presented for the eigenvalues in Sect. 7.5.2 remain valid for heterogeneous beams too except one thing: the product IE should be replaced by $I_{ey} = I_{e\eta}$. The latter equality reflects that the cross section and the material distribution over the cross section are the same for the straight and curved beams.

Let $\hat{\lambda}_i$ and $\hat{\omega}_i$ be the i-th eigenvalue and eigenfrequency ($i = 1, \ldots, 4$) of a straight beam (pinned-pinned, fixed-fixed, or pinned-fixed). The first eigenfrequency of the pinned-pinned beam is, however, denoted by $\hat{\omega}_{1\,pp}$. For our later considerations, we determine the quotient $C_{i\,\text{char}} = \hat{\omega}_i/\hat{\omega}_{1\,pp}$. The calculations are based on equation (7.109b) and the last column of Table 7.5:

$$C_{i\,\text{char}} = \frac{\hat{\omega}_i}{\hat{\omega}_{1\,pp}} = \left(\frac{\beta_i}{\beta_{1\,pp}}\right)^2 = \left(\frac{\beta_i \ell}{\beta_{1\,pp}\ell}\right)^2 = \frac{\beta_i^2}{\pi^2}. \tag{10.74}$$

The results are gathered in Table 10.2:

Table 10.2 Typical values of $C_{i,\text{char}}$

Beams	$i = 1$	$i = 2$	$i = 3$	$i = 4$
Pinned-pinned beams	1	4	9	16
Fixed-fixed beams	2.260	6.265	12.25	20.25
Pinned-fixed beams	1.562	5.062	10.56	18.06

Since

$$\hat{\omega}_{1\,pp} = \frac{\pi^2}{\ell^2 \sqrt{\frac{\rho_a A}{I_{e\eta}}}}, \tag{10.75}$$

it also holds that

$$\hat{\omega}_i = C_{i\,\text{char}}\, \hat{\omega}_{1\,pp} = \frac{C_{i\,\text{char}}\, \pi^2}{\ell^2 \sqrt{\frac{\rho_a A}{I_{e\eta}}}}. \tag{10.76}$$

The eigenvalue problem governed by the Fredholm integral equation (10.71) is solved numerically. The computations are based on the algorithm presented in Sect. 9.2.14) and coded in Fortran 90. Recalling that (10.31b) gives ω_i in terms of λ_i and taking into account that $\ell = 2\vartheta = \bar{\vartheta}$ Figs. 10.4, 10.5, and 10.6 show the quotient

$$C_{i,\text{char}} \frac{\omega_i}{\hat{\omega}_i} = \frac{\dfrac{\sqrt{\lambda_i}}{\sqrt{\frac{\rho_a A}{I_{e\eta}}}R^2}}{\dfrac{\pi^2}{\sqrt{\frac{\rho_a A}{I_{e\eta}}}\ell^2}} = \frac{\bar{\vartheta}^2 \sqrt{\lambda_i}}{\pi^2} \tag{10.77}$$

against $\bar{\vartheta} = 2\vartheta$ and m is a parameter.

The eigenfrequencies of the curved beams we have considered are therefore compared to the first eigenfrequency of straight beams with the same length and same material composition. It is worth emphasizing that the material composition is incorporated into the model via the parameter m—the figures clearly represent its effect—$I_{e\eta}$ and ρ_a.

For the quotients (10.77) depicted in Figs. 10.4, 10.5, and 10.6, the dimensionless quantity m serves as a parameter. The selected values of m are 750, 1 000, 1 300,, 1 750, 2 400, 3 400, 5 000, 7 500, 12 000, 20 000, 35 000, 60 000, 100 000, and 200 000. It should also be mentioned that Figs. 10.4, 10.5, and 10.6 are taken

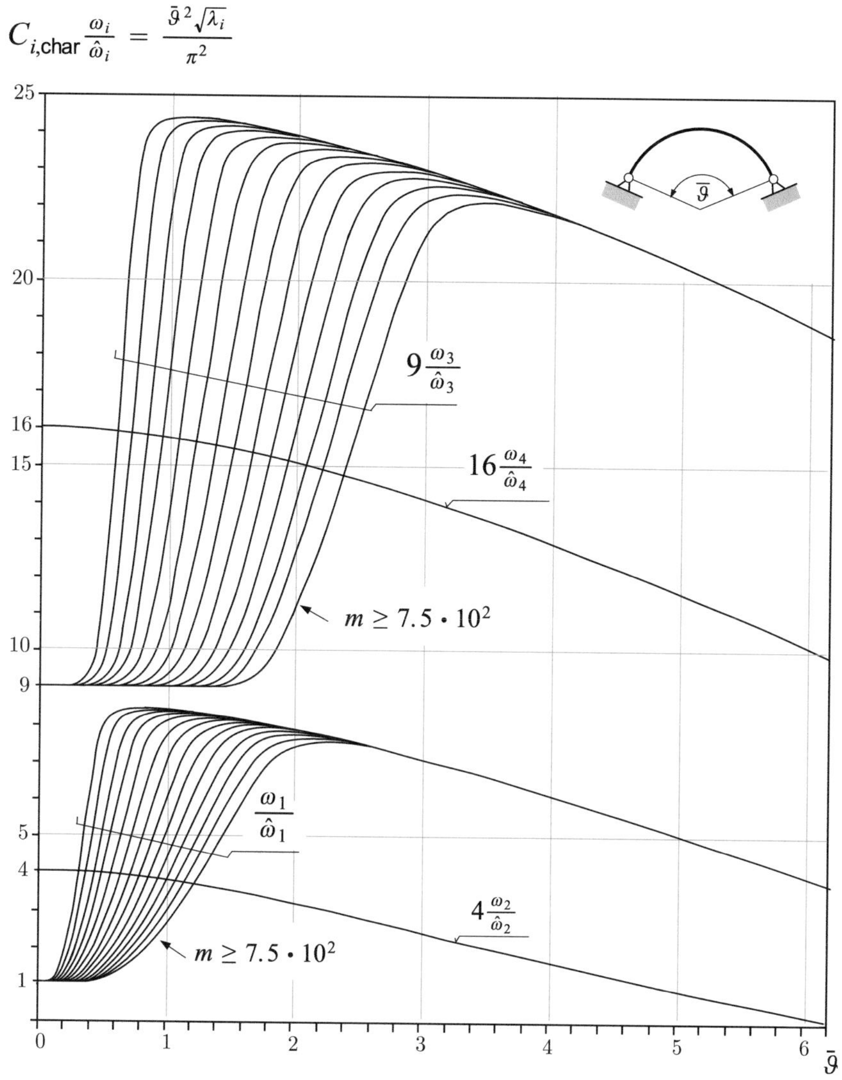

Fig. 10.4 The first four natural circular frequencies against the central angle for pinned-pinned curved beam

from paper [14] with a permission from the Miskolc University Press. The results obtained for pinned-pinned and fixed-fixed beams are identical to those published in [15] for homogeneous beams. In this work, the eigenfrequencies are, however, computed by utilizing the frequency determinant.

For $\bar{\vartheta} = 2\pi$, the eigenfrequency ω_2 of the pinned-pinned beam is zero. This result reflects the fact that the pinned-pinned beam can rotate freely about the two supports which, in this case, coincide.

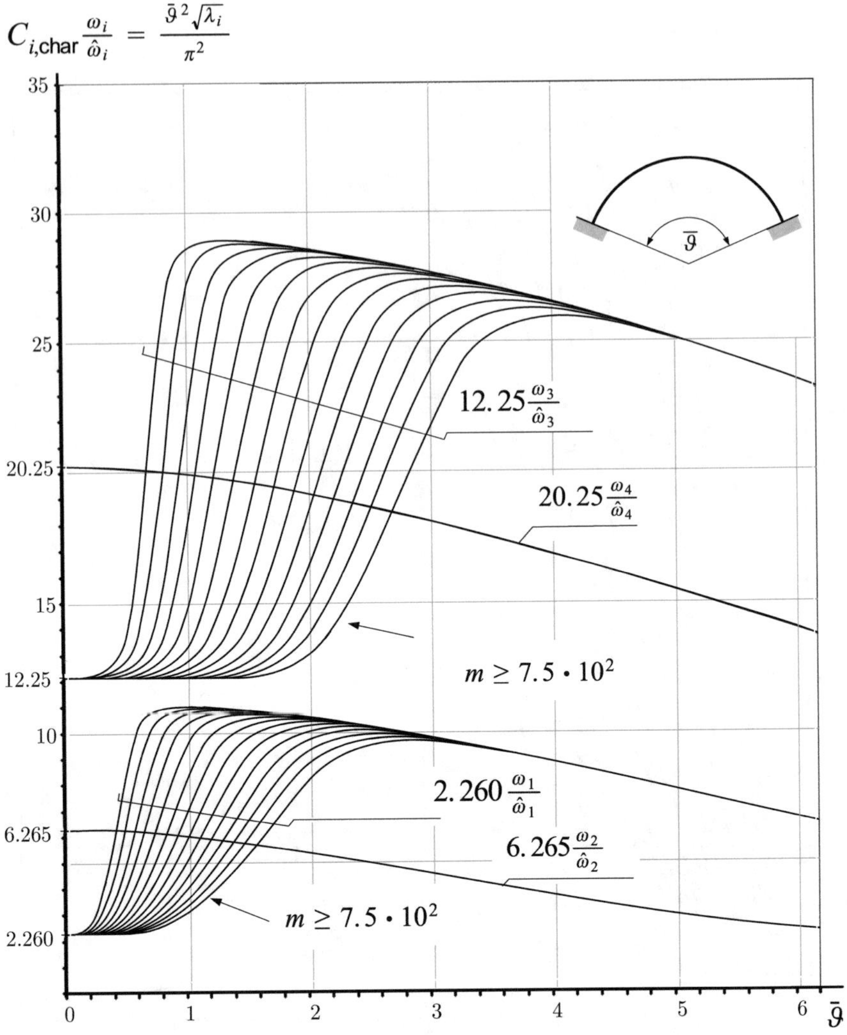

Fig. 10.5 The first four natural circular frequencies against the central angle for fixed-fixed curved beam

It is also an interesting phenomenon that the ratios for the even eigenfrequencies do not depend on m. Another important property is that a frequency shift can be observed: in terms of magnitude, the first/third frequency becomes the second/fourth one if the central angle is sufficiently great.

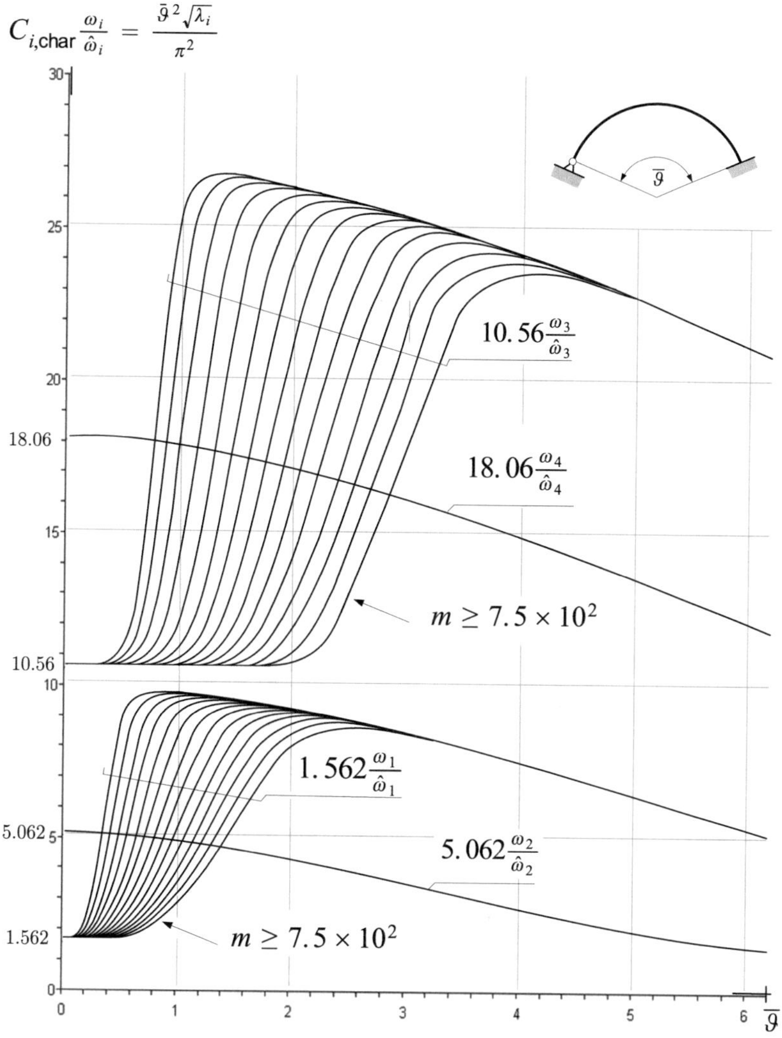

Fig. 10.6 The first four natural circular frequencies against the central angle for pinned-fixed curved beam

Note that the results obtained for pinned-fixed beams are larger than those for pinned-pinned beams and smaller than the results for fixed-fixed beams. This difference follows from the fact that the rigidity of the pinned-fixed beams is larger than that of the pinned-pinned beams and is smaller than the rigidity of fixed-fixed beams.

Table 10.3 Data comparison for pinned-pinned beam

2ϑ		Current model	Ref. [7]
$\pi/2$	ω_1	38.41	38.28
$\pi/2$	ω_2	89.77	89.08
$\pi/2$	ω_3	172.18	169.75
$\pi/2$	ω_4	245.82	243.05
π	ω_1	6.33	6.32
π	ω_2	19.32	19.28
π	ω_3	39.05	38.87
π	ω_4	63.79	63.29

Table 10.4 Data comparison for fixed-fixed beam

2ϑ		Current model	Ref. [7]
$\pi/2$	ω_1	63.1	62.62
$\pi/2$	ω_2	117.5	115.85
$\pi/2$	ω_3	218.2	213.28
$\pi/2$	ω_4	249.8	247.96
π	ω_1	12.24	12.21
π	ω_2	26.92	26.80
π	ω_3	50.07	49.70
π	ω_4	76.85	75.95

The numerical results obtained and presented graphically in Figs. 10.4, 10.5, and 10.6 are compared with the findings of [7]. This article incorporates axial extension, rotatory inertia, and transverse shear effects into the model as well. Homogeneous, rectangular cross section is assumed with the following data: $\rho_a = 7\,800\ \text{kg/m}^3$, $A = 0.01\ \text{m}^2$, $I_{e\eta} = 1.66 \cdot 10^6\ \text{Nm}^2$. The natural circular frequencies are compared in Table 10.3 for pinned-pinned and in Table 10.4 for fixed-fixed supports as long as $m = 10\,000$. The correlation happens to be really good.

If there is a vertical load applied to the curved beam at the crown point, the eigenfrequencies of the vibrations will change. The effect of this load is clarified partly in thesis [13] and in papers [17–19]. Measurement data for the free vibrations of fixed-fixed beams are also published in [17]. They coincide with the computed values with a good accuracy.

10.3 Problems

Problem 10.1 Prove that the eigenvalue problems defined by differential equation (10.35) and boundary conditions (10.36), (10.37), and (10.38) are all self-adjoint.

Problem 10.2 Find the Rayleigh quotient for the eigenvalue problems considered in the previous problem.

Problem 10.3 Consider the pinned-fixed curved beams and find the equation system for the non-zero elements of the matrices $\mathbf{A}_\ell$ which are included in the Green function matrix. Determine the solutions too.

Problem 10.4 Find the Rayleigh quotient by utilizing the solution of Problem 10.2 if

$$u_1 = \frac{2x}{\pi} - \frac{2\vartheta}{\pi}\sin\frac{\pi}{2\vartheta}x\,, \qquad u_2 = \cos\frac{\pi}{2\vartheta}x\,. \tag{10.78}$$

These functions are comparative for pinned-pinned curved beams.

Problem 10.5 Show that the free vibrations of straight beams with cross-sectional heterogeneity are described by differential equation (10.72).

References

1. J.P. Den Hartog, *Mechanical Vibrations*. Civil, Mechanical and Other Engineering Series, 1st edn.: 1934 (Dover Publications, Mineola, 1985)
2. J.P. Den Hartog, Vibration of frames of electrical machines. J. Appl. Mech. **50**, 1–6 (1828)
3. E. Volterra, J.D. Morrel, On the fundamental frequencies of curved beams. Bull. Polytech. Inst. Jassy **7**(11), 1–2 (1961)
4. K. Federhofer, *Dynamik des Bogenträgers und Kreisringes [Dynamics of Arches and Circular Rings]* (Springer, Wien, 1950)
5. M.S. Qatu, A.A. Elsharkawy, Vibration of laminated composite arches with deep curvature and arbitrary boundaries. Comput. Struct. **47**(2), 305–311 (1993). https://doi.org/10.1016/0045-7949(93)90381-M
6. K. Kang, C.W. Bert, A.G. Striz, Vibration analysis of shear deformable circular arches by the differential quadrature method. J. Sound Vib. **181**(2), 353–360 (1995). https://doi.org/10.1006/jsvi.1995.0258
7. E. Tüfekçi, A. Arpaci, Exact solution of in-plane vibrations of circular arches with account taken of axial extension, transverse shear and rotatory inertia affects. J. Sound Vib. **209**(5), 845–856 (1997). https://doi.org/10.1006/jsvi.1997.1290
8. A. Krishnan, Y.J. Suresh, A simple cubic linear element for static and free vibration analyses of curved beams. Comput. Struct. **68**(5), 473–489 (1998). https://doi.org/10.1016/S0045-7949(98)00091-1
9. I. Ecsedi, K. Dluhi, A linear model for the static and dynamic analysis of non-homogeneous curved beams. Appl. Math. Model. **29**(12), 1211–1231 (2006). https://doi.org/10.1016/j.apm.2005.03.006
10. F.F. Çalim, Forced vibration of curved beams on two-parameter elastic foundation. Appl. Math. Model. **36**(3), 964–973 (2012). https://doi.org/10.1016/j.apm.2011.07.066
11. M. Hajianmaleki, M.S. Qatu, Vibrations of straight and curved composite beams: a review. Compos. Struct. **100**, 218–232 (2013). https://doi.org/10.1016/j.compstruct.2013.01.001.340
12. B. Kovács, Vibration analysis of layered curved arch. J. Sound Vib. **332**(18), 4223–4240 (2013). https://doi.org/10.1016/j.jsv.2013.03.011
13. L.P. Kiss, Vibrations and stability of heterogeneous curved beams. Ph.D. thesis. Institute of Applied Mechanics, University of Miskolc, Hungary (2015), 142 pp
14. L.P. Kiss, Green's functions for nonhomogeneous curved beams with applications to vibration problems. J. Comput. Appl. Mech. **12**(1), 19–41 (2017). https://doi.org/10.32973/jcam.2017.002

15. G. Szeidl, Effect of change in length on the natural frequencies and stability of circular beams in Hungarian. Ph.D. thesis. Department of Mechanics, University of Miskolc, Hungary (1975), 158 pp
16. J.G. Obádovics, On the boundary and initial value problems of differential equation systems. Ph.D. thesis. Hungarian Academy of Sciences (1967) (in Hungarian)
17. L. Kiss et al., Vibrations of fixed-fixed heterogeneous curved beams loaded by a central force at the crown point. Int. J. Eng. Model. **27**(3–4), 85–100 (2014)
18. L. Kiss, G. Szeidl, Vibrations of pinned-pinned heterogeneous circular beams subjected to a radial force at the crown point. Mech. Based Des. Struct. Mach.: Int. J. **43**(4), 424–449 (2015). https://doi.org/10.1080/15397734.2015.1022659
19. L.P. Kiss, G. Szeidl, Vibrations of pinned-fixed heterogeneous circular beams pre-loaded by a vertical force at the crown point. J. Sound Vib. **393**, 92–113 (2017). https://doi.org/10.1016/j.jsv.2016.12.032

Appendix A
A Short Introduction to Tensor Algebra

A.1 Notational Conventions

A.1.1 Some Vector Operations

Let $\mathbf{a}$ and $\mathbf{b}$ be two non-zero vectors. The scalar product s of the two vectors is denoted in the usual way:

$$s = \mathbf{a} \cdot \mathbf{b}\,. \tag{A.1.1}$$

The operation sign is the reason for calling it dot product.

The scalar product is a commutative vector operation: $\mathbf{a} \cdot \mathbf{b} = \mathbf{b} \cdot \mathbf{a}$.

If $\mathbf{a} \cdot \mathbf{b} = 0$, then $\mathbf{a}$ and $\mathbf{b}$ are perpendicular to each other. This condition is known as condition of perpendicularity.

The zero vector is perpendicular to any other vector.

The cross product $\mathbf{c}$ of the vectors $\mathbf{a}$ and $\mathbf{b}$ is also denoted in the usual way:

$$\mathbf{c} = \mathbf{a} \times \mathbf{b}\,. \tag{A.1.2}$$

The cross product is not commutative: $\mathbf{a} \times \mathbf{b} = -\mathbf{b} \times \mathbf{a}$.

If $\mathbf{c} = \mathbf{a} \times \mathbf{b} = \mathbf{0}$, then $\mathbf{a}$ and $\mathbf{b}$ are parallel to each other. This condition is that of the parallelism for $\mathbf{a}$ and $\mathbf{b}$.

The zero vector is parallel to any other vector.

The dyadic product of two vectors is denoted by

$$\boldsymbol{W} = \mathbf{a} \circ \mathbf{b}\,. \tag{A.1.3}$$

The left side of this equation is the name we have given to the product. We remark that any name can be given to a dyadic product. These names are, in general, typeset by boldface and italic letters.

G. Szeidl and L. P. Kiss, *Mechanical Vibrations*, Foundations of Engineering Mechanics,
https://doi.org/10.1007/978-3-030-45074-8

Properties:

1. If we dot multiply it from left or right by $\mathbf{c}$, we get

$$\boldsymbol{W} \cdot \mathbf{c} = (\mathbf{a} \circ \mathbf{b}) \cdot \mathbf{c} = \mathbf{a} \, (\mathbf{b} \cdot \mathbf{c}) \; , \tag{A.1.4a}$$

$$\mathbf{c} \cdot \boldsymbol{W} = \mathbf{c} \cdot (\mathbf{a} \circ \mathbf{b}) = (\mathbf{c} \cdot \mathbf{a}) \, \mathbf{b} \, . \tag{A.1.4b}$$

Let $\mathbf{d}$ be a non-zero vector. With regard to (A.1.4a) and (A.1.4b), it is clear that the product

$$\mathbf{d} \cdot \boldsymbol{W} \cdot \mathbf{c} = \mathbf{d} \cdot (\mathbf{a} \circ \mathbf{b}) \cdot \mathbf{c} = (\mathbf{d} \cdot \mathbf{a}) \, (\mathbf{b} \cdot \mathbf{c}) \tag{A.1.4c}$$

is a scalar.

2. Equations (A.1.4) show that the dyadic product is not a commutative operation.
3. The dot product of the two dyadic products $\mathbf{a} \circ \mathbf{b}$ and $\mathbf{c} \circ \mathbf{d}$ is also a dyadic product defined by the relation

$$(\mathbf{a} \circ \mathbf{b}) \cdot (\mathbf{c} \circ \mathbf{d}) = (\mathbf{a} \circ \mathbf{d}) \, \mathbf{b} \cdot \mathbf{c} \, . \tag{A.1.5}$$

4. The inner product $\mathbf{a} \circ \mathbf{b}$ and $\mathbf{c} \circ \mathbf{d}$ is a scalar:

$$(\mathbf{a} \circ \mathbf{b}) \cdot \cdot \, (\mathbf{c} \circ \mathbf{d}) = (\mathbf{a} \cdot \mathbf{c}) \, (\mathbf{b} \cdot \mathbf{d}) \; . \tag{A.1.6}$$

5. If we cross multiply $\boldsymbol{W}$ from left or right by $\mathbf{c}$, we get two dyadic products:

$$\boldsymbol{W} \times \mathbf{c} = (\mathbf{a} \circ \mathbf{b}) \times \mathbf{c} = \mathbf{a} \circ (\mathbf{b} \times \mathbf{c}) \; , \tag{A.1.7a}$$

$$\mathbf{c} \times \boldsymbol{W} = \mathbf{c} \times (\mathbf{a} \circ \mathbf{b}) = (\mathbf{c} \times \mathbf{a}) \circ \mathbf{b} \, . \tag{A.1.7b}$$

The box product of the vectors $\mathbf{a}$, $\mathbf{b}$, and $\mathbf{c}$ is defined by the following equation:

$$[\mathbf{abc}] = (\mathbf{a} \times \mathbf{b}) \cdot \mathbf{c}. \tag{A.1.8}$$

The box product has the following properties:

1. The placement of the operation signs is interchangeable:

$$(\mathbf{a} \times \mathbf{b}) \cdot \mathbf{c} = \mathbf{a} \cdot (\mathbf{b} \times \mathbf{c}) \; . \tag{A.1.9a}$$

2. Cyclic interchangeability:

$$[\mathbf{abc}] = [\mathbf{bca}] = [\mathbf{cab}] = -\,[\mathbf{bac}] = -\,[\mathbf{cba}] = -\,[\mathbf{acb}] \; . \tag{A.1.9b}$$

If the three vectors are measured from the same point, the box product results in the volume of the parallelepiped determined by the three vectors.

The box product is zero if the three vectors lie in the same plane.

The triple cross products of the vectors $\mathbf{a}$, $\mathbf{b}$, and $\mathbf{c}$ can be expended as

$$\mathbf{a} \times (\mathbf{b} \times \mathbf{c}) = (\mathbf{a} \cdot \mathbf{c}) \, \mathbf{b} - (\mathbf{a} \cdot \mathbf{b}) \, \mathbf{c} \; , \qquad (\mathbf{a} \times \mathbf{b}) \times \mathbf{c} = (\mathbf{a} \cdot \mathbf{c}) \, \mathbf{b} - (\mathbf{b} \cdot \mathbf{c}) \, \mathbf{a} \; . \tag{A.1.10}$$

A.1.2 Base Vectors

Let $\mathbf{g}_1$, $\mathbf{g}_2$, and $\mathbf{g}_3$ be three non-zero vectors. It will be assumed that

$$\left[\mathbf{g}_1\mathbf{g}_2\mathbf{g}_3\right] = \gamma_\circ \neq 0, \qquad \gamma_\circ^* = 1/\gamma_\circ\,. \tag{A.1.11}$$

Then, the three vectors (if measured from the same point) have no common plane. It is obvious that any vector $\mathbf{v}$ can be given in the form

$$\mathbf{v} = v^1\mathbf{g}^1 + v^2\mathbf{g}_2 + v^3\mathbf{g}_3, \tag{A.1.12}$$

where v^ℓ $(\ell = 1, 2, 3)$ is the component of $\mathbf{v}$ in the direction $\mathbf{g}_\ell$. This statement follows form the fact that the above equation system, which is detailed here

$$\begin{bmatrix} v_x \\ v_y \\ v_z \end{bmatrix} = \begin{bmatrix} g_{x1} & g_{x2} & g_{x3} \\ g_{y1} & g_{y2} & g_{y3} \\ g_{z1} & g_{z2} & g_{z3} \end{bmatrix} \begin{bmatrix} v^1 \\ v^2 \\ v^3 \end{bmatrix}, \tag{A.1.13}$$

always has a unique solution for v^ℓ. The triplet $\mathbf{g}_\ell$ is called basis in the 3D space since any vector $\mathbf{v}$ can be given in terms of $\mathbf{g}_\ell$. The dual basis (The dual base vectors) is (are) defined by the following equations:

$$\mathbf{g}_1^* = \frac{\mathbf{g}_2 \times \mathbf{g}_3}{\gamma_\circ}, \quad \mathbf{g}_2^* = \frac{\mathbf{g}_3 \times \mathbf{g}_1}{\gamma_\circ}, \quad \mathbf{g}_3^* = \frac{\mathbf{g}_1 \times \mathbf{g}_2}{\gamma_\circ}. \tag{A.1.14a}$$

It can be shown that

$$\mathbf{g}_1 = \frac{\mathbf{g}_2^* \times \mathbf{g}_3^*}{\gamma_\circ^*}, \quad \mathbf{g}_2 = \frac{\mathbf{g}_3^* \times \mathbf{g}_1^*}{\gamma_\circ^*}, \quad \mathbf{g}_3 = \frac{\mathbf{g}_1^* \times \mathbf{g}_2^*}{\gamma_\circ^*}. \tag{A.1.14b}$$

The base vectors $\mathbf{g}_k$ and $\mathbf{g}_\ell^*$ have the following properties:

$$\mathbf{g}_k \cdot \mathbf{g}_\ell^* = \begin{cases} 1 \text{ if } k = \ell\,, \\ 0 \text{ if } k \neq \ell\,. \end{cases} \qquad k, \ell = 1, 2, 3 \tag{A.1.15}$$

For example

$$\mathbf{g}_1 \cdot \mathbf{g}_1^* = \mathbf{g}_1 \cdot \frac{\mathbf{g}_2 \times \mathbf{g}_3}{\gamma_\circ} = \frac{\left[\mathbf{g}_1\mathbf{g}_2\mathbf{g}_3\right]}{\gamma_\circ} = \frac{\gamma_\circ}{\gamma_\circ} = 1\,,$$

or

$$\mathbf{g}_2 \cdot \mathbf{g}_3^* = \mathbf{g}_2 \cdot \frac{\mathbf{g}_1 \times \mathbf{g}_2}{\gamma_\circ} = \frac{\mathbf{g}_1 \times \mathbf{g}_2}{\gamma_\circ} \cdot \mathbf{g}_2 = \frac{\left[\mathbf{g}_1\mathbf{g}_2\mathbf{a}_2\right]}{\gamma_\circ} = \frac{\mathbf{g}_1 \cdot (\mathbf{g}_2 \times \mathbf{g}_2)}{\gamma_\circ} = 0\,.$$

Since

$$\mathbf{i}_x \times \mathbf{i}_y = \mathbf{i}_z\,, \quad \mathbf{i}_y \times \mathbf{i}_z = \mathbf{i}_x\,, \quad \mathbf{i}_z \times \mathbf{i}_x = \mathbf{i}_y, \quad \text{and} \quad \left[\mathbf{i}_x \mathbf{i}_y \mathbf{i}_z\right] = 1, \tag{A.1.16a}$$

it follows that

$$\mathbf{i}_x = \mathbf{i}_x^*\,, \quad \mathbf{i}_y = \mathbf{i}_y^*\,, \quad \mathbf{i}_z = \mathbf{i}_z^*\,. \tag{A.1.16b}$$

This means that the basis and the dual basis is the same in the Cartesian coordinate system. If we utilize relation (A.1.15), we get from Eq. (A.1.12) that

$$v_\ell = \mathbf{v} \cdot \mathbf{g}_\ell^*. \tag{A.1.17}$$

In the Cartesian coordinate system (xyz), it also holds that

$$v_m = \mathbf{v} \cdot \mathbf{i}_m\,, \qquad m = x, y, z\,. \tag{A.1.18}$$

A.2 Tensors

A.2.1 Classification

A scalar field, which describes a physical state of a body, is said to be a tensor field of order zero. (For example, the temperature field in a solid body.) A vector field, which describes a physical state of a body, is said to be a tensor field of order one. (For example, the displacement field of a solid body.)

A common property of tensors of order zero and one is that they are independent of the coordinate system, i.e., they do not change if we translate and rotate the coordinate system (if the coordinate system performs a rigid body motion).

In the following subsection, we shall deal with tensors of order two.

A.2.2 Homogeneous Linear Vector-Vector Functions: Tensors of Order Two

Consider the vector-vector function

$$\mathbf{w} = \mathbf{f}(\mathbf{v}), \tag{A.2.19}$$

in which $\mathbf{w}$ is the dependent variable and $\mathbf{f}$ denotes the prescription made on the independent variable $\mathbf{v}$. Function (A.2.19) is a mapping of the vectors $\mathbf{v}$ (measured from the origin O_v) onto the vectors $\mathbf{w}$ (measured from the origin O_w—see Fig. A.1):

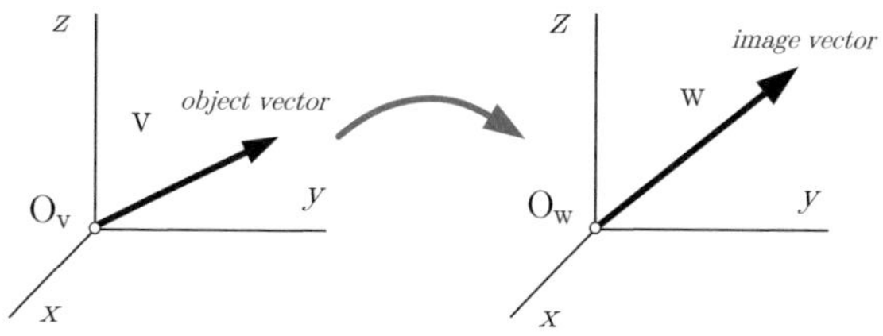

Fig. A.1 Mapping of **v** onto **w**

The vector-vector function $\mathbf{w} = \mathbf{f}(\mathbf{v})$ is a homogeneous linear function if the following functional equation is satisfied:

$$\mathbf{w} = \mathbf{f}(\mathbf{v}) = \mathbf{f}(v_x\mathbf{i}_x + v_y\mathbf{i}_y + v_y\mathbf{i}_z) = v_x\underbrace{\mathbf{f}(\mathbf{i}_x)}_{\mathbf{w}_x} + v_y\underbrace{\mathbf{f}(\mathbf{i}_y)}_{\mathbf{w}_y} + v_z\underbrace{\mathbf{f}(\mathbf{i}_z)}_{\mathbf{w}_z} =$$
$$= \mathbf{w}_x\, v_x + \mathbf{w}_y\, v_y + \mathbf{w}_z\, v_z, \qquad \text{(A.2.20)}$$

where

$$\begin{aligned} \mathbf{w}_x &= w_{xx}\mathbf{i}_x + w_{yx}\mathbf{i}_y + w_{zx}\mathbf{i}_z\,, \\ \mathbf{w}_y &= w_{xy}\mathbf{i}_x + w_{yy}\mathbf{i}_y + w_{zy}\mathbf{i}_z\,, \\ \mathbf{w}_z &= w_{xz}\mathbf{i}_x + w_{yz}\mathbf{i}_y + w_{zz}\mathbf{i}_z \end{aligned} \qquad \text{(A.2.21)}$$

are the images of the base vectors $\mathbf{i}_x$, $\mathbf{i}_y$, and $\mathbf{i}_z$.

Mapping (A.2.20) is uniquely determined if we know the nine scalars w_{xx}, $w_{yx}, \ldots, w_{zz}$.

Mapping (A.2.20) is degenerated if the $3D$ space (the vectors **v**) is (are) mapped onto a plane, or a straight line, or the origin O_w.

Since

$$\begin{aligned} v_x\mathbf{w}_x &= \mathbf{w}_x\underbrace{(\mathbf{i}_x \cdot \mathbf{v})}_{v_x} = (\mathbf{w}_x \circ \mathbf{i}_x) \cdot \mathbf{v}\,, \\ v_y\mathbf{w}_y &= \mathbf{w}_y\underbrace{(\mathbf{i}_y \cdot \mathbf{v})}_{v_y} = (\mathbf{w}_y \circ \mathbf{i}_y) \cdot \mathbf{v}\,, \\ v_x\mathbf{w}_z &= \mathbf{w}_z\underbrace{(\mathbf{i}_z \cdot \mathbf{v})}_{v_z} = (\mathbf{w}_z \circ \mathbf{i}_z) \cdot \mathbf{v}, \end{aligned} \qquad \text{(A.2.22)}$$

we can rewrite Eq. (A.2.20) in the following form:

$$\mathbf{w} = \big(\mathbf{w}_x \circ \mathbf{i}_x + \mathbf{w}_y \circ \mathbf{i}_y + \mathbf{w}_z \circ \mathbf{i}_z\big) \cdot \mathbf{v}, \qquad \text{(A.2.23)}$$

where the sum

$$\boxed{\boldsymbol{W} = \mathbf{w}_x \circ \mathbf{i}_x + \mathbf{w}_y \circ \mathbf{i}_y + \mathbf{w}_z \circ \mathbf{i}_z} \qquad \text{(A.2.24)}$$

is called tensor of order two (or simply tensor). With the knowledge of $\boldsymbol{W}$ mapping (A.2.20) takes the form

$$\boxed{\mathbf{w} = \boldsymbol{W} \cdot \mathbf{v}\,.} \qquad \text{(A.2.25)}$$

Because of its definition mapping (A.2.20) is independent of the coordinate system. Hence, so is the tensor $\boldsymbol{W}$. The sum (A.2.24) is, however, the presentation of the tensor $\boldsymbol{W}$ in the Cartesian coordinate system (xyz).

It is obvious that the image vectors $\mathbf{w}_x$, $\mathbf{w}_y$, and $\mathbf{w}_z$ can be given in terms of $\boldsymbol{W}$ as

$$\mathbf{w}_x = \boldsymbol{W} \cdot \mathbf{i}_x\,, \qquad \mathbf{w}_y = \boldsymbol{W} \cdot \mathbf{i}_y\,, \qquad \mathbf{w}_z = \boldsymbol{W} \cdot \mathbf{i}_z\,. \tag{A.2.26}$$

If we dot multiply these equations by $\mathbf{i}_m$ and take relations (A.1.4c) and (A.2.21) into account, we get

$$\boxed{w_{mn} = \mathbf{i}_m \cdot \boldsymbol{W} \cdot \mathbf{i}_n\,, \qquad m, n = x, y, z\,.} \tag{A.2.27}$$

We remark that the simple dyad (A.1.3) is also a tensor:

$$\boldsymbol{A} = \mathbf{a} \circ \mathbf{b} = \mathbf{a} \circ \left(b_x \mathbf{i}_x + b_y \mathbf{i}_y + b_z \mathbf{i}_z\right) = \underbrace{\mathbf{a} b_x}_{\text{image of } \mathbf{i}_x} \circ\, \mathbf{i}_x + \underbrace{\mathbf{a} b_y}_{\text{image of } \mathbf{i}_y} \circ\, \mathbf{i}_y + \underbrace{\mathbf{a} b_z}_{\text{image of } \mathbf{i}_z} \circ\, \mathbf{i}_z\,. \tag{A.2.28}$$

Observe that tensor $\mathbf{a} \circ \mathbf{b}$ is degenerated since the image vectors $\boldsymbol{A} \cdot \mathbf{c} = (\mathbf{a} \circ \mathbf{b}) \cdot \mathbf{c} = \mathbf{a}\,(\mathbf{b} \cdot \mathbf{c})$ are all parallel to $\mathbf{a}$ independently of the value $\mathbf{c}$ has.

A.2.3 Special Tensors

The zero tensor $\boldsymbol{0}$, for which the images of the unit vectors $\mathbf{i}_x$, $\mathbf{i}_y$, and $\mathbf{i}_z$ are all zero vectors, maps any vector onto a zero vector.

The unit tensor $\boldsymbol{1}$ maps each vector onto itself. It follows from the transformation

$$\begin{aligned}\mathbf{v} = v_x \mathbf{i}_x + v_y \mathbf{i}_y + v_z \mathbf{i}_z = \mathbf{i}_x\,(\mathbf{i}_x \cdot \mathbf{v}) + \mathbf{i}_y\left(\mathbf{i}_y \cdot \mathbf{v}\right) + \mathbf{i}_z\,(\mathbf{i}_z \cdot \mathbf{v}) = \\ = \left(\mathbf{i}_x \circ \mathbf{i}_x + \mathbf{i}_y \circ \mathbf{i}_y + \mathbf{i}_z \circ \mathbf{i}_z\right) \cdot \mathbf{v}\end{aligned}$$

that

$$\boxed{\boldsymbol{1} = \mathbf{i}_x \circ \mathbf{i}_x + \mathbf{i}_y \circ \mathbf{i}_y + \mathbf{i}_z \circ \mathbf{i}_z\,.} \tag{A.2.29}$$

Consider the vector triplet

$$\mathbf{n}_\ell\,, \qquad |\mathbf{n}_\ell| = 1\,, \qquad \mathbf{n}_k \cdot \mathbf{n}_\ell = \begin{cases} 1 & \text{if } k = \ell\,, \\ 0 & \text{if } k \neq \ell \end{cases} \qquad k, \ell = 1, 2, 3\,. \tag{A.2.30}$$

Since the unit vectors $\mathbf{n}_1$, $\mathbf{n}_2$, and $\mathbf{n}_3$ are mutually perpendicular to each other, they are said to [constitute] (determine) [an orthonormal basis] (a Cartesian coordinate system) in the 3D space. It is obvious that the unit tensor assumes the form

$$\boldsymbol{1} = \mathbf{n}_1 \circ \mathbf{n}_1 + \mathbf{n}_2 \circ \mathbf{n}_2 + \mathbf{n}_3 \circ \mathbf{n}_3 \tag{A.2.31}$$

in this basis.

The inverse of the tensor

$$\boldsymbol{W} = \mathbf{w}_x \circ \mathbf{i}_x + \mathbf{w}_y \circ \mathbf{i}_y + \mathbf{w}_z \circ \mathbf{i}_z\,, \qquad w_o = \left[\mathbf{w}_x \mathbf{w}_y \mathbf{w}_z\right] \neq 0$$

is denoted by $\boldsymbol{W}^{-1}$ and is defined by the following equation:

$$\boldsymbol{W} \cdot \boldsymbol{W}^{-1} = \boldsymbol{W}^{-1} \cdot \boldsymbol{W} = \boldsymbol{1}\,. \tag{A.2.32}$$

Since $w_o = \left[\mathbf{w}_x \mathbf{w}_y \mathbf{w}_z\right] \neq 0$, there is a dual base for the image vectors $\mathbf{w}_x$, $\mathbf{w}_y$, and $\mathbf{w}_z$:

$$\mathbf{w}_x^* = \frac{\mathbf{w}_y \times \mathbf{w}_z}{w_o}\,, \quad \mathbf{w}_y^* = \frac{\mathbf{w}_z \times \mathbf{w}_x}{w_o}\,, \quad \mathbf{w}_z^* = \frac{\mathbf{w}_x \times \mathbf{w}_y}{w_o}\,. \tag{A.2.33}$$

The inverse $\boldsymbol{W}^{-1}$ can be given in terms of the dual base:

$$\boxed{\boldsymbol{W}^{-1} = \mathbf{i}_x \circ \mathbf{w}_x^* + \mathbf{i}_y \circ \mathbf{w}_y^* + \mathbf{i}_z \circ \mathbf{w}_z^*\,.} \tag{A.2.34}$$

Proof

$$\begin{aligned}
\boldsymbol{W}^{-1}\cdot\boldsymbol{W} &= \left(\mathbf{i}_x \circ \mathbf{w}_x^* + \mathbf{i}_y \circ \mathbf{w}_y^* + \mathbf{i}_z \circ \mathbf{w}_z^*\right) \cdot \left(\mathbf{w}_x \circ \mathbf{i}_x + \mathbf{w}_y \circ \mathbf{i}_y + \mathbf{w}_z \circ \mathbf{i}_z\right) = \\
&= \mathbf{i}_x \circ \mathbf{i}_x \underbrace{\left(\mathbf{w}_x^* \cdot \mathbf{w}_x\right)}_{=1} + \mathbf{i}_x \circ \mathbf{i}_y \underbrace{\left(\mathbf{w}_x^* \cdot \mathbf{w}_y\right)}_{=0} + \mathbf{i}_x \circ \mathbf{i}_z \underbrace{\left(\mathbf{w}_x^* \cdot \mathbf{w}_z\right)}_{=0} + \\
&+ \mathbf{i}_y \circ \mathbf{i}_x \underbrace{\left(\mathbf{w}_y^* \cdot \mathbf{w}_x\right)}_{=0} + \mathbf{i}_y \circ \mathbf{i}_y \underbrace{\left(\mathbf{w}_y^* \cdot \mathbf{w}_y\right)}_{=1} + \mathbf{i}_y \circ \mathbf{i}_z \underbrace{\left(\mathbf{w}_y^* \cdot \mathbf{w}_z\right)}_{=0} + \\
&+ \mathbf{i}_z \circ \mathbf{i}_x \underbrace{\left(\mathbf{w}_z^* \cdot \mathbf{w}_x\right)}_{=0} + \mathbf{i}_z \circ \mathbf{i}_y \underbrace{\left(\mathbf{w}_z^* \cdot \mathbf{w}_y\right)}_{=0} + \mathbf{i}_z \circ \mathbf{i}_z \underbrace{\left(\mathbf{w}_z^* \cdot \mathbf{w}_z\right)}_{=1} = \\
&= \mathbf{i}_x \circ \mathbf{i}_x + \mathbf{i}_y \circ \mathbf{i}_y + \mathbf{i}_z \circ \mathbf{i}_z = \boldsymbol{1}\,.
\end{aligned}$$

The transpose of the tensor $\boldsymbol{W}$ is defined by the following equation:

$$\boxed{\boldsymbol{W}^T = \mathbf{i}_x \circ \mathbf{w}_x + \mathbf{i}_y \circ \mathbf{w}_y + \mathbf{i}_z \circ \mathbf{w}_z\,.} \tag{A.2.35}$$

Note that

$$\begin{aligned}
\boldsymbol{W} \cdot \mathbf{v} &= \mathbf{w}_x \left(\mathbf{i}_x \cdot \mathbf{v}\right) + \mathbf{w}_y \left(\mathbf{i}_y \cdot \mathbf{v}\right) + \mathbf{w}_y \left(\mathbf{i}_y \cdot \mathbf{v}\right) = \\
&= \left(\mathbf{v} \cdot \mathbf{i}_x\right) \mathbf{w}_x + \left(\mathbf{v} \cdot \mathbf{i}_y\right) \mathbf{w}_y + \left(\mathbf{v} \cdot \mathbf{i}_z\right) \mathbf{w}_z = \mathbf{v} \cdot \boldsymbol{W}^T\,.
\end{aligned}$$

If we dot multiply the above equation by the vector $\mathbf{u}$, we get the following relation:

$$\boxed{\mathbf{u} \cdot \boldsymbol{W} \cdot \mathbf{v} = \mathbf{v} \cdot \boldsymbol{W}^T \cdot \mathbf{u},} \tag{A.2.36}$$

which should hold for any $\mathbf{v}$ and $\mathbf{u}$.

Let **p** and **q** be two non-zero vectors. Further, let $\boldsymbol{S} = \mathbf{u} \circ \mathbf{v}$ and $\boldsymbol{T} = \mathbf{p} \circ \mathbf{q}$ be two tensors. Some properties of the transpose are detailed below:

$$\left(\boldsymbol{W}^T\right)^T = \boldsymbol{W}\,, \tag{A.2.37a}$$

$$\boxed{(\boldsymbol{S} \cdot \boldsymbol{T})^T = \boldsymbol{T}^T \cdot \boldsymbol{S}^T\,.} \tag{A.2.37b}$$

Property (A.2.37a) is obvious. As regards property (A.2.37a), compare the following two equations:

$$(\boldsymbol{S} \cdot \boldsymbol{T})^T = \left[(\mathbf{u} \circ \mathbf{v}) \cdot (\mathbf{p} \circ \mathbf{q})\right]^T = \left[(\mathbf{v} \cdot \mathbf{p})\, \mathbf{u} \circ \mathbf{q}\right]^T = (\mathbf{v} \cdot \mathbf{p})\, \mathbf{q} \circ \mathbf{u}\,,$$

$$\boldsymbol{T}^T \cdot \boldsymbol{S}^T = (\mathbf{q} \circ \mathbf{p}) \cdot (\mathbf{v} \circ \mathbf{u}) = (\mathbf{v} \cdot \mathbf{p})\, \mathbf{q} \circ \mathbf{u}\,.$$

The tensor $\boldsymbol{W}$ is symmetric if

$$\boldsymbol{W} = \boldsymbol{W}^T\,. \tag{A.2.38}$$

For a symmetric $\boldsymbol{W}$, Eq. (A.2.36) yields

$$\mathbf{v} \cdot \boldsymbol{W} \cdot \mathbf{u} = \mathbf{u} \cdot \boldsymbol{W} \cdot \mathbf{v}\,. \tag{A.2.39}$$

If we write $\mathbf{i}_m$ for **v**, $\mathbf{i}_n$ for **u** $(m, n = x, y, z)$ in the above equation and take (A.2.27) into account, we get for the symmetric tensor $\boldsymbol{W}$ that

$$w_{mn} = \mathbf{i}_m \cdot \boldsymbol{W} \cdot \mathbf{i}_n = \mathbf{i}_n \cdot \boldsymbol{W} \cdot \mathbf{i}_m = w_{nm}\,. \tag{A.2.40}$$

The tensor $\boldsymbol{W}$ is skew if

$$\boldsymbol{W} = -\boldsymbol{W}^T\,. \tag{A.2.41}$$

For the skew tensor $\boldsymbol{W}$, it follows from Eq. (A.2.36) that

$$\mathbf{v} \cdot \boldsymbol{W} \cdot \mathbf{u} = -\mathbf{u} \cdot \boldsymbol{W} \cdot \mathbf{v}\,. \tag{A.2.42}$$

If we now write $\mathbf{i}_m$ for **v**, $\mathbf{i}_n$ for **u** $(m, n = x, y, z)$ in (A.2.42), and take (A.2.27) into account, we obtain for the skew tensor $\boldsymbol{W}$ that

$$w_{mn} = \mathbf{i}_m \cdot \boldsymbol{W} \cdot \mathbf{i}_n = -\mathbf{i}_n \cdot \boldsymbol{W} \cdot \mathbf{i}_m = -w_{nm} \quad \text{hence} \quad w_{mm} = 0\,. \tag{A.2.43}$$

The tensor $\boldsymbol{W}$ is said to be

$$\left.\begin{array}{l}\text{positive definite}\\ \text{positive semidefinite}\\ \text{negative semidefinite}\\ \text{negative definite}\end{array}\right\} \quad \text{if, for any } \mathbf{u}, \quad \left\{\begin{array}{l}\mathbf{u} \cdot \boldsymbol{W} \cdot \mathbf{u} > 0\\ \mathbf{u} \cdot \boldsymbol{W} \cdot \mathbf{u} \ge 0\\ \mathbf{u} \cdot \boldsymbol{W} \cdot \mathbf{u} \le 0\\ \mathbf{u} \cdot \boldsymbol{W} \cdot \mathbf{u} < 0\end{array}\right. .$$

and it is indefinite if none of the above relations is satisfied.

It follows from the identity

$$W = \frac{1}{2}\left(W + W^T\right) + \frac{1}{2}\left(W - W^T\right) \tag{A.2.44}$$

that any tensor can be resolved into symmetric and skew parts:

$$\boxed{\begin{array}{c} W = W_{\text{sym}} + W_{\text{skew}} \\ W_{\text{sym}} = \frac{1}{2}\left(W + W^T\right), \qquad W_{\text{skew}} = \frac{1}{2}\left(W - W^T\right). \end{array}} \tag{A.2.45}$$

This theorem is that of the additive resolution.

A.2.4 Axial Vector

Consider the mapping that belongs to W_{skew}. We can write

$$\begin{aligned} W_{\text{skew}} \cdot \mathbf{v} &= \frac{1}{2}\left(W - W^T\right) \cdot \mathbf{v} = \\ &= \frac{1}{2}\left(\mathbf{w}_x \circ \mathbf{i}_x + \mathbf{w}_y \circ \mathbf{i}_y + \mathbf{w}_z \circ \mathbf{i}_z - \mathbf{i}_x \circ \mathbf{w}_x + \mathbf{i}_y \circ \mathbf{w}_y + \mathbf{i}_z \circ \mathbf{w}_z\right) \cdot \mathbf{v} = \\ &= -\frac{1}{2}\left[\mathbf{i}_x\left(\mathbf{w}_x \cdot \mathbf{v}\right) - \mathbf{w}_x\left(\mathbf{i}_x \cdot \mathbf{v}\right) + \mathbf{i}_y\left(\mathbf{w}_y \cdot \mathbf{v}\right) - \mathbf{w}_y\left(\mathbf{i}_y \cdot \mathbf{v}\right) + \mathbf{i}_z\left(\mathbf{w}_z \cdot \mathbf{v}\right) - \mathbf{w}_z\left(\mathbf{i}_z \cdot \mathbf{v}\right)\right] = \\ &\underset{(A.1.10)}{=\uparrow=} -\frac{1}{2}\left(\mathbf{w}_x \times \mathbf{i}_x + \mathbf{w}_y \times \mathbf{i}_y + \mathbf{w}_z \times \mathbf{i}_z\right) \times \mathbf{v}. \end{aligned} \tag{A.2.46}$$

Here the vector

$$\boxed{\mathbf{w}_a = -\frac{1}{2}\left(\mathbf{w}_x \times \mathbf{i}_x + \mathbf{w}_y \times \mathbf{i}_y + \mathbf{w}_z \times \mathbf{i}_z\right)} \tag{A.2.47}$$

is called axial vector. With $\mathbf{w}_a$ it holds that

$$\boxed{W_{\text{skew}} \cdot \mathbf{v} = \mathbf{w}_a \times \mathbf{v}.} \tag{A.2.48}$$

It is obvious that

(a) the axial vector $\mathbf{w}_a$ is independent of the coordinate system,
(b) mapping (A.2.48) is degenerated since the image vector is always perpendicular to the axial vector $\mathbf{w}_a$.

By utilizing the cross products

$$\mathbf{w}_x \times \mathbf{i}_x = w_{zx}\mathbf{i}_y - w_{yx}\mathbf{i}_z\,, \quad \mathbf{w}_y \times \mathbf{i}_y = w_{xy}\mathbf{i}_z - w_{zy}\mathbf{i}_x\,, \quad \mathbf{w}_z \times \mathbf{i}_z = w_{yz}\mathbf{i}_x - w_{xz}\mathbf{i}_y$$

we get

$$\boxed{\begin{gathered} \mathbf{w}_a = w_{ax}\mathbf{i}_x + w_{ay}\mathbf{i}_y + w_{az}\mathbf{i}_z\,, \\ w_{ax} = -\frac{1}{2}\left(w_{yz} - w_{zy}\right)\,, \quad w_{ay} = -\frac{1}{2}\left(w_{zx} - w_{xz}\right)\,, \quad w_{az} = -\frac{1}{2}\left(w_{xy} - w_{yx}\right)\,. \end{gathered}} \tag{A.2.49}$$

If the tensor $\boldsymbol{W}$ is symmetric $w_{mn} = w_{nm}$ $(m, n = x, y, z)$—see (A.2.40). Hence, it follows from (A.2.49) that the axial vector of a symmetric tensor is zero vector.

A.2.5 Tensors and Matrices

Recalling Eqs. (A.2.25) and (A.2.20), we can write

$$\boldsymbol{W} \cdot \mathbf{v} = \mathbf{w}_x v_x + \mathbf{w}_y v_y + \mathbf{w}_z v_z\,. \tag{A.2.50}$$

By introducing the column matrices

$$\underset{(3\times1)}{\underline{\mathbf{w}}} = \begin{bmatrix} w_x \\ w_y \\ w_z \end{bmatrix}, \quad \underset{(3\times1)}{\underline{\mathbf{w}}_x} = \begin{bmatrix} w_{xx} \\ w_{yx} \\ w_{zx} \end{bmatrix}, \quad \underset{(3\times1)}{\underline{\mathbf{w}}_y} = \begin{bmatrix} w_{xy} \\ w_{yy} \\ w_{zy} \end{bmatrix}, \quad \underset{(3\times1)}{\underline{\mathbf{w}}_z} = \begin{bmatrix} w_{xz} \\ w_{yz} \\ w_{zz} \end{bmatrix}, \tag{A.2.51}$$

we can manipulate Eq. (A.2.50) into the following form:

$$\begin{aligned} \underset{(3\times1)}{\underline{\mathbf{w}}} = \begin{bmatrix} w_x \\ w_y \\ w_z \end{bmatrix} &= \begin{bmatrix} w_{xx} \\ w_{yx} \\ w_{zx} \end{bmatrix} v_x + \begin{bmatrix} w_{xy} \\ w_{yy} \\ w_{zy} \end{bmatrix} v_y + \begin{bmatrix} w_{xz} \\ w_{yz} \\ w_{zz} \end{bmatrix} v_z = \\ &= \underbrace{\begin{bmatrix} w_{xx} & w_{xy} & w_{xz} \\ w_{yx} & w_{yy} & w_{yz} \\ w_{zx} & w_{zy} & w_{zz} \end{bmatrix}}_{\underset{(3\times3)}{\underline{\mathbf{W}}}} \underbrace{\begin{bmatrix} v_x \\ v_y \\ v_z \end{bmatrix}}_{\underset{(3\times1)}{\underline{\mathbf{v}}}} \end{aligned} \tag{A.2.52}$$

or shortly

$$\underset{(3\times1)}{\underline{\mathbf{w}}} = \underbrace{\left[\, \underset{(3\times1)}{\underline{\mathbf{w}}_x} \mid \underset{(3\times1)}{\underline{\mathbf{w}}_y} \mid \underset{(3\times1)}{\underline{\mathbf{w}}_z} \,\right]}_{\underset{(3\times3)}{\underline{\mathbf{W}}}} \underset{(3\times1)}{\underline{\mathbf{v}}} = \underset{(3\times3)}{\underline{\mathbf{W}}}\ \underset{(3\times1)}{\underline{\mathbf{v}}}\,, \tag{A.2.53}$$

where the matrix $\underline{\mathbf{W}}$ is that of the tensor $\boldsymbol{W}$ in the Cartesian coordinate system (xyz). It follows from (A.2.28) that the matrix of the tensor $\boldsymbol{A} = \mathbf{a} \circ \mathbf{b}$ is of the form

$$\underset{(3\times3)}{\underline{\mathbf{A}}} = \begin{bmatrix} a_x b_x & a_x b_y & a_x b_z \\ a_y b_x & a_y b_y & a_y b_z \\ a_z b_x & a_z b_y & a_z b_z \end{bmatrix} = \begin{bmatrix} a_x \\ a_y \\ a_z \end{bmatrix} \left[b_x \,|\, b_y \,|\, b_z \right] = \underset{(3\times1)}{\underline{\mathbf{a}}} \; \underset{(1\times3)}{\underline{\mathbf{b}}^T} = \underline{\mathbf{a}}\,\underline{\mathbf{b}}^T. \tag{A.2.54}$$

This means that the matrix product

$$\underset{(3\times1)}{\underline{\mathbf{a}}} \; \underset{(1\times3)}{\underline{\mathbf{b}}^T} \quad \text{is equivalent to the dyadic product} \quad \mathbf{a} \circ \mathbf{b} \tag{A.2.55}$$

if we use matrix notation in the Cartesian coordinate system (xyz).

Consider now the dot product

$$\mathbf{v} = \mathbf{u} \cdot \boldsymbol{W} = \underset{(\mathrm{A.2.24})}{\uparrow} = \underbrace{\mathbf{u} \cdot \mathbf{w}_x}_{v_x} \mathbf{i}_x + \underbrace{\mathbf{u} \cdot \mathbf{w}_y}_{v_y} \mathbf{i}_y + \underbrace{\mathbf{u} \cdot \mathbf{w}_z}_{v_z} \mathbf{i}_z \,. \tag{A.2.56}$$

It can be checked with ease that the matrix product

$$\underset{(1\times3)}{\underline{\mathbf{v}}^T} = \left[v_x \,|\, v_y \,| v_z \right] = \left[u_x \,|\, u_y \,| u_z \right] \begin{bmatrix} w_{xx} & w_{xy} & w_{xz} \\ w_{yx} & w_{yy} & w_{yz} \\ w_{zx} & w_{zy} & w_{zz} \end{bmatrix} = \underset{(1\times3)}{\underline{\mathbf{u}}^T} \; \underset{(3\times3)}{\underline{\mathbf{W}}} \tag{A.2.57}$$

yields the very same result as a row matrix.

We shall proceed with the cross product

$$\mathbf{u} \times \boldsymbol{W} = \underset{(\mathrm{A.1.7b})}{\uparrow} = \underbrace{\mathbf{u} \times \mathbf{w}_x}_{\text{image of } \mathbf{i}_x} \circ \mathbf{i}_x + \underbrace{\mathbf{u} \times \mathbf{w}_y}_{\text{image of } \mathbf{i}_y} \circ \mathbf{i}_y + \underbrace{\mathbf{u} \times \mathbf{w}_z}_{\text{image of } \mathbf{i}_z} \circ \mathbf{i}_z \tag{A.2.58}$$

which results in a tensor. If we recall Eqs. (1.14) and introduce the matrix

$$\underset{(3\times3)}{\underline{\mathbf{u}}_{\times}} = \begin{bmatrix} 0 & -u_z & u_y \\ u_z & 0 & -u_x \\ -u_y & u_x & 0 \end{bmatrix}, \tag{A.2.59}$$

we arrive at the result that the matrix of the tensor

$$\mathbf{u} \times \boldsymbol{W} \tag{A.2.60a}$$

can be obtained by performing the following matrix product:

$$\boxed{\underset{(3\times3)}{\underline{\mathbf{u}\times \boldsymbol{W}}} = \underset{(3\times3)}{\underline{\mathbf{u}}_\times}\,\underset{(3\times3)}{\underline{\mathbf{W}}} = \begin{bmatrix} 0 & -u_z & u_y \\ u_z & 0 & -u_x \\ -u_y & u_x & 0 \end{bmatrix} \begin{bmatrix} w_{xx} & w_{xy} & w_{xz} \\ w_{yx} & w_{yy} & w_{yz} \\ w_{zx} & w_{zy} & w_{zz} \end{bmatrix}.} \tag{A.2.60b}$$

Let $\boldsymbol{T}$ and $\boldsymbol{S}$ be two tensors:

$$\boldsymbol{T} = \mathbf{t}_x \circ \mathbf{i}_x + \mathbf{t}_y \circ \mathbf{i}_y + \mathbf{t}_y \circ \mathbf{i}_y\,, \qquad \boldsymbol{S} = \mathbf{s}_x \circ \mathbf{i}_x + \mathbf{s}_y \circ \mathbf{i}_y + \mathbf{s}_y \circ \mathbf{i}_y\,. \tag{A.2.61}$$

Making use of (A.1.5), we can determine the dot product of these tensors:

$$\boldsymbol{T}\cdot\boldsymbol{S} = \underbrace{\left(\mathbf{t}_x\, s_{xx} + \mathbf{t}_y\, s_{yx} + \mathbf{t}_z\, s_{zx}\right)}_{\text{the first column in the matrix } \underset{(3\times3)}{\underline{\boldsymbol{T}\cdot\boldsymbol{S}}}} \circ \mathbf{i}_x + \underbrace{\left(\mathbf{t}_x\, s_{xy} + \mathbf{t}_y\, s_{yy} + \mathbf{t}_z\, s_{zy}\right)}_{\text{the second column in the matrix } \underset{(3\times3)}{\underline{\boldsymbol{T}\cdot\boldsymbol{S}}}} \circ \mathbf{i}_y + $$
$$+ \underbrace{\left(\mathbf{t}_x\, s_{xz} + \mathbf{t}_y\, s_{yz} + \mathbf{t}_z\, s_{zz}\right)}_{\text{the third column in the matrix } \underset{(3\times3)}{\underline{\boldsymbol{T}\cdot\boldsymbol{S}}}} \circ \mathbf{i}_z\,. \tag{A.2.62}$$

We can read off from this equation that the matrix of the dot product $\boldsymbol{T}\cdot\boldsymbol{S}$ is given by the following matrix product:

$$\boxed{\underset{(3\times3)}{\underline{\boldsymbol{T}\cdot\boldsymbol{S}}} = \underset{(3\times3)}{\underline{\boldsymbol{T}}}\,\underset{(3\times3)}{\underline{\boldsymbol{S}}} = \begin{bmatrix} t_{xx} & t_{xy} & t_{xz} \\ t_{yx} & t_{yy} & t_{yz} \\ t_{zx} & t_{zy} & t_{zz} \end{bmatrix} \begin{bmatrix} s_{xx} & s_{xy} & s_{xz} \\ s_{yx} & s_{yy} & s_{yz} \\ s_{zx} & s_{zy} & s_{zz} \end{bmatrix}.} \tag{A.2.63}$$

It also holds that

$$\underset{(3\times3)}{\underline{(\boldsymbol{T}\cdot\boldsymbol{S})^T}} = \Big(\underset{(3\times3)}{\underline{\boldsymbol{T}}}\,\underset{(3\times3)}{\underline{\boldsymbol{S}}}\Big)^T = \underset{(1\times3)}{\underline{\mathbf{S}}^T}\,\underset{(1\times3)}{\underline{\mathbf{T}}^T}. \tag{A.2.64}$$

The determinant of a tensor is that of its matrix.

A.2.6 Eigenvalue Problem of Symmetric Tensors

Let $\boldsymbol{W}$ be a symmetric tensor. Find the direction n for which it holds that the unit vector (Fig. A.2)

$$\mathbf{n} = n_x\mathbf{i}_x + n_y\mathbf{i}_y + n_z\mathbf{i}_z; \qquad n_x^2 + n_y^2 + n_z^2 = 1, \tag{A.2.65}$$

which belongs to the direction n satisfies the relation

$$\mathbf{w}_n = \boldsymbol{W} \cdot \mathbf{n} = \lambda \mathbf{n}. \tag{A.2.66}$$

which means that the object vector and the image vector are parallel to each other. Note that here the scalar λ—which is referred to as eigenvalue—and the vector $\mathbf{n}$ (n_x, n_y and n_z)—which is called eigenvector—are the unknowns.

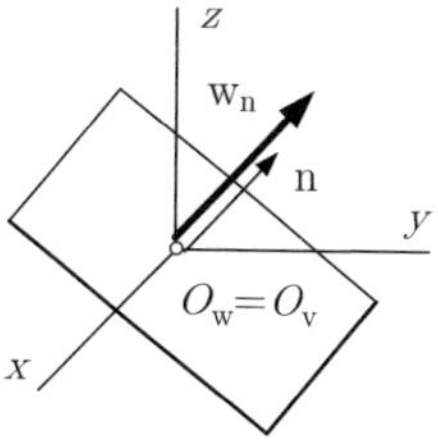

Fig. A.2 Parallel object and image vectors

Since the unit tensor $\boldsymbol{1}$ maps each vector onto itself, we can rewrite Eq. (A.2.66) in the following form:

$$(\boldsymbol{W} - \lambda \boldsymbol{1}) \cdot \mathbf{n} = \mathbf{0}. \tag{A.2.67}$$

If we now utilize the matrices of $\boldsymbol{W}$, $\boldsymbol{1}$, and $\mathbf{n}$, we get a homogeneous linear equation system:

$$\underbrace{\begin{bmatrix} w_{xx} - \lambda & w_{xy} & w_{xz} \\ w_{yx} & w_{yy} - \lambda & w_{yz} \\ w_{zx} & w_{zy} & w_{zz} - \lambda \end{bmatrix}}_{\underline{\mathbf{W}} - \lambda \underline{\mathbf{1}}} \underbrace{\begin{bmatrix} n_x \\ n_y \\ n_z \end{bmatrix}}_{\underline{\mathbf{n}}} = 0. \tag{A.2.68}$$

Let $P_3(\lambda) = -\det\left(\underline{\mathbf{W}} - \lambda \underline{\mathbf{1}}\right)$. This function is a cubic polynomial of λ. Solution for $\mathbf{n}$ different from the trivial one exists if and only if

$$P_3(\lambda) = -\begin{vmatrix} w_{xx} - \lambda & w_{xy} & w_{xz} \\ w_{yx} & w_{yy} - \lambda & w_{yz} \\ w_{zx} & w_{zy} & w_{zz} - \lambda \end{vmatrix} = \lambda^3 - W_I \lambda^2 + W_{II} \lambda - W_{III} = 0, \tag{A.2.69}$$

where

$$W_I = w_{xx} + w_{yy} + w_{zz}\,, \tag{A.2.70a}$$

$$W_{II} = \begin{vmatrix} w_{xx} & w_{xy} \\ w_{yx} & w_{yy} \end{vmatrix} + \begin{vmatrix} w_{xx} & w_{xz} \\ w_{zx} & w_{zz} \end{vmatrix} + \begin{vmatrix} w_{yy} & w_{yz} \\ w_{zy} & w_{zz} \end{vmatrix} \tag{A.2.70b}$$

and

$$W_{III} = \begin{vmatrix} w_{xx} & w_{xy} & w_{xz} \\ w_{yx} & w_{yy} & w_{yz} \\ w_{zx} & w_{zy} & w_{zz} \end{vmatrix}. \tag{A.2.70c}$$

Since mapping (A.2.66) is independent of the coordinate system, the solutions for λ should also be coordinate system independent quantities. This requirement is satisfied if and only if the coefficients W_I, W_{II}, and W_{III} are also independent of the coordinate system. For this reason, they are called scalar invariants of the tensor $\boldsymbol{W}$.

The most important results for the eigenvalue problem considered are as follows [1]:

Assume that the three eigenvalues are ordered: $\lambda_1 \geq \lambda_2 \geq \lambda_3$. Assume further the eigenvectors are attached to the same point, i.e., to the point where the tensor $\boldsymbol{W}$ is defined.

(a) If the three eigenvalues are different, there exist three eigenvectors $\mathbf{n}_1$, $\mathbf{n}_2$, and $\mathbf{n}_3$ which are mutually perpendicular to each other. Therefore, they constitute an orthonormal basis in the 3D space in which

$$\boldsymbol{W} = \sum_{\ell=1}^{3} \lambda_\ell \, \mathbf{n}_\ell \circ \mathbf{n}_\ell \tag{A.2.71}$$

is the form of the tensor.

(b) If λ_1 is a single root and λ_2 is a double root, then the eigenvector $\mathbf{n}_1$ is unique while any direction in the plane perpendicular to $\mathbf{n}_1$ is a principal direction. Select the vectors $\mathbf{n}_2$ and $\mathbf{n}_3$ perpendicularly to each other and $\mathbf{n}_1$. Then the vectors $\mathbf{n}_1$, $\mathbf{n}_2$, and $\mathbf{n}_3$ constitute a basis in the 3D space and

$$\begin{aligned} \boldsymbol{W} &= \lambda_1 \, \mathbf{n}_1 \circ \mathbf{n}_1 + \lambda_2 \, (\mathbf{n}_2 \circ \mathbf{n}_2 + \mathbf{n}_3 \circ \mathbf{n}_3) = \underset{\text{(A.2.31)}}{\uparrow} = \\ &= (\lambda_1 - \lambda_2) \, \mathbf{n}_1 \circ \mathbf{n}_1 + \lambda_2 \, \underbrace{(\mathbf{n}_1 \circ \mathbf{n}_1 + \mathbf{n}_2 \circ \mathbf{n}_2 + \mathbf{n}_3 \circ \mathbf{n}_3)}_{\boldsymbol{I}} = \\ &= (\lambda_1 - \lambda_2) \, \mathbf{n}_1 \circ \mathbf{n}_1 + \lambda_2 \boldsymbol{I} \end{aligned} \tag{A.2.72}$$

is the form of the tensor $\boldsymbol{W}$ in the basis mentioned. Here, we have not assumed that the roots are ordered.

(c) If the three roots are the same denoted, say, by λ_1, then every direction is a principal direction. This means that we can select again an orthonormal basis $\mathbf{n}_1$, $\mathbf{n}_2$, and $\mathbf{n}_3$ in which

$$\boldsymbol{W} = \lambda_1 \, (\mathbf{n}_1 \circ \mathbf{n}_1 + \mathbf{n}_2 \circ \mathbf{n}_2 + \mathbf{n}_3 \circ \mathbf{n}_3) = \lambda_1 \, \boldsymbol{I} \tag{A.2.73}$$

is the form of the tensor $\boldsymbol{W}$.

Appendix B
Some Useful Relationships

B.1 The Symbolic Dirac Function and Its Most Important Properties

B.1.1 The Dirac Function in One Variable

Let $f(x)$ be a continuous function in the neighborhood of the point P_1 with coordinate $x = \ell_1$. Further let $a < b$. The Dirac function of the variable x is defined by the following equations [2, 3] :

$$\delta(x - \ell_1) = \begin{cases} 0 & \text{if} \quad x \neq \ell_1\,, \\ \infty & \text{if} \quad x - \ell_1 = 0 \end{cases} \tag{B.1.1}$$

and

$$\int_a^b f(x)\,\delta(x - \ell_1)\,\mathrm{d}x = \begin{cases} 0 & \text{if} \quad \ell_1 \notin [a, b]\,, \\ \dfrac{1}{2} f(\ell_1) & \text{if} \quad \ell_1 = a \quad \text{or} \quad \ell_1 = b\,, \\ f(\ell_1) & \text{if} \quad \ell_1 \in (a, b)\,, \end{cases} \tag{B.1.2}$$

which shows that the Dirac[1] functions tend to infinity for $x = \ell_1$ in such a way that relations (B.1.2) are also satisfied.

The Dirac function is not a real function, and its integrals do not exist in the traditional sense of the integral concept of the classical analysis. Instead, it is such a generalized function, or symbol, which can be used to calculate the value $f(\ell_1)$ of the function $f(x)$ at $x = \ell_1$ by performing a formal integral transformation. The latter property of the Dirac function is very useful since it makes possible to model a concentrated load as if it were a distributed one.

---

[1] Paul Adrien Maurice Dirac (1902–1984).

G. Szeidl and L. P. Kiss, *Mechanical Vibrations*, Foundations of Engineering Mechanics,
https://doi.org/10.1007/978-3-030-45074-8

Since the independent variable x belongs to a point on the x axis, denoted, say, by P—the point P is naturally not fixed—and the coordinate $\ell_1 = x_Q$ belongs to a fixed point denoted, say, by Q we can rewrite Eq. (B.1.2) in the following form:

$$\int_a^b f(P)\,\delta(P-Q)\,\mathrm{d}x = \begin{cases} 0 & \text{if } \; Q \notin [a,b] \\ \dfrac{1}{2} f(Q) & \text{if } \; Q \text{ coincides with } a \text{ or } b\,, \\ f(Q) & \text{if } \; \xi_{1Q} \in (a,b) \end{cases} \tag{B.1.3}$$

where the arc element ds_P taken at the point P corresponds to $\mathrm{d}x$.

B.1.2 The Dirac Function in Two Variables

Let P and Q be two points in the coordinate plane (x, y) not necessarily different from each other. The corresponding position vectors are denoted by $\mathbf{r}_P = x\mathbf{i}_x + y\mathbf{i}_y$ and $\mathbf{r}_Q = x_Q\mathbf{i}_x + y_Q\mathbf{i}_y$. The Dirac delta function in the coordinate plane (x, y) is defined by the following equation:

$$\delta(P-Q) = \delta(x-x_Q)\delta(y-y_Q)\,. \tag{B.1.4}$$

Let S be a plane region. Further let $f(x, y) = f(P)$ be a continuous function in two variables in the coordinate plane (x, y). Then it follows from (B.1.2) that

$$\int_S f(P)\,\delta(P-Q)\,\mathrm{s}A_P = \begin{cases} 0 & \text{if } \; Q \notin S\,, \\ f(Q) & \text{if } \; Q \in S\,. \end{cases} \tag{B.1.5}$$

B.1.3 The Dirac Function in the 3D Space

Definition of the Dirac function in the 3D space is basically the same as its definition in the 2D space. Let P and Q two points in the 3D space with coordinates x, y, z and x_Q, y_Q, z_Q. Further let $f(Q)$ be a continuous function in the 3D space. Then it holds that

$$\int_V f(P)\,\delta(P-Q)\,\mathrm{d}V_P = \begin{cases} 0 & \text{if } \; Q \notin V\,, \\ f(Q) & \text{if } \; Q \in V \end{cases} \tag{B.1.6}$$

where V_P is a volume region of the 3D space.

B.2 Bessel Functions—Fundamental Relations

Let us denote the Bessel[2] functions $J_n(x)$ and $Y_n(x)$ uniformly by $H_n(x)$. The following relations hold:

$$\frac{dH_0(x)}{dx} = -H_1\,(x)\ , \tag{B.2.7a}$$

$$\frac{dH_1(x)}{dx} = \frac{1}{x}\,(H_0\,(x)\,x - H_1\,(x)) = H_0\,(x) - \frac{1}{x}H_1\,(x) = \frac{1}{2}\,(H_0(x) - H_2(x))\,, \tag{B.2.7b}$$

$$\frac{dH_n(x)}{dx} = \frac{1}{2}\,(H_{n-1}(x) - H_{n+1}(x))\,, \tag{B.2.7c}$$

$$\frac{2n}{x}H_n(x) = H_{n-1}(x) + H_{n+1}(x) \tag{B.2.7d}$$

and

$$Y_n(x)J_{n+1}(x) - Y_{n+1}(x)J_n(x) = \frac{2}{\pi x}\ . \tag{B.2.7e}$$

See [4, 1977a] a for more details.

As regards the Bessel functions $I_n(x)$ and $K_n(x)$ use has been made of the following relations:

$$\frac{dI_0(x)}{dx} = I_1(x)\ , \qquad \frac{dK_0(x)}{dx} = -K_1(x)\ , \tag{B.2.8a}$$

$$\frac{dI_1(x)}{dx} = \frac{1}{x}\,(x\,I_0\,(x) - I_1\,(x)) = I_0\,(x) - \frac{1}{x}I_1\,(x) = \frac{1}{2}\,(I_0(x) + I_2(x))\ , \tag{B.2.8b}$$

$$\frac{dI_n(x)}{dx} = \frac{1}{2}\left(I_{n-1}(x) + I_{n+1}(x)\right)\ , \tag{B.2.8c}$$

$$\frac{2n}{x}I_n(x) = I_{n-1}(x) - I_{n+1}(x)\ , \tag{B.2.8d}$$

$$\frac{dK_1(x)}{dx} = -\frac{1}{x}\,(x\,K_0\,(x) + K_1\,(x)) = -K_0\,(x) - \frac{1}{x}K_1\,(x) = -\frac{1}{2}\,(K_0(x) + K_2(x))\ , \tag{B.2.8e}$$

$$\frac{dK_n(x)}{dx} = -\frac{1}{2}\left(K_{n-1}(x) + K_{n+1}(x)\right)\ , \tag{B.2.8f}$$

$$-\frac{2n}{x}I_n(x) = K_{n-1}(x) - K_{n+1}(x) \tag{B.2.8g}$$

and

$$I_n(x)K_{n+1}(x) + I_{n+1}(x)K_n(x) = \frac{1}{x}\ . \tag{B.2.8h}$$

See [4, 1977b] for more details.

[2]Friedrich Wilhelm Bessel (1784–1846).

B.3 Gaussian Quadrature Abscissae and Weights

This section contains the tabulations for calculating integral approximations in the following form[3]:

$$J(\phi) = \int_{\eta=-1}^{\eta=1} \phi(\eta)\,\mathrm{d}\eta = \sum_{j=1}^{n} w_j \phi(\eta_j)\,, \tag{B.3.9}$$

in which w_j is the weight and η_j is the abscissa where the function to be integrated should be taken. The data w_j and η_j in Tables B.1, B.2, B.3, B.4, B.5 and B.6 are presented for $n = 4, 5, 6, 7, 8, 10$.

Table B.1 Weights and abscissae of the Gauss points if $n = 4$

j	Abscissa η_i	Weight w_i
1	0.6521451548625461	-0.3399810435848563
2	0.6521451548625461	0.3399810435848563
3	0.3478548451374538	-0.8611363115940526
4	0.3478548451374538	0.8611363115940526

Table B.2 Weights and abscissae of the Gauss points if $n = 5$

j	Abscissa η_i	Weight w_i
1	-0.9061798459386640	0.2369268850561891
2	-0.5384693101056831	0.4786286704993665
3	0.0000000000000000	0.5688888888888889
4	0.5384693101056831	0.4786286704993665
5	0.9061798459386640	0.2369268850561891

Table B.3 Weights and abscissae of the Gauss points if $n = 6$

j	Abscissa η_i	Weight w_i
1	-0.9324695142031521	0.1713244923791704
2	-0.6612093864662645	0.3607615730481386
3	-0.2386191860831969	0.4679139345726910
4	0.2386191860831969	0.4679139345726910
5	0.6612093864662645	0.3607615730481386
6	0.9324695142031521	0.1713244923791704

[3]These data are taken from the homepage https://pomax.github.io/bezierinfo/legendre-gauss.html.

Table B.4 Weights and abscissae of the Gauss points if $n = 7$

j	Abscissa η_i	Weight w_i
1	-0.9491079123427585	0.129484966168869
2	-0.7415311855993945	0.2797053914892766
3	-0.4058451513773972	0.3818300505051189
4	0.000000000000000	0.4179591836734694
5	0.4058451513773972	0.3818300505051189
6	0.7415311855993945	0.2797053914892766
7	0.9491079123427585	0.129484966168869

Table B.5 Weights and abscissae of the Gauss points if $n = 8$

j	Abscissa η_i	Weight w_i
1	-0.9602898564975363	0.1012285362903763
2	-0.7966664774136267	0.2223810344533745
3	-0.5255324099163290	0.3137066458778873
4	-0.1834346424956498	0.3626837833783620
5	0.1834346424956498	0.3626837833783620
6	0.5255324099163290	0.3137066458778873
7	0.7966664774136267	0.2223810344533745
8	0.9602898564975363	0.1012285362903763

Table B.6 Weights and abscissae of the Gauss points if $n = 10$

j	Abscissa η_i	Weight w_i
1	-0.9739065285171717	0.0666713443086881
2	-0.8650633666889845	0.1494513491505806
3	-0.6794095682990244	0.2190863625159820
4	-0.4333953941292472	0.2692667193099963
5	-0.1488743389816312	0.2955242247147529
6	0.1488743389816312	0.2955242247147529
7	0.4333953941292472	0.2692667193099963
8	0.6794095682990244	0.2190863625159820
9	0.8650633666889845	0.1494513491505806
10	0.9739065285171717	0.0666713443086881

B.4 Shear Correction Factor

A few shear correction factors are presented on the basis of [5, 6] in Table B.7 where ν is the Poisson number:

Table B.7 Shear correction factors

Cross section	Name	κ_y
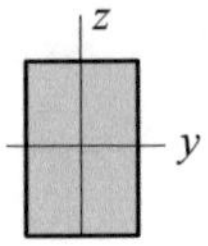	Rectangle	$\dfrac{10(1+\nu)}{12+11\nu}$
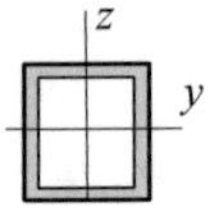	Thin-walled Square Tube	$\dfrac{20(1+\nu)}{48+39\nu}$
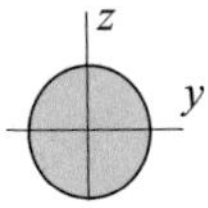	Circle	$\dfrac{6(1+\nu)}{7+6\nu}$
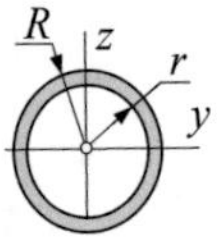	Hollow Cylinder	$\dfrac{6(1+\nu)(1+r/R)^2}{(7+6\nu)(1+r/R)^2+4(5+3\nu)(r/R)^2}$
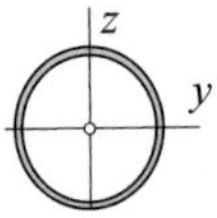	Thin-walled Circular Tube	$\dfrac{2(1+\nu)}{4+\nu}$
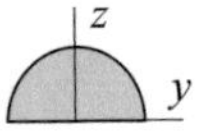	Semi-circle	$\dfrac{1+\nu}{1.305+1.273\nu}$

Appendix C
Solutions to Selected Problems

C.1 Problems in Chap. 1

Problem 1.1 The proof is presented by using tensorial notations. Figure C.1 shows the mass center G, the position vector $\boldsymbol{\rho}$ of the mass element with respect to the mass center G, the direction vectors $\mathbf{n}$ and $\mathbf{m}$ for which it holds that $|\mathbf{n}| = |\mathbf{m}| = 1$, $\mathbf{n} \cdot \mathbf{m} = 0$, $\boldsymbol{\rho} \cdot \mathbf{n} = \rho_n$ and $\boldsymbol{\rho} \cdot \mathbf{m} = \rho_m$.

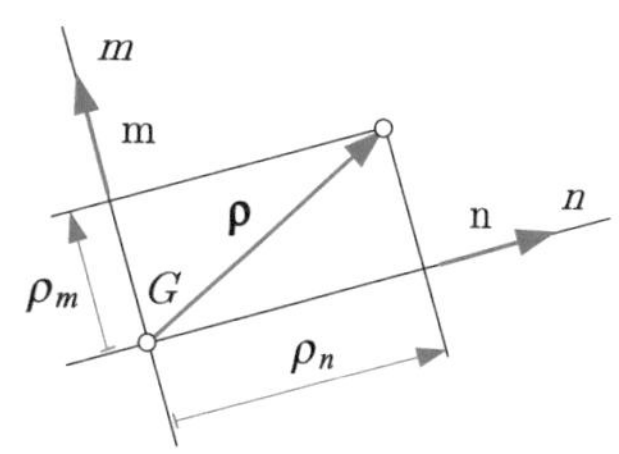

Fig. C.1 Resolution of $\boldsymbol{\rho}$ into two components

Recalling (1.22) for the product $(1.17\text{d})_1$ we can write

$$\mathbf{n} \cdot \boldsymbol{J}_G \cdot \mathbf{n} = \int_V \mathbf{n} \cdot \left(\rho^2 \boldsymbol{1} - \boldsymbol{\rho} \circ \boldsymbol{\rho}\right) \cdot \mathbf{n} \underbrace{\rho \mathrm{d}V}_{\mathrm{d}m} = \int_V \Big(\rho^2 \underbrace{\mathbf{n} \cdot \boldsymbol{1} \cdot \mathbf{n}}_{\mathbf{n}} - \underbrace{\mathbf{n} \cdot \boldsymbol{\rho}}_{\rho_n} \circ \underbrace{\boldsymbol{\rho} \cdot \mathbf{n}}_{\rho_n}\Big)\, \mathrm{d}m = \underset{\mathbf{n}\cdot\mathbf{n}=1}{\uparrow} =$$

$$= \int_V \left(\rho^2 - \rho_n^2\right) \mathrm{d}m = \underset{\rho^2 - \rho_n^2 = \rho_m^2}{\uparrow} = \int_V \rho_m^2 \,\mathrm{d}m = J_n,$$

which is really the moment of inertia with respect to the axis n.

For the product $(1.17\text{d})_2$, a similar line of thought yields

$$-\mathbf{m} \cdot \boldsymbol{J}_G \cdot \mathbf{n} = -\int_V \mathbf{m} \cdot \left(\rho^2 \boldsymbol{1} - \boldsymbol{\rho} \circ \boldsymbol{\rho}\right) \cdot \mathbf{n} \underbrace{\rho \mathrm{d}V}_{\mathrm{d}m} =$$

$$= -\int_V \Big(\rho^2 \underbrace{\mathbf{m} \cdot \boldsymbol{1} \cdot \mathbf{n}}_{\mathbf{n}} - \underbrace{\mathbf{m} \cdot \boldsymbol{\rho}}_{\rho_m} \circ \underbrace{\boldsymbol{\rho} \cdot \mathbf{n}}_{\rho_n}\Big)\, \mathrm{d}m = \underset{\mathbf{m}\cdot\mathbf{n}=0}{\uparrow} = \int_V \rho_m \rho_n \,\mathrm{d}m = J_{mn},$$

G. Szeidl and L. P. Kiss, *Mechanical Vibrations*, Foundations of Engineering Mechanics,
https://doi.org/10.1007/978-3-030-45074-8

which is the product of inertia with respect to the axes n and m.
Problem 1.2 The moment of momentum about the point A is

$$\mathbf{H}_A = \int_V \hat{\boldsymbol{\rho}} \times \mathbf{v}\, \rho \mathrm{d}V$$

in which

$$\mathbf{v} = \mathbf{v}_A + \boldsymbol{\omega} \times \hat{\boldsymbol{\rho}}\,.$$

Hence

$$\mathbf{H}_A = \int_V \hat{\boldsymbol{\rho}} \times \left(\mathbf{v}_A + \boldsymbol{\omega} \times \hat{\boldsymbol{\rho}}\right) \rho \mathrm{d}V = \int_V \hat{\boldsymbol{\rho}}\, \rho \mathrm{d}V \times \mathbf{v}_A + \int_V \hat{\boldsymbol{\rho}} \times \left(\boldsymbol{\omega} \times \hat{\boldsymbol{\rho}}\right) \rho \mathrm{d}V, \tag{C.1.1}$$

where

$$\int_V \hat{\boldsymbol{\rho}}\, \rho \mathrm{d}V \times \mathbf{v}_A \underset{(1.7)}{=\uparrow=} \underbrace{\mathbf{Q}_A}_{m\mathbf{r}_{AG}} \times \mathbf{v}_A = m\mathbf{r}_{AG} \times \mathbf{v}_A \tag{C.1.2a}$$

and

$$\int_V \hat{\boldsymbol{\rho}} \times \left(\boldsymbol{\omega} \times \hat{\boldsymbol{\rho}}\right) \rho \mathrm{d}V = \int_V \left(\hat{\rho}^2 \boldsymbol{\omega} - \hat{\boldsymbol{\rho}}\left(\hat{\boldsymbol{\rho}} \cdot \boldsymbol{\omega}\right)\right) \rho \mathrm{d}V =$$
$$= \int_V \left(\hat{\rho}^2 \boldsymbol{I} - \hat{\boldsymbol{\rho}} \circ \hat{\boldsymbol{\rho}}\right) \rho \mathrm{d}V \cdot \boldsymbol{\omega}\,. \tag{C.1.2b}$$

In this equation, $\boldsymbol{I}$ is the unit tensor and $\circ$ stands for the operation sign of a dyadic product. Integral

$$\boldsymbol{J}_A = \int_V \left(\hat{\rho}^2 \boldsymbol{I} - \hat{\boldsymbol{\rho}} \circ \hat{\boldsymbol{\rho}}\right) \rho \mathrm{d}V$$

defines the tensor of inertia at A. Using the tensor of inertia $\boldsymbol{J}_A$, integral (C.1.2b) can be given in the form

$$\int_V \hat{\boldsymbol{\rho}} \times \left(\boldsymbol{\omega} \times \hat{\boldsymbol{\rho}}\right) \rho \mathrm{d}V = \boldsymbol{J}_A \cdot \boldsymbol{\omega}\,. \tag{C.1.2c}$$

Comparison of Eqs. (C.1.1), (C.1.2a) and (C.1.2c) yields the equation to be proved in tensorial notions:

$$\mathbf{H}_A = m\mathbf{r}_{AG} \times \mathbf{v}_A + \boldsymbol{J}_A \cdot \boldsymbol{\omega}. \tag{C.1.3}$$

Problem 1.3 To prove the parallel axis theorem rewrite Eq. (C.1.2c) by taking in account that $\hat{\boldsymbol{\rho}} = \mathbf{r}_{AG} + \boldsymbol{\rho}$. We obtain

$$\boldsymbol{J}_A \cdot \boldsymbol{\omega} = \int_V \hat{\boldsymbol{\rho}} \times \left(\boldsymbol{\omega} \times \hat{\boldsymbol{\rho}}\right) \rho \mathrm{d}V = \int_V (\mathbf{r}_{AG} + \boldsymbol{\rho}) \times [\boldsymbol{\omega} \times (\mathbf{r}_{AG} + \boldsymbol{\rho})] \rho \mathrm{d}V =$$

$$= \int_V \boldsymbol{\rho} \times (\boldsymbol{\omega} \times \boldsymbol{\rho}) \rho \mathrm{d}V + \mathbf{r}_{AG} \times (\boldsymbol{\omega} \times \mathbf{r}_{AG}) \underbrace{\int_V \rho \mathrm{d}V}_{m} +$$

$$+ \mathbf{r}_{AG} \times \left(\boldsymbol{\omega} \times \underbrace{\int_V \boldsymbol{\rho} \rho \mathrm{d}V}_{\mathbf{Q}_G = \mathbf{0}}\right) + \underbrace{\int_V \boldsymbol{\rho} \rho \mathrm{d}V}_{\mathbf{Q}_G = \mathbf{0}} \times (\boldsymbol{\omega} \times \mathbf{r}_{AG}) \, .$$

Here, we utilized that (a) the first moment with respect to the mass center G is zero, and (b) the integral $\int_V \rho \, \mathrm{d}V$ is the mass of the body. Consequently, it holds that

$$\boldsymbol{J}_A \cdot \boldsymbol{\omega} = \int_V \boldsymbol{\rho} \times (\boldsymbol{\omega} \times \boldsymbol{\rho}) \rho \mathrm{d}V + m \mathbf{r}_{AG} \times (\boldsymbol{\omega} \times \mathbf{r}_{AG}) \, . \tag{C.1.4}$$

If we now recall (2.1) and (1.23), we get

$$\int_V \boldsymbol{\rho} \times (\boldsymbol{\omega} \times \boldsymbol{\rho}) \rho \mathrm{d}V = \boldsymbol{J}_G \cdot \boldsymbol{\omega} \, . \tag{C.1.5a}$$

After expanding the triple vector product $m\mathbf{r}_{AG} \times (\boldsymbol{\omega} \times \mathbf{r}_{AG})$, we find

$$m\mathbf{r}_{AG} \times (\boldsymbol{\omega} \times \mathbf{r}_{AG}) = m \left(\mathbf{r}_{AG}^2 \boldsymbol{\omega} - \mathbf{r}_{AG} \left(\mathbf{r}_{AG} \cdot \boldsymbol{\omega}\right) \rho\right) =$$

$$= m \underbrace{\left(\mathbf{r}_{AG}^2 \boldsymbol{I} - \mathbf{r}_{AG} \circ \mathbf{r}_{AG}\right)}_{\boldsymbol{J}_{AG}} \cdot \boldsymbol{\omega} = m \boldsymbol{J}_{AG} \cdot \boldsymbol{\omega},$$

where

$$\boldsymbol{J}_{AG} = \mathbf{r}_{AG}^2 \boldsymbol{I} - \mathbf{r}_{AG} \circ \mathbf{r}_{AG} \tag{C.1.5b}$$

is a tensor of order two. Using the tensor $\boldsymbol{J}_{AG}$, we can rewrite the triple cross product:

$$m\mathbf{r}_{AG} \times (\boldsymbol{\omega} \times \mathbf{r}_{AG}) = m \boldsymbol{J}_{AG} \cdot \boldsymbol{\omega} \, . \tag{C.1.5c}$$

After substituting (C.1.5a) and (C.1.5c) into (C.1.4), we arrive at the following result:

$$\boldsymbol{J}_A \cdot \boldsymbol{\omega} = (\boldsymbol{J}_G + m \, \boldsymbol{J}_{AG}) \cdot \boldsymbol{\omega},$$

which should hold for any $\boldsymbol{\omega}$. Hence

$$\boldsymbol{J}_A = \boldsymbol{J}_G + m \, \boldsymbol{J}_{AG}, \tag{C.1.6}$$

which is the parallel axis theorem in tensorial form.

Problem 1.4 The proof needs some preparation. Figure C.2 shows two coordinate systems. The coordinate system (xyz) is at rest (this is the absolute coordinate system). The coordinate system $(\xi\eta\zeta)$ is in motion: it moves as if it were a rigid body.

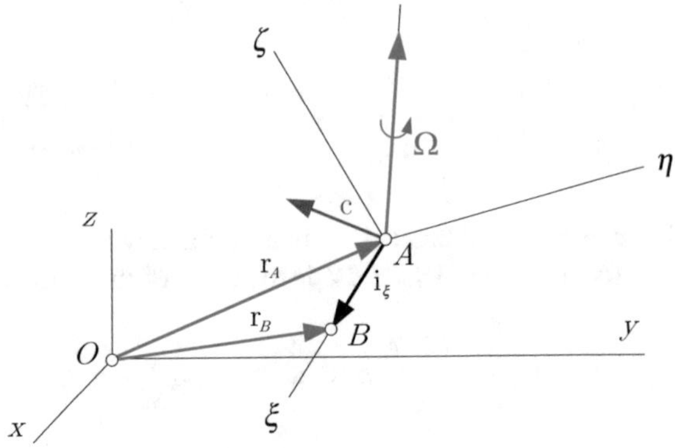

Fig. C.2 Absolute and moving coordinate systems

The unit vectors in the two coordinate systems are denoted by $\mathbf{i}_x$, $\mathbf{i}_y$, $\mathbf{i}_z$ and $\mathbf{i}_\xi$, $\mathbf{i}_\eta$, $\mathbf{i}_\zeta$. We remark that Fig. C.2 shows the unit vector $\mathbf{i}_\xi$ only. For the observer at the origin O (in the coordinate system (xyz)), the unit vectors $\mathbf{i}_\xi$, $\mathbf{i}_\eta$, $\mathbf{i}_\zeta$ depend on time, i.e.,

$$\mathbf{i}_\xi = \mathbf{i}_\xi(t) , \qquad \mathbf{i}_\eta = \mathbf{i}_\eta(t) , \qquad \mathbf{i}_\zeta = \mathbf{i}_\zeta(t) .$$

For the observer at the origin A (in the coordinate system $(\xi\eta\zeta)$), the unit vectors $\mathbf{i}_\xi$, $\mathbf{i}_\eta$, $\mathbf{i}_\zeta$ are, however, constants (are independent of time). Assume that the angular velocity of the Greek coordinate system (the moving coordinate system) $\boldsymbol{\Omega} = \boldsymbol{\Omega}(t)$ is known. Assume further that we also know the position vectors $\mathbf{r}_A = \mathbf{r}_A(t)$ and $\mathbf{r}_B = \mathbf{r}_B(t)$. Since $\mathbf{r}_B - \mathbf{r}_A = \mathbf{r}_{AB} = \mathbf{i}_\xi$, it holds—$\boldsymbol{\Omega}$ corresponds to $\boldsymbol{\omega}$ in (1.5)—that

$$\frac{\mathrm{d}\mathbf{i}_\xi}{\mathrm{d}t} = \frac{\mathrm{d}}{\mathrm{d}t}(\mathbf{r}_B - \mathbf{r}_A) = \mathbf{v}_B - \mathbf{v}_A \underset{(1.5)}{=\uparrow=} \boldsymbol{\Omega} \times \mathbf{r}_{AB} = \boldsymbol{\Omega} \times (\mathbf{r}_B - \mathbf{r}_A) = \boldsymbol{\Omega} \times \mathbf{i}_\xi .$$

On the base of this equation, we can come to the conclusion that the time derivatives of the unit vectors $\mathbf{i}_\xi$, $\mathbf{i}_\eta$, $\mathbf{i}_\zeta$ in the Latin (the absolute) coordinate system are given by

$$\frac{\mathrm{d}\mathbf{i}_\xi}{\mathrm{d}t} = \boldsymbol{\Omega} \times \mathbf{i}_\xi , \qquad \frac{\mathrm{d}\mathbf{i}_\eta}{\mathrm{d}t} = \boldsymbol{\Omega} \times \mathbf{i}_\eta , \qquad \frac{\mathrm{d}\mathbf{i}_\zeta}{\mathrm{d}t} = \boldsymbol{\Omega} \times \mathbf{i}_\zeta . \tag{C.1.7}$$

In what follows the time derivative of a quantity in the absolute coordinate system will be denoted by

$$\left.\frac{\mathrm{d}(\ldots)}{\mathrm{d}t}\right|_{(xyz)} \qquad \text{or} \qquad (\ldots)^{\cdot} . \tag{C.1.8a}$$

To make a difference, the time derivative of a quantity in the Greek coordinate system (in the moving coordinate system) will, however, be denoted by

$$\left.\frac{\mathrm{d}(\ldots)}{\mathrm{d}t}\right|_{(\xi\eta\zeta)} \qquad \text{or} \qquad (\ldots)^* \,. \tag{C.1.8b}$$

If the quantity considered is scalar, there is no difference between the two time derivatives:

$$(\ldots)^{\cdot} = (\ldots)^* \,. \tag{C.1.8c}$$

Let us now examine the time rate of the vector $\mathbf{c}(t)$, which can be given in both coordinate systems:

$$\mathbf{c}(t) = c_x\,\mathbf{i}_x + c_y\,\mathbf{i}_y + c_z\,\mathbf{i}_z = c_\xi\,\mathbf{i}_\xi + c_\eta\,\mathbf{i}_\eta + c_\zeta\,\mathbf{i}_\zeta \,.$$

For the observer in the Greek coordinate system

$$\left.\frac{\mathrm{d}\mathbf{c}}{\mathrm{d}t}\right|_{(\xi\eta\zeta)} = \overset{*}{\mathbf{c}} = \overset{*}{c}_\xi\,\mathbf{i}_\xi + \overset{*}{c}_\eta\,\mathbf{i}_\eta + \overset{*}{c}_\zeta\,\mathbf{i}_\zeta = \dot{c}_\xi\,\mathbf{i}_\xi + \dot{c}_\eta\,\mathbf{i}_\eta + \dot{c}_\zeta\,\mathbf{i}_\zeta \tag{C.1.9}$$

is the time rate sought. For the observer in the absolute coordinate system, it holds that

$$\left.\frac{\mathrm{d}\mathbf{c}}{\mathrm{d}t}\right|_{(xyz)} = \dot{\mathbf{c}} = \dot{c}_x\,\mathbf{i}_x + \dot{c}_y\,\mathbf{i}_y + \dot{c}_z\,\mathbf{i}_z = \underbrace{\dot{c}_\xi\,\mathbf{i}_\xi + \dot{c}_\eta\,\mathbf{i}_\eta + \dot{c}_\zeta\,\mathbf{i}_\zeta}_{\overset{*}{\mathbf{c}}} + c_\xi\,\overset{*}{\mathbf{i}}_\xi + c_\eta\,\overset{*}{\mathbf{i}}_\eta + c_\zeta\,\overset{*}{\mathbf{i}}_\zeta = \underset{(\text{C.1.7})}{\uparrow} =$$

$$= \overset{*}{\mathbf{c}} + \boldsymbol{\Omega}\times c_\xi\,\mathbf{i}_\xi + \boldsymbol{\Omega}\times c_\eta\,\mathbf{i}_\eta + \boldsymbol{\Omega}\times c_\zeta\,\mathbf{i}_\zeta = \overset{*}{\mathbf{c}} + \boldsymbol{\Omega}\times\mathbf{c}\,. \tag{C.1.10}$$

The result

$$\boxed{\dot{\mathbf{c}} = \overset{*}{\mathbf{c}} + \boldsymbol{\Omega}\times\mathbf{c}} \tag{C.1.11}$$

we have just arrived at relates the time rate of the vector field $\mathbf{c}$ in the absolute coordinate system to that of the vector field $\mathbf{c}$ considered in the moving coordinate system. This result is known as the Coriolis theorem [7, 8].

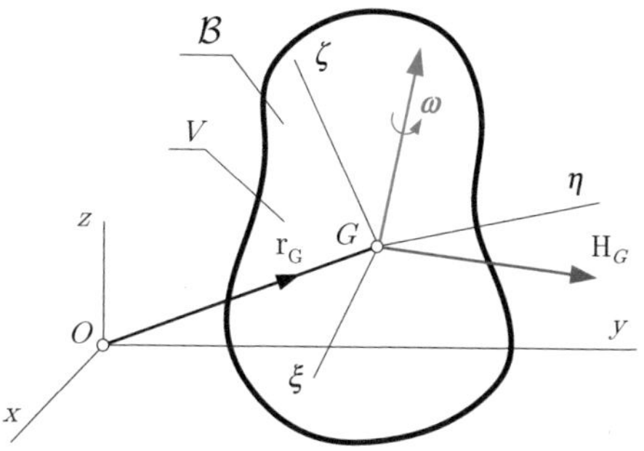

Fig. C.3 Moving body with the moment of momentum vector

Figure C.3 shows the moving body $\mathcal{B}$. The moment of momentum $\mathbf{H}_G$ is also shown at the mass center G. The origin of the Greek coordinate system $(\xi\eta\zeta)$ always coincides with the mass center and we shall assume that this coordinate system moves together with the body, which means that it is rigidly attached to the body. Hence, its angular velocity is the same as that of the body: $\boldsymbol{\Omega} = \boldsymbol{\omega}$.

For the time derivative of the angular velocity, it holds—write $\boldsymbol{\omega}$ for $\mathbf{c}$ in (C.1.11)—that

$$\dot{\boldsymbol{\omega}} = \overset{*}{\boldsymbol{\omega}} + \underbrace{\boldsymbol{\omega} \times \boldsymbol{\omega}}_{=\mathbf{0}} = \overset{*}{\boldsymbol{\omega}} = \boldsymbol{\alpha}, \tag{C.1.12}$$

where $\boldsymbol{\alpha}$ is the angular acceleration.

The moment of momentum about G is given by

$$\mathbf{H}_G = \boldsymbol{J}_G \cdot \boldsymbol{\omega},$$

in which the tensor of inertia $\boldsymbol{J}_G$ is constant in the coordinate system $(\xi\eta\zeta)$. If we apply the rule (C.1.11) to $\mathbf{H}_G$, we get

$$\left.\frac{\mathrm{d}\mathbf{H}_G}{\mathrm{d}t}\right|_{(xyz)} = \dot{\mathbf{H}}_G = \overset{*}{\mathbf{H}}_G + \boldsymbol{\omega} \times \mathbf{H}_G = \underbrace{\overset{*}{\boldsymbol{J}}_G}_{\text{zero tensor}} \cdot \boldsymbol{\omega} + \boldsymbol{J}_G \cdot \underbrace{\overset{*}{\boldsymbol{\omega}}}_{\boldsymbol{\alpha}} + \boldsymbol{\omega} \times \boldsymbol{J}_G \cdot \boldsymbol{\omega} =$$

$$= \boldsymbol{J}_G \cdot \boldsymbol{\alpha} + \boldsymbol{\omega} \times \boldsymbol{J}_G \cdot \boldsymbol{\omega} = \underset{(1.38)}{\uparrow} = \boldsymbol{\mathcal{D}}_G\,.$$

That was to be proved.

C.2 Problems in Chap. 2

Problem 2.1 The kinetic energy of the two bodies before and after impact are given by

$$\mathcal{E}_b = \frac{1}{2} m_1 \left(v_{1n}^2 + v_{1t}^2\right) + \frac{1}{2} m_2 \left(v_{2n}^2 + v_{2t}^2\right)$$

and

$$\mathcal{E}_a = \frac{1}{2} m_1 \left(\left(v'_{1n}\right)^2 + v_{1t}^2\right) + \frac{1}{2} m_2 \left(\left(v'_{2n}\right)^2 + v_{2t}^2\right).$$

Hence

$$\mathcal{E}_{loss} = \mathcal{E}_b - \mathcal{E}_a = \frac{1}{2} m_1 \left(v_{1n}^2 - \left(v'_{1n}\right)^2\right) + \frac{1}{2} m_2 \left(v_{2n}^2 - \left(v'_{2n}\right)^2\right)$$

is the energy loss. After substituting solutions (2.12) and then c_n from (2.9b), we get

$$\begin{aligned}
\mathcal{E}_{loss} &= \frac{1}{2}m_1\left(v_{1n}^2-\left(v'_{1n}\right)^2\right)+\frac{1}{2}m_2\left(v_{2n}^2-\left(v'_{2n}\right)^2\right)= \\
&=\frac{1}{2}m_1\left(v_{1n}^2-c_n^2+2c_ne\frac{m_2}{m_1+m_2}\left(v_{1n}-v_{2n}\right)-e^2\frac{m_2^2}{\left(m_1+m_2\right)^2}\left(v_{1n}-v_{2n}\right)^2\right)+ \\
&+\frac{1}{2}m_2\left(v_{2n}^2-c_n^2-2c_ne\frac{m_1}{m_1+m_2}\left(v_{1n}-v_{2n}\right)-e^2\frac{m_1^2}{\left(m_1+m_2\right)^2}\left(v_{1n}-v_{2n}\right)^2\right)= \\
&=\frac{1}{2}\left[m_1v_{1n}^2+m_2v_{2n}^2-\frac{\left(m_1v_{1n}+m_2v_{2n}\right)^2}{m_1+m_2}-e^2\frac{m_1m_2}{m_1+m_2}\left(v_{1n}-v_{2n}\right)^2\right]= \\
&=\frac{1}{2}\frac{1}{m_1+m_2}\left[m_1v_{1n}^2\left(m_1+m_2\right)+m_2v_{2n}^2\left(m_1+m_2\right)-\left(m_1v_{1n}+m_2v_{2n}\right)^2-\right. \\
&\left.-e^2m_1m_2\left(v_{1n}-v_{2n}\right)^2\right]=\frac{1}{2}\frac{1}{m_1+m_2}\left[\left(v_{1n}-v_{2n}\right)^2-e^2m_1m_2\left(v_{1n}-v_{2n}\right)^2\right]
\end{aligned}$$

or

$$\mathcal{E}_{loss}=\frac{1}{2}\left(1-e^2\right)\frac{m_1m_2}{m_1+m_2}\left(v_{1n}-v_{2n}\right)^2 .$$

Problem 2.2 With the closed-form solutions (see Eqs. (2.9a), (2.10), and (2.12) for details)

$$\begin{aligned}
v'_{1t}&=v_{1t}\,,\quad v'_{2t}=v_{2t}\,, \\
c_n&=\frac{m_1v_{1n}+m_2v_{2n}}{m_1+m_2}
\end{aligned}$$

and

$$v'_{1n}=c_n-e\frac{m_2}{m_1+m_2}\left(v_{1n}-v_{2n}\right)\,,\qquad v'_{2n}=c_n+e\frac{m_1}{m_1+m_2}\left(v_{1n}-v_{2n}\right),$$

we get if $e=0.6$ that

$$\begin{aligned}
v'_{1t}&=v'_{1y}=v_{1y}=4\ [\text{m/s}]\,,\quad v'_{2t}=v'_{2y}=v_{2y}=-4\ [\text{m/s}]\,, \\
c_n&=\frac{m_1v_{1n}+m_2v_{2n}}{m_1+m_2}=\frac{m_1v_{1x}+m_2v_{2x}}{m_1+m_2}=\frac{2\cdot4+6\cdot0}{2+6}=1\ [\text{m/s}]\,, \\
v'_{1n}&=c_n-e\frac{m_2}{m_1+m_2}\left(v_{1n}-v_{2n}\right)=1-0.6\cdot\frac{6}{2+6}\left(4-0\right)=-0.8\ [\text{m/s}]\,, \\
v'_{2n}&=c_n+e\frac{m_1}{m_1+m_2}\left(v_{1n}-v_{2n}\right)=1+0.6\cdot\frac{2}{2+6}\left(4-0\right)=1.6\ [\text{m/s}]\,.
\end{aligned}$$

For $e=1$ and $e=0$, the calculations can now be performed with ease.

Problem 2.3 We have to solve the equation system

$$\begin{aligned}
m_1v'_{1n}+m_2v'_{2n}&=m_1v_{1n}+m_2v_{2n}\,, \\
v'_{2n}-v'_{1n}&=e\left(v_{1n}-v_{2n}\right)
\end{aligned}$$

for v_{1n} and v_{2n}:

$$v_{1n} = \frac{m_1 v'_{1n} + m_2 v'_{2n}}{m_1 + m_2} + \frac{m_2}{e(m_1 + m_2)}(v'_{2n} - v'_{1n}) = c_n + \frac{m_2}{e(m_1 + m_2)}(v'_{2n} - v'_{1n}),$$
$$v_{2n} = \frac{m_1 v'_{1n} + m_2 v'_{2n}}{m_1 + m_2} - \frac{m_1}{e(m_1 + m_2)}(v'_{2n} - v'_{1n}) = c_n - \frac{m_1}{e(m_1 + m_2)}(v'_{2n} - v'_{1n}).$$

After substituting the data of the problem into the above equations, we obtain the solutions sought:

$$v_{1n} = \frac{2 \cdot (-2) + 6 \cdot 2}{8} + \frac{6}{8}(2+2) = 1 + 3 = 4\,\text{m/s}.$$
$$v_{2n} = \frac{2 \cdot (-2) - 2 \cdot 2}{8} - \frac{2}{8}(2+2) = 1 - 1 = 0\,\text{m/s}.$$

Problem 2.4 The velocities $\mathbf{v}'_1$ and $\mathbf{v}'_1$ the pucks have after impact are shown in the Maxwell diagram—see Fig. C.4 for the details.

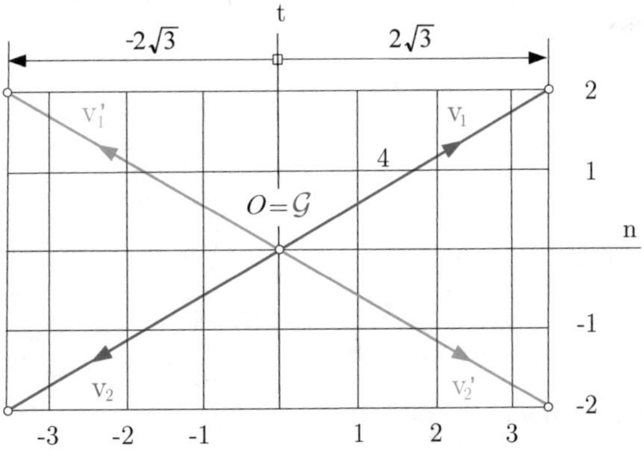

Fig. C.4 Solutions for the velocities $\mathbf{v}'_1$ and $\mathbf{v}'_2$

It is clear from Fig. C.4 that

$$\mathbf{v}'_1 = -2\sqrt{3}\mathbf{t} + 2\mathbf{t}\ [\text{m/s}]\,, \quad \mathbf{v}'_2 = 2\sqrt{3}\mathbf{t} - 2\mathbf{t}\ [\text{m/s}]\,.$$

Problem 2.7 The solution is based on Sect. 2.3.1.

(i) Figure C.5 shows the rod and support B before collision. The notations are basically the same as those for Figs. 2.14 and 2.15.
Support B (i.e., body $\mathcal{B}_1$) is at rest before, during and after impact. Hence

$$\mathbf{v}_{T_1} = \mathbf{v}'_{T_1} = \boldsymbol{\omega}_1 = \boldsymbol{\omega}'_1 = \mathbf{0}\,.$$

As regards the rod (i.e., body $\mathcal{B}_2$), we may write

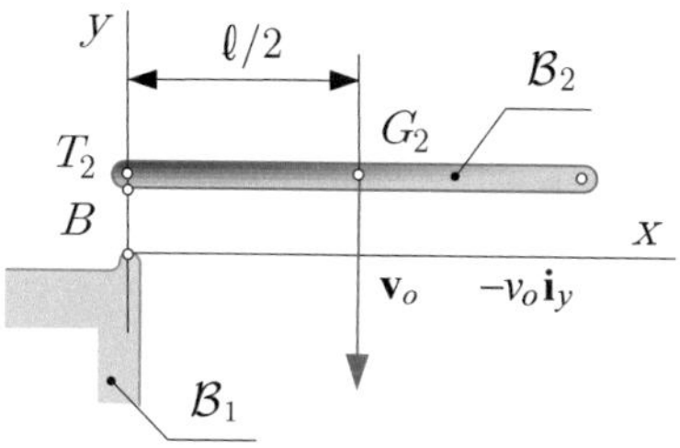

Fig. C.5 Rod and support B before collision

$$m_2 = m\,, \qquad v_{T_2x} = v'_{T_2x} = 0\,, \qquad v_{T_2y} = -v_o\,.$$

Then it follows from Eq. (2.29c) that

$$v'_{T_2y} - \underbrace{v'_{T_1y}}_{=0} = e\big(\underbrace{v_{T_1n}}_{=0} - \underbrace{v_{T_2y}}_{=-v_o}\big)\,.$$

Thus

$$v'_{T_2y} = v_o\,.$$

It is obvious that

$$J_{G_2} = \frac{m\ell^2}{12}\,, \qquad J_{T_2} = \frac{m\ell^2}{3}\,, \qquad \tilde{m}_2 = m_2\frac{J_{G_2}}{J_{T_2}} = \frac{m}{4}.$$

Since the rod is slender, we may assume that $T_2 \approx D$. The angular velocity can be calculated by using Eq. (2.31). We get

$$\boldsymbol{\omega}'_2 = \underbrace{\boldsymbol{\omega}_2}_{=\mathbf{0}} + \frac{\tilde{m}_2}{J_{G_2}}\mathbf{r}_{G_2A} \times \left(\mathbf{v}'_{T_2} - \mathbf{v}_{T_2}\right) = \frac{\frac{m}{4}}{\frac{m\ell^2}{12}}\left(-\frac{\ell}{2}\mathbf{i}_x\right) \times (v_o + v_o)\,\mathbf{i}_y = -\frac{3v_o}{\ell}\mathbf{i}_z\,.$$

With $\boldsymbol{\omega}'_2$, it is found that

$$\mathbf{v}_{G_2} = \mathbf{v}_{T_2} + \boldsymbol{\omega}'_2 \times \mathbf{r}_{T_2G_2} = v_o\mathbf{i}_y - \frac{3v_o}{\ell}\mathbf{i}_z \times \frac{\ell}{2}\mathbf{i}x = -\frac{v_o}{2}\mathbf{i}_y$$

and

$$\mathbf{v}_D = \mathbf{v}_{T_2} + \boldsymbol{\omega}'_2 \times \mathbf{r}_{T_2D} = v_o\mathbf{i}_y - \frac{3v_o}{\ell}\mathbf{i}_z \times \ell\mathbf{i}x = -2v_o\mathbf{i}_y$$

are the velocities at the mass center and right end of the rod.

(ii) The solutions for the part Problem (ii) are based on Fig. C.6 and the previous line of thought.
Without entering into details, we get

Fig. C.6 The rod before striking point D of the support

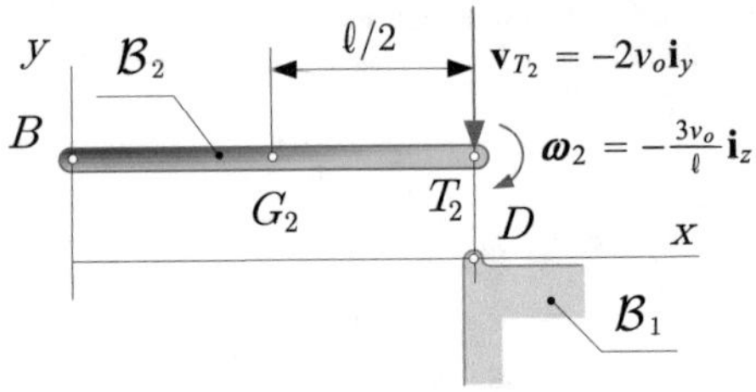

$$v'_{T_2y} = 2v_o\,, \qquad \omega'_2 = \frac{3v_o}{\ell}\mathbf{i}_z\,, \qquad \mathbf{v}_{G_2} = \frac{v_o}{2}\mathbf{i}_y \qquad \mathbf{v}_B = -v_o\mathbf{i}_y\,.$$

(iii) The solutions are given by the following equations:

$$\mathbf{v}'_{T_2y} = \mathbf{v}_B = v_o\mathbf{i}_y\,, \qquad \omega'_2 = \mathbf{0}\,, \qquad \mathbf{v}_{G_2} = \mathbf{v}_D = v_o\mathbf{i}_y\,.$$

C.3 Problems in Chap. 3

Problem 3.1 It follows from (3.6b) and (3.5a) that

$$\omega_n = 2\pi f_n = 60\pi \ 1/\mathrm{s}\,,$$

$$|D| = \frac{a_{\max}}{\omega_n^2} = \frac{50}{(60\pi)^2} = 1.4072387 \cdot 10^{-3}\,\mathrm{m} = 1.40724\,\mathrm{mm}\,.$$

Hence $v_{\max} = D\omega_n = 1.40724 \cdot 60\pi = 265.258$ mm/s.

Problem 3.2
(a) For $v_o = 0$ Eq. (3.21) yields

$$x = \frac{x_o}{2\mu}\left[(\beta+\mu)e^{-(\beta-\mu)t} - (\beta-\mu)e^{-(\beta+\mu)t}\right] = \frac{x_o}{2\mu}(\beta+\mu)e^{-(\beta-\mu)t}\left[1 - \frac{\beta-\mu}{\beta+\mu}e^{-2\mu t}\right],$$

where

$$\frac{x_o}{2\mu}(\beta+\mu)e^{-(\beta-\mu)t}$$

is [positive](negative) if x_o is [positive](negative). Since μ and t are greater than 0 and $\beta > \mu$, we get that

$$0 < 1 - \frac{\beta-\mu}{\beta+\mu}e^{-2\mu t} < 1\,.$$

Consequently, x does not change its sign.

(b) If $x_o = 0$ solution (3.21) simplifies to the form

$$x = \frac{v_o}{2\mu}\left[e^{-(\beta-\mu)t} - e^{-(\beta+\mu)t}\right] = \frac{x_o}{2\mu}(\beta+\mu)e^{-(\beta-\mu)t}\left[1 - e^{-2\mu t}\right],$$

which is [positive](negative) if v_o is [positive](negative), i.e., there is again no sign change.

Problem 3.3 We can manipulate formula (3.27) for the logarithmic decrement into the given form with ease:

$$\delta = \tau_d \frac{c}{2m} = \underset{(3.26),\,(3.23c)}{\uparrow} = \frac{2\pi}{\omega_n}\frac{c}{2m}\frac{1}{\sqrt{1-(c/c_{\text{crit}})^2}} = \underset{(3.20)}{\uparrow} = \frac{2\pi c/c_n}{\sqrt{1-(c/c_n)^2}}\,.$$

Problem 3.4 The line of thought leading to (3.27) yields

$$\delta = \frac{1}{k}\ln\frac{x\,(t)}{x\,(t+k\tau_d)} = \frac{1}{k}\ln\frac{e^{-\beta t}D\sin\left(\omega_d t+\phi\right)}{e^{-\beta(t+k\tau_d)}D\sin\left[\omega_d\left(t+k\tau_d\right)+\phi\right]} = \underset{\omega_d\tau_d=2\pi}{\uparrow} =$$

$$= \frac{1}{k}\ln\frac{e^{-\beta t}D\sin\left(\omega_d t+\phi\right)}{e^{-\beta(t+k\tau_d)}D\sin\left(\omega_d t+\phi+2\pi k\right)} = \frac{1}{k}\ln e^{-\beta t}e^{\beta(t+k\tau_d)} = \frac{1}{k}\ln e^{\frac{c}{2m}k\tau_d} = \frac{c}{2m}\tau_d.$$

Problem 3.5 If we recall Exercise 3.4, we can check with ease that the equations of motion for systems (1) and (2) are as follows:

$$\underbrace{\frac{1}{3}m\ell^2}_{m_g}\ddot{\varphi} + \underbrace{c\ell^2}_{c_{g1}}\dot{\varphi} + \underbrace{k\left(\frac{3\ell}{4}\right)^2}_{k_{g1}}\varphi = \underbrace{F_{fo}\frac{\ell}{2}}_{F_{go}}\cos\omega_f t\,,$$

$$\underbrace{\frac{1}{3}m\ell^2}_{m_g}\ddot{\varphi} + \underbrace{c\left(\frac{3\ell}{4}\right)^2}_{c_{g2}}\dot{\varphi} + \underbrace{k\ell^2}_{k_{g2}}\varphi = \underbrace{F_{fo}\frac{\ell}{2}}_{F_{go}}\cos\omega_f t,$$

where m is the mass of the rod and φ is the polar angle. Making use of the notations

$$\omega_{no}^2 = \frac{k}{m}, \quad \delta_{fo} = \frac{F_{fo}}{k}, \quad 2\beta_o = \frac{c}{m}, \quad \zeta_o = \frac{\beta_o}{\omega_{no}}, \quad \eta_o = \frac{\omega_f}{\omega_{no}},$$

we can write

$$\omega_{n1}^2 = \frac{k_{g1}}{m_g} = \frac{27}{16}\frac{k}{m} = \frac{27}{16}\omega_{no}^2\,, \; \delta_{f1} = \frac{F_{go}}{k_{g1}} = \frac{8}{9\ell}\frac{F_{fo}}{k} = \frac{8}{9\ell}\delta_{fo}\,,$$

$$2\beta_{g1} = \frac{c_{g1}}{m_g} = 3\frac{c}{m} = 3\cdot 2\beta_o,$$

$$\omega_{n2}^2 = \frac{k_{g2}}{m_g} = \frac{3k}{m} = 3\omega_{no}^2\,,\,\delta_{f2} = \frac{F_{go}}{k_{g2}} = \frac{1}{2\ell}\frac{F_{fo}}{k} = \frac{1}{2\ell}\delta_{fo}\,,$$

$$2\beta_{g2} = \frac{c_{g2}}{m_g} = \frac{27}{16}\frac{c}{m} = \frac{27}{16}\cdot 2\beta_o\,.$$

It follows from (3.37) that the amplitudes of the steady-state motions are given by

$$\ell\frac{x_{m1}}{\delta_{fo}} = \frac{8}{9}\frac{1}{\sqrt{\left(1-\frac{\omega_f^2}{\frac{27}{16}\omega_{no}^2}\right)^2+\left(\frac{3\cdot 2\beta_o}{\frac{27}{16}\omega_{no}^2}\omega_f\right)^2}} = \frac{8}{9}\frac{1}{\sqrt{\left(1-\frac{16}{27}\eta_o^2\right)^2+(4\zeta_o\eta_o)^2}}\,,$$

$$\ell\frac{x_{m2}}{\delta_{fo}} = \frac{1}{2}\frac{1}{\sqrt{\left(1-\frac{\omega_f^2}{3\omega_{no}^2}\right)^2+\left(\frac{\frac{27}{16}\cdot 2\beta_o}{3\omega_{no}^2}\omega_f\right)^2}} = \frac{1}{2}\frac{1}{\sqrt{\left(1-\frac{1}{3}\eta_o^2\right)^2+(\zeta_o\eta_o)^2}}\,.$$

Consequently,

$$\frac{x_{m2}}{x_{m1}} = \frac{9}{16}\frac{\sqrt{\left(1-\frac{16}{27}\eta_o^2\right)^2+(4\zeta_o\eta_o)^2}}{\sqrt{\left(1-\frac{1}{3}\eta_o^2\right)^2+(\zeta_o\eta_o)^2}}\,.$$

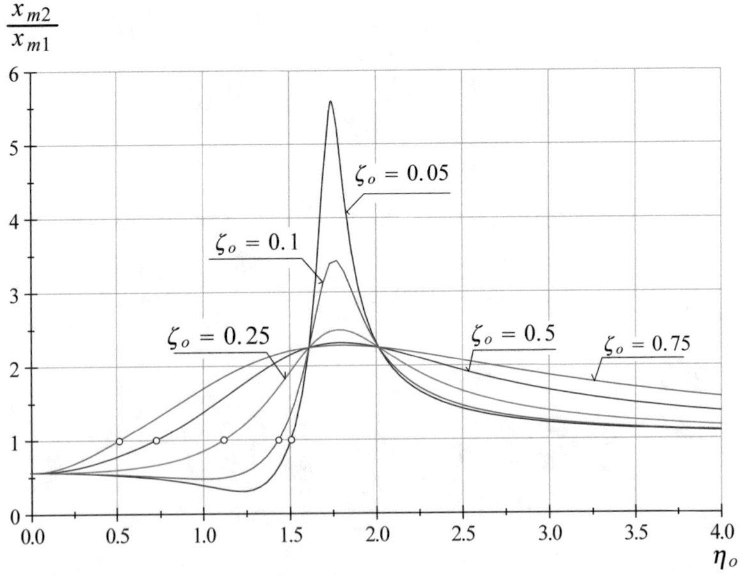

Fig. C.7 Quotient of amplitudes

Figure C.7 shows the quotient x_{m2}/x_{m1} against η_o. If, for instance, $\zeta_o = 0.1$ and $\eta_o \in [0, 1.0)$ then $x_{m2} < x_{m1}$, etc.

Problem 3.6 The solution is based on Exercise 3.7. When the impact begins

$$v_1 = \sqrt{2gh} = \sqrt{2 \cdot 9.81 \cdot 2.5} = 7.0036 \text{ [m/s]}$$

is the velocity of mass m_1. When the impact ends

$$c_n = \frac{m_1 v_{1n} + m_2 v_{2n}}{m_1 + m_2} = \underset{v_{1n}=v_1}{\uparrow} \; \underset{v_{2n}=0}{\uparrow} = \frac{m_1 v_1}{m_1 + m_2} = \frac{20 \cdot 7.0036}{30} = 4.6691 \text{ [m/s]}$$

is the common velocity of the masses m_1 and m_2. To find the maximum deflection x_d of the pan, we have to apply the principle of work an energy

$$\mathcal{E}_2 - \mathcal{E}_1 = W_{12}$$

which yields—see formula (3.92) in Exercise 3.7—the following equation:

$$x_d = \delta_f \underbrace{\left(1 + \sqrt{1 + 2\frac{h}{\delta_f}\frac{1}{1 + \frac{m_2}{m_1}}}\right)}_{\nu},$$

where

$$\delta_f = \frac{m_1 g}{k} = \frac{20 \cdot 9.81 \frac{\text{kg m}}{\text{s}^2}}{20 \frac{\text{kN}}{\text{m}}} = \frac{20 \cdot 9.81 \quad \frac{\text{kg m}}{\text{s}^2}}{20 \cdot 10^3 \; \frac{\text{kg m}}{\text{s}^2}\frac{1}{\text{m}}} = 0.009\,81 \text{ m}.$$

Hence

$$x_d = 0.009\,81 \cdot \left(1 + \sqrt{1 + 2 \cdot \frac{2.5}{0.009\,81} \cdot \frac{1}{1 + \frac{10}{20}}}\right) \simeq 0.190\,907 \text{ m}.$$

C.4 Problems in Chap. 4

Problem 4.2 The kinetic energy of the system is given by

$$\mathcal{E} = \frac{1}{2} m_1 \left(\mathbf{v}_{G_1}\right)^2 + \frac{1}{2} m_2 \left(\mathbf{v}_{G_2}\right)^2 + \frac{1}{2} J_{G_2} \omega_2^2,$$

where

$$\mathbf{v}_{G_1} = \dot{q}\,\mathbf{i}_x, \quad \mathbf{v}_{G_2} = \dot{q}\,\mathbf{i}_x + \dot{\vartheta} L \left(\cos\vartheta\,\mathbf{i}_x + \sin\vartheta\,\mathbf{i}_y\right), \quad \boldsymbol{\omega} = \dot{\vartheta}\,\mathbf{i}_z, \quad J_{G_2} = \frac{1}{12} m_2 (2L)^2 .$$

Hence

$$\mathcal{E} = \frac{1}{2}m_1\dot{q}^2 + \frac{1}{2}m_2\left(\dot{q}\,\mathbf{i}_x + \dot{\vartheta}L\left(\cos\vartheta\,\mathbf{i}_x + \sin\vartheta\,\mathbf{i}_y\right)\right)^2 + \frac{1}{2}J_{G_2}\dot{\vartheta}^2 =$$

$$= \frac{1}{2}\left(m_1 + m_2\right)\dot{q}^2 + \frac{1}{2}2m_2\dot{q}\dot{\vartheta}L\cos\vartheta + \frac{1}{2}\left(m_2L^2 + J_{G_2}\right)\dot{\vartheta}^2 =$$

$$\frac{1}{2}\left(m_1 + m_2\right)\dot{q}^2 + m_2\dot{q}\dot{\vartheta}L\cos\vartheta + \frac{1}{2}\underbrace{\frac{4}{3}m_2L^2}_{J_{G_1}}\dot{\vartheta}^2 .$$

The generalized damping force Q_{c1} and the generalized force of excitation Q_{f1} are given by the following equations:

$$Q_{c1} = -c\,\dot{q}, \qquad Q_{f1} = F_1\cos\omega_f t .$$

The potential energy stored in the spring is

$$\mathcal{U} = \frac{1}{2}kq^2 .$$

Making use of the above relations from the Lagrange equations

$$\frac{\partial}{\partial t}\frac{\partial\mathcal{E}}{\partial\dot{q}} - \frac{\partial\mathcal{E}}{\partial q} = -\frac{\partial\mathcal{U}}{\partial q} + Q_{c1} + Q_{f1} ,$$
$$\frac{\partial}{\partial t}\frac{\partial\mathcal{E}}{\partial\dot{\vartheta}} - \frac{\partial\mathcal{E}}{\partial\vartheta} = -\frac{\partial\mathcal{U}}{\partial\vartheta},$$

we obtain the following equations of motion:

$$\left(m_1 + m_2\right)\ddot{q} + m_2\ddot{\vartheta}L\cos\vartheta = -kq - c\,\dot{q} + F_1\cos\omega_f t ,$$
$$m_2\ddot{q}L\cos\vartheta + \frac{4}{3}m_2L^2\ddot{\vartheta} = 0$$

or

$$\begin{bmatrix} m_1 + m_2 & m_2L\cos\vartheta \\ m_2L\cos\vartheta & \frac{4}{3}m_2L^2 \end{bmatrix}\begin{bmatrix}\ddot{q}\\ \ddot{\vartheta}\end{bmatrix} + \begin{bmatrix} c & 0 \\ 0 & 0\end{bmatrix}\begin{bmatrix}\dot{q}\\ \dot{\vartheta}\end{bmatrix} +$$
$$+\begin{bmatrix} k & 0 \\ 0 & 0\end{bmatrix}\begin{bmatrix}q\\ \vartheta\end{bmatrix} = \begin{bmatrix}F_1\\ 0\end{bmatrix}\cos\omega_f t .$$

If there is no damping and ϑ is very small, we get a system of linear differential equations:

$$\begin{bmatrix} m_1 + m_2 & m_2L \\ m_2L & \frac{4}{3}m_2L^2 \end{bmatrix}\begin{bmatrix}\ddot{q}\\ \ddot{\vartheta}\end{bmatrix} + \begin{bmatrix} k & 0 \\ 0 & 0\end{bmatrix}\begin{bmatrix}q\\ \vartheta\end{bmatrix} = \begin{bmatrix}F_1\\ 0\end{bmatrix}\cos\omega_f t .$$

Problem 4.3 The eigenvalue problem to be solved is

$$\left\{\begin{bmatrix} 2 & -1 \\ -1 & 2 \end{bmatrix} - \underbrace{\frac{m}{k}\omega^2}_{\lambda}\begin{bmatrix} 1 & 0 \\ 0 & 1 \end{bmatrix}\right\}\begin{bmatrix} A_1 \\ A_2 \end{bmatrix} = \begin{bmatrix} 0 \\ 0 \end{bmatrix}.$$

It can be checked with ease that

$$\lambda_1 = 1\,, \quad \underline{\mathbf{A}}_1 = \begin{bmatrix} 1 \\ 1 \end{bmatrix} \quad \text{and} \quad \lambda_2 = 3\,, \quad \underline{\mathbf{A}}_2 = \begin{bmatrix} -1 \\ 1 \end{bmatrix}.$$

C.5 Problems in Chap. 5

Problem 5.1 The stiffness matrix and the equations of motion are given by the following relations:

$$\underset{(3\times 3)}{\underline{\mathbf{K}}} = \frac{3IE}{7a^3}\begin{bmatrix} 23 & -22 & 9 \\ -22 & 32 & -22 \\ 9 & -22 & 23 \end{bmatrix}$$

and

$$\underbrace{\begin{bmatrix} m_1 & 0 & 0 \\ 0 & m_2 & 0 \\ 0 & 0 & m_3 \end{bmatrix}}_{\underset{(3\times 3)}{\underline{\mathbf{M}}}}\underbrace{\begin{bmatrix} \ddot{q}_1 \\ \ddot{q}_2 \\ \ddot{q}_3 \end{bmatrix}}_{\underset{(3\times 1)}{\underline{\ddot{\mathbf{q}}}}} + \underbrace{\frac{3IE}{7a^3}\begin{bmatrix} 23 & -22 & 9 \\ -22 & 32 & -22 \\ 9 & -22 & 23 \end{bmatrix}}_{\underset{(3\times 3)}{\underline{\mathbf{K}}}}\underbrace{\begin{bmatrix} q_1 \\ q_2 \\ q_3 \end{bmatrix}}_{\underset{(3\times 1)}{\underline{\mathbf{q}}}} = \underbrace{\begin{bmatrix} 0 \\ 0 \\ 0 \end{bmatrix}}_{\underset{(3\times 1)}{\underline{\mathbf{0}}}}. \tag{C.5.13}$$

Problem 5.2 Eigenvalues:

$$\begin{aligned} \lambda_{n1} &= \omega_{n1}^2 = \frac{IE}{a^3 m}\left(\frac{96}{7} - \frac{66}{7}\sqrt{2}\right) \approx 0.380\,272\,1262\frac{IE}{a^3 m}\,, \\ \lambda_{n2} &= \omega_{n2}^2 = 6\frac{IE}{a^3 m}\,, \\ \lambda_{n3} &= \omega_{n3}^2 = \frac{IE}{a^3 m}\left(\frac{96}{7} + \frac{66}{7}\sqrt{2}\right) \approx 27.048\,299\,3\frac{IE}{a^3 m}\,. \end{aligned} \tag{C.5.14a}$$

Eigenvectors:

$$\underline{\mathbf{A}}_1 = \begin{bmatrix} 1 \\ \sqrt{2} \\ 1 \end{bmatrix}, \quad \underline{\mathbf{A}}_2 = \begin{bmatrix} -1 \\ 0 \\ 1 \end{bmatrix}, \quad \underline{\mathbf{A}}_3 = \begin{bmatrix} -1 \\ \sqrt{2} \\ 1 \end{bmatrix}. \tag{C.5.14b}$$

Problem 5.3 Equations of motion:

$$\underbrace{\begin{bmatrix} m_1 & 0 & 0 \\ 0 & m_2 & 0 \\ 0 & 0 & m_3 \end{bmatrix}}_{\underset{(3\times3)}{\underline{\mathbf{M}}}} \underbrace{\begin{bmatrix} \ddot{q}_1 \\ \ddot{q}_2 \\ \ddot{q}_3 \end{bmatrix}}_{\underset{(3\times1)}{\ddot{\underline{\mathbf{q}}}}} + \underbrace{\frac{3EI}{13a^3}\begin{bmatrix} 80 & -46 & 12 \\ -46 & 44 & -16 \\ 12 & -16 & 7 \end{bmatrix}}_{\underset{(3\times3)}{\underline{\mathbf{K}}}} \underbrace{\begin{bmatrix} q_1 \\ q_2 \\ q_3 \end{bmatrix}}_{\underset{(3\times1)}{\underline{\mathbf{q}}}} = \underbrace{\begin{bmatrix} 0 \\ 0 \\ 0 \end{bmatrix}}_{\underset{(3\times1)}{\underline{\mathbf{0}}}}. \tag{C.5.15}$$

Problem 5.4 Eigenvalues:

$$\begin{aligned} \lambda_{n1} &= \omega_{n1}^2 \approx 0.08554\,621\frac{EI}{a^3m}, \\ \lambda_{n2} &= \omega_{n2}^2 \approx 3.667\,782\frac{EI}{a^3m}, \\ \lambda_{n3} &= \omega_{n3}^2 \approx 26.477\,440\frac{EI}{a^3m}. \end{aligned} \tag{C.5.16a}$$

Eigenvectors:

$$\underline{\mathbf{A}}_1 = \begin{bmatrix} 0.156\,422 \\ 0.531\,648 \\ 1.0 \end{bmatrix}, \quad \underline{\mathbf{A}}_2 = \begin{bmatrix} -1.268\,969 \\ -1.507\,585 \\ 1.0 \end{bmatrix}, \quad \underline{\mathbf{A}}_3 = \begin{bmatrix} 4.647\,029 \\ -3.248\,201 \\ 1.0 \end{bmatrix}. \tag{C.5.16b}$$

Problem 5.5 The equation of motion is of the form

$$\begin{bmatrix} m & 0 & 0 \\ 0 & m & 0 \\ 0 & 0 & m \end{bmatrix} \begin{bmatrix} \ddot{x}_1 \\ \ddot{x}_2 \\ \ddot{x}_3 \end{bmatrix} + \begin{bmatrix} 3k & -k & 0 \\ -k & 2k & -k \\ 0 & -k & k \end{bmatrix} \begin{bmatrix} x_1 \\ x_2 \\ x_3 \end{bmatrix} = \begin{bmatrix} 0 \\ 0 \\ 0 \end{bmatrix}$$

If we divide throughout by k we obtain

$$\underbrace{\frac{m}{k}\begin{bmatrix} 1 & 0 & 0 \\ 0 & 1 & 0 \\ 0 & 0 & 1 \end{bmatrix}}_{\underset{(3\times3)}{\underline{\mathbf{M}}}} \underbrace{\begin{bmatrix} \ddot{x}_1 \\ \ddot{x}_2 \\ \ddot{x}_3 \end{bmatrix}}_{\underset{(3\times1)}{\underline{\mathbf{R}}}} + \underbrace{\begin{bmatrix} 3 & -1 & 0 \\ -1 & 2 & -1 \\ 0 & -1 & 1 \end{bmatrix}}_{\underset{(3\times3)}{\underline{\mathbf{K}}}} \underbrace{\begin{bmatrix} x_1 \\ x_2 \\ x_3 \end{bmatrix}}_{\underset{(3\times1)}{\underline{\mathbf{q}}}} = \underbrace{\begin{bmatrix} 0 \\ 0 \\ 0 \end{bmatrix}}_{\underset{(3\times1)}{\underline{\mathbf{0}}}}.$$

This equation defines the mass and stiffness matrices for the matrix iteration detailed in Sect. 5.3.2. On the base of the above equation

$$\lambda = \frac{\omega^2 m}{k}$$

is the eigenvalue we seek when we apply the matrix iteration.

We remark that the exact solutions for λ and the corresponding eigenvectors are known and are given below:

$$\lambda_1 = 2 - \sqrt{3} = 0.267\,949\,19, \quad \underline{\mathbf{A}}_1 = \begin{bmatrix} 1 \\ 1+\sqrt{3} \\ 2+\sqrt{3} \end{bmatrix} = \begin{bmatrix} 1 \\ 2.732\,050\,8 \\ 3.732\,050\,8 \end{bmatrix},$$

$$\lambda_2 = 2, \quad \underline{\mathbf{A}}_2 = \begin{bmatrix} -1 \\ -1 \\ 1 \end{bmatrix},$$

$$\lambda_3 = 2 + \sqrt{3} = 3.732\,050\,8, \quad \underline{\mathbf{A}}_3 = \begin{bmatrix} 1 \\ 1-\sqrt{3} \\ 2-\sqrt{3} \end{bmatrix} = \begin{bmatrix} 1 \\ -0.732\,050\,81 \\ 0.267\,949\,19 \end{bmatrix}.$$

Now we proceed with the matrix iteration. Let

$$\underset{(0)}{\underline{\mathbf{A}}_1^T} = [1\ 1\ 1]$$

be the initial value for the first eigenvector. Though the details are omitted, the results for the first seven iteration steps are presented in the table below (Tables C.1 and C.2):

Table C.1 Values of $\underset{(\ell)}{\underline{\varphi}_1^T}$ and $\underset{(\ell)}{\lambda_1}$

Results of the matrix iteration		
ℓ	$\underset{(\ell)}{\underline{\varphi}_1^T}$	$\underset{(\ell)}{\lambda_1} = m \underset{(\ell)}{\omega_1^2}/k$
0	$\frac{1}{\sqrt{3}}[1\ 1\ 1]$	0.666 666 67
1	[0.254 456 68 0.593 732 25 0.763 370 04]	0.273 381 30
2	[0.216 202 43 0.580 332 85 0.785 156 2]	0.268 030 55
3	[0.211 911 17 0.577 800 97 0.788 187 61]	0.267 950 57
4	[0.211 398 56 0.577 414 25 0.788 608 56]	0.267 949 22
5	[0.211 334 39 0.577 359 09 0.788 666 12]	0.267 949 19
6	[0.211 326 12 0.577 351 48 0.788 673 92]	0.267 949 19
7	[0.211 325 03 0.577 350 43 0.788 674 97]	0.267 949 19

For the sake of a comparison, we have normalized the first element in the seventh approximation of the first eigenvector $\underset{(7)}{\underline{\varphi}_1}$ to unity. Row A contains the exact values, and row B the approximations:

Table C.2 The first eigenvectors

Comparison			
A	$\underline{\mathbf{A}}_{1\,\text{exact}}$	$\begin{bmatrix} 1.0 & 2.732\,050\,8 & 3.732\,050\,8 \end{bmatrix}$	0.267 949 19
B	$\underline{\mathbf{A}}_{1\,\text{approx}}$	$\begin{bmatrix} 1.0 & 2.732\,049\,4 & 2.732\,047\,1 \end{bmatrix}$	0.267 949 19

It is obvious from these tables that the approximative values are very good.

For finding the second eigenvector and eigenvalue, we have to determine the matrices $\underline{\mathbf{M}}_2$ and $\underline{\mathbf{K}}_2$. Making use of Eqs. (5.75) and (5.76), we get

$$\hat{\underline{\mathbf{A}}}_1 = \frac{\underline{\mathbf{A}}_1}{|\underline{\mathbf{A}}_1|} = \frac{1}{\sqrt{1+\left(1+\sqrt{3}\right)^2+\left(2+\sqrt{3}\right)^2}}\begin{bmatrix} 1 \\ 1+\sqrt{3} \\ 2+\sqrt{3} \end{bmatrix} = \begin{bmatrix} 0.211\,324\,87 \\ 0.577\,350\,27 \\ 0.788\,675\,13 \end{bmatrix},$$

$$\underline{\mathbf{M}}_2 = \underline{\mathbf{M}}_1 - \underline{\mathbf{M}}_1\hat{\underline{\mathbf{A}}}_1\hat{\underline{\mathbf{A}}}_1^T = \\ = \begin{bmatrix} 0.955\,341\,80 & -0.122\,008\,47 & -0.166\,666\,67 \\ -0.122\,008\,47 & 0.666\,666\,67 & -0.455\,341\,80 \\ -0.166\,666\,67 & -0.455\,341\,80 & 0.377\,991\,54 \end{bmatrix}$$

and

$$\underline{\mathbf{K}}_2 = \underline{\mathbf{K}}_1 - \underline{\mathbf{K}}_1\hat{\underline{\mathbf{A}}}_1\hat{\underline{\mathbf{A}}}_1^T = \\ = \begin{bmatrix} 2.988\,033\,9 & -1.032\,692\,1 & -4.465\,820\,9\cdot 10^{-2} \\ -1.032\,692\,1 & 1.910\,683\,6 & -1.122\,008\,5 \\ -4.465\,819\,9\cdot 10^{-2} & -1.122\,008\,5 & 0.833\,333\,34 \end{bmatrix}.$$

Let

$$\underset{(0)}{\underline{\mathbf{A}}_2^T} = \begin{bmatrix} 1 \\ 1 \\ 1 \end{bmatrix}$$

be the very first approximation of the second eigenvector. Note that this is the same as that for the first eigenvector. The table below shows the results for the first eleven iteration steps (Tables C.3 and C.4).

Table C.3 Values of $\underset{(\ell)}{\underline{\varphi}}{}_2^T$ and $\underset{(\ell)}{\lambda}{}_2$

Results of the matrix iteration		
ℓ	$\underset{(\ell)}{\underline{\varphi}}{}_2^T$	$\underset{(\ell)}{\lambda}{}_2 = m\underset{(\ell)}{\omega}{}_2^2/k$
0	$\left[1.3975887\ 1.3975887\ 1.3975887\right]$	2.6043387
1	$\left[0.61715404\ -0.24259405\ -1.2378518\right]$	2.2310142
2	$\left[0.81152168\ 0.67642475\ -0.20705079\right]$	2.0733144
3	$\left[0.61036327\ 0.36795934\ -0.74283897\right]$	2.0217100
4	$\left[0.60021431\ 0.47698072\ -0.6517074\right]$	2.0062911
5	$\left[0.53388662\ 0.37077667\ -0.82648159\right]$	2.0018114
6	$\left[0.59914909\ 0.58969086\ -0.5429505\right]$	2.0005205
7	$\left[0.54432794\ 0.46178912\ -0.72585895\right]$	2.0001495
8	$\left[0.56729628\ 0.53629015\ -0.62844288\right]$	2.0000429
9	$\left[0.57326179\ 0.55889318\ -0.59988966\right]$	2.0000123
10	$\left[0.56038348\ 0.52709008\ -0.64457546\right]$	2.0000035
11	$\left[0.57771445\ 0.57625169\ -0.57808417\right]$	2.0000011

To make a comparison, we normalized the first element in the eleventh approximation of the second eigenvector $\underset{(11)}{\underline{\varphi}}{}_2$ to unity. Row A contains the exact values, and row B the approximation:

Table C.4 The second eigenvectors

Comparison			
A	$\underline{\mathbf{A}}_{2\,\text{exact}}$	$\left[1.0000000\ 1.0000000\ -1.00000000\right]$	2.0000000
B	$\underline{\mathbf{A}}_{2\,\text{approx}}$	$\left[1.0000000\ 0.99746802\ -1.00064000\right]$	2.0000011

As regards the third iteration step with

$$\underline{\mathbf{A}}_2 = \begin{bmatrix} 1 \\ 1 \\ -1 \end{bmatrix}, \quad |\underline{\mathbf{A}}_2| = \sqrt{3}, \quad \hat{\underline{\mathbf{A}}}_2 = \frac{\underline{\mathbf{A}}_1}{|\underline{\mathbf{A}}_1|} = \frac{1}{\sqrt{3}}\begin{bmatrix} 1 \\ 1 \\ -1 \end{bmatrix} = \begin{bmatrix} 0.57735027 \\ 0.57735027 \\ -0.57735027 \end{bmatrix},$$

we get

$$\underline{\mathbf{M}}_3 = \underline{\mathbf{M}}_2 - \underline{\mathbf{M}}_2\hat{\underline{\mathbf{A}}}_2\hat{\underline{\mathbf{A}}}_2^T = \begin{bmatrix} 0.62200846 & -0.45534180 & 0.16666666 \\ -0.45534180 & 0.33333334 & -0.12200847 \\ 0.16666667 & -0.12200846 & 0.04465820 \end{bmatrix},$$

$$\underline{\mathbf{K}}_3 = \underline{\mathbf{K}}_2 - \underline{\mathbf{K}}_2 \underline{\hat{\mathbf{A}}}_2 \underline{\hat{\mathbf{A}}}_2^T = \begin{bmatrix} 2.321\,367\,23 & -1.699\,358\,77 & 0.622\,008\,46 \\ -1.699\,358\,77 & 1.244\,016\,93 & -0.455\,341\,83 \\ 0.622\,008\,48 & -0.455\,341\,82 & 0.166\,666\,66 \end{bmatrix}.$$

Assume that

$$\underset{(0)}{\underline{\mathbf{A}}_3^T} = \begin{bmatrix} 1.00 & -0.65 & 0.25 \end{bmatrix}.$$

Then

$$\left| \underset{(0)}{\underline{\mathbf{A}}_3} \right| = \sqrt{1.480\,563\,37} = 1.216\,784\,03,$$

$$\underset{(0)}{\underline{\varphi}_3} = \frac{\underset{(0)}{\underline{\mathbf{A}}_3}}{\left| \underset{(0)}{\underline{\mathbf{A}}_3} \right|} = \begin{bmatrix} 0.821\,838\,53 \\ -0.534\,195\,05 \\ 0.205\,459\,63 \end{bmatrix}$$

and

$$\underset{(0)}{\lambda_3} = \left(\underset{(0)}{\underline{\varphi}_2}, \underset{(0)}{\underline{\varphi}_2} \right)_{K_3} = 3.732\,050\,85\,.$$

With $\underset{(0)}{\underline{\varphi}_3}$, we obtain

$$\underline{\mathbf{M}}_3 \underset{(0)}{\underline{\varphi}_3} = \begin{bmatrix} 0.788\,675\,13 \\ -0.577\,350\,27 \\ 0.211\,324\,86 \end{bmatrix}.$$

If we normalize the first element in this column matrix to unity

$$\underline{\tilde{\mathbf{A}}}_3 = \begin{bmatrix} 1.000\,000\,00 \\ -0.732\,050\,81 \\ 0.267\,949\,19 \end{bmatrix}$$

is the result. Observe that

$$\underset{(0)}{\lambda_3} = \lambda_3 \qquad \text{and} \qquad \underline{\tilde{\mathbf{A}}}_3 = \underline{\mathbf{A}}_3\,.$$

These results show that there is no need for an iteration. The reason for this is simple: the product

$$\underline{\mathbf{M}}_3 \underset{(0)}{\underline{\varphi}_3} \tag{C.5.17}$$

is always orthogonal to $\underline{\varphi}_1$ and $\underline{\varphi}_2$ no matter what value the approximation $\underset{(0)}{\underline{\varphi}_3}$ has. In the 3D space, there is, however, only one direction orthogonal to $\underline{\varphi}_1$ and $\underline{\varphi}_2$. Consequently, product (C.5.17) must be the solution for the eigenvector we seek. And with the solution for the eigenvector, the Rayleigh quotient yields the exact eigenvalue.

C.6 Problems in Chap. 6

Problem 6.1 Figure C.8 shows the shaft in question before and during deformation.

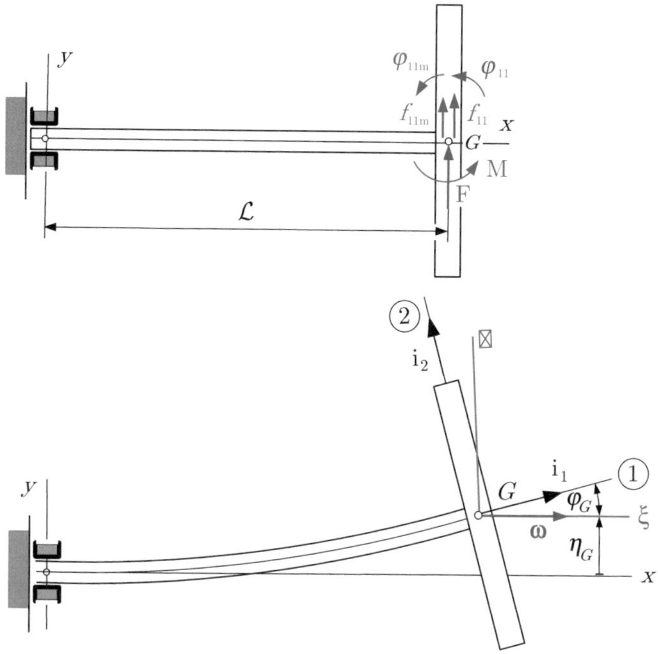

Fig. C.8 Rotating shaft fixed at the left end

If we recall the line of thought leading to Eqs. (6.37), we can come to the conclusion that the displacement η_G and angle of rotation φ_G can be calculated in the same way as for the simply supported rotating shaft shown in Fig. 6.10. Hence

$$\eta_G = f_{11}F + f_{11m}M \quad \text{and} \quad \varphi_G = \varphi_{11}F + \varphi_{11m}M,$$

where now

$$f_{11} = \frac{\mathcal{L}^3}{3IE}, \qquad f_{11m} = \varphi_{11} = \frac{\mathcal{L}^2}{2IE}, \qquad \varphi_{11m} = \frac{\mathcal{L}}{IE}$$

and

$$F = m\eta_G\omega^2, \qquad M = -\omega^2 J_d\,\varphi_G$$

while m is the disk mass and $J_d = J_1$ is the moment of inertia of the disk. Hence

$$\eta_G = f_{11}F + f_{11m}\,M = f_{11}m\omega^2\eta_G - f_{11m}\omega^2 J_d\,\varphi_G\,,$$
$$\varphi_G = \varphi_{11}F + \varphi_{11m}\,M = \varphi_{11}m\omega^2\eta_G - \varphi_{11m}\,\omega^2 J_d\,\varphi_G\,.$$

The determinant of this homogeneous equation system, which is given by Eq. (6.44b), should vanish:

$$(\varphi_{11}\varphi_{11m} - f_{11}\varphi_{11m})\, J_d\chi^4 + \left(f_{11}\varphi_{11m}J_d - f_{11}^2 m\right)\chi^2 + f_{11}^2 m = 0\,, \quad \chi^2 = f_{11}m\omega^2.$$

If we substitute the flexibilities f_{11}, $f_{11m} = \varphi_{11}$, φ_{11m} and multiply the result by $-(IE)^2/12\mathcal{L}^4$, we get

$$J_d\chi^4 - 4\left(J_d - \frac{\mathcal{L}^2 m}{3}\right) - \frac{4}{3}\mathcal{L}^2 m = 0\,.$$

Using the notation $\gamma = J_d/\mathcal{L}^2 m$, we can rewrite this equation in the following form:

$$\frac{3}{4}\gamma\chi^4 + (1 - 3\gamma)\,\chi^2 - 1 = 0\,.$$

Thus

$$\chi^2 = f_{11}m\omega^2 = \frac{\omega^2}{\omega_{\text{cr}}^2} = \frac{2}{3\gamma}\left(3\gamma - 1 \pm \sqrt{9\gamma^2 - 3\gamma + 1}\right)\,.$$

Figure C.9 shows the graphs of the two roots against γ.

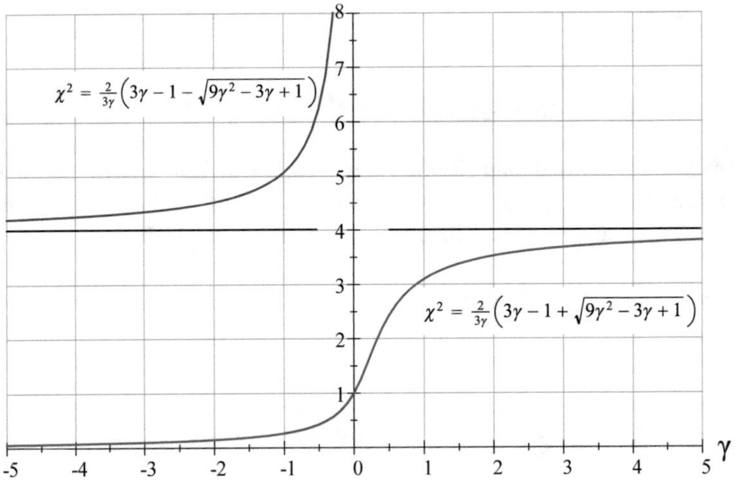

Fig. C.9 χ^2 as function of γ

C.7 Problems in Chap. 7

Problem 7.3 The positiveness of λ (or ω^2) for the support arrangements shown in Table 7.5 is a consequence of Eq. (7.131).

Problem 7.4

(a) Fixed-fixed beam.
Boundary conditions:

$$W(\beta x)|_{x=0} = \mathcal{A}_1 = 0\,, \quad \left.\frac{\mathrm{d}W(\beta x)}{\mathrm{d}x}\right|_{x=0} = \beta\mathcal{A}_2 = 0\,,$$
$$W(\beta x)|_{x=\ell} = \mathcal{A}_3\,U(\beta\ell) + \mathcal{A}_4\,V(\beta\ell) = 0\,,$$
$$\left.\frac{\mathrm{d}W(\beta x)}{\mathrm{d}x}\right|_{x=\ell} = \beta\,(\mathcal{A}_3\,T(\beta\ell) + \mathcal{A}_4\,U(\beta\ell)) = 0\,.$$

Equation system for the non-zero integration constants:

$$\mathcal{A}_3\,U(\beta\ell) + \mathcal{A}_4\,V(\beta\ell) = 0\,,$$
$$\mathcal{A}_3\,T(\beta\ell) + \mathcal{A}_4\,U(\beta\ell) = 0\,.$$

Frequency equation:

$$\begin{vmatrix} U(\beta\ell) & V(\beta\ell) \\ T(\beta\ell) & U(\beta\ell) \end{vmatrix} = U^2(\beta\ell) - V(\beta\ell)T(\beta\ell) = \frac{1}{2}\,(1 - \cosh\beta\ell\cos\beta\ell) = 0\,.$$

Hence,

$$\cosh\beta_k\ell\cos\beta_k\ell = 1\,.$$

Relation between the integration constants:

$$\mathcal{A}_{4k} = -\mathcal{A}_{3k}\frac{U(\beta_k\ell)}{V(\beta_k\ell)} = -\mathcal{A}_{3k}\frac{\cosh\beta_k\ell - \cos\beta_k\ell}{\sinh\beta_k\ell - \sin\beta_k\ell}\,.$$

Eigenfunctions:

$$W_k(x) = U(\beta_k x) - \frac{U(\beta_k\ell)}{V(\beta_k\ell)}\,V(\beta_k x) =$$
$$= \cosh\beta_k x - \cos\beta_k x - b_k\,(\sinh\beta_k x - \sin\beta_k x)\;,$$

$$b_k = \frac{\cosh\beta_k\ell - \cos\beta_k\ell}{\sinh\beta_k\ell - \sin\beta_k\ell}\,.$$

(b) Fixed-free beam.
Boundary conditions:

$$W(\beta x)|_{x=0} = \mathcal{A}_1 = 0\,, \quad \left.\frac{\mathrm{d}W(\beta x)}{\mathrm{d}x}\right|_{x=0} = \beta\mathcal{A}_2 = 0\,,$$

$$\left.\frac{\mathrm{d}^2W(\beta x)}{\mathrm{d}x^2}\right|_{x=\ell} = \beta^2\left(\mathcal{A}_3\,S(\beta\ell) + \mathcal{A}_4\,T(\beta\ell)\right) = 0\,,$$

$$\left.\frac{\mathrm{d}^3W(\beta x)}{\mathrm{d}x^3}\right|_{x=\ell} = \beta^3\left(\mathcal{A}_3\,V(\beta\ell) + \mathcal{A}_4\,S(\beta\ell)\right) = 0\,.$$

Equation system for the non-zero integration constants:

$$\mathcal{A}_3\,S(\beta\ell) + \mathcal{A}_4\,T(\beta\ell) = 0\,,$$
$$\mathcal{A}_3\,V(\beta\ell) + \mathcal{A}_4\,S(\beta\ell) = 0\,.$$

Frequency equation:

$$\begin{vmatrix} S(\beta\ell) & T(\beta\ell) \\ V(\beta\ell) & S(\beta\ell) \end{vmatrix} = S^2(\beta\ell) - T(\beta\ell)V(\beta\ell) = \frac{1}{2}\left(\cosh\beta\ell\cos\beta\ell + 1\right) = 0\,.$$

Hence,

$$\cosh\beta_k\ell\cos\beta_k\ell = -1\,.$$

Relation between the integration constants:

$$\mathcal{A}_{4k} = -\mathcal{A}_{3k}\frac{S(\beta_k\ell)}{T(\beta_k\ell)} = -\mathcal{A}_{3k}\frac{\cosh\beta_k\ell + \cos\beta_k\ell}{\sinh\beta_k\ell + \sin\beta_k\ell}\,.$$

Eigenfunctions:

$$W_k(x) = U(\beta_k x) - \frac{S(\beta_k\ell)}{T(\beta_k\ell)}\,V(\beta_k x) =$$
$$= \cosh\beta_k x - \cos\beta_k x - b_k\left(\sinh\beta_k x - \sin\beta_k x\right)\,,$$

$$b_k = \frac{\cosh\beta_k\ell + \cos\beta_k\ell}{\sinh\beta_k\ell + \sin\beta_k\ell}\,.$$

(c) Pinned-free beam.
Boundary conditions:

$$W(\beta x)|_{x=0} = \mathcal{A}_1 = 0\,, \quad \left.\frac{\mathrm{d}^2W(\beta x)}{\mathrm{d}x^2}\right|_{x=0} = \beta^2\mathcal{A}_3 = 0\,,$$

$$\left.\frac{\mathrm{d}^2 W(\beta x)}{\mathrm{d}x^2}\right|_{x=\ell} = \beta^2 \left(\mathcal{A}_2\, V(\beta\ell) + \mathcal{A}_4\, T(\beta\ell)\right) = 0\,,$$

$$\left.\frac{\mathrm{d}^3 W(\beta x)}{\mathrm{d}x^3}\right|_{x=\ell} = \beta^3 \left(\mathcal{A}_2\, U(\beta\ell) + \mathcal{A}_4\, S(\beta\ell)\right) = 0\,.$$

Equation system for the non-zero integration constants:

$$\mathcal{A}_2\, V(\beta\ell) + \mathcal{A}_4\, T(\beta\ell) = 0\,,$$
$$\mathcal{A}_2\, U(\beta\ell) + \mathcal{A}_4\, S(\beta\ell) = 0\,.$$

Frequency equation:

$$\begin{vmatrix} V(\beta\ell) & T(\beta\ell) \\ U(\beta\ell) & S(\beta\ell) \end{vmatrix} = V(\beta\ell)S(\beta\ell) - T(\beta\ell)U(\beta\ell) =$$
$$= \frac{1}{2}\left(\cos\beta\ell\sinh\beta\ell - \cosh\beta\ell\sin\beta\ell\right) = 0\,.$$

Hence,

$$\tan\beta_k\ell = \tanh\beta_k\ell\,.$$

Relation between the integration constants:

$$\mathcal{A}_{4k} = -\mathcal{A}_{2k}\frac{V(\beta_k\ell)}{T(\beta_k\ell)} = -\mathcal{A}_{2k}\frac{\sinh\beta_k\ell - \sin\beta_k\ell}{\sinh\beta_k\ell + \sin\beta_k\ell}\,.$$

Eigenfunctions:

$$T(\beta_k x) - \frac{V(\beta_k\ell)}{T(\beta_k\ell)}\, V(\beta_k x) =$$
$$= \sinh\beta_k x + \sin\beta_k x - \frac{\sinh\beta_k\ell - \sin\beta_k\ell}{\sinh\beta_k\ell + \sin\beta_k\ell}\left(\sinh\beta_k x - \sin\beta_k x\right) =$$
$$= 2\frac{\sinh\beta_k x\sin\beta_k\ell + \sin\beta_k x\sinh\beta_k\ell}{\sinh\beta_k\ell + \sin\beta_k\ell}$$

from where it follows that

$$W_k(x) = \sinh\beta_k x + \frac{\sinh\beta_k\ell}{\sin\beta_k\ell}\sin\beta_k x\,.$$

(d) Free-free beam.
Boundary conditions:

$$\left.\frac{\mathrm{d}^2 W(\beta x)}{\mathrm{d}x^2}\right|_{x=0} = \beta^2\mathcal{A}_3 = 0\,, \quad \left.\frac{\mathrm{d}^3 W(\beta x)}{\mathrm{d}x^3}\right|_{x=0} = \beta^3\mathcal{A}_4 = 0\,,$$

$$\left.\frac{\mathrm{d}^2 W(\beta x)}{\mathrm{d}x^2}\right|_{x=\ell} = \beta^2 \left(\mathcal{A}_1\, U(\beta\ell) + \mathcal{A}_2\, S(\beta\ell)\right) = 0\,,$$
$$\left.\frac{\mathrm{d}^3 W(\beta x)}{\mathrm{d}x^3}\right|_{x=\ell} = \beta^3 \left(\mathcal{A}_1\, T(\beta\ell) + \mathcal{A}_4\, U(\beta\ell)\right) = 0\,.$$

Equation system for the non-zero integration constants:

$$\mathcal{A}_1\, U(\beta\ell) + \mathcal{A}_2\, V(\beta\ell) = 0\,,$$
$$\mathcal{A}_1\, T(\beta\ell) + \mathcal{A}_4\, U(\beta\ell) = 0\,.$$

Frequency equation:

$$\begin{vmatrix} U(\beta\ell) & V(\beta\ell) \\ T(\beta\ell) & U(\beta\ell) \end{vmatrix} = U^{\hat{}2}(\beta\ell) - V(\beta\ell)T(\beta\ell) = \frac{1}{2}\left(1 - \cosh\beta\ell\cos\beta\ell\right) = 0\,.$$

Hence,

$$\cosh\beta_k\ell\cos\beta_k\ell = 1\,.$$

Relation between the integration constants:

$$\mathcal{A}_{2k} = -\mathcal{A}_{1k}\frac{U(\beta_k\ell)}{V(\beta_k\ell)} = -\mathcal{A}_{2k}\frac{\cosh\beta_k\ell - \cos\beta_k\ell}{\sinh\beta_k\ell - \sin\beta_k\ell}\,.$$

Eigenfunctions:

$$W_k(x) = S(\beta_k x) - \frac{U(\beta_k\ell)}{V(\beta_k\ell)}\,T(\beta_k x) =$$
$$= \cosh\beta_k x + \cos\beta_k x - b_k\left(\sinh\beta_k x + \sin\beta_k x\right)\,,$$

$$b_k = \frac{\cosh\beta_k\ell - \cos\beta_k\ell}{\sinh\beta_k\ell - \sin\beta_k\ell}\,.$$

Problem 7.5 It can be checked with paper-and-pencil calculations that the inverse of the flexibility matrix is

$$\underset{(3\times3)}{\underline{\mathbf{f}}}^{-1} = \frac{IE}{\ell^3}\begin{bmatrix} 680.4 & -421.2 & 194.4 \\ -421.2 & 453.6 & -421.2 \\ 194.4 & -421.2 & 680.4 \end{bmatrix}$$

With the inverse, we get an eigenvalue problem for the three degree of freedom system shown in Fig. 7.15:

$$\underbrace{3\frac{\ell^3}{IE}\underset{(3\times3)}{\underline{\mathbf{f}}}^{-1}}_{\underline{\mathbf{K}}} - \omega^2\underbrace{\frac{\rho A\ell^4}{IE}\underset{(3\times3)}{\underline{\mathbf{I}}}}_{\underline{\mathbf{M}}} = \underset{(3\times3)}{\underline{\mathbf{0}}}\,,$$

where

$$\underset{(3\times3)}{\mathbf{K}} = \begin{bmatrix} 2041.2 & -1263.6 & 583.2 \\ -1263.6 & 1360.8 & -1263.6 \\ 583.2 & -1263.6 & 2041.2 \end{bmatrix}$$

and

$$\underset{(3\times3)}{\mathbf{M}} = \frac{\rho A\ell^4}{IE}\begin{bmatrix} 1 & 0 & 0 \\ 0 & 1 & 0 \\ 0 & 0 & 1 \end{bmatrix}.$$

We have determined the smallest eigenvalue by using the program MAPLE. The result is

$$\omega_1^2 = 97.200\frac{IE}{\rho A\ell^4}, \qquad \omega_1 = 9.8590\sqrt{\frac{IE}{\rho A\ell^4}}.$$

The exact value for the continuous beam is given by the following equation:

$$\omega_1^2 = 97.409\frac{IE}{\rho A\ell^4}, \qquad \omega_1 = 9.8696\sqrt{\frac{IE}{\rho A\ell^4}}.$$

The relative error for ω_1 is

$$\delta = \left|1 - \frac{9.8590}{9.8696}\right| = 0.107\%.$$

This result is in accordance with the known thumb rule which says that a continuous beam system should be replaced by a finite degree of freedom system with a degree of freedom of at least $2n + 1$ if we want to get an acceptable approximation for the first n eigenfrequencies by solving the corresponding finite eigenvalue problem.

C.8 Problems in Chap. 8

Problem 8.1 The coefficients $b_1(\xi), \ldots, b_4(\xi)$ given by Eq. (8.65) are all independent of the boundary conditions. This means that we should determine the boundary condition-dependent coefficients $a_1(\xi), \ldots, a_4(\xi)$ only and then we have to utilize Eq. (8.53) to find the Green functions sought.

(a) Fixed-pinned beam:

Boundary conditions

$$w(0) = w^{(1)}(0) = w(\ell) = w^{(2)}(\ell) = 0$$

lead to the equation system

$$
\begin{aligned}
a_1w_1(0)+a_2w_2(0)+a_3w_3(0)+a_4w_4(0)&=\\
&=-b_1w_1(0)-b_2w_2(0)-b_3w_3(0)-b_4w_4(0),\\
a_1w_1^{(1)}(0)+a_2w_2^{(1)}(0)+a_3w_3^{(1)}(0)+a_4w_4^{(1)}(0)&=\\
&=-b_1w_1^{(1)}(0)-b_2w_2^{(1)}(0)-b_3w_3^{(1)}(0)-b_4w_4^{(1)}(0),\\
a_1w_1(\ell)+a_2w_2(\ell)+a_3w_3(\ell)+a_4w_4(\ell)&=\\
&=b_1w_1(\ell)+b_2w_2(\ell)+b_3w_3(\ell)+b_4w_4(\ell),\\
a_1w_1^{(2)}(\ell)+a_2w_2^{(2)}(\ell)+a_3w_3^{(2)}(\ell)+a_4w_4^{(2)}(\ell)&=\\
&=b_1w_1^{(2)}(\ell)+b_2w_2^{(2)}(\ell)+b_3w_3^{(2)}(\ell)+b_4w_4^{(2)}(\ell)
\end{aligned}
$$

from where after substituting the linearly independent particular solutions

$$w_1(x),\dots,w_4(x)\quad\text{and the values of}\quad b_1(\xi),\dots,b_4(\xi)$$

we get

$$
\begin{bmatrix}1&0&0&0\\0&1&0&0\\1&\ell&\ell^2&\ell^3\\0&0&2&6\ell\end{bmatrix}
\begin{bmatrix}a_1\\a_2\\a_3\\a_4\end{bmatrix}=\frac{1}{12}
\begin{bmatrix}-\xi^3\\3\xi^2\\\xi^3-3\xi^2\ell+3\xi\ell^2-\ell^3\\6\xi-6\ell\end{bmatrix}.
$$

Thus

$$
\begin{bmatrix}a_1\\a_2\\a_3\\a_4\end{bmatrix}=\frac{1}{12}
\begin{bmatrix}-\xi^3\\3\xi^2\\3\xi\frac{\xi^2-3\xi\ell+\ell^2}{\ell^2}\\-\frac{\xi^3-3\xi^2\ell+\ell^3}{\ell^3}\end{bmatrix}
$$

and

$$
\begin{aligned}
\mathcal{G}(x,\xi)&=\frac{1}{IE}\left[(a_1+b_1)+(a_2+b_2)\,x+(a_3+b_3)\,x^2+(a_4+b_4)\,x^3\right]=\\
&=\frac{1}{12IE}\left[\left(3\xi\frac{\xi^2-3\xi\ell+\ell^2}{\ell^2}+3\xi\right)x^2+\left(-\frac{\xi^3-3\xi^2\ell+\ell^3}{\ell^3}-1\right)x^3\right]=\\
&=\frac{x^2}{12IE\ell^3}(\ell-\xi)\left(6\xi\ell^2-2\ell^2x-3\xi^2\ell-2\ell\xi x+x\xi^2\right)\qquad x\le\xi,
\end{aligned}
\tag{C.8.18a}
$$

$$
\mathcal{G}(x,\xi)=\frac{x^2}{12IE\ell^3}(\ell-x)\left(6x\ell^2-2\ell^2\xi-3x^2\ell-2\ell x\xi+\xi x^2\right)\qquad x\ge\xi.
\tag{C.8.18b}
$$

(b) Fixed-fixed beam:

Boundary conditions

$$w(0) = w^{(1)}(0) = w(\ell) = w^{(1)}(\ell) = 0$$

lead to the equation system

$$\begin{bmatrix} 1 & 0 & 0 & 0 \\ 0 & 1 & 0 & 0 \\ 1 & \ell & \ell^2 & \ell^3 \\ 0 & 1 & 2\ell & 3\ell^2 \end{bmatrix} \begin{bmatrix} a_1 \\ a_2 \\ a_3 \\ a_4 \end{bmatrix} = \frac{1}{12} \begin{bmatrix} \xi^3 \\ 3\xi^2 \\ \xi^3 - 3\xi^2\ell + 3\xi\ell^2 - \ell^3 \\ -3\xi^2 + 6\ell\xi - 3\ell^2 \end{bmatrix}$$

from where

$$\begin{bmatrix} a_1 \\ a_2 \\ a_3 \\ a_4 \end{bmatrix} = \frac{1}{12\ell^3} \begin{bmatrix} -\xi^3\ell^3 \\ 3\xi^2\ell^3 \\ 3\xi\ell\left(2\xi^2 - 4\xi\ell + \ell^2\right) \\ -4\xi^3 + 6\xi^2\ell - \ell^3 \end{bmatrix}$$

and

$$\begin{aligned} \mathcal{G}(x,\xi) &= \frac{1}{IE}\left[(a_1+b_1) + (a_2+b_2)\,x + (a_3+b_3)\,x^2 + (a_4+b_4)\,x^3\right] = \\ &= \frac{2x^2}{12IE\ell^3}\left(3\xi^3\ell - 6\xi^2\ell^2 + 3\xi\ell^3 - 2x\xi^3 + 3x\xi^2\ell - x\ell^3\right) = \\ &= \frac{x^2}{6IE\ell^3}\,(\ell-\xi)^2\,(3\xi\ell - \ell x - 2\xi x) \quad x \le \xi\,, \end{aligned} \tag{C.8.19a}$$

$$\mathcal{G}(x,\xi) = \frac{\xi^2}{6IE\ell^3}\,(\ell-x)^2\,(3x\ell - \ell\xi - 2x\xi) \quad x \ge \xi\,. \tag{C.8.19b}$$

(c) Fixed-free beam:

Boundary conditions

$$w(0) = w^{(1)}(0) = w^{(2)}(\ell) = w^{(3)}(\ell) = 0$$

yield

$$\begin{bmatrix} 1 & 0 & 0 & 0 \\ 0 & 1 & 0 & 0 \\ 0 & 0 & 2 & 6\ell \\ 0 & 0 & 0 & 6 \end{bmatrix} \begin{bmatrix} a_1 \\ a_2 \\ a_3 \\ a_4 \end{bmatrix} = \frac{1}{12} \begin{bmatrix} -\xi^3 \\ 3\xi^2 \\ 6\xi - 6\ell \\ -6 \end{bmatrix}.$$

Hence,

$$\begin{bmatrix} a_1 \\ a_2 \\ a_3 \\ a_4 \end{bmatrix} = \frac{1}{12} \begin{bmatrix} -\xi^3 \\ 3\xi^2 \\ 3\xi \\ -1 \end{bmatrix}$$

and now we get in the same manner as above that

$$\mathcal{G}(x,\xi) = \frac{1}{6IE}x^2(3\xi - x), \quad x \le \xi, \tag{C.8.20a}$$

$$\mathcal{G}(x,\xi) = \frac{1}{6IE}\xi^2(3x - \xi), \quad x \ge \xi. \tag{C.8.20b}$$

Since the three boundary value problems considered are all self-adjoint, the Green function is symmetric. Consequently, we have used symmetry Eq. (8.61) to give the three Green functions if $x \ge \xi$, i.e., we wrote ξ for x and x for ξ in the formulae we established for $\mathcal{G}(x,\xi)$ $x \le \xi$ so that we can get $\mathcal{G}(x,\xi)$ $x \ge \xi$.

Problem 8.2 It follows on the basis of Remark 8.8 and/or the first row of Table 8.1 that

$$f_{11} = \mathcal{G}(x,\xi)|_{x=\ell/6,\xi=\ell/6} = \frac{\xi}{6IE\ell}(\ell - x)\left(2\ell x - \xi^2 - x^2\right)\Big|_{x=\ell/6,\xi=\ell/6} =$$

$$= \frac{1}{36IE}\left(\frac{6\ell}{6} - \frac{\ell}{6}\right)\left(2\ell\frac{\ell}{6} - \frac{\ell^2}{36} - \frac{\ell^2}{36}\right) = \frac{\ell^3}{36\cdot 1.5\cdot 36IE}\cdot 5\cdot(12-2) = \frac{12.5\ell^3}{1944IE}$$

or

$$f_{13} = f_{31} = \mathcal{G}(x,\xi)|_{x=\ell/6,\xi=5\ell/6} = \frac{x}{6IE\ell}(\ell - \xi)\left(2\ell\xi - x^2 - \xi^2\right)\Big|_{x=\ell/6,\xi=5\ell/6} =$$

$$= \frac{1}{36IE}\left(\frac{6\ell}{6} - \frac{5\ell}{6}\right)\left(2\ell\frac{5\ell}{6} - \frac{\ell^2}{36} - \frac{25\ell^2}{36}\right) = \frac{\ell^3}{36\cdot 1.5\cdot 36IE}\cdot\frac{(60-1-25)}{4} =$$

$$= \frac{8.5\ell^3}{1944IE}.$$

The line of thought is the same for the flexibilities f_{22}, f_{33}, $f_{12} = f_{21}$ and $f_{23} = f_{32}$.

Problem 8.3 The solution is based on Sect. 8.11.2.

Let the axial force be a compressive one. Boundary conditions (8.73a), $\mathcal{G}^{(1)}(0,\xi) = 0$, continuity and discontinuity conditions (8.74), and boundary conditions (8.75a), $\mathcal{G}^{(1)}(\ell,\xi) = 0$ lead to the following equation system:

$$\underline{\mathcal{A}} = \begin{bmatrix} 1 & 0 & 1 & 0 & 0 & 0 & 0 & 0 \\ 0 & 1 & 0 & p & 0 & 0 & 0 & 0 \\ 1 & \xi & \cos p\xi & \sin p\xi & -1 & -\xi & -\cos p\xi & -\sin p\xi \\ 0 & 1 & -p\sin p\xi & p\cos p\xi & 0 & -1 & p\sin p\xi & -p\cos p\xi \\ 0 & 0 & -p^2\cos p\xi & -p^2\sin p\xi & 0 & 0 & p^2\cos p\xi & p^2\sin p\xi \\ 0 & 0 & p^3\sin p\xi & -p^3\cos p\xi & 0 & 0 & -p^3\sin p\xi & p^3\cos p\xi \\ 0 & 0 & 0 & 0 & 1 & \ell & \cos p\ell & \sin p\ell \\ 0 & 0 & 0 & 0 & 0 & 1 & -p\sin p\ell & p\cos p\ell \end{bmatrix},$$

$$\underline{\boldsymbol{\mathcal{X}}}^T = [\,\mathcal{A}_1|\mathcal{A}_2|\mathcal{A}_3|\mathcal{A}_4|\,\mathcal{B}_1|\mathcal{B}_2|\mathcal{B}_3|\mathcal{B}_4\,],$$
$$\underline{\mathcal{F}}^T = [\,0|0|0|0|0|-1/IE|0|0\,],$$

$$\underline{\boldsymbol{\mathcal{A}}}\,\underline{\boldsymbol{\mathcal{X}}} = \underline{\mathcal{F}}. \tag{C.8.21}$$

Making use of the solutions

$$\frac{A_1}{\mathcal{D}_c} = -\xi p + \sin p\,(\xi-\ell) + \ell p \cos p\,(\xi-\ell) - \sin p\xi + \sin p\ell + (\xi-\ell)\,p\cos p\ell\,,$$
$$\frac{A_2}{p\mathcal{D}_c} = 1 - \cos p\,(\xi-\ell) + (\xi-\ell)\,p\sin p\ell - \cos p\ell + \cos p\xi = -\frac{A_4}{\mathcal{D}_c}\,,$$
$$\frac{A_3}{\mathcal{D}_c} = \xi p - \sin p\,(\xi-\ell) - \ell p\cos p\,(\xi-\ell) + \sin p\xi - \sin p\ell - (\xi-\ell)\,p\cos p\ell = -\frac{A_1}{\mathcal{D}_c}\,,$$
$$\frac{A_4}{\mathcal{D}_c} = -1 + \cos p\,(\xi-\ell) - (\xi-\ell)\,p\sin p\ell + \cos p\ell - \cos p\xi = -\frac{A_2}{\mathcal{D}_c p}$$

$$\mathcal{D}_c = 1/p^3 IE\,(2\cos p\ell + \ell p\sin p\ell - 2)$$

for $\mathcal{A}_1(\xi), \ldots, \mathcal{A}_4(\xi)$ we can give the Green function in the following form:

$$\begin{aligned}\mathcal{G}(x,\xi) &=\\ &= \mathcal{A}_1 + \mathcal{A}_2 x + \mathcal{A}_3\cos px + \mathcal{A}_4\sin px = \mathcal{A}_1\,(1-\cos px) + \mathcal{A}_4\,(\sin px - px) =\\ &= \mathcal{D}_c\,\{[\sin p\,(\xi-\ell) + \ell p\cos p\,(\xi-\ell) - \xi p - \sin p\xi + \sin p\ell + (\xi-\ell)\,p\cos p\ell]\,(1-\cos px) -\\ &\quad -[1-\cos p\,(\xi-\ell) + (\xi-\ell)\,p\sin p\ell - \cos p\ell + \cos p\xi]\,(\sin px - px)\}\,,\quad x\le\xi\,.\end{aligned} \tag{C.8.22a}$$

Since the boundary value problem considered is self-adjoint write ξ for x and x for ξ to get the Green function if $x \ge \xi$:

$$\begin{aligned}\mathcal{G}(x,\xi) &=\\ &= \mathcal{D}_c\,\{[\sin p\,(x-\ell) + \ell p\cos p\,(x-\ell) - xp - \sin px + \sin p\ell + (x-\ell)\,p\cos p\ell]\,(1-\cos p\xi) -\\ &\quad -[1-\cos p\,(x-\ell) + (x-\ell)\,p\sin p\ell - \cos p\ell + \cos px]\,(\sin p\xi - p\xi)\}\,,\quad x\ge\xi\,.\end{aligned} \tag{C.8.22b}$$

If the axial force is tensile

$$\underline{\boldsymbol{\mathcal{A}}} = \begin{bmatrix} 1 & 0 & 1 & 0 & 0 & 0 & 0 & 0\\ 0 & 1 & 0 & p & 0 & 0 & 0 & 0\\ 1 & \xi & \cosh p\xi & \sinh p\xi & -1 & -\xi & -\cosh p\xi & -\sinh p\xi\\ 0 & 1 & p\sinh p\xi & p\cosh p\xi & 0 & -1 & -p\sinh p\xi & -p\cosh p\xi\\ 0 & 0 & p^2\cosh p\xi & p^2\sinh p\xi & 0 & 0 & -p^2\cosh p\xi & -p^2\sinh p\xi\\ 0 & 0 & p^3\sinh p\xi & p^3\cosh p\xi & 0 & 0 & -p^3\sinh p\xi & -p^3\cosh p\xi\\ 0 & 0 & 0 & 0 & 1 & \ell & \cosh p\ell & \sinh p\ell\\ 0 & 0 & 0 & 0 & 0 & 1 & p\sinh p\ell & p\cosh p\ell \end{bmatrix}$$

is the coefficient matrix in equation system (C.8.21). Utilizing the solutions

$$\frac{A_1}{\mathcal{D}_t} = \sinh p\,(\xi-\ell)+\ell p\cosh p\,(\xi-\ell)-\xi p-$$
$$-\sinh p\xi+\sinh p\ell+(\xi-\ell)\,p\cosh p\ell,$$
$$\frac{A_2}{p\mathcal{D}_t} = 1-\cosh p\,(\xi-\ell)-(\xi-\ell)\,p\sinh p\ell-\cosh p\ell+\cosh p\xi,$$
$$\frac{A_3}{\mathcal{D}_t} = \xi p-\sinh p\,(\xi-\ell)-\ell p\cosh p\,(\xi-\ell)+\sinh p\xi-$$
$$-\sinh p\ell-(\xi-\ell)\,p\cosh p\ell=-\frac{A_1}{\mathcal{D}_t},$$
$$\frac{A_4}{\mathcal{D}_t} = -1+\cos p\,(\xi-\ell)-(\xi-\ell)\,p\sin p\ell+\cos p\ell-\cos p\xi=-\frac{A_2}{p\mathcal{D}_t}$$

$$\mathcal{D}_t = 1/p^3 IE\,(2-2\cosh p\ell+\ell p\sinh p\ell)$$

for $\mathcal{A}_1(\xi),\dots,\mathcal{A}_4(\xi)$ we get

$$\begin{aligned}&\mathcal{G}(x,\xi)=\\&=\mathcal{A}_1+\mathcal{A}_2x+\mathcal{A}_3\cosh px+\mathcal{A}_4\sinh px=\mathcal{A}_1\,(1-\cosh px)+\mathcal{A}_4\,(\sinh px-px)=\\&=\mathcal{D}_t\big\{[\sinh p\,(\xi-\ell)+\ell p\cosh p\,(\xi-\ell)-\xi p-\sinh p\xi+\sinh p\ell+(\xi-\ell)\,p\cosh p\ell]\,(1-\cosh px)-\\&\quad-[1-\cosh p\,(\xi-\ell)-(\xi-\ell)\,p\sinh p\ell-\cosh p\ell+\cosh p\xi]\,(\sinh px-px)\big\},\qquad x\le\xi\end{aligned} \tag{C.8.23a}$$

and

$$\begin{aligned}&\mathcal{G}(x,\xi)=\\&=\mathcal{D}_t\big\{[\sinh p\,(x-\ell)+\ell p\cosh p\,(x-\ell)-xp-\sinh px+\sinh p\ell+(x-\ell)\,p\cosh p\ell]\,(1-\cosh p\xi)-\\&\quad-[1-\cosh p\,(x-\ell)-(x-\ell)\,p\sinh p\ell-\cosh p\ell+\cosh px]\,(\sinh p\xi-p\xi)\big\},\qquad x\ge\xi\end{aligned} \tag{C.8.23b}$$

Problem 8.4 The solution is based again on Sect. 8.11.2.

Let the axial force be a compressive one. The equation system to be solved contains the following equations: boundary conditions (8.73a), $\mathcal{G}^{(1)}(0,\xi)=0$, continuity and discontinuity conditions (8.74), and boundary conditions (8.75a), $\mathcal{G}^{(2)}(\ell,\xi)=0$. If we compare these equations and the equation system (C.8.21), it turns out that only the last equation, i.e., the boundary condition $\mathcal{G}^{(2)}(\ell,\xi)=0$, is different. It coincides with the last equation in (8.76). For this reason, the equation system for the unknowns $\mathcal{A}_1(\xi),\dots,\mathcal{A}_4(\xi)$ and $\mathcal{B}_1(\xi),\dots,\mathcal{B}_4(\xi)$ is not detailed here. Instead, the solutions that can be used for calculating the Green functions are presented only:

$$\mathcal{A}_1=\mathcal{A}_3=p\mathcal{D}_c\,[(\xi-\ell)\sin p\ell-\ell\sin p\,(\xi-\ell)]\,,$$
$$\mathcal{A}_2=-p\mathcal{A}_4=p\mathcal{D}_c\,[p\,(\xi-\ell)\cos p\ell-\sin p\,(\xi-\ell)]\,,$$
$$\mathcal{D}_c=1/p^3IE\,(p\ell\cos p\ell-\sin p\ell)\,.$$

With $\mathcal{A}_1(\xi), \ldots, \mathcal{A}_4(\xi)$ we have

$$\begin{aligned}\mathcal{G}(x,\xi) &= \mathcal{A}_1 + \mathcal{A}_2 x + \mathcal{A}_3 \cos px + \mathcal{A}_4 \sin px = \mathcal{A}_1\,(1+\cos px) + \mathcal{A}_4\,(\sin px - px) = \\ &= \mathcal{D}_c\big\{p\,[(\xi-\ell)\sin p\ell - \ell \sin p\,(\xi-\ell)]\,(1+\cos px) - \\ &- [p\,(\xi-\ell)\cos p\ell - \sin p\,(\xi-\ell)]\,(\sin px - px)\big\}, \quad x \le \xi \end{aligned} \tag{C.8.24}$$

and

$$\begin{aligned}\mathcal{G}(x,\xi) &= \mathcal{D}_c\big\{p\,[(x-\ell)\sin p\ell - \ell\sin p\,(x-\ell)]\,(1+\cos p\xi) - \\ &- [p\,(x-\ell)\cos p\ell - \sin p\,(x-\ell)]\,(\sin p\xi - p\xi)\big\}, \quad x \ge \xi\,. \end{aligned} \tag{C.8.25}$$

If the axial force is a tensile one, a similar line of thought yields

$$\begin{aligned}\mathcal{A}_1 &= -\mathcal{A}_3 = p\mathcal{D}_t\,[(\xi-\ell)\sinh p\ell - \ell\sinh p\,(\xi-\ell)]\,,\\ \mathcal{A}_2 &= -\mathcal{A}_4 p = p\mathcal{D}_t\,[\sinh p\,(\xi-\ell) - p\,(\xi-\ell)\cosh p\ell]\,,\\ \mathcal{D}_t &= 1/p^3 IE\,(p\ell\cosh p\ell - \sinh p\ell)\end{aligned}$$

and

$$\begin{aligned}\mathcal{G}(x,\xi) &= \mathcal{A}_1 + \mathcal{A}_2 x + \mathcal{A}_3\cosh px + \mathcal{A}_4\sinh px = \\ &= \mathcal{A}_1\,(1-\cosh px) + \mathcal{A}_4\,(\sinh px - px) = \\ &= \mathcal{D}_t\big\{p\,[(\xi-\ell)\sinh p\ell - \ell\sinh p\,(\xi-\ell)]\,(1-\cosh px) - \\ &- [\sinh p\,(\xi-\ell) - p\,(\xi-\ell)\cosh p\ell]\,(\sinh px - px)\big\}, \quad x \le \xi, \end{aligned} \tag{C.8.26}$$

$$\begin{aligned}\mathcal{G}(x,\xi) &= \mathcal{D}_t\big\{p\,[(x-\ell)\sinh p\ell - \ell\sinh p\,(x-\ell)]\,(1-\cosh p\xi) - \\ &- [\sinh p\,(x-\ell) - p\,(x-\ell)\cosh p\ell]\,(\sinh p\xi - p\xi)\big\}, \quad x \ge \xi\,. \end{aligned} \tag{C.8.27}$$

C.9 Problems in Chap. 9

Problem 9.1 According to Eq. (9.52b), the product $\left(\underline{\mathbf{u}}, \underline{\mathbf{v}}\right)_M$ is commutative. It follows from (9.52a) that

$$\begin{aligned}\left(\underline{\mathbf{u}}, \underline{\mathbf{v}}\right)_K &= \underbrace{\left(\hat{\kappa} v_1 u_1^{(1)} + v_2 u_2^{(1)}\right)\Big|_0^\ell - \hat{\kappa} u_1 v_2\Big|_0^\ell}_{=0} - \\ &-\hat{\kappa}\int_0^\ell \left(u_1^{(1)} v_2 + v_1^{(1)} u_2\right)\mathrm{d}x - \int_0^\ell \left(\hat{\kappa} v_1^{(1)} u_1^{(1)} + v_2^{(1)} u_2^{(1)}\right)\mathrm{d}x - \int_0^\ell \hat{\kappa} v_2 u_2 \mathrm{d}x = \\ &= \left(\underline{\mathbf{v}}, \underline{\mathbf{u}}\right)_K\,,\end{aligned}$$

in which the terms underbraced on the right side vanish if we substitute the boundary conditions in question. Hence, the product $\left(\underline{\mathbf{u}}, \underline{\mathbf{v}}\right)_K$ is also commutative. This means that the eigenvalue problems defined by differential equation (9.34a) and the boundary conditions in Row 2, 3 and 4 of Table 9.1 are all self-adjoint.

Problem 9.2 It is worth following the solution steps presented in Exercise 9.5. By using the Green function given in the last row of Table 8.1, we get

$$f_{11} = \int_0^{\ell} \mathcal{G}\left(\ell, \xi\right) \delta\left(\ell - \xi\right) \mathrm{d}\xi = \int_0^{\ell} \frac{1}{6IE} \ell^2 \left(3\xi - \ell\right) \delta\left(\ell - \xi\right) \mathrm{d}\xi = \frac{\ell^3}{3IE},$$

$$k_{11} = \frac{1}{f_{11}} = \frac{3IE}{\ell^3} \quad \text{and} \quad \omega^2 = \frac{k_{11}}{m} = \frac{3IE}{\ell^3 m}.$$

With $x_\ell = \ell/\ell = 1$, $r_1\left(\xi\right) = -\delta\left(x_\ell - \xi\right)\ell^3/IE$, $r_2\left(\xi\right) = 0$ and $\hat{\xi} = \ell\xi$, it holds within the framework of the Timoshenko beam theory that

$$\begin{aligned}
\begin{bmatrix} y_1\left(x_\ell\right) \\ y_2\left(x_\ell\right) \end{bmatrix} &= \int_0^1 \begin{bmatrix} G_{11}\left(x_\ell, \xi\right) & G_{12}\left(x_\ell, \xi\right) \\ G_{21}\left(x_\ell, \xi\right) & G_{22}\left(x_\ell, \xi\right) \end{bmatrix} \begin{bmatrix} r_1\left(\xi\right) \\ r_2\left(\xi\right) \end{bmatrix} \mathrm{d}\xi = \\
&= -\frac{\ell^3}{IE} \int_0^{\ell} \begin{bmatrix} G_{11}\left(x_\ell, \frac{\hat{\xi}}{\ell}\right) & G_{12}\left(x_\ell, \frac{\hat{\xi}}{\ell}\right) \\ G_{21}\left(x_\ell, \frac{\hat{\xi}}{\ell}\right) & G_{22}\left(x_\ell, \frac{\hat{\xi}}{\ell}\right) \end{bmatrix} \begin{bmatrix} \delta\left(x_\ell - \frac{\hat{\xi}}{\ell}\right) \\ 0 \end{bmatrix} \frac{\mathrm{d}\hat{\xi}}{\ell}.
\end{aligned}$$

Substituting now the elements of the Green function matrix from (9.107) yields

$$\begin{aligned}
f_{11} = w\left(x_\ell\right) = \ell y_1\left(x_\ell\right) &= -\frac{\ell^3}{IE} \int_0^{\ell} G_{11}\left(x_\ell, \frac{\hat{\xi}}{\ell}\right) \delta\left(x_\ell - \frac{\hat{\xi}}{\ell}\right) \mathrm{d}\hat{\xi} = \\
= -\frac{\ell^3}{IE} \left\{ \left[1\; x_\ell\right] \begin{bmatrix} 0 \\ -\frac{1}{4\chi}\left(\chi x_\ell^2 + 2\right) + \frac{1}{4}x_\ell^2 - \frac{1}{2\chi} \end{bmatrix} - \cdot \left[\tfrac{1}{2}x_\ell^2\; \tfrac{1}{3}x_\ell^3\right] \begin{bmatrix} x_\ell \\ -\frac{1}{2} \end{bmatrix} \right\} &\underset{x_\ell = 1}{=\uparrow=} \\
&= \frac{\ell^3}{3IE}\left(1 + \frac{3}{\chi}\right).
\end{aligned}$$

Hence,

$$\omega^2 = \frac{1}{f_{11} m} = \frac{k_{11}}{m} = \frac{3IE}{\ell^3\left(1 + \frac{3}{\chi}\right)} \underset{\chi = \frac{AG\kappa_y \ell^2}{IE}\;\; E = 2G(1+\nu)\;\; r_y^2 = \frac{I}{A}}{=\uparrow\quad\uparrow\quad\uparrow=} \frac{3IE}{\ell^3\left(1 + \frac{2r_y^2}{\ell^2}\frac{(1+\nu)}{\kappa_y}\right)}.$$

Problem 9.3 The main steps of the solutions are presented here only.

(a) If the axial force is compressive, it follows from Eq. (9.76) by utilizing the solutions for $\underline{\mathbf{Y}}_\ell$ that the matrices $\underline{\mathbf{B}}_\ell$ should satisfy the following equations:

$$\sum_{\ell=1}^{2} \underline{\mathbf{Y}}_\ell(\xi)\,\underline{\mathbf{B}}_\ell(\xi) = \begin{bmatrix} 1 & \xi \\ 0 & -\underbrace{\frac{1}{\chi}(\mathcal{N}+\chi)}_{\kappa}\end{bmatrix}\begin{bmatrix}\overset{1}{B}_{11} & \overset{1}{B}_{12}\\ \overset{1}{B}_{21} & \overset{1}{B}_{22}\end{bmatrix} + $$

$$+\begin{bmatrix}\frac{1}{n}\cos n\xi & -\frac{1}{n}\sin n\xi \\ \sin n\xi & \cos n\xi\end{bmatrix}\begin{bmatrix}\overset{2}{B}_{11} & \overset{2}{B}_{12}\\ \overset{2}{B}_{21} & \overset{2}{B}_{22}\end{bmatrix} = \begin{bmatrix}0 & 0\\ 0 & 0\end{bmatrix},$$

$$\sum_{\ell=1}^{2} \underline{\mathbf{Y}}_\ell^{(1)}(\xi)\,\underline{\mathbf{B}}_\ell(\xi) = \begin{bmatrix} 0 & 1 \\ 0 & 0\end{bmatrix}\begin{bmatrix}\overset{1}{B}_{11} & \overset{1}{B}_{12}\\ \overset{1}{B}_{21} & \overset{1}{B}_{22}\end{bmatrix} + $$

$$+\begin{bmatrix}-\sin n\xi & -\cos n\xi \\ n\cos n\xi & -n\sin n\xi\end{bmatrix}\begin{bmatrix}\overset{2}{B}_{11} & \overset{2}{B}_{12}\\ \overset{2}{B}_{21} & \overset{2}{B}_{22}\end{bmatrix} = -\frac{1}{2}\begin{bmatrix}\frac{1}{\chi} & 0\\ 0 & 1\end{bmatrix} = -\frac{1}{2}\underline{\mathbf{P}}_2^{-1}.$$

The solutions of the previous linear equation systems are given by

$$\begin{bmatrix}\overset{1}{B}_{11} & \overset{1}{B}_{12}\\ \overset{1}{B}_{21} & \overset{1}{B}_{22}\end{bmatrix} = \begin{bmatrix} -\frac{\xi}{2\chi(\kappa-1)} & -\frac{1}{2n^2} \\ \frac{1}{2\chi(\kappa-1)} & 0\end{bmatrix}, \tag{C.9.28a}$$

$$\begin{bmatrix}\overset{2}{B}_{11} & \overset{2}{B}_{12}\\ \overset{2}{B}_{21} & \overset{2}{B}_{22}\end{bmatrix} = \begin{bmatrix} \frac{\kappa\sin n\xi}{2\chi(\kappa-1)} & -\frac{\cos n\xi}{2n} \\ \frac{\kappa\cos n\xi}{2\chi(\kappa-1)} & \frac{\sin n\xi}{2n}\end{bmatrix}. \tag{C.9.28b}$$

Note that the matrices $\underline{\mathbf{B}}_\ell$ are independent of the boundary conditions.

Let us assume that the matrices $\underline{\mathbf{Y}}_\ell$, $\underline{\mathbf{A}}_\ell$, and $\underline{\mathbf{B}}_\ell$ are partitioned in the same manner as shown by Eqs. (9.94). Since the boundary conditions are the same as those of the pinned-pinned Timoshenko beam, it follows that the elements of the matrices $\underline{\mathbf{A}}_\ell$ should satisfy the Eqs. (9.98) and (9.99). Take into account that the coefficient matrices on the left sides of these equations are the same and substitute then $\underline{\mathbf{Y}}_\ell$ and $\underline{\mathbf{Y}}_\ell^{(1)}$. We arrive at the following linear equations system:

$$\begin{bmatrix} 1 & 0 & \frac{1}{n} & 0 \\ 1 & 1 & \frac{\cos n}{n} & -\frac{\sin n}{n} \\ 0 & 0 & n & 0 \\ 0 & 0 & n\cos n & n\sin n\end{bmatrix}\begin{bmatrix}\overset{1}{A}_{1i}\\ \overset{1}{A}_{2i}\\ \overset{2}{A}_{1i}\\ \overset{2}{A}_{2i}\end{bmatrix} = \begin{bmatrix} -\overset{1}{B}_{1i} - \frac{\overset{2}{B}_{1i}}{n} \\ \overset{1}{B}_{1i} + \overset{1}{B}_{2i} + \frac{\overset{2}{B}_{1i}}{n}\cos n - \frac{\overset{2}{B}_{2i}}{n}\sin n \\ -n\overset{2}{B}_{1i} \\ \overset{2}{B}_{1i} n\cos n + \overset{2}{B}_{2i} n\sin n\end{bmatrix}, \quad i = 1, 2.$$

The solutions are as follows:

$$
\begin{gathered}
\overset{1}{A}_{1i} = \overset{1}{B}_{1i}, \qquad \overset{1}{A}_{2i} = 2\overset{1}{B}_{1i} + \overset{1}{B}_{2i}, \qquad \overset{2}{A}_{1i} = \overset{1}{B}_{1i}, \\
\overset{2}{A}_{2i} = -\frac{1}{\sin n}\left(2\overset{2}{B}_{1i}\cos n - \overset{2}{B}_{2i}\sin n\right).
\end{gathered}
\tag{C.9.29}
$$

Substituting the solutions (C.9.28) for the elements of $\underline{\mathbf{B}}_\ell$ yields

$$
\begin{bmatrix} \overset{1}{A}_{11} & \overset{1}{A}_{12} \\ \overset{1}{A}_{21} & \overset{1}{A}_{22} \end{bmatrix} = \begin{bmatrix} \frac{\xi}{2\chi(\kappa-1)} & -\frac{1}{2n^2} \\ -\frac{2\xi-1}{2\chi(\kappa-1)} & \frac{1}{n^2} \end{bmatrix},
\tag{C.9.30a}
$$

$$
\begin{bmatrix} \overset{2}{A}_{11} & \overset{2}{A}_{12} \\ \overset{2}{A}_{21} & \overset{2}{A}_{22} \end{bmatrix} = \begin{bmatrix} -\frac{\kappa\sin n\xi}{2\chi(\kappa-1)} & \frac{\cos n\xi}{2n} \\ \frac{\kappa(2\sin n\xi\cos n + \cos n\xi\sin n)}{2\chi(\kappa-1)\sin n} & \frac{2\cos n\cos n\xi + \sin n\sin n\xi}{2n\sin n} \end{bmatrix}.
\tag{C.9.30b}
$$

Making use of Eqs. (C.9.28), (C.9.30), it is now easy to check the correctness of the Green function matrix (9.112).

(b) If the axial force is tensile, the solution steps are the same as those for the case of a compressive force. In the sequel, the main solution steps (with the corresponding results) are presented only.

Equation system for the matrices $\underline{\mathbf{B}}_\ell$:

$$
\begin{gathered}
\begin{bmatrix} 1 & \xi \\ 0 & \underbrace{-\frac{1}{\chi}(\chi-\mathcal{N})}_{\kappa} \end{bmatrix} \begin{bmatrix} \overset{1}{B}_{11} & \overset{1}{B}_{12} \\ \overset{1}{B}_{21} & \overset{1}{B}_{22} \end{bmatrix} + \\
+ \begin{bmatrix} -\frac{1}{n}\cosh n\xi & -\frac{1}{n}\sinh n\xi \\ \sinh n\xi & \cosh n\xi \end{bmatrix} \begin{bmatrix} \overset{2}{B}_{11} & \overset{2}{B}_{12} \\ \overset{2}{B}_{21} & \overset{2}{B}_{22} \end{bmatrix} = \begin{bmatrix} 0 & 0 \\ 0 & 0 \end{bmatrix},
\end{gathered}
$$

$$
\begin{gathered}
\begin{bmatrix} 0 & 1 \\ 0 & 0 \end{bmatrix} \begin{bmatrix} \overset{1}{B}_{11} & \overset{1}{B}_{12} \\ \overset{1}{B}_{21} & \overset{1}{B}_{22} \end{bmatrix} + \\
+ \begin{bmatrix} -\sinh n\xi & -\cosh n\xi \\ n\cosh n\xi & n\sinh n\xi \end{bmatrix} \begin{bmatrix} \overset{2}{B}_{11} & \overset{2}{B}_{12} \\ \overset{2}{B}_{21} & \overset{2}{B}_{22} \end{bmatrix} = -\frac{1}{2}\begin{bmatrix} \frac{1}{\chi} & 0 \\ 0 & 1 \end{bmatrix}.
\end{gathered}
$$

The solutions are given by the following equations:

$$
\begin{bmatrix} \overset{1}{B}_{11} & \overset{1}{B}_{12} \\ \overset{1}{B}_{21} & \overset{1}{B}_{22} \end{bmatrix} = \begin{bmatrix} -\frac{\xi}{2\chi(\kappa-1)} & -\frac{1}{2n^2} \\ \frac{1}{2\chi(\kappa-1)} & 0 \end{bmatrix},
\tag{C.9.31a}
$$

$$\begin{bmatrix} \overset{2}{B}_{11} & \overset{2}{B}_{12} \\ \overset{2}{B}_{21} & \overset{2}{B}_{22} \end{bmatrix} = \begin{bmatrix} -\frac{\kappa \sinh n\xi}{2\chi(\kappa-1)} & -\frac{\cosh n\xi}{2n} \\ \frac{\kappa \cosh n\xi}{2\chi(\kappa-1)} & \frac{\sinh n\xi}{2n} \end{bmatrix}. \tag{C.9.31b}$$

The boundary conditions lead to the following equation system:

$$\begin{bmatrix} 1\,0 & -\frac{1}{n} & 0 \\ 1\,1 & -\frac{\cosh n}{n} & -\frac{\sinh n}{n} \\ 0\,0 & n & 0 \\ 0\,0\,n\cosh n & n\sinh n \end{bmatrix} \begin{bmatrix} \overset{1}{A}_{1i} \\ \overset{1}{A}_{2i} \\ \overset{2}{A}_{1i} \\ \overset{2}{A}_{2i} \end{bmatrix} = \begin{bmatrix} -\overset{1}{B}_{1i} + \frac{\overset{2}{B}_{1i}}{n} \\ \overset{1}{B}_{1i} + \overset{1}{B}_{2i} - \frac{\overset{2}{B}_{1i}}{n}\cosh n - \frac{\overset{2}{B}_{2i}}{n}\sinh n \\ -n\overset{2}{B}_{1i} \\ \overset{2}{B}_{1i} n\cosh n + \overset{2}{B}_{2i} n \sinh n \end{bmatrix}, \quad i = 1, 2$$

from where

$$\overset{1}{A}_{1i} = -\overset{1}{B}_{1i}, \qquad \overset{1}{A}_{2i} = 2\overset{1}{B}_{1i}, +\overset{1}{B}_{2i}, \qquad \overset{2}{A}_{1i} = -n\overset{2}{B}_{1i},$$
$$\overset{2}{A}_{2i} = \frac{1}{\sinh n}\left(2\overset{2}{B}_{1i}\cosh n + \overset{2}{B}_{2i}\sinh n\right). \tag{C.9.32}$$

After substituting (C.9.31) for the elements of $\underline{\mathbf{B}}_\ell$, we have

$$\begin{bmatrix} \overset{1}{A}_{11} & \overset{1}{A}_{12} \\ \overset{1}{A}_{21} & \overset{1}{A}_{22} \end{bmatrix} = \begin{bmatrix} \frac{\xi}{2\chi(\kappa-1)} & \frac{1}{2n^2} \\ -\frac{2\xi-1}{2\chi(\kappa-1)} & -\frac{1}{n^2} \end{bmatrix}, \tag{C.9.33a}$$

$$\begin{bmatrix} \overset{2}{A}_{11} & \overset{2}{A}_{12} \\ \overset{2}{A}_{21} & \overset{2}{A}_{22} \end{bmatrix} = \begin{bmatrix} \frac{\kappa \sinh n\xi}{2\chi(\kappa-1)} & \frac{\cosh n\xi}{2n} \\ \frac{\kappa(\cosh n\xi \sinh n - 2\sinh n\xi \cosh n)}{2\chi(\kappa-1)\sinh n} & \frac{\sinh n\xi \sinh n - 2\cosh n\xi \cosh n}{n \sinh n} \end{bmatrix}. \tag{C.9.33b}$$

Problem 9.4 The solution is based on Eqs. (9.149) which, for completeness, are repeated here:

$$\chi\left(\frac{d^2 y_1}{dx^2} + \frac{dy_2}{dx}\right) = -\lambda y_1, \qquad \frac{d^2 y_2}{dx^2} - \chi\left(\frac{dy_1}{dx} + y_2\right) = -\lambda \mathfrak{r}^2 y_2. \tag{C.9.34}$$

It is obvious that

$$y_2^{(1)} = -\frac{\lambda}{\chi} y_1 - y_1^{(2)}. \tag{C.9.35}$$

Deriving Eq. $(C.9.34)_2$ with respect to x and substituting then $(C.9.34)_1$, we obtain

$$y_2^{(2)} + \lambda y_1 = -\lambda \mathfrak{r}^2 y_2.$$

Hence,

$$y_2^{(3)} + \lambda y_1^{(1)} = -\lambda \mathfrak{r}^2 y_2^{(1)}.$$

Substitution of (C.9.35) into this equation yields the elimination of y_2:

$$y_1^{(4)} + \lambda \mathfrak{r}^2 \left(1 + \frac{1}{\chi \mathfrak{r}^2}\right) y_1^{(2)} - \lambda y_1 + \lambda^2 \frac{\mathfrak{r}^2}{\chi} y_1 = 0 . \tag{C.9.36}$$

It will be assumed that

$$y_1(x) = e^{\varkappa x}, \tag{C.9.37}$$

where $\varkappa$ is an unknown parameter. The above equation is a solution if

$$\varkappa^4 + \lambda \mathfrak{r}^2 \left(1 + \frac{1}{\chi \mathfrak{r}^2}\right) \varkappa^2 - \lambda + \lambda^2 \frac{\mathfrak{r}^2}{\chi} = 0 . \tag{C.9.38}$$

Consequently,

$$\varkappa^2 = -\lambda \frac{\mathfrak{r}^2}{2} \left(1 + \frac{1}{\chi \mathfrak{r}^2}\right) \pm 1 \sqrt{\left(\lambda \frac{\mathfrak{r}^2}{2}\right)^2 \left(1 + \frac{1}{\chi \mathfrak{r}^2}\right)^2 + \lambda - \lambda^2 \frac{\mathfrak{r}^2}{\chi}} =$$
$$= -\frac{1}{2} \frac{\lambda}{\chi} \left(\mathfrak{r}^2 \chi + 1\right) \pm \frac{1}{2} \sqrt{\frac{\lambda}{\chi^2} \left(\lambda \left(\mathfrak{r}^2 \chi - 1\right)^2 + 4\chi^2\right)},$$

where the discriminant is always positive. It will be assumed that $\lambda - \lambda^2 \frac{\mathfrak{r}^2}{\chi}$ is negative—according to numerical experiences, this assumption is in general true. Then $\varkappa^2$ is negative; therefore, the general solution can be given in the form

$$y_1 = \mathcal{A}_1 \sin \varkappa_1 x + \mathcal{A}_2 \cos \varkappa_1 x + \mathcal{A}_3 \sin \varkappa_2 x + \mathcal{A}_4 \cos \varkappa_2 x, \tag{C.9.39}$$

where the undetermined integration constants are denoted by $\mathcal{A}_k$ $(k = 1, \dots, 4)$ and

$$\begin{aligned} \varkappa_1^2 &= \lambda \frac{\mathfrak{r}^2}{2} \left(1 + \frac{1}{\chi \mathfrak{r}^2}\right) - \sqrt{\left(\lambda \frac{\mathfrak{r}^2}{2}\right)^2 \left(1 + \frac{1}{\chi \mathfrak{r}^2}\right)^2 + \lambda - \lambda^2 \frac{\mathfrak{r}^2}{\chi}}, \\ \varkappa_2^2 &= \lambda \frac{\mathfrak{r}^2}{2} \left(1 + \frac{1}{\chi \mathfrak{r}^2}\right) + \sqrt{\left(\lambda \frac{\mathfrak{r}^2}{2}\right)^2 \left(1 + \frac{1}{\chi \mathfrak{r}^2}\right)^2 + \lambda - \lambda^2 \frac{\mathfrak{r}^2}{\chi}}. \end{aligned} \tag{C.9.40}$$

With y_1 Eq. (C.9.35) leads to the following result:

$$\begin{aligned} y_2^{(1)} = -\frac{\lambda}{\chi} y - y^{(2)} = -\mathcal{A}_1 \left(\frac{\lambda}{\chi} - \varkappa_1^2\right) \sin \varkappa_1 x - \mathcal{A}_2 \left(\frac{\lambda}{\chi} - \varkappa_1^2\right) \cos \varkappa_1 x - \\ -\mathcal{A}_3 \left(\frac{\lambda}{\chi} - \varkappa_2^2\right) \sin \varkappa_2 x - \mathcal{A}_4 \left(\frac{\lambda}{\chi} - \varkappa_2^2\right) \cos \varkappa_2 x . \end{aligned} \tag{C.9.41}$$

Utilizing the boundary conditions

$$y_1(0) = y_1(1) = y_2^{(1)}(0) = y_2^{(1)}(1) = 0, \tag{C.9.42}$$

which are valid for pinned-pinned beams, it can be shown that the undetermined integration constants should satisfy the following homogeneous equation system:

$$\mathcal{A}_2 + \mathcal{A}_4 = 0, \tag{C.9.43a}$$

$$\mathcal{A}_1 \sin \varkappa_1 + \mathcal{A}_2 \cos \varkappa_1 + \mathcal{A}_3 \sin \varkappa_2 + \mathcal{A}_4 \cos \varkappa_2 = 0, \tag{C.9.43b}$$

$$A_2 \left(\frac{\lambda}{\chi} - \varkappa_1^2\right) - A_4 \left(\frac{\lambda}{\chi} - \varkappa_2^2\right) = 0, \tag{C.9.43c}$$

$$\begin{aligned}\mathcal{A}_1 \left(\frac{\lambda}{\chi} - \varkappa_1^2\right) \sin \varkappa_1 + \mathcal{A}_2 \left(\frac{\lambda}{\chi} - \varkappa_1^2\right) \cos \varkappa_1 + \\ + \mathcal{A}_3 \left(\frac{\lambda}{\chi} - \varkappa_2^2\right) \sin \varkappa_2 + \mathcal{A}_4 \left(\frac{\lambda}{\chi} - \varkappa_2^2\right) \cos \varkappa_2 = 0.\end{aligned} \tag{C.9.43d}$$

Solutions for $\mathcal{A}_k$ $(k = 1, \ldots, 4)$ exist if the determinant of this equation system is zero, i.e., it holds that

$$\begin{aligned}\varkappa_1^4 \sin \varkappa_1 \sin \varkappa_2 + \varkappa_2^4 \sin \varkappa_1 \sin \varkappa_2 - 2\varkappa_1^2 \varkappa_2^2 \sin \varkappa_1 \sin \varkappa_2 = \\ (\sin \varkappa_2 \sin \varkappa_1)(\varkappa_1 - \varkappa_2)^2 (\varkappa_1 + \varkappa_2)^2 = 0\,.\end{aligned} \tag{C.9.44}$$

This determinant is zero if

$$\varkappa_1^2 = (k\pi)^2 \qquad \text{and} \qquad \varkappa_2^2 = (k\pi)^2 \qquad (k = 1, 2, 3, \ldots), \tag{C.9.45}$$

or

$$\varkappa_2^2 \pm 2\varkappa_1 \varkappa_2 + \varkappa_1^2 = 0. \tag{C.9.46}$$

Substituting $(\text{C.9.40})_1$ into Eq. $(\text{C.9.45})_1$ yields

$$\varkappa_2^2 = (k\pi)^2 = \lambda \frac{\mathfrak{r}^2}{2}\left(1 + \frac{1}{\chi \mathfrak{r}^2}\right) - \sqrt{\left(\lambda \frac{\mathfrak{r}^2}{2}\right)^2 \left(1 + \frac{1}{\chi \mathfrak{r}^2}\right)^2 + \lambda - \lambda^2 \frac{\mathfrak{r}^2}{\chi}},$$

from where the following equation can be obtained for the eigenvalue λ:

$$\lambda^2 - \frac{\chi}{\mathfrak{r}^2}\left[1 + \mathfrak{r}^2 \left(1 + \frac{1}{\chi \mathfrak{r}^2}\right)(k\pi)^2\right]\lambda + \frac{\chi}{\mathfrak{r}^2}(k\pi)^4 = 0\,. \tag{C.9.47}$$

Hence,

$$\begin{aligned}\lambda = \frac{\chi}{2\mathfrak{r}^2}\left[1 + \mathfrak{r}^2 \left(1 + \frac{1}{\chi \mathfrak{r}^2}\right)(k\pi)^2\right] \pm \\ \pm \sqrt{\left(\frac{\chi}{2\mathfrak{r}^2}\right)^2 \left[1 + \mathfrak{r}^2 \left(1 + \frac{1}{\chi \mathfrak{r}^2}\right)(k\pi)^2\right]^2 - \frac{\chi}{\mathfrak{r}^2}(k\pi)^4}.\end{aligned} \tag{C.9.48}$$

This solution is, in fact, the same as the solution given by Eq. (9.43). It should be mentioned that Eq. (C.9.45)$_2$ would result in the same solution for λ.

There are, however, other possibilities for calculating λ. If, for example,

$$\lambda = \frac{\chi}{\mathfrak{r}^2} \tag{C.9.49}$$

Equation (C.9.40) yields

$$\varkappa_1 = 0, \qquad \varkappa_1 = \sqrt{\chi\left(1 + \frac{1}{\chi\mathfrak{r}^2}\right)}. \tag{C.9.50}$$

Thus, determinant (C.9.44) is really zero. Under these conditions, the general solution assumes the form

$$y_1 = \mathcal{A}_1 + \mathcal{A}_2 x + \mathcal{A}_3 \sin \varkappa_2 x + \mathcal{A}_4 \cos \varkappa_2 x$$

and it can be shown by making use of boundary conditions (C.9.42) that $\mathcal{A}_1 = \mathcal{A}_2 = \mathcal{A}_3 = \mathcal{A}_4 = 0$, i.e., the corresponding eigenfunctions vanish.

C.10 Problems in Chap. 10

Problem 10.1 If the eigenvalue problems defined by differential equation (10.35) and boundary conditions (10.36)–(10.38) are self-adjoint then

$$(\underline{\mathbf{u}}, \underline{\mathbf{v}})_K = (\underline{\mathbf{v}}, \underline{\mathbf{u}})_K \quad \text{and} \quad (\underline{\mathbf{u}}, \underline{\mathbf{v}})_M = (\underline{\mathbf{v}}, \underline{\mathbf{u}})_M, \tag{C.10.51}$$

where $\underline{\mathbf{v}}$ and $\underline{\mathbf{u}}$ are comparative vectors. Making use of (10.35), Eq. (C.10.51)$_1$ can be written as

$$(\underline{\mathbf{u}}, \underline{\mathbf{v}})_K = \int_{-\vartheta}^{\vartheta} \underline{\mathbf{u}}^T(x) \sum_{\nu=0}^{4} \underline{\mathbf{K}}_\nu(x) \underline{\mathbf{v}}^\nu(x)\, \mathrm{d}x =$$
$$= \int_{-\vartheta}^{\vartheta} \left[u_2 v_2^{(4)} - m u_1 v_1^{(2)} + 2u_2 v_2^{(2)} - m u_1 v_2^{(1)} + m u_2 v_1^{(1)} + (m+1) u_2 v_2\right] \mathrm{d}x, \tag{C.10.52}$$

where

$$\int_{-\vartheta}^{\vartheta} u_2 v_2^{(4)}\, \mathrm{d}x = \underbrace{u_2 v_2^{(3)}\Big|_{-\vartheta}^{\vartheta} - u_2^{(1)} v_2^{(2)}\Big|_{-\vartheta}^{\vartheta}}_{=0} + \int_{-\vartheta}^{\vartheta} u_2^{(2)} v_2^{(2)}\, \mathrm{d}x = \int_{-\vartheta}^{\vartheta} u_2^{(2)} v_2^{(2)}\, \mathrm{d}x\,,$$

$$-m\int_{-\vartheta}^{\vartheta} u_1 v_1^{(2)}\,\mathrm{d}x = \underbrace{-m u_1 v_1^{(1)}\Big|_{-\vartheta}^{\vartheta}}_{=0} + m\int_{-\vartheta}^{\vartheta} u_1^{(1)} v_1^{(1)}\,\mathrm{d}x = m\int_{-\vartheta}^{\vartheta} u_1^{(1)} v_1^{(1)}\,\mathrm{d}x\,,$$

$$2\int_{-\vartheta}^{\vartheta} u_2 v_2^{(2)}\,\mathrm{d}x = \underbrace{2u_2 v_2^{(1)}\Big|_{-\vartheta}^{\vartheta}}_{=0} - 2\int_{-\vartheta}^{\vartheta} u_2^{(1)} v_2^{(1)}\,\mathrm{d}x = 2\int_{-\vartheta}^{\vartheta} u_2^{(1)} v_2^{(1)}\,\mathrm{d}x\,,$$

$$\int_{-\vartheta}^{\vartheta}\left(-m u_1 v_2^{(1)} + m u_2 v_1^{(1)}\right)\mathrm{d}x = \underbrace{-m u_1 v_2|_{-\vartheta}^{\vartheta}}_{=0} + \int_{-\vartheta}^{\vartheta}\left(m u_1^{(1)} v_2 + m v_1^{(1)} u_2\right)\mathrm{d}x\,.$$

Note that these formulae are obtained by performing partial integration and taking boundary conditions (10.36)–(10.38) into account. Substituting the above integrals into (C.10.52), we have

$$\begin{aligned}\left(\underline{\mathbf{u}},\underline{\mathbf{v}}\right)_K = \int_{-\vartheta}^{\vartheta}\Big[&u_2^{(2)} v_2^{(2)} + m u_1^{(1)} v_1^{(1)} - 2u_2^{(1)} v_2^{(1)} + \\ &+ m\left(u_1^{(1)} v_2 + v_1^{(1)} u_2\right) + (m+1)\,u_2 v_2\Big]\,\mathrm{d}x = \left(\underline{\mathbf{v}},\underline{\mathbf{u}}\right)_K\,,\end{aligned} \tag{C.10.53}$$

which shows that the product $\left(\underline{\mathbf{u}},\underline{\mathbf{v}}\right)_K$ is commutative. It can be checked with ease that the product $\left(\underline{\mathbf{u}},\underline{\mathbf{v}}\right)_M$ is also commutative. Hence, the considered eigenvalue problems are all self-adjoint.

Problem 10.2 Write u for v in (C.10.53) and take into account that

$$(\underline{\mathbf{u}},\underline{\mathbf{u}})_M = \int_{-\vartheta}^{\vartheta}(u_1^2+u_2^2)\,\mathrm{d}x\,.$$

The result is

$$\begin{aligned}\mathcal{R}[\underline{\mathbf{u}}(x)] &= \frac{\left(\underline{\mathbf{u}},\underline{\mathbf{u}}\right)_K}{\left(\underline{\mathbf{u}},\underline{\mathbf{u}}\right)_M} = \\ &= \frac{\int_{-\vartheta}^{\vartheta}\left[(u_2^{(2)})^2 + m(u_1^{(1)})^2 - 2(u_2^{(1)})^2 + 2m u_1^{(1)} u_2 + (m+1)\,(u_2)^2\right]\mathrm{d}x}{\int_{-\vartheta}^{\vartheta}(u_1^2+u_2^2)\,\mathrm{d}x}\,.\end{aligned} \tag{C.10.54}$$

Problem 10.3 As regards the case of the pinned-fixed beam, only one boundary condition differs from those valid for the fixed-fixed beam. This means that the boundary condition $y_2^{(1)}(x)|_{x=-\vartheta}=0$ should be changed to $y_2^{(2)}(x)|_{x=\vartheta}=0$. The effect of the boundary condition $y_2^{(1)}(x)|_{x=-\vartheta}=0$ appears in the fifth row of the equation system (10.65) used for finding the non-zero elements of the matrices $\underline{\mathbf{A}}_\ell$ for fixed-fixed beams. Recalling that the fifth row in equation system (10.62) valid for pinned-pinned beams is in fact the boundary condition $y_2^{(2)}(x)|_{x=\vartheta} = 0$, it is sufficient to rewrite the fifth row in equation system (10.65) by the fifth row in equation system (10.62) for establishing the equation system valid for pinned-fixed beams. For this reason, we do not present the whole equation system here. Instead, the solutions sought are presented only. For simplicity in writing, they are given in terms of $a, b, \ldots, f$ here. With

$$\begin{aligned} D = & -4m + 11m\cos^2\vartheta - 7m\cos^4\vartheta - 4m\vartheta\sin\vartheta\cos^3\vartheta - \\ & - 2m\vartheta\sin\vartheta\cos\vartheta + 2\vartheta\cos\vartheta\sin\vartheta - 4\vartheta\cos^3\vartheta\sin\vartheta + 3m\vartheta^2 + 2\vartheta^2, \end{aligned} \tag{C.10.55}$$

the solutions in question are as follows:

$$\begin{aligned} \overset{1}{A}_{1i} = & -\frac{1}{D}\{a\left[-2\vartheta^2(m+1)\cos^2\vartheta + 2m\vartheta\cos\vartheta\sin\vartheta\right] + \\ & + b\left[-2\vartheta^2 m\sin\vartheta\cos\vartheta - 2\vartheta^2\sin\vartheta\cos\vartheta - m\vartheta\cos^2\vartheta + 4m\vartheta\cos^4\vartheta - \right. \\ & \left. -7m\sin\vartheta\cos^3\vartheta + 4m\sin\vartheta\cos\vartheta - 2\vartheta\cos^2\vartheta + 4\vartheta\cos^4\vartheta\right] + \\ & c\left[-2\vartheta^3 - 2\vartheta^3(m+1)\cos^2\vartheta - 3\vartheta^3 m + 4m\vartheta\sin^2\vartheta + m\vartheta^2\sin\vartheta\cos\vartheta - 2\vartheta^2\sin\vartheta\cos\vartheta\right] \\ & + d\left[m(m+1)\vartheta^3\cos\vartheta - 4m^2\vartheta\cos\vartheta + m^2\vartheta\cos^3\vartheta + 2\vartheta m\cos\vartheta - \right. \\ & - 4\vartheta m\cos^3\vartheta - 4m^2\sin\vartheta + 7m^2\sin\vartheta\cos^2\vartheta + \\ & \left. + 3m(m+1)\vartheta^2\sin\vartheta\cos^2\vartheta + 2m\vartheta^2\sin\vartheta + 3m^2\vartheta^2\sin\vartheta\right] + \\ & + e\left[-2(m+1)\vartheta^3\sin\vartheta\cos\vartheta - 4\vartheta m\cos\vartheta(\vartheta\cos\vartheta - \sin\vartheta) - 2\vartheta^2\cos^2\vartheta\right] + \\ & + f m\vartheta\cos\vartheta[\vartheta - \sin\vartheta\cos\vartheta]\}, \end{aligned} \tag{C.10.56a}$$

$$\begin{aligned} \overset{2}{A}_{1i} = & -\frac{1}{D}\{a\left[-2\vartheta - m\vartheta + 6\vartheta\cos^2\vartheta - 4(m+1)\vartheta\cos^4\vartheta + 3m\vartheta\cos^2\vartheta - \right. \\ & \left. - 5m\cos\vartheta\sin\vartheta + 7m\sin\vartheta\cos^3\vartheta - 2\vartheta^2(m+1)\sin\vartheta\cos\vartheta\right] + \\ & + b\left[2m\sin^2\vartheta - 2m\vartheta\sin\vartheta\cos\vartheta - 2(m+1)\vartheta^2\sin^2\vartheta\right] + \\ & + c\left[4m\vartheta\sin\vartheta\cos\vartheta + 2m\cos^2\vartheta - 2\vartheta^2\cos^2\vartheta - 2mc - \right. \\ & \left. 2(m+1)\vartheta^3\sin\vartheta\cos\vartheta + 2m\vartheta^2 - 4m\vartheta^2\cos^2\vartheta + 2\vartheta^2\right] + \\ & + d\left[m\vartheta^2\cos\vartheta\sin^2\vartheta + 2m^2\vartheta^2\cos\vartheta\sin^2\vartheta - m^2\vartheta\sin\vartheta\sin^2\vartheta - \right. \\ & \left. -m^2\cos\vartheta\sin^2\vartheta + m(m+1)\vartheta^3\sin\vartheta\right] + \\ & + e\left[6m\vartheta - 5m\vartheta^3 + 2\vartheta^2\sin\vartheta\cos\vartheta + 2m\vartheta^3\cos^2\vartheta - 4\vartheta^3 - \right. \\ & \left. -2m\sin\vartheta\cos\vartheta + 3m\vartheta^2\cos\vartheta\sin\vartheta - 4m\vartheta\cos^2\vartheta + 2\vartheta^3\cos^2\vartheta\right] + \\ & + f\left[-2m\sin\vartheta + 4m\sin\vartheta\cos^2\vartheta - 5m\vartheta\cos\vartheta + 3m\vartheta\cos^3\vartheta + \vartheta^2 m\sin\vartheta\right]\}, \end{aligned} \tag{C.10.56b}$$

$$\begin{aligned} \overset{3}{A}_{1i} = & \frac{1}{D}\{a\left[2\vartheta - 2\vartheta(m+1)\cos^2\vartheta - m\cos\vartheta\sin\vartheta + 3m\vartheta\right] + \\ & + b\left[2m\sin^2\vartheta - 2(m+1)\vartheta\sin\vartheta\cos\vartheta\right] + \\ & + c\left[4m\vartheta\sin\vartheta\cos\vartheta - 2m\sin^2\vartheta - 2(m+1)\vartheta^2\cos^2\vartheta + 2\vartheta\sin\vartheta\cos\vartheta\right] + \\ & + d\left[m(m+1)\vartheta\sin\vartheta\cos^2\vartheta + m(m+1)\vartheta^2\cos\vartheta - m^2\vartheta\sin\vartheta - m^2\sin^2\vartheta\cos\vartheta\right] + \end{aligned}$$

$$+e\left[m\vartheta\cos^2\vartheta-4m\vartheta\cos^4\vartheta-2\left(m+1\right)\vartheta^2\sin\vartheta\cos\vartheta-\right.$$
$$-6m\sin\vartheta\cos\vartheta+7m\sin\vartheta\cos^3\vartheta+2m\vartheta+4\vartheta\cos^2\vartheta\sin^2\vartheta\left.\right]+$$
$$\left.+f\left[2m\sin\vartheta-3m\sin\vartheta\cos^2\vartheta+m\vartheta\cos\vartheta\right]\right\}, \tag{C.10.56c}$$

$$\overset{3}{A}_{2i}=\frac{1}{D}\left\{2a\left(-3\sin\vartheta\cos^2\vartheta+2\sin\vartheta+\vartheta\cos\vartheta\right)+2b\left(-\cos\vartheta+\cos^3\vartheta+\vartheta\sin\vartheta\right)+\right.$$
$$+2c\left(-\vartheta\sin\vartheta\cos^2\vartheta+\vartheta^2\cos\vartheta-\vartheta\sin\vartheta-\cos^3\vartheta+\cos\vartheta\right)-$$
$$-dm\left(\vartheta^2-\cos^2\vartheta+\cos^4\vartheta\right)+2e\left(3\vartheta\cos^3\vartheta-4\vartheta\cos\vartheta+\vartheta^2\sin\vartheta+\sin\vartheta\cos^2\vartheta\right)+$$
$$\left.+2f\left(-2\sin\vartheta\cos^3\vartheta+\sin\vartheta\cos\vartheta+\vartheta\right)\right\}, \tag{C.10.56d}$$

$$\overset{4}{A}_{1i}=\frac{1}{D}\left\{-2a\left(-m\sin^2\vartheta+(1+m)\vartheta\sin\vartheta\cos\vartheta\right)+\right.$$
$$+b\left(2m\vartheta\cos^2\vartheta-3m\sin\vartheta\cos\vartheta+m\vartheta+2\vartheta\cos^2\vartheta\right)-$$
$$-c\left[m\vartheta-7m\sin^3\vartheta\cos\vartheta+3m\vartheta\cos^2\vartheta-4m\vartheta\cos^4\vartheta+\right.$$
$$\left.+2\left(m+1\right)\vartheta^2\sin\vartheta\cos\vartheta+4\vartheta\cos^2\vartheta\sin^2\vartheta\right]+$$
$$+dm\left(\vartheta\cos\vartheta-2m\sin\vartheta+5m\sin\vartheta\cos^2\vartheta+(m+1)\vartheta^2\sin\vartheta-3\left(m+1\right)\vartheta\cos^3\vartheta\right)+$$
$$+2e\left(-\vartheta^2\sin^2\vartheta+2m\sin^2\vartheta-m\vartheta^2\sin^2\vartheta-\vartheta\sin\vartheta\cos\vartheta-2m\vartheta\sin\vartheta\cos\vartheta\right)+$$
$$\left.+fm\left(\vartheta\sin\vartheta-\cos\vartheta+\cos^3\vartheta\right)\right\}, \tag{C.10.56e}$$

$$\overset{4}{A}_{2i}=\frac{1}{D}a\left[2m\sin^2\cos\vartheta-2\vartheta^2\left(m+1\right)\cos\vartheta-2\vartheta^2\cos\vartheta+\right.$$
$$\left.+4m\vartheta^2\cos^5\vartheta+2m\vartheta\sin\vartheta-2\left(m+1\right)\vartheta\sin\vartheta\cos^2\vartheta\right]+$$
$$+b\left[-2\vartheta\cos\vartheta+6b\vartheta\cos^3\vartheta-10m\sin\vartheta\cos^2\vartheta-\right.$$
$$\left.-2m\vartheta^2\sin\vartheta+6m\vartheta\cos^3\vartheta-2\vartheta^2\sin\vartheta+4m\sin\vartheta\right]+$$
$$+c\left[-2\left(m+1\right)\vartheta^3\cos\vartheta-2\vartheta\sin^2\vartheta\cos\vartheta-2\vartheta^2\sin\vartheta-4m\vartheta^2\sin\vartheta+4m\sin\vartheta-\right.$$
$$\left.-4m\sin\vartheta\cos^2\vartheta+8m\vartheta\cos\vartheta\sin^2\vartheta-6\left(m+1\right)\vartheta^2\sin\vartheta\cos^2\vartheta\right]+$$
$$+d\left[2\vartheta^3+4m\vartheta\left(m\vartheta^2-1\right)+6m\vartheta\left(\vartheta^2-m\right)+14m\left(m+1\right)\vartheta\sin^2\vartheta\cos^2\vartheta+\right.$$
$$+12m^2\sin\vartheta\cos^3\vartheta-4\vartheta^2\left(m+1\right)^2\sin\vartheta\cos^3\vartheta+$$
$$\left.+2\left(\vartheta^2m^2-3m^2+\vartheta^2+2m\vartheta^2\right)\cos\vartheta\sin\vartheta\right]+$$
$$+e\left[2\vartheta^3\sin\vartheta+2\left(m+1\right)\vartheta^2\cos^3\vartheta-4\vartheta^2\cos\vartheta+4m\cos\vartheta\sin^2\vartheta-\right.$$
$$\left.-4m\vartheta\sin\vartheta\cos^2\vartheta-2\vartheta\sin\vartheta\cos^2\vartheta+4m\vartheta\sin\vartheta-6m\vartheta^2\cos\vartheta-2\vartheta^3m\sin\vartheta\right]+$$
$$+fm\left(\vartheta^2-\cos^2\vartheta+\cos^4\vartheta\right). \tag{C.10.56f}$$

Problem 10.4 Substituting

$$u_1=\frac{2x}{\pi}-\frac{2\vartheta}{\pi}\sin\frac{\pi}{2\vartheta}x,\qquad u_2=\cos\frac{\pi}{2\vartheta}x,$$

$$\frac{\mathrm{d}u_1}{\mathrm{d}x}=\frac{2}{\pi}-\cos\frac{\pi}{2\vartheta}x,\qquad \frac{\mathrm{d}u_2}{\mathrm{d}x}=-\frac{\pi}{2\vartheta}\sin\frac{\pi}{2\vartheta}x$$

and

$$\frac{\mathrm{d}^2 u_2}{\mathrm{d}x^2} = -\frac{\pi^2}{4\vartheta^2}\cos\frac{\pi}{2\vartheta}x$$

into the numerator of the Rayleigh quotient (C.10.54) yields

$$\begin{aligned}
(\underline{\mathbf{u}}, \underline{\mathbf{u}})_K &= \int_{-\vartheta}^{\vartheta} \underline{\mathbf{u}}^T \underline{\mathbf{K}}\,\underline{\mathbf{u}}\,\mathrm{d}x = \\
&= \int_{-\vartheta}^{\vartheta} \left[(u_2^{(2)})^2 + m(u_1^{(1)})^2 - 2(u_2^{(1)})^2 + 2mu_1^{(1)}u_2 + (m+1)\,(u_2)^2\right] \mathrm{d}x = \\
&= \int_{-\vartheta}^{\vartheta} \left[\left(-\frac{\pi^2}{4\vartheta^2}\cos\frac{\pi}{2\vartheta}x\right)^2 + m\left(\frac{2}{\pi} - \cos\frac{\pi}{2\vartheta}x\right)^2 - 2\left(-\frac{\pi}{2\vartheta}\sin\frac{\pi}{2\vartheta}x\right)^2 + \right. \\
&\quad \left. +2m\left(\frac{2}{\pi} - \cos\frac{\pi}{2\vartheta}x\right)\cos\frac{\pi}{2\vartheta}x + (m+1)\left(\cos\frac{\pi}{2\vartheta}x\right)^2\right] \mathrm{d}x = \\
&= \frac{1}{16}\frac{\pi^6 + 128m\vartheta^4 - 8\pi^4\vartheta^2 + 16\vartheta^4\pi^2}{\vartheta^3\pi^2}.
\end{aligned}$$

For the denominator, we get in the same way:

$$\begin{aligned}
(\underline{\mathbf{u}}, \underline{\mathbf{u}})_M &= \int_{-\vartheta}^{\vartheta} \underline{\mathbf{u}}^T \underline{\mathbf{M}}\,\underline{\mathbf{u}}\,\mathrm{d}x = \int_{-\vartheta}^{\vartheta} (u_1^2 + u_2^2)\,\mathrm{d}x = \\
&\int_{-\vartheta}^{\vartheta} \left[\left(\frac{2x}{\pi} - \frac{2\vartheta}{\pi}\sin\frac{\pi}{2\vartheta}x\right)^2 + \left(\cos\frac{\pi}{2\vartheta}x\right)^2\right] \mathrm{d}x = \\
&= \frac{1}{3}\vartheta\frac{20\vartheta^2\pi^2 - 192\vartheta^2 + 3\pi^4}{\pi^4}.
\end{aligned}$$

Hence,

$$\mathcal{R}[\underline{\mathbf{u}}(x)] = \frac{(\underline{\mathbf{u}}, \underline{\mathbf{u}})_K}{(\underline{\mathbf{u}}, \underline{\mathbf{u}})_M} = \frac{3}{16}\frac{\pi^2\left(16\pi^2\vartheta^4 + 128m\vartheta^4 - 8\pi^4\vartheta^2 + \pi^6\right)}{\vartheta^4\left(20\vartheta^2\pi^2 - 192\vartheta^2 + 3\pi^4\right)}.$$

Table C.5 Results for the quotient $\omega_1/\hat{\omega}_1$

		$\omega_1/\hat{\omega}_1$	
m	$\bar{\vartheta}$	Numerical sol.	Rayleigh approx.
10 000	0.2	1.060	1.061
10 000	0.4	1.758	1.759
10 000	0.6	3.389	3.419
10 000	0.8	5.673	5.904
10 000	1.0	7.564	9.145

Table C.5 shows the same comparable results for the quotient $\omega_1/\hat{\omega}_1$. Note that the approximation is very good if $\omega_1 < \omega_2$ (or if $\bar{\vartheta} < 0.8$). After the order of the eigenfrequencies changes, the mode shapes will also change; therefore, the above approximation will significantly overestimate the actual values.

Problem 10.5 It follows from Eq. (10.16) that

$$M = -\frac{I_{e\eta}}{R^2}\left(w_o^{(2)} + w_o\right) = -I_{e\eta}\left(\frac{\mathrm{d}^2 w_o}{\mathrm{d}s^2} + \frac{w_o}{R^2}\right).$$

For heterogeneous straight beams $M = M_y$, $w_o = w$, $s = x$, $I_{e\eta} = I_{ey}$, and R tends to infinity. Hence

$$M_y = -I_{ey}\frac{\mathrm{d}^2 w}{\mathrm{d}x^2}.$$

According to Eqs. (7.72) ($N = 0$) and (7.78), the motion equation of heterogeneous straight beams is

$$\frac{\partial^2 M_y}{\partial x^2} - \rho_a A\frac{\partial^2 w}{\partial t^2} = 0, \qquad \text{or} \qquad \frac{\partial^2}{\partial x^2}\left(I_{ey}\frac{\partial^2 w}{\partial x^2}\right) + \rho_a A\frac{\partial^2 w}{\partial t^2} = 0$$

from where by performing the steps leading to $(7.82)_1$ we get Eq. (10.72).

References

1. M.E. Gurtin, *An Introduction to Continuum Mechanics* (Academic, New York, 1981)
2. P. Dirac, *The Principles of Quantum Mechanics*, 4th edn. (Clarendon Press, Oxford, 1958)
3. I.M. Gelfand, G.E. Silov, *Generalized Functions* (Academic, New York, 1964)
4. E. Janke, F. Emde, F. Lösch, Tafelen Höheren Funktionen. Nauka, Moscow (in Russian) 1977 (a) p. 241–242 and (b) 245–246
5. G.R. Cowper, The shear coefficient in Timoshenko's beam theory. J. Appl. Mech. **33**(2), 335–340 (1966). https://doi.org/10.1115/1.3625046
6. S. Timoshenko, J.M. Gere, *Theory of Elastic Stability*, 2nd edn. (McGraw-Hill, New York, 1961)
7. G.G. Coriolis, Sur les équations du mouvement relatif des systémes de corps. J. De l'Ecole Royale Polytechnique **15**, 144–154 (1835)
8. J.L. Meriam, L.G. Kraige, *Engineering Dynamics*, 6th edn. (Wiley, New York, 1981)

Bibliography

9. S. Timoshenko, *Vibration Problems in Engineering*, 2nd edn. 1st edn: 1928 (D. Van Nostran Company, New York, 1937), 470 pp
10. E.J. Nestroides, *A Handbook on Torsional Vibrations*, 2nd edn.: 2011 (Cambridge University Press, Cambridge, 1958)
11. F. Gantmacher, *Lectures in Analytical Mechanics* (Mir Publishers, Moscow, 1975)
12. P. Srinivasulu, C.V. Vaidyanathan, *Handbook of Machine Foundations* (McGraw-Hill Education, New York, 1976)
13. R.S. Rao, *Mechanical Vibrations* (Addison Wesley Publishing Company, Boston, 1986)
14. W.T. Thomson, *Theory of Vibration with Applications*, 4th edn. (Georg Allen & Unwin, London, 1993)
15. D.J. Inman, *Engineering Vibration* (Prentice Hall, Upper Saddle River, 1995), 560 pp
16. L. Meirovitch, *Fundamentals of Vibration*, International edn. (McGraw-Hill, New York, 2001), 806 pp
17. V.A. Svetlitsky, A.S. Lidvansky, R.A. Mukhamedshin, *Engineering Vibration Analysis: Worked Problems 2*. Engineering Vibration Analysis (Springer, Berlin, 2004)
18. V.A. Chechin, V.A. Svetlitsky, G.I. Merzon, *Engineering Vibration Analysis: Worked Problems 1*. Foundations of Engineering Mechanics (Springer, Berlin, 2012)
19. G. Csernák, G. Stépán, *Fundamentals of Mechanical Vibrations* (in Hungarian). An Electronic edn. (2012)
20. P.L. Gatti, *Applied Structural and Mechanical Vibrations: Theory and Methods*, 2nd edn. (CRC Press, Taylor & Francis Group, Boca Raton, 2014)
21. H. Baruh, *Applied Dynamics* (CRC Press, Taylor & Francis Group, Boca Raton, 2015)

Index

G. Szeidl and L. P. Kiss, *Mechanical Vibrations*, Foundations of Engineering Mechanics,
https://doi.org/10.1007/978-3-030-45074-8

Printed in the United States
by Baker & Taylor Publisher Services